UNITEXT – La Matematica per il 3+2

Volume 87

More information about this series at
http://www.springer.com/series/5418

Sandro Salsa · Gianmaria Verzini

Partial Differential Equations in Action

Complements and Exercises

 Springer

Sandro Salsa
Dipartimento di Matematica
Politecnico di Milano
Milano, Italy

Gianmaria Verzini
Dipartimento di Matematica
Politecnico di Milano
Milano, Italy

Translated by Simon G. Chiossi, UFBA – Universidade Federal da Bahia, Salvador (Brazil).

Translation from the Italian language edition: *Equazioni a derivate parziali. Complementi ed esercizi*, Sandro Salsa e Gianmaria Verzini, © Springer-Verlag Italia, Milano 2005. All rights reserved.

UNITEXT – La Matematica per il 3+2
ISSN 2038-5722   ISSN 2038-5757 (electronic)
ISBN 978-3-319-15415-2   ISBN 978-3-319-15416-9 (eBook)
DOI 10.1007/978-3-319-15416-9
Springer Cham Heidelberg New York Dordrecht London

Library of Congress Control Number: 2015930285

Cover Design: Simona Colombo, Giochi di Grafica, Milano, Italy
Typesetting with LaTeX: PTP-Berlin, Protago TeX-Production GmbH, Germany (www.ptp-berlin.eu)

Springer is a part of Springer Science+Business Media (www.springer.com)

Preface

This book is designed for advanced undergraduate students from various disciplines, including applied mathematics, physics, and engineering. It evolved during the PDE courses that both authors have taught during recent decades at the Politecnico di Milano, and consists of problems of various types and difficulties.

In the first part of the book, while much emphasis is placed on the most common methods of resolution, such as separation of variables or the method of characteristics, we also invite the student to handle the basic theoretical tools and properties of the solutions to the fundamental equations of mathematical physics.

The second part is slightly more advanced and requires basic tools from functional analysis. A small number of exercises aims to familiarize the student with the first elements of the theory of distributions and of the Hilbertian Sobolev spaces. The focus then switches to the variational formulation of the most common boundary value problems for uniformly elliptic equations. A substantial number of problems is devoted to the use of the Riesz representation and the Lax-Milgram theorems together with Fredholm alternative to analyse well posedness or solvability of those problems. Next, a number of problems addresses the analysis of weak solutions to initial-boundary value problems for the heat or the wave equation.

The text is completed by two short appendixes, the first dealing with Sturm-Liouville problems and Bessel functions and the second listing frequently used formulas.

Each chapter begins with a brief review of the main theoretical concepts and tools that constitute necessary prerequisites for a proper understanding. The text Partial Differential Equation in Action [18], by S. Salsa, is the natural theoretical reference.

Within each chapter, the problems are divided into two sections. In the first one we present detailed solutions and comments to provide the student with a reasonably complete guide. In the second section, we propose a set of problems that each student should try to solve by him- or herself. In each case, a solution can be found at the end of the chapter. Some problems are proposed as theoretical complements and may prove particularly challenging; this is especially true of those marked with one or two asterisks.

Milano, January 2015   Sandro Salsa
                                                                     Gianmaria Verzini

Contents

1

Diffusion

1.1 Backgrounds

We shall recall a few notions and results that crop up frequently concerning the diffusion equation

$$u_t - D\Delta u = f$$

defined on a cylindrical domain $Q_T = \Omega \times (0, T)$, where Ω is a domain (connected, open subset) of $\mathbb{R}^n$. Here $u = u(\mathbf{x}, t)$, and the Laplacian Δ is taken with respect to the spatial variables $\mathbf{x}$ only.

• *Parabolic boundary.* The union of the base of Q_T (i.e. $\Omega \times \{0\}$) and the lateral surface $S_T = \partial\Omega \times [0, T]$ is the *parabolic boundary of* Q_T, which we denote by $\partial_p Q_T$. In well-posed problems for the diffusion equation this is where one must assign the data.

• *Maximum principles.* Let Ω be bounded and $w \in C^{2,1}(Q_T) \cap C(\overline{Q}_T)$ a sub-solution (or super-solution), that is

$$w_t - D\Delta w = q \le 0 \qquad (\text{resp. } \ge 0) \qquad \text{in } Q_T.$$

Then w reaches its maximum value (resp. minimum) on the parabolic boundary $\partial_p Q_T$ of Q_T:

$$\max_{\overline{Q}_T} w = \max_{\partial_p Q_T} w$$

(*weak maximum principle*). In particular, if w is negative (resp. positive) on $\partial_p Q_T$, then it is negative (resp. positive) over all of Q_T. If, further, $w(\mathbf{x}_0, t_0) = \max_{\overline{Q}_T}$ with $\mathbf{x}_0 \in \Omega$, then w is constant on $\Omega \times [0, t_0]$ (*strong maximum principle*).

• *Fundamental solution and global Cauchy problem.* The function

$$\Gamma_D(\mathbf{x}, t) = \frac{1}{(4\pi D t)^{n/2}} e^{-|\mathbf{x}|^2/(4Dt)}, \qquad t > 0,$$

is called *fundamental solution* to the diffusion equation; when $t > 0$ it solves

© Springer International Publishing Switzerland 2015
S. Salsa, G. Verzini, *Partial Differential Equations in Action. Complements and Exercises*,
UNITEXT – La Matematica per il 3+2 87, DOI 10.1007/978-3-319-15416-9_1

$u_t - D\Delta u = 0$ and is the unique function satisfying

$$\lim_{t\downarrow 0} \Gamma_D(\mathbf{x}, t) = \delta_n(\mathbf{x}), \qquad \int_{\mathbb{R}^n} \Gamma_D(\mathbf{x}, t)\, d\mathbf{x} = 1 \text{ for any } t > 0,$$

where $\delta_n(\mathbf{x})$ denotes the n-dimensional *Dirac's delta function*.

The fundamental solution enables to construct the general solution to the global Cauchy problem

$$\begin{cases} u_t - D\Delta u = f(\mathbf{x}, t) & \text{in } \mathbb{R}^n \times (0, \infty) \\ u(\mathbf{x}, 0) = g(\mathbf{x}) & \text{in } \mathbb{R}^n, \end{cases}$$

by means of the formula

$$u(\mathbf{x}, t) = \int_{\mathbb{R}^n} \Gamma_D(\mathbf{x} - \mathbf{y}, t)\, g(\mathbf{y})\, d\mathbf{y} + \int_0^t \int_{\mathbb{R}^n} \Gamma(\mathbf{x} - \mathbf{y}, t - s)\, f(\mathbf{y}, s)\, d\mathbf{y} ds.$$

The latter holds, for example, when $|g(\mathbf{x})| \le c e^{A|\mathbf{x}|}$, f is bounded and f, f_t, f_{x_j}, $f_{x_i x_j}$ are continuous on $\mathbb{R}^n \times (0, +\infty)$. At a point $\mathbf{x}_0$ of continuity of g we have

$$u(\mathbf{x}, t) \to g(\mathbf{x}_0) \qquad \text{as } (\mathbf{y}, t) \to (\mathbf{x}_0, 0), \, t > 0.$$

• *Random walk and fundamental solution* $(n = 1)$. Let us consider a particle of unit mass moving along the x-axis as follows.

i) In a time interval τ the particle advances by h, starting from $x = 0$.

ii) It moves leftwards or rightwards with probability $p = 1/2$, each time independently of the previous action.

At time $t = N\tau$, i.e. after N iterations, the particle will have reached point $x = mh$, where N is a natural number and m an integer. The probability $p(x, t)$ that it will be in x at time t is the solution of the discrete problem

$$p(x, t + \tau) = \frac{1}{2} p(x - h, t) + \frac{1}{2} p(x + h, t), \qquad (1.1)$$

with initial conditions

$$p(0, 0) = 1 \text{ and } p(x, 0) = 0 \quad \text{if } x \ne 0.$$

Passing to the limit in (1.1) for $h, \tau \to 0$, whilst keeping $h^2/\tau = 2D = \text{constant}$ and interpreting p as a probability density, gives the equation $p_t = D p_{xx}$, and the initial conditions now read

$$\lim_{t\downarrow 0} p(x, t) = \delta \qquad (\delta_1 = \delta).$$

We have already noticed that the unique solution with unit mass is the fundamental solution to the diffusion equation:

$$p(x, t) = \Gamma_D(x, t).$$

1.2 Solved Problems

- **1.2.1 – 1.2.7** : The method of separation of variables.
- **1.2.8 – 1.2.11** : Use of the maximum principle.
- **1.2.12 – 1.2.18** : Applying the notion of fundamental solution.
- **1.2.19 – 1.2.22** : Use of Fourier and Laplace transforms.
- **1.2.23 – 1.2.26** : Problems in dimension higher than one.

1.2.1 The method of separation of variables

Problem 1.2.1 (Cauchy-Dirichlet). *Let $D > 0$ be a constant and $g \in C^1([0, \pi])$, with $g(0) = g(\pi) = 0$. Solve, by separating the variables, the problem:*

$$\begin{cases} u_t(x,t) - Du_{xx}(x,t) = 0 & 0 < x < \pi, \, t > 0 \\ u(x,0) = g(x) & 0 \le x \le \pi \\ u(0,t) = u(\pi,t) = 0 & t > 0. \end{cases}$$

Discuss uniqueness and continuous dependence on the initial data.

Solution. We start with two preliminary observations. First of all, the choice of $[0, \pi]$ as domain for the space variable is just to keep the formulas simpler. In case the space variable x varies between 0 and $L > 0$ we can use Fourier series on suitable intervals, or reduce to $[0, \pi]$ by the change $y = x\pi/L$, $v(y,t) = u(Ly/\pi,t)$, which would give the following problem for v:

$$\begin{cases} v_t - \frac{D\pi^2}{L^2}v_{yy} = 0 & 0 < y < \pi, \, t > 0 \\ v(y,0) = g(Ly/\pi) & 0 \le y \le \pi \\ v(0,t) = v(\pi,t) = 0 & t > 0. \end{cases}$$

Notice that the boundary condition is of Dirichlet type and *homogeneous*. The first step of the method consists in seeking *non-zero* solutions of the form

$$u(x,t) = v(x)w(t).$$

Substituting into the equation gives

$$v(x)w'(t) - Dv''(x)w(t) = 0.$$

Dividing by $v(x)w(t)$ and rearranging terms we find:

$$\frac{1}{D}\frac{w'(t)}{w(t)} = \frac{v''(x)}{v(x)}. \tag{1.2}$$

This is an identity between two members depending on different variables. Consequently they must both be equal to some constant $\lambda \in \mathbb{R}$. Thus we can split (1.2) into the two equations

$$w'(t) - \lambda D w(t) = 0,$$

solved by

$$w(t) = C e^{\lambda D t}, \quad C \in \mathbb{R}, \tag{1.3}$$

and

$$v''(x) - \lambda v(x) = 0. \tag{1.4}$$

The Dirichlet conditions force $v(0)w(t) = v(\pi)w(t) = 0$ for any $t > 0$, i.e.

$$v(0) = v(\pi) = 0. \tag{1.5}$$

The boundary-value problem (1.4), (1.5) has non-trivial solutions only for special values of λ, called *eigenvalues*. The corresponding solutions are said *eigenfunctions*. We distinguish three cases.

Case $\lambda = \mu^2 > 0$. The general integral of (1.4) is

$$v(x) = C_1 e^{\mu x} + C_2 e^{-\mu x}.$$

By imposing the boundary conditions we find

$$\begin{cases} C_1 + C_2 = 0 \\ e^{\mu \pi} C_1 + e^{-\mu \pi} C_2 = 0, \end{cases}$$

so $C_1 = C_2 = 0$. This gives the zero solution only.

Case $\lambda = 0$. This situations is essentially the same as the above one. From

$$v(x) = C_1 + C_2 x,$$

the Dirichlet constraints immediately force $C_1 = C_2 = 0$.

Case $\lambda = -\mu^2 < 0$. Now we have

$$v(x) = C_1 \cos \mu x + C_2 \sin \mu x, \qquad v(0) = v(\pi) = 0.$$

From $v(0) = 0$ we deduce $C_1 = 0$; from $v(\pi) = 0$ we get

$$C_2 \sin \mu \pi = 0 \implies \mu = k \text{ positive integer and } C_2 \text{ arbitrary.}$$

Then, the eigenvalues are $\lambda_k = -k^2$ and the eigenfunctions $v_k(x) = \sin kx$. Recalling (1.3), we have the infinitely-many solutions

$$\varphi_k(x, t) = C e^{-k^2 D t} \sin kx, \quad k = 1, 2, \ldots,$$

fulfilling $\varphi_k(0) = \varphi_k(\pi) = 0$. None of these functions satisfies the condition $u(x, 0) = g(x)$, except when $g(x) = C \sin mx$ and m is an integer. The idea is then to exploit the problem linearity by assembling the v_k into a linear combination, and trying to determine the coefficients so to satisfy the initial condition. Then our candidate solution has the form:

$$u(x, t) = \sum_{k=1}^{\infty} c_k e^{-k^2 Dt} \sin kx,$$

and we seek the constants c_k by imposing

$$u(x, 0) = \sum_{k=1}^{\infty} c_k \sin kx = g(x). \tag{1.6}$$

Notice that $u(x, 0)$ is a sines-Fourier series; therefore we extend g on $[-\pi, \pi]$ as an odd function and expand it in a sines-Fourier series:

$$g(x) = \sum_{k=1}^{\infty} g_k \sin kx, \qquad g_k = \frac{2}{\pi} \int_0^{\pi} g(x) \sin kx \, dx.$$

By comparison with (1.6) we have $c_k = g_k$, and thus

$$u(x, t) = \sum_{k=1}^{\infty} g_k e^{-k^2 Dt} \sin kx \tag{1.7}$$

is the (formal) solution.

• *Analysis of (1.7)*. The function g is $C^1([0, \pi])$ and vanishes at the endpoints, so its odd prolongation on $[-\pi, \pi]$ is $C^1([-\pi, \pi])$. The theory of Fourier series guarantees that $\sum_{k=1}^{\infty} |g_k|$ converges. Since

$$\left| g_k e^{-k^2 Dt} \right| \leq |g_k|,$$

the function (1.6) converges uniformly on the entire strip $[0, \pi] \times [0, \infty)$ and we may swap the sum with the limit. This ensures that (1.7) is continuous on $[0, \pi] \times [0, \infty)$. On the other hand if $t \geq t_0 > 0$ the fast convergence rate of the exponential as $k \to \infty$ allows to differentiate term-wise (to any order), and in particular

$$u_t - Du_{xx} = \sum_{k=1}^{\infty} g_k [(u_k)_t - D(u_k)_{xx}] = 0,$$

so (1.7) solves the differential equation inside the strip.

• *Uniqueness and continuous dependence on initial data.* The uniqueness of a solution, continuous on $[0, \pi] \times [0, \infty)$, and the fact it depends in a continuous manner upon the initial data both follow from the maximum principle: indeed, if u_g is a solution corresponding to the datum g, we have

$$\max_{[0,\pi]\times[0,\infty)} |u_{g_1} - u_{g_2}| \leq \max_{[0,\pi]} |g_1 - g_2|.$$

Problem 1.2.2 (Cauchy-Neumann). *Let $D > 0$ be a constant and $g \in C^1([0,\pi])$ such that $g'(0) = g'(\pi) = 0$. Solve by separation of variables:*

$$\begin{cases} u_t(x,t) - Du_{xx}(x,t) = 0 & 0 < x < \pi, t > 0 \\ u(x,0) = g(x) & 0 \leq x \leq \pi \\ u_x(0,t) = 0, \; u_x(\pi,t) = 0 & t > 0. \end{cases}$$

Discuss uniqueness and continuous dependence on the initial data.

Solution. Since the Neumann conditions are homogeneous, we proceed by seeking *non-zero* solutions of the form

$$u(x,t) = v(x)w(t).$$

As in the previous problem, we obtain for w the equation

$$w'(t) - \lambda D w(t) = 0,$$

with general solution

$$w(t) = Ce^{\lambda Dt}, \; C \in \mathbb{R}. \tag{1.8}$$

For v we have the eigenvalue problem

$$\begin{cases} v''(x) - \lambda v(x) = 0 \\ v'(0) = v'(\pi) = 0, \end{cases}$$

where λ is a real number. As usual, we must distinguish three cases.

Case $\lambda = \mu^2 > 0$. The general integral reads

$$v(x) = C_1 e^{\mu x} + C_2 e^{-\mu x}.$$

The Neumann conditions impose $v'(0) = v'(\pi) = 0$, so

$$\begin{cases} \mu C_1 - \mu C_2 = 0 \\ e^{\mu \pi} C_1 - e^{-\mu \pi} C_2 = 0, \end{cases}$$

and then $C_1 = C_2 = 0$ because $\mu(e^{-\mu\pi} + e^{\mu\pi}) \neq 0$. The only solution is trivial.

Case $\lambda = 0$. From

$$v(x) = C_1 + C_2 x,$$

and the Neumann conditions, we deduce immediately $C_2 = 0$ and C_1 arbitrary. Now the eigenfunctions are constant functions.

Case $\lambda = -\mu^2 < 0$. We have

$$v(x) = C_1 \cos \mu x + C_2 \sin \mu x, \qquad v'(0) = v'(\pi) = 0.$$

Since

$$v'(x) = -\mu C_1 \sin \mu x + \mu C_2 \cos \mu x,$$

from $v'(0) = 0$ we deduce $C_2 = 0$; from $v'(\pi) = 0$ we infer

$$C_1 \sin \mu \pi = 0 \implies \mu = k \in \mathbb{N}, C_2 \text{ arbitrary.}$$

The eigenvalues are then $\lambda_k = -k^2$ and the eigenfunctions $v_k(x) = \cos kx$.

Recalling (1.8), we have infinitely-many solutions

$$\varphi_k(x,t) = Ce^{-k^2 Dt} \cos kx, \quad k \in \mathbb{N}$$

satisfying $\varphi'_k(0) = \varphi'_k(\pi) = 0$. None fulfils $u(x,0) = g(x)$ except when $g(x) = C \cos mx$, m integer. So let us set

$$u(x,t) = \sum_{k=0}^{\infty} c_k e^{-k^2 Dt} \cos kx$$

as candidate solution (in particular, for $k = 0$ we also obtain the constant solutions of case $\lambda = 0$). The coefficients c_k must be chosen so that

$$u(x,0) = \sum_{k=0}^{\infty} c_k \cos kx = g(x). \tag{1.9}$$

Since $u(x,0)$ is a cosines-Fourier series, we extend g as even function on $[-\pi, \pi]$ and expand it in cosines-Fourier series:

$$g(x) = \frac{g_0}{2} + \sum_{k=1}^{\infty} g_k \cos kx, \qquad g_k = \frac{2}{\pi} \int_0^{\pi} g(x) \cos kx \, dx.$$

Notice that $g_0/2$ is the mean value of the datum g on the interval $[0, \pi]$. After comparison with (1.9) we must have $c_0 = g_0/2$, $c_k = g_k$, giving the (formal) solution

$$u(x,t) = \frac{g_0}{2} + \sum_{k=1}^{\infty} g_k u_k(x,t) = \frac{g_0}{2} + \sum_{k=1}^{\infty} g_k e^{-k^2 Dt} \cos kx. \tag{1.10}$$

• *Analysis of* (1.10). The function g belongs to $C^1([0, \pi])$ and has null derivative at the endpoints, whence its even prolongation on $[-\pi, \pi]$ is in $C^1([-\pi, \pi])$. The theory of

Fourier series guarantees that $\sum_{k=1}^{\infty} |g_k|$ converges. As

$$\left| g_k e^{-k^2 Dt} \right| \leq |g_k|,$$

the function (1.10) converges uniformly on the strip $[0, \pi] \times [0, \infty)$ and we may swap the limit and the sum. This makes sure that (1.10) is continuous on $[0, \pi] \times [0, \infty)$. Now let us check the Neumann conditions on the boundary. Fix $t_0 > 0$; for t close to t_0 we can differentiate term by term, so

$$u_x(x, t) = -\sum_{k=1}^{\infty} k g_k e^{-k^2 Dt} \sin kx. \tag{1.11}$$

Since[1]

$$\left| k g_k e^{-k^2 Dt} \right| \leq \frac{1}{\sqrt{2eDt}} |g_k|,$$

the series (1.11) converges uniformly on $[0, \pi] \times [t_0, \infty)$ for any $t_0 > 0$. In particular

$$\lim_{(x,t)\to(0,t_0)} u_x(x, t) = -\sum_{k=1}^{\infty} k g_k \lim_{(x,t)\to(0,t_0)} [e^{-k^2 Dt} \sin kx] = 0$$

$$\lim_{(x,t)\to(\pi,t_0)} u_x(x, t) = -\sum_{k=1}^{\infty} k g_k \lim_{(x,t)\to(\pi,t_0)} [e^{-k^2 Dt} \sin kx] = 0.$$

The function is therefore C^1 on any strip $[0, \pi] \times [t_0, \infty)$. Similar computations show that if $t \geq t_0 > 0$ the fast convergence to zero of the exponential as $k \to \infty$ allows to differentiate (to any order) each term separately. In particular

$$u_t - Du_{xx} = \sum_{k=1}^{\infty} g_k [(u_k)_t - D(u_k)_{xx}] = 0,$$

so (1.10) is indeed a solution on the strip $[0, \pi] \times (0, \infty)$.

• *Uniqueness and continuous dependence on the data.* We use an *energy method*. Suppose there exist two solutions u and v of the same problem, defined on $[0, \pi] \times [0, \infty)$ and C^1 on $[0, \pi] \times (0, \infty)$. Set $w = u - v$ and

$$E(t) = \int_0^\pi w^2(x, t) \, dx.$$

Then $E(t) \geq 0$, $E(0) = \lim_{t \downarrow 0} E(t) = 0$, and for $t > 0$ also

$$E'(t) = 2\int_0^\pi w w_t \, dx = 2D \int_0^\pi w w_{xx} \, dx.$$

---

[1] Maximise the function $f(x) = xe^{-x^2 Dt}$.

We integrate by parts and recall that w_x vanishes at the endpoints:

$$E'(t) = -2D \int_0^\pi (w_x)^2 \, dx \le 0.$$

Consequently E decreases and therefore $E = 0$ for any $t \ge 0$. As w is continuous, $w(x,t) \equiv 0$.

By Bessel's equality, moreover,

$$\sup_{t>0} \|u(\cdot,t)\|_{L^2(0,\pi)}^2 = \sup_{t>0} \int_0^\pi u^2(x,t) \, dx \le \pi \sum_{k=0}^\infty |g_k|^2 = \pi \|g\|_{L^2(0,\pi)}^2,$$

showing that the solution depends continuously (in L^2) on the initial datum.

Problem 1.2.3 (Stationary state and asymptotic behaviour). *Let u be a solution to:*

$$\begin{cases} u_t(x,t) = D u_{xx}(x,t) & 0 < x < L, \, t > 0 \\ u(x,0) = g(x) & 0 \le x \le L \\ u_x(0,t) = u_x(L,t) = 0 & t > 0. \end{cases}$$

a) *Interpret the problem supposing u is the concentration (mass per unit of length) of a substance under the diffusion. Explain heuristically why*

$$u(x,t) \to U \quad \text{(constant)} \quad \text{as } t \to +\infty.$$

Find the value of U by integrating the equation suitably.

b) *Assume $g \in C([0,L])$, u continuous on $[0,L] \times [0,\infty)$ and C^1 on $[0,L] \times [t_0,\infty)$ for any $t_0 > 0$. Show that $u(x,t) \to U$ as $t \to +\infty$ in $L^2(0,L)$, i.e.*

$$\int_0^L (u(x,t) - U)^2 dx \to 0 \quad \text{as } t \to \infty.$$

c) *Let $g \in C^1([0,L])$ with $g'(0) = g'(\pi) = 0$. Using the formula for u of Problem 1.2.2, show that $u(x,t) \to U$ uniformly on $[0,L]$ as $t \to +\infty$.*

Solution. a) If the equation governs the (one-dimensional) diffusion of a concentrated substance then the Neumann boundary conditions tell us that the flow across the endpoints is zero. Thus it is natural to expect that the total mass is preserved, and that the substance will tend to distribute uniformly, eventually reaching a state of constant density. At the same time, neglecting the initial condition, the only stationary solutions (independent of t) of the problem are exactly the constant solutions. Integrating the equation in x over $[0,L]$ and using the Neumann conditions gives

$$\int_0^L u_t(x,t) \, dx = \int_0^L D u_{xx}(x,t) \, dx = D u_x(L,t) - D u_x(0,t) = 0,$$

whence

$$\frac{d}{dt}\int_0^L u(x,t)\,dx = 0,\tag{1.12}$$

which is precisely the conservation of the mass. As u is continuous on $[0, L]\times[0,\infty)$, we have

$$\int_0^L u(x,t)\,dx \to \int_0^L g(x)\,dx \qquad \text{for } t \to 0.$$

Therefore

$$\int_0^L u(x,t)\,dx \equiv \int_0^L g(x)\,dx.$$

Then, if $u(x,t)\to U$ as $t\to+\infty$, it must be

$$U = \frac{1}{L}\int_0^L g(x)\,dx.\tag{1.13}$$

b) We shall prove that, indeed, u converges in $L^2(0, L)$ to the constant U just defined. Set $w(x,t) = u(x,t) - U$. Clearly $w(x,0) = g(x) - U$, and by (1.13) the function w has zero spatial mean: $\int_0^L w(x,t)\,dx \equiv 0$, for any $t \ge 0$. As in Problem 1.2.2, if

$$E(t) = \int_0^L w^2(x,t)\,dx$$

it follows that

$$E'(t) = -2D\int_0^L w_x^2(x,t)\,dx.$$

At the very end we shall prove that[2]

$$\int_0^L w_x^2(x,t)\,dx \ge \frac{E(t)}{L}.\tag{1.14}$$

So we have an (ordinary) *differential inequality*:

$$E'(t) \le -\frac{2D}{L}E(t),$$

that is,

$$\frac{d}{dt}\log E(t) \le -\frac{2D}{L}.$$

Integrating from 0 to t

$$\log E(t) - \log E(0) \le -\frac{2D}{L}t$$

and thus

$$E(t) \le E(0)e^{-\frac{2D}{L}t}.$$

---

[2] Poincaré's inequality.

Therefore as $t \to +\infty$, $E(t)$ tends to 0, that is

$$\int_0^L (u(x,t) - U)^2 dx \to 0 \qquad \text{as } t \to \infty,$$

i.e. $u(\cdot, t)$ tends to U in L^2-norm as $t \to \infty$.

c) Take $g \in C^1([0, L])$ with $g'(0) = g'(L) = 0$. We use the explicit expression of the solution obtained in Problem 1.2.2. Passing from $[0, \pi]$ to $[0, L]$ we find

$$u(x,t) = \frac{a_0}{2} + \sum_{n=1}^{\infty} a_n e^{-D\lambda_n^2 t} \cos \lambda_n x,$$

where $\lambda_n = n\pi/L$, $a_n = \frac{2}{L} \int_0^L g(x) \cos \lambda_n x \, dx$. Now if $n \geq 1, t > 0$,

$$\left| a_n e^{-D\lambda_n^2 t} \cos \lambda_n x \right| \leq e^{-D\lambda_1^2 t} |a_n|$$

and since $g \in C^1([0, L])$, $g'(0) = g'(L) = 0$,

$$\sum_{n=1}^{\infty} |a_n| = S < \infty.$$

Hence

$$\left| \sum_{n=1}^{\infty} a_n e^{-D\lambda_n^2 t} \cos \lambda_n x \right| \leq e^{-D\lambda_1^2 t} \sum_{n=1}^{\infty} |a_n| = e^{-D\lambda_1^2 t} S \to 0 \qquad \text{as } t \to +\infty.$$

Therefore, as $t \to +\infty$, $u(x,t)$ tends (exponentially) to $a_0/2 = U$, uniformly in x.

• *Proof of* (1.14). By the intermediate value theorem, for any $t > 0$ one can find $x(t)$ such that

$$w(x(t), t) = \frac{1}{L} \int_0^L w(x,t) \, dx = 0.$$

The fundamental theorem of calculus then implies, for any $x \in [0, L]$:

$$w(x,t) = \int_{x(t)}^x w_x(s,t) \, ds$$

and by Schwarz's inequality:

$$|w(x,t)| = \left| \int_{x(t)}^x w_x(s,t) \, ds \right| \leq \int_0^L |w_x(s,t)| \, ds \leq \sqrt{L} \left(\int_0^L w_x^2(s,t) \, ds \right)^{1/2}.$$

Squaring both sides and integrating over $[0, L]$ gives

$$\int_0^L w^2(s,t)\, ds \le L \int_0^L w_x^2(s,t)\, ds.$$

Problem 1.2.4 (Cauchy-Neumann; non-homogeneous equation). *Solve, using separation of variables, the problem:*

$$\begin{cases} u_t(x,t) - u_{xx}(x,t) = tx & 0 < x < \pi,\, t > 0 \\ u(x,0) = 1 & 0 \le x \le \pi \\ u_x(0,t) = u_x(\pi,t) = 0 & t > 0. \end{cases}$$

Solution. This is a *non-homogeneous* Neumann problem with homogeneous boundary conditions. In order to be able to separate the variables it is convenient to consider first the homogeneous equation, and in particular the associated eigenvalue problem:

$$\begin{cases} v''(x) - \lambda v(x) = 0 \\ v'(0) = v'(\pi) = 0. \end{cases}$$

In Problem 1.2.2 we found the eigenvalues $\lambda_k = -k^2$ and the eigenfunctions $v_k(x) = \cos kx$. Let us write the candidate solution as

$$u(x,t) = \sum_{k=0}^{\infty} c_k(t) \cos kx$$

and impose (recall that $v_k'' = -k^2 v_k$):

$$u_t - u_{xx} = \sum_{k=0}^{\infty} \left[c_k'(t) + k^2 c_k(t) \right] \cos kx = tx$$

with

$$u(x,0) = \sum_{k=0}^{\infty} c_k(0) \cos kx = 1.$$

We expand $f(x) = x$ in cosines-Fourier series:

$$x = \frac{\pi}{2} - \frac{4}{\pi} \sum_{k=0}^{\infty} \frac{\cos\left[(2k+1)x\right]}{(2k+1)^2},$$

uniformly convergent on $[0, \pi]$. Comparing the last three equations, the coefficients $c_k(t)$ must solve the following Cauchy problems:

$$c_0'(t) = \tfrac{\pi}{2}t, \qquad\qquad\qquad\qquad c_0(0) = 1;$$
$$c_{2k}'(t) + 4k^2 c_{2k}(t) = 0, \qquad\qquad c_{2k}(0) = 0,\, k \ge 1;$$
$$c_{2k+1}'(t) + (2k+1)^2 c_{2k+1}(t) = -\tfrac{4}{\pi}\tfrac{1}{(2k+1)^2}t, \quad c_{2k+1}(0) = 0,\, k \ge 0.$$

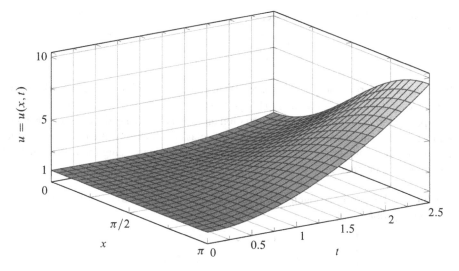

Fig. 1.1 Solution to Problem 1.2.4 for $0 < x < \pi, 0 < t < 2.5$

Solving the ODE problem, we find

$$c_0(t) = \tfrac{\pi}{4}t^2 + 1;$$

$$c_{2k}(t) = 0, \qquad\qquad\qquad\qquad k \geq 1;$$

$$c_{2k+1}(t) = -\frac{4}{\pi(2k+1)^4}\left[t + \frac{1}{(2k+1)^2}\left(e^{-(2k+1)^2 t} - 1\right)\right], \quad k \geq 0.$$

Hence the (formal) solution reads (Fig. 1.1):

$$u(x,t) = \frac{\pi}{4}t^2 + 1 + \sum_{k=0}^{\infty} c_{2k+1}(t)\cos\left[(2k+1)x\right]. \tag{1.15}$$

• *Analysis of* (1.15). As

$$\left|\left(e^{-(2k+1)^2 t} - 1\right)\cos(2k+1)x\right| \leq 2$$

we deduce that the series

$$\sum_{k=0}^{\infty} \frac{e^{-(2k+1)^2 t} - 1}{(2k+1)^6}\cos(2k+1)x,$$

the series of first and second partial derivatives in x, given by

$$-\sum_{k=0}^{\infty} \frac{e^{-(2k+1)^2 t} - 1}{(2k+1)^5}\sin(2k+1)x, \qquad -\sum_{k=0}^{\infty} \frac{e^{-(2k+1)^2 t} - 1}{(2k+1)^4}\cos(2k+1)x,$$

and the series of derivatives in t, given by

$$-\sum_{k=0}^{\infty} \frac{e^{-(2k+1)^2 t}}{(2k+1)^4} \cos (2k+1) x,$$

all converge uniformly on $[0, \pi] \times [0, \infty)$. Hence the derivatives can be carried into the sum, making u a C^2 function on $[0, \pi] \times [0, \infty)$.

In particular u solves the diffusion equation on $(0, \pi) \times (0, \infty)$ and it assumes the given boundary values in pointwise sense. Within this class of functions, it can be proved that the solution is unique and it depends continuously on the data using the energy method (Problem 1.2.2 on page 6).

Problem 1.2.5 (Non-homogeneous Neumann). *Solve by separation of variables the following problem:*

$$\begin{cases} u_t(x,t) - u_{xx}(x,t) = 0 & 0 < x < \pi, t > 0 \\ u(x,0) = 0 & 0 \le x \le \pi \\ u_x(0,t) = 0, \ u_x(\pi,t) = U & t > 0. \end{cases}$$

If $U \neq 0$, can there be a stationary solution $u_\infty = u_\infty(x)$?

Solution. This Neumann problem has non-homogeneous boundary conditions. Let us observe immediately that $U \neq 0$ prevents the existence of stationary solutions $u_\infty = u_\infty(x)$, for otherwise we would have $u_\infty''(x) = 0$, $u_\infty'(0) = 0$, $u_\infty'(\pi) = U$, which is impossible.

To separate variables we reduce to homogeneous conditions by setting

$$w(x,t) = u(x,t) - v(x)$$

where $v_x(0) = 0$, $v_x(\pi) = U$. For example we can choose

$$v(x) = \frac{Ux^2}{2\pi}.$$

The function w solves the problem

$$\begin{cases} w_t(x,t) - w_{xx}(x,t) = U/\pi & 0 < x < \pi, t > 0 \\ w(x,0) = -Ux^2/2\pi & 0 \le x \le \pi \\ w_x(0,t) = 0, \ w_x(\pi,t) = 0 & t > 0. \end{cases}$$

As in Problem 1.2.4, given the homogeneous Neumann conditions, we write

$$w(x,t) = \frac{c_0(t)}{2} + \sum_{k=1}^{\infty} c_k(t) \cos kx$$

so that the Neumann conditions are (formally) satisfied. We have to find $c_k(t)$ so that

$$w_t - w_{xx} = \frac{c_0'(t)}{2} + \sum_{k=1}^{\infty} [c_k'(t) + k^2 c_k(t)] \cos kx = \frac{U}{\pi}$$

and

$$w(x,0) = \frac{c_0(0)}{2} + \sum_{k=1}^{\infty} c_k(0) \cos kx = -\frac{Ux^2}{2\pi}.$$

Let us expand $g(x) = \frac{Ux^2}{2\pi}$ in cosines Fourier series:

$$\frac{Ux^2}{2\pi} = \frac{U}{2\pi} \left\{ \frac{\pi^2}{3} + 4 \sum_{k=1}^{\infty} \frac{(-1)^k}{k^2} \cos kx \right\},$$

which is uniformly convergent on $[0, \pi]$. The comparison of the last three formulas forces the coefficients $c_k(t)$ to solve the Cauchy problems:

$$c_0'(t) = \frac{2U}{\pi}, \qquad c_0(0) = -\frac{U\pi}{3};$$

$$c_k'(t) + k^2 c_k(t) = 0, \quad c_k(0) = \frac{2U}{\pi} \frac{(-1)^{k+1}}{k^2}, \quad k \geq 1.$$

We find

$$c_0(t) = \frac{2U}{\pi} t - \frac{U\pi}{3}, \qquad c_k(t) = \frac{2U}{\pi} \frac{(-1)^{k+1}}{k^2} e^{-k^2 t}, \quad k \geq 1,$$

and the solution reads (Fig. 1.2):

$$u(x,t) = \frac{U}{\pi} t + \frac{Ux^2}{2\pi} - \frac{U\pi}{6} + \frac{2U}{\pi} \sum_{k=1}^{\infty} \frac{(-1)^{k+1}}{k^2} e^{-k^2 t} \cos kx. \tag{1.16}$$

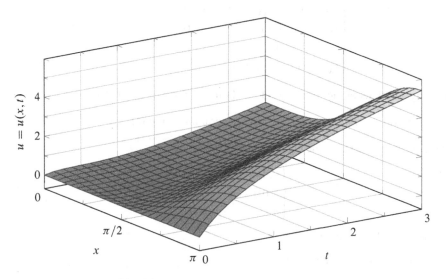

Fig. 1.2 Solution to Problem 1.2.5 ($U = \pi$)

• *Analysis of* (1.16). The series is uniformly convergent on $[0, \pi] \times [0, \infty)$, so u is continuous there. All derivatives pass into the sum on $[0, \pi] \times [t_0, \infty)$ for any $t_0 > 0$. Thus u solves the diffusion equation on $(0, \pi) \times (0, \infty)$.

Problem 1.2.6 (Mixed Neumann-Robin). *A bar made of a homogeneous material, of length L, is insulated on its sides and at its left end in $x = 0$. The other end is subject to a linear Newton's law of cooling (the heat flow at this end is a multiple of the difference between the temperature u and the ambient temperature U).*

a) Write the mathematical model for the evolution of u when $t > 0$.

b) Solve the problem by separation of variables, assuming that the initial temperature profile is continuous in $[0, L]$. Study the behaviour of the solution as $t \to +\infty$.

Solution. a) Call D the diffusivity coefficient, $\gamma > 0$ the coefficient appearing in Newton's law at $x = L$, and g the temperature function at time $t = 0$, with $g \in C(\mathbb{R})$ and L-periodic. The ensuing problem is of Neumann-Robin type:

$$\begin{cases} u_t(x,t) - Du_{xx}(x,t) = 0 & 0 < x < L, t > 0 \\ u(x,0) = g(x) & 0 \le x \le L \\ u_x(0,t) = 0 & t > 0 \\ u_x(L,t) = -\gamma(u(L,t) - U) & t > 0. \end{cases}$$

b) Since the Robin conditions at $x = L$ is not homogeneous, we set $z(x,t) = u(x,t) - U$. The new unknown satisfies the homogeneous problem

$$\begin{cases} z_t(x,t) - Dz_{xx}(x,t) = 0 & 0 < x < L, t > 0 \\ z(x,0) = g(x) - U & 0 \le x \le L \\ z_x(0,t) = 0 & t > 0 \\ z_x(L,t) = -\gamma z(L,t) & t > 0. \end{cases}$$

Seeking solutions of the form $z(x,t) = v(x)w(t)$, the equation for w is

$$w'(t) - \lambda Dw(t) = 0, \qquad \text{with general integral } w(t) = Ce^{\lambda Dt}, C \in \mathbb{R},$$

while for v we are led to the eigenvalue problem

$$\begin{cases} v''(x) + \lambda^2 v(x) = 0 \\ v'(0) = 0 \\ v'(L) = -\gamma v(L). \end{cases} \tag{1.17}$$

We distinguish three cases:

Case $\lambda = \mu^2 > 0$. We have

$$v(x) = C_1 e^{\mu x} + C_2 e^{-\mu x}$$

and then

$$\begin{cases} \mu\, C_1 - \mu C_2 = 0 \\ (\mu + \gamma)e^{\mu L}\, C_1 - (\mu - \gamma)e^{-\mu L}\, C_2 = 0. \end{cases}$$

Since[3] $\mu\left[(\mu + \gamma)e^{\mu L} - (\mu - \gamma)e^{-\mu L}\right] \neq 0$, we find $C_1 = C_2 = 0$.

Case $\lambda = 0$. From

$$v(x) = C_1 + C_2 x,$$

the Robin conditions force $C_1 = C_2 = 0$, hence again the trivial solution.

Case $\lambda = -\mu^2 < 0$. We have

$$v(x) = C_1 \cos \mu x + C_2 \sin \mu x.$$

As

$$v'(x) = -\mu C_1 \sin \mu x + \mu C_2 \cos \mu x,$$

$v'(0) = 0$ implies $C_2 = 0$, and $v'(L) = -\gamma v(L)$ implies

$$\mu \sin \mu L = \gamma \cos \mu L, \qquad \text{i.e. } \tan \mu L = \frac{\gamma}{\mu}. \tag{1.18}$$

Define $s = \mu L$, so (1.18) reads $\tan s = \gamma L/s$. In Fig. 1.3 we see that when $s > 0$ the graphs of the tangent function $y = \tan s$ and the hyperbola $y = \gamma L/s$ cross infinitely many times, for $0 < s_1 = \mu_1 L < s_2 = \mu_2 L < \ldots$, say. Note that

$$(n-1)\pi < \mu_n L < n\pi,$$

so $\mu_n \sim n\pi/L$ as $n \to \infty$. Therefore also $\tan \mu_n L$ and $\sin \mu_n L$ tend to zero as $n \to \infty$. The eigenvalues are

$$\lambda_n = -\mu_n^2 = -\frac{s_n^2}{L^2}$$

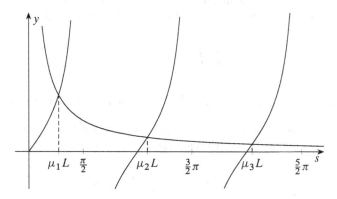

Fig. 1.3 Intersections between the graphs of $y = \tan s$ and $y = \gamma L/s$

---

[3] In fact, $e^{2\mu L} > (\mu - \gamma)/(\mu + \gamma)$ for any $\mu, \gamma > 0$.

and the eigenfunctions $v_n(x) = \cos \mu_n x$. The candidate solution is

$$z(x,t) = \sum_{n=1}^{\infty} a_n e^{-D\mu_n^2 t} \cos \mu_n x.$$

The initial condition requires

$$z(x,0) = \sum_{n=1}^{\infty} a_n \cos \mu_n x = g(x) - U, \qquad 0 \le x \le L.$$

Problem (1.17) is a *regular Sturm-Liouville*[4] problem, and therefore the eigenfunctions $v_n(x) = \cos \mu_n x$ satisfy

$$\int_0^L v_n(x) v_m(x)\, dx = \begin{cases} 0 & m \ne n \\ \frac{L}{2} + \frac{\sin(2\mu_n L)}{4\mu_n} \equiv \beta_n & m = n. \end{cases}$$

If $g \in C([0, L])$ we can write

$$g(x) - U = \sum_{n=1}^{\infty} g_n \cos \mu_n x,$$

with convergence in $L^2(0, L)$, where

$$g_n = \frac{1}{\beta_n} \int_0^L [g(x) - U] \cos \mu_n x\, dx.$$

This gives

$$z(x,t) = \sum_{n=1}^{\infty} g_n e^{-D\mu_n^2 t} \cos \mu_n x \qquad (1.19)$$

and so $u(x,t) = z(x,t) + U$.

• *Analysis of* (1.19). We remind that $\mu_n \sim n\pi/L$, so that $\tan \mu_n L$ and $\sin \mu_n L$ tend to zero as $n \to \infty$. Moreover, $\beta_n \to L/2$ if $n \to \infty$. This implies, given the assumptions on g,

$$|g_n| \le \frac{1}{\beta_n} \int_0^L |g(x) - U|\, dx \le M$$

and

$$\left| g_n e^{-D\mu_n^2 t} \cos \mu_n x \right| \le M e^{-D\mu_n^2 t}, \qquad \text{for a suitable constant } M.$$

The series (1.19) is uniformly convergent on $[0, \pi] \times [t_0, \infty)$ and all derivatives can be carried under the sum, for any $t_0 > 0$. Thus u solves the diffusion equation on $(0, \pi) \times (0, \infty)$. When $t \to +\infty$, we have

$$z(x,t) \to 0, \qquad \text{uniformly on } [0, \pi],$$

hence $u(x,t) \to U$.

---

[4] Appendix A.

Problem 1.2.7 (Ill posed problem; *backward* equation). *Let u be the solution to the end-value problem:*

$$\begin{cases} u_t(x,t) - u_{xx}(x,t) = 0 & 0 < x < \pi, 0 < t < T \\ u(x,T) = g(x) & 0 \leq x \leq \pi \\ u(0,t) = u(\pi,t) = 0 & 0 < t < T. \end{cases} \quad (1.20)$$

a) *Show that the variable change* $t = T - s$ *transforms problem (1.20) for a forward equation into an initial-value problem for a backward equation.*

b) *Solve (formally) by separation of variables, specifying which hypotheses on g guarantee that the expression found is indeed a solution.*

c) *Show the solution does not depend continuously on the data, by taking* $g_n(x) = \frac{1}{n} \sin nx$.

Solution. a) Set $t = T - s$ and $v(x,s) = u(x, T - s)$, so that

$$v_s(x,s) = -u_t(T-s) \quad \text{and} \quad v(x,0) = u(x,T),$$

resulting in the following problem for v:

$$\begin{cases} v_s(x,s) + v_{xx}(x,s) = 0 & 0 < x < \pi, 0 < s < T \\ v(x,0) = g(x) & 0 \leq x \leq \pi \\ v(0,s) = u(1,s) = 0 & 0 < s < T, \end{cases}$$

which is a *backward in time* equation with initial datum g.

b) Exactly as in Problem 1.2.1 (page 3) we find the following expression for the solution:

$$u(x,t) = \sum_{k=1}^{\infty} c_k v_k(x,t) = \sum_{k=1}^{\infty} c_k e^{-k^2 t} \sin kx.$$

This time, though, the coefficients c_k must be chosen so that $u(x,T) = g(x)$, in other words

$$c_k = g_k e^{k^2 T}$$

by writing

$$g(x) = \sum_{k=1}^{\infty} g_k \sin kx.$$

This gives

$$u(x,t) = \sum_{k=1}^{\infty} c_k v_k(x,t) = \sum_{k=1}^{\infty} g_k e^{k^2(T-t)} \sin kx. \quad (1.21)$$

If we compare to the initial-value situation, now the exponential "fights against" convergence, for the exponent is positive for $t < T$.

To be able to differentiate inside the sum, thus ensuring that (1.21) really is a solution, we need the coefficients g_k to tend to zero *very rapidly*. For instance, it suffices to have

$$\sum_{k=1}^{\infty} |g_k| e^{k^2 T} < \infty. \tag{1.22}$$

In particular, (1.22) implies $g_k = o(k^{-m})$ for any $m \geq 1$, i.e. g *must* have a 2π-periodic odd prolongation at least of class $C^\infty (\mathbb{R})$. This is not so surprising if we think that in the *forward* problem the solution becomes $C^\infty (\mathbb{R})$ for $t > 0$, at least on $0 < x < \pi$, even if we start with a non-regular datum at $t = 0$. The datum has to take into account the *strong regularising* effect of the diffusion equation.

c) Let us pick

$$g_n (x) = \frac{1}{n} \sin nx$$

as final datum. Equation (1.21) gives

$$u_n (x, t) = \frac{1}{n} e^{n^2 (T-t)} \sin nx.$$

Since

$$|g_n (x)| = \frac{1}{n} |\sin nx| \leq \frac{1}{n},$$

$g_n \to 0$ uniformly on $\mathbb{R}$ as $n \to \infty$; on the other hand for $n = 2m + 1$

$$\left| u_n \left(\frac{\pi}{2}, 0 \right) \right| = \frac{1}{n} e^{n^2 T} \to \infty$$

as $n \to \infty$. This implies that the solution does not depend on the data in a continuous way, and the problem is ill-posed.

1.2.2 Use of the maximum principle

> **Problem 1.2.8** (Maximum principle). *Let u be a solution to*
>
> $$\begin{cases} u_t(x,t) - u_{xx}(x,t) = 0 & 0 < x < 1, t > 0 \\ u(x,0) = \sin \pi x & 0 \leq x \leq 1 \\ u(0,t) = 2te^{1-t}, u(1,t) = 1 - \cos \pi t & t > 0, \end{cases}$$
>
> *that is continuous[a] on the closure of the half-strip $S = (0, 1) \times (0, \infty)$.*
>
> *a) Prove that u is non-negative.*
>
> *b) Determine an upper bound for $u \left(\frac{1}{2}, \frac{1}{2} \right)$ and $u \left(\frac{1}{2}, 3 \right)$.*
>
> ──────────
>
> [a] Note how the boundary values agree on $(0,0)$ and $(1,0)$.

Solution. a) The parabolic boundary $\partial_p S$ of the strip is the union of the half-lines $x = 0$, $x = 1$, $t > 0$ and the segment $0 \leq x \leq 1$ on the x-axis ($t = 0$). By the maximum principle u is non-negative on the entire strip provided $u \geq 0$ on $\partial_p S$. On the boundary

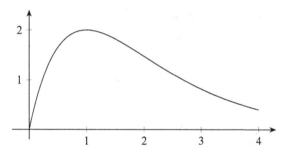

Fig. 1.4 Graph of $t \mapsto 2te^{1-t}$

half-lines $2te^{-t} \geq 0$, and $1 - \cos \pi x \geq 0$; also the datum $\sin \pi x$ is non-negative. Hence $u \geq 0$ on S.

b) By the maximum principle, the value $u \left(1/2, 1/8\right)$ does not exceed the data on the parabolic boundary of the strip $S_{1/8} = (0, 1) \times (0, 1/8)$, which is

$$\{0 \leq x \leq 1, t = 0\} \cup \left\{x = 0,\ 0 \leq t \leq \frac{1}{8}\right\} \cup \left\{x = 1,\ 0 \leq t \leq \frac{1}{8}\right\}.$$

The maximum value of the datum and of $1-\cos \pi t$ is 1. The graph of $2te^{1-t}$ (see Fig. 1.4) has an absolute maximum 2 at $t = 1$; on $[0, 1/8]$ the maximum is $e^{7/8}/4 \simeq 0.599\,72 < 1$. We can only say $u \left(1/2, 1/8\right) < 1$ (the strict inequality holds by the strong maximum principle). Similarly, $u \left(\frac{1}{2}, 3\right) < 2$. Actually, we can also say that $u \left(x, t\right) < 2$ on S.

Problem 1.2.9 (Asymptotic behaviour). *Let u be a continuous solution on the closure of $S = (0, 1) \times (0, \infty)$ to the problem*

$$\begin{cases} u_t(x,t) - u_{xx}(x,t) = 0 & 0 < x < 1,\ t > 0 \\ u(x,0) = x(1-x) & 0 \leq x \leq 1 \\ u(0,t) = u(1,t) = 0 & t > 0. \end{cases}$$

After proving that u is non-negative, determine positive numbers α, β, so that

$$u(x,t) \leq w(x,t) \equiv \alpha x(1-x)e^{-\beta t}.$$

Deduce that $u(x,t) \rightarrow 0$ uniformly on $[0, L]$ as $t \rightarrow +\infty$.

Solution. We begin by proving that $u \geq 0$. For this we recall that by the maximum principle is enough to have $u \geq 0$ on the parabolic boundary $\partial_p S$ of the strip S. For $t = 0$, in fact, $u(x,0) = x(1-x)$ is nonnegative on $[0, 1]$. Moreover $u = 0$ along the sides $x = 0, x = 1$. Hence $u \geq 0$. The idea, to have w larger that u, is to use the maximum principle for the continuous function $v = w - u$. More precisely, we look for α, β so that

$v \geq 0$ on $\partial_p S$ and v is a *super-solution* (i.e. $v_t - v_{xx} \geq 0$):

$$w(x,0) = \alpha x(1-x)$$
$$w_t(x,t) = -\alpha\beta x(1-x)e^{-\beta t}$$
$$w_{xx}(x,t) = -2\alpha e^{-\beta t},$$

so

$$\begin{cases} v_t(x,t) - v_{xx}(x,t) = \alpha(2 - \beta x(1-x))e^{-\beta t} & 0 < x < 1,\ t > 0 \\ v(x,0) = (\alpha - 1)x(1-x) & 0 \leq x \leq 1 \\ v(0,t) = v(1,t) = 0 & t > 0. \end{cases}$$

Let us find $\beta > 0$ so that $2 - \beta x(1-x) \geq 0$. As $x(1-x) \leq \frac{1}{4}$,

$$2 - \beta x(1-x) \geq 2 - \frac{1}{4}\beta$$

and then is suffices to choose $0 < \beta \leq 8$. Let us find the sign of v on $\partial_p S$. Along the sides of S we have $v = 0$, while $v(x,0) = (\alpha - 1)x(1-x) \geq 0$, for $t = 0$, provided $\alpha \geq 1$. Hence $v \geq 0$ on $\partial_p S$ if $\alpha \geq 1$. To sum up, for $\alpha \geq 1$ and $0 < \beta \leq 8$, v is a non-negative super-solution. We may then use the maximum principle on v, obtaining non-negativity, and so

$$0 \leq u(x,t) \leq \alpha x(1-x)e^{-\beta t} \leq \frac{\alpha}{4}e^{-\beta t}$$

because $x(1-x) \leq 1/4$. As $\beta > 0$, $e^{-\beta t} \to 0$ for $t \to \infty$, so $u(x,t) \to 0$ uniformly on $[0,1]$ as $t \to +\infty$.

Problem 1.2.10 (Stationary state and asymptotic behaviour). *Consider the problem*

$$\begin{cases} u_t(x,t) - u_{xx}(x,t) = 1 & 0 < x < 1,\ t > 0 \\ u(x,0) = 0 & 0 \leq x \leq 1 \\ u(0,t) = u(1,t) = 0 & t > 0. \end{cases}$$

a) *Determine the stationary solution $u^s = u^s(x)$ that satisfies the boundary conditions.*

b) *Show $u(x,t) \leq u^s(x)$ when $t > 0$.*

c) *Find $\beta > 0$ so that $u(x,t) \geq (1 - e^{-\beta t})u^s(x)$.*

d) *Deduce $u(x,t)$ tends to $u^s(x)$ as $t \to +\infty$, uniformly on $[0,1]$.*

e) *Double-check the result by solving the problem by separation of variables.*

Solution. a) We recall that a stationary solution does not depend on the time variable ($u_t(x,t) \equiv 0$). Thus, we are asked to find $u^s(x) = \psi(x)$ such that

$$\begin{cases} -\psi''(x) = 1, & 0 < x < 1, \\ \psi(0) = \psi(1) = 0. \end{cases}$$

The solution is the parabola

$$u^s(x) = \frac{1}{2}x(1-x).$$

b) Set $v(x,t) = u^s(x) - u(x,t)$. Then

$$v_t(x,t) - v_{xx}(x,t) = 0, \qquad 0 < x < 1, \, t > 0.$$

As u is continuous on the closure of $S = [0,1] \times (0,\infty)$, to prove the non-negativity of v we can show that the boundary data are non-negative on the parabolic boundary, by the maximum principle. In fact $v(0,t) = v(1,t) = 0$ for $t > 0$, and

$$v(x,0) = \frac{1}{2}x(1-x) \geq 0,$$

if $0 \leq x \leq 1$. The maximum principle implies $v \geq 0$, i.e. $u^s \geq u$.

c) As for Problem 1.2.9 we set $w(x,t) = u(x,t) - (1 - e^{-\beta t})u^s(x)$ and look for $\beta > 0$ so that w is a super-solution and non-negative on $\partial_p S$. Since

$$\partial_t[(1 - e^{-\beta t})u^s(x)] = \frac{\beta}{2}x(1-x)e^{-\beta t}$$

$$\partial_{xx}[(1 - e^{-\beta t})u^s(x)] = -1 + e^{-\beta t},$$

we have

$$\begin{cases} w_t(x,t) - w_{xx}(x,t) = e^{-\beta t}(1 - \frac{\beta}{2}x + \frac{\beta}{2}x^2) & 0 < x < 1, \, t > 0 \\ w(x,0) = 0 & 0 \leq x \leq 1 \\ w(0,t) = w(1,t) = 0 & t > 0. \end{cases}$$

So, we have to find β rendering the right-hand side non-negative. After that, the maximum principle allows to conclude. Now the right side of the equation is non-negative if

$$\beta x^2 - \beta x + 2 \geq 0,$$

so it suffices to demand $\beta \leq 8$.

d) Take $\beta \leq 8$, so that:

$$(1 - e^{-\beta t})u^s(x) \leq u(x,t) \leq u^s(x),$$

i.e.

$$0 \leq u^s(x) - u(x,t) \leq e^{-\beta t}u^s(x).$$

Then

$$\sup_{x \in [0,1]} |u^s(x) - u(x,t)| \leq \sup_{x \in [0,1]} e^{-\beta t}u^s(x) \leq \frac{1}{8}e^{-\beta t} \to 0 \qquad \text{as } t \to +\infty.$$

Therefore $u(x,t) \to 0$ uniformly on $[0,1]$ as $t \to +\infty$.

e) The solution has the form

$$w(x,t) = \frac{1}{2}x(1-x) + \sum_{k=1}^{\infty} c_k e^{-k^2\pi^2 t} \sin k\pi x$$

where the c_k are chosen so that

$$w(x,0) = \frac{1}{2}x(1-x) + \sum_{k=1}^{\infty} c_k \sin k\pi x = 0.$$

Since $x(1-x)$ vanishes at $x = 0, 1$, we have that $S = \sum_1^{\infty} |c_k| < \infty$. As a consequence,

$$\left| w(x,t) - \frac{1}{2}x(1-x) \right| \le S e^{-\pi^2 t} \to 0, \qquad \text{as } t \to \infty.$$

**Problem 1.2.11 (Hopf's principle). *Let u solve*

$$u_t(x,t) - u_{xx}(x,t) = 0$$

on the rectangle $Q_T = (0,1) \times (0,T)$, and assume that u is $C^1(\overline{Q}_T)$.

a) Given $0 < t_0 \le T$ suppose

$$u(x,t) > m, \qquad \text{for } 0 \le x \le 1, \ 0 < t < t_0,$$

and $u(0,t_0) = m$. Prove that $u_x(0,t_0) > 0$ (it cannot vanish)[a].

b) Deduce that the Neumann problem

$$\begin{cases} u_t(x,t) - u_{xx}(x,t) = 0 & \text{in } Q_T \\ u(x,0) = g(x) & 0 \le x \le 1 \\ u_x(0,t) = u_x(1,t) = 0 & 0 < t \le T, \end{cases}$$

has a unique solution u, and

$$\min_{[0,1]} g = \min_{\overline{Q}_T} u \le \max_{\overline{Q}_T} u = \max_{[0,1]} g.$$

---

[a] *Hint.* Compare u with $z(x,t) = e^x - 1$. Note that $z(0,t) = 0$, $z_x(0,t) = 1 > 0$.

Solution. a) First, set $v = u - m$ so that $v(0,t_0) = 0$ and $v > 0$ when $0 \le x \le 1$, $0 < t < t_0$. The idea is to find a function w smaller than v on a neighbourhood of $(0,t_0)$, that vanishes at $(0,t_0)$ and with $w_x(0,t_0) > 0$. Once w is found we can write

$$\frac{v(h,t_0) - v(0,t_0)}{h} > \frac{w(h,t_0) - w(0,t_0)}{h},$$

and passing to the limit for $h \to 0^+$ we obtain

$$v_x(0,t_0) \ge w_x(0,t_0) > 0.$$

We now construct w. As a neighbourhood of $(0,t_0)$ we choose the rectangle

$$R = \left(0, \frac{1}{2}\right) \times \left(\frac{t_0}{2}, t_0\right).$$

Note that the function

$$z(x,t) = e^x - 1$$

is zero at $x = 0$. On R, moreover,

$$0 \le z \le \sqrt{e} - 1 \equiv a, \qquad z_x(0, t_0) = 1 > 0, \qquad z_t - z_{xx} = -e^x < 0.$$

Call $m_0 \, (> 0)$ the minimum of v along the sides $x = 1/2$ and $t = t_0/2$ of ∂R. The function

$$w(x) = \frac{m_0}{a} z(x)$$

is a non-negative sub-solution of the diffusion equation; it is smaller than v on the parabolic boundary of R, it vanishes at $x = 0$ and $w_x(0, t_0) = \frac{m_0}{a} \frac{1}{e} > 0$. These are the properties we wanted.

b) We shall prove $\min_{[0,1]} g = \min_{\overline{Q}_T} u$. The claim about the maximum is similar. Suppose $\min_{[0,1]} g > \min_{\overline{Q}_T} u$, so the maximum principle implies u has minimum on $\overline{Q}_T$ at points on the sides of Q_T. If either $(0, t_0)$ or $(1, t_0)$, $0 < t_0 \le T$, were the minimum point with smallest t-coordinate, by part a) the spatial derivative should be non-zero there; but this would contradict the homogeneous Neumann conditions.

Finally, the uniqueness of the solution comes from the fact that the difference of any two solutions with same data would have zero initial datum.

1.2.3 Applying the notion of fundamental solution

Problem 1.2.12 (Non-instantaneous point source). *A polluting agent with concentration* $u = u(x, t)$ *(mass per unity of length) diffuses, with coefficient* D, *along a narrow channel (the x-axis). At* $x = 0$ *the quantity* $q = q(t)$ *of pollutant (mass per second per unity of length) enters, where*

$$q(t) = \begin{cases} Q & 0 < t < T \\ 0 & t > T. \end{cases}$$

Determine the pollutant concentration at $x = 0$ *and time* t, *and its asymptotic behaviour as* $t \to \infty$.

Solution. Recall that, if $t > s$, $\Gamma_D(0, t - s)$ indicates the concentration at $x = 0$ and time t due to a point source with unit intensity placed at $x = 0$ at time $t = s$. If the source at $t = s$ produces $q(s)$, the contribution to the concentration at $x = 0$ and time t is

$$\Gamma_D(0, t - s)q(s) = \begin{cases} 0 & t < s \\ \dfrac{1}{\sqrt{4\pi D(t - s)}} q(s) & t > s. \end{cases}$$

The sum of all contributions as $s < t$ varies gives

$$u(0, t) = \int_0^t \Gamma_D(0, t - s)q(s)\, ds.$$

Therefore, for $t < T$,

$$u(0, t) = \int_0^t Q\Gamma_D(0, t - s)\, ds = \frac{Q}{\sqrt{4\pi D}} \int_0^t \frac{1}{\sqrt{t - s}}\, ds = \frac{Q}{\sqrt{\pi D}} \sqrt{t},$$

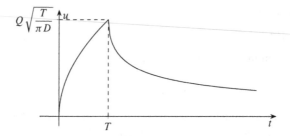

Fig. 1.5 The concentration $u(0,t)$ in Problem 1.2.12

while, for $t > T$,

$$u(0,t) = \int_0^T \Gamma_D(0, t-s)\, ds = \frac{Q}{\sqrt{4\pi D}} \int_0^T \frac{1}{\sqrt{t-s}}\, ds = \frac{Q}{\sqrt{\pi D}}\left[\sqrt{t} - \sqrt{t-T}\right].$$

This means that when $t \to \infty$ we have (Fig. 1.5):

$$u(0,t) = \frac{Q}{\sqrt{\pi D}}\left[\sqrt{t} - \sqrt{t-T}\right] = \frac{Q}{\sqrt{\pi D}} \frac{T}{\sqrt{t} + \sqrt{t-T}} \sim \frac{2QT}{\sqrt{\pi D}} \frac{1}{\sqrt{t}}.$$

Problem 1.2.13 (The error function). *Find all solutions to $u_t(x,t) - D u_{xx}(x,t) = 0$ of the form*

$$u(x,t) = v\left(\frac{x}{\sqrt{t}}\right).$$

Use the result to recover the function $\Gamma_D(x,t)$.

Solution. We just have to substitute the expression into the equation. To simplify the notation we set

$$\xi = \frac{x}{\sqrt{t}} \qquad \text{whence} \qquad \frac{\partial \xi}{\partial t} = -\frac{x}{2t\sqrt{t}}, \quad \frac{\partial \xi}{\partial x} = \frac{1}{\sqrt{t}}, \quad \frac{\partial^2 \xi}{\partial x^2} = 0.$$

So,

$$u_t(x,t) = -\frac{x}{2t\sqrt{t}} v'(\xi), \qquad u_x(x,t) = \frac{1}{\sqrt{t}} v'(\xi), \qquad u_{xx}(x,t) = \frac{1}{t} v''(\xi)$$

and then

$$-\frac{x}{2t\sqrt{t}} v'(\xi) - D \frac{1}{t} v''(\xi) = 0,$$

or

$$\frac{\xi}{2D} v'(\xi) + v''(\xi) = 0.$$

This is a first order linear ODE for v', that gives $v'(\xi) = C \exp\left(-\xi^2/4D\right)$, and, by a further integration,

$$v(\xi) = C_1 + C_2 \int \exp\left(-\frac{\xi^2}{4D}\right) = C_1 + C_2 \int_0^{\frac{\xi}{\sqrt{4D}}} e^{-z^2}\, dz.$$

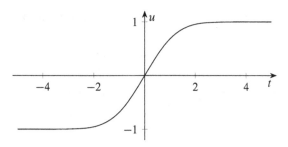

Fig. 1.6 The function erf (x)

Introducing the *error function* (Fig. 1.6)

$$\text{erf}(x) = \frac{2}{\sqrt{\pi}} \int_0^x e^{-z^2} \, dz \qquad \text{(Gauss' error function)},$$

we can write

$$u(x,t) = C_1 + C_2 \, \text{erf} \left(\frac{x}{\sqrt{4Dt}} \right).$$

Choosing $C_2 = \frac{1}{\pi}$ we find $\Gamma_D(x,t) = u_x(x,t)$.

Remark. The error function erf is invariant under the parabolic dilation $x \mapsto \lambda x, t \mapsto \lambda^2 t$; solutions of this type are called *self-similar* and are useful when the domain and the data are invariant under parabolic dilations (see Exercise 1.3.11 on page 52).

Problem 1.2.14 (Absorbing barriers and Dirichlet conditions; method of images). *Consider a one-dimensional symmetric random walk [a] of a particle of unit mass initially placed at the origin. Let h and τ be the space and time steps, and $p = p(x,t)$ the transition probability[b]. Suppose to have an absorbing barrier at $L = \overline{m}h > 0$, which means that if the particle is at $L - h$ at time t and moves to the right, at time $t + \tau$ it will be absorbed and stop at L.*

a) *Which problem does $p = p(x,t)$ solve when we pass to the limit $h, \tau \to 0$, maintaining $h^2/\tau = 1$?*

b) *Find the explicit expression of p.*

c) *Show that in the limit the particle reaches L in finite time with probability 1. Observe, in particular, that for $t > 0$*

$$\int_{-\infty}^L p(x,t) \, dx < 1.$$

---

[a] [18, Chap. 2, Sect. 4].
[b] The probability that the particle will reach x at time t.

Solution. a) This being a symmetric walk, the limit density p solves

$$p_t - \frac{1}{2} p_{xx} = 0$$

for $x < L$ and $t > 0$; as the particle starts from the origin, we have

$$p(x,0) = \delta(x)$$

on $(-\infty, L)$. To understand what happens at $x = L$, let us use the absorbing condition. If the particle is located at $L-h$ at time $t+\tau$, it can only come from $L-2h$, with probability $1/2$. By the law of total probability

$$p(L-h, t+\tau) = \frac{1}{2} p(L-2h, t). \tag{1.23}$$

Taking the limit as $h, \tau \to 0$ gives $p(L,t) = 0$, which is a *homogeneous Dirichlet condition*.

b) In order to find $p(x,t)$ we shall use the so-called *method of images*, which consists in starting another random walk at $2L$, the symmetric point to the origin with respect to L. Then using the linearity of the heat equation, we consider the difference of the fundamental solutions with initial conditions $\delta(x)$ and $\delta(x-2L)$. Let

$$p^A(x,t) = \Gamma_{1/2}(x,t) - \Gamma_{1/2}(x-2L,t) = \Gamma_{1/2}(x,t) - \Gamma_{1/2}(2L-x,t). \tag{1.24}$$

Thus p^A is the required solution, for

$$p^A(x,0) = \Gamma_{1/2}(x,0) - \Gamma_{1/2}(2L-x,0) = \delta(x)$$

and

$$p^A(L,t) = \Gamma_{1/2}(L,t) - \Gamma_{1/2}(L,t) = 0$$

when $-\infty < x < L$.

c) We indicate by $X(t)$ the position at time t and by T_L *the first instant at which the particle reaches L. T_L* is a random variable, defined by

$$T_L = \inf_s \{X(s) = L\}.$$

We claim that the probability of the event $\{T_L < \infty\}$ is 1. This follows if we show

$$\text{Prob}\{T_L > t\} \to 0 \text{ as } t \to \infty.$$

Now, $\{T_L > t\}$ occurs if and only if at time t the particle is inside the interval $(-\infty, L)$, (in particular it will not have yet reached L). The probability of being in $(-\infty, L)$ is

$$
\text{Prob}\{T_L > t\} = \int_{-\infty}^{L} p^A(x,t)\,dx = \int_{-\infty}^{L} \left[\Gamma_{1/2}(x,t) - \Gamma_{1/2}(x - 2L, t)\right] dx
$$

$$
= \frac{1}{\sqrt{2\pi t}} \int_{-\infty}^{L} \left[e^{-\frac{x^2}{2t}} - e^{-\frac{(x-2L)^2}{2t}}\right] dx = \frac{1}{\sqrt{2\pi t}} \int_{-L}^{L} e^{-\frac{x^2}{2t}}\,dx
$$

$$
= \frac{1}{\sqrt{\pi}} \int_{-L/\sqrt{2t}}^{L/\sqrt{2t}} e^{-y^2}\,dy
$$

(in the last line we substituted $x = \sqrt{2t}\,y$). If now $t \to \infty$, $\text{Prob}\{T_L > t\} \to 0$.

Problem 1.2.15 (Problems on the half-line; reflection method). *Let* $g : [0, +\infty) \to \mathbb{R}$ *be a continuous and bounded function.*

a) Find a formula for the solution to

$$
\begin{cases}
u_t(x,t) - Du_{xx}(x,t) = 0 & x > 0, \ t > 0 \\
u(x,0) = g(x) & x > 0 \\
u(0,t) = 0 & t > 0.
\end{cases}
$$

Hint. *Extend the initial datum, for* $x < 0$, *to an odd function and use the formula for the global Cauchy problem.*

b) Find a formula for the solution to

$$
\begin{cases}
u_t(x,t) - Du_{xx}(x,t) = 0 & x > 0, \ t > 0 \\
u(x,0) = g(x) & x > 0 \\
u_x(0,t) = 0 & t > 0.
\end{cases}
$$

Hint. *Extend the initial datum, for* $x < 0$, *to an even function and use the formula for the global Cauchy problem.*

c) Show that either formula provides the unique bounded solution to the respective problem.

Solution. a) In the first case we extend g in an odd way:

$$
\tilde{g}(x) = \begin{cases} g(x) & x \geq 0 \\ -g(-x) & x < 0 \end{cases} \qquad \text{(odd reflection)}.
$$

The new function is continuous on $\mathbb{R}$ only if $g(0) = 0$. Consider the global Cauchy problem

$$
\begin{cases}
u_t(x,t) - Du_{xx}(x,t) = 0 & x \in \mathbb{R}, \ t > 0 \\
u(x,0) = \tilde{g}(x) & x \in \mathbb{R}.
\end{cases}
$$

For any $x \in \mathbb{R}, t > 0$, the solution reads

$$
\begin{aligned}
\tilde{u}(x,t) &= \int_R \Gamma_D(x - y, t)\tilde{g}(y)\, dy \\
&= \int_0^{+\infty} \Gamma_D(x - y, t)g(y)\, dy - \int_{-\infty}^0 \Gamma_D(x - y, t)g(-y)\, dy \\
&= \int_0^{+\infty} \Gamma_D(x - y, t)g(y)\, dy - \int_0^{+\infty} \Gamma_D(x + y, t)g(y)\, dy
\end{aligned}
$$

where in the last term we wrote y instead of $-y$ and swapped endpoints. Let $u(x,t)$ denote the restriction of $\tilde{u}$ to the first quadrant. The previous computation tells

$$
u(x,t) = \int_0^{+\infty} [\Gamma_D(x - y, t) - \Gamma_D(x + y, t)]\, g(y)\, dy. \tag{1.25}
$$

• *Analysis of (1.25).* Clearly u is bounded and solves the heat equation on the quadrant $x > 0, t > 0$. Since Γ_D is even in the spatial variable we obtain[5]

$$
u(0,t) = \int_0^{+\infty} [\Gamma_D(-y, t) - \Gamma_D(y, t)]\, g(y)\, dy = 0 \qquad \text{for any } t > 0.
$$

Hence u fulfils the Dirichlet condition of the half-line $x = 0$. Write

$$
g^+(x) = \begin{cases} g(x) & x \geq 0 \\ 0 & x < 0 \end{cases} \quad \text{and} \quad g^-(x) = \begin{cases} 0 & x \geq 0 \\ g(-x) & x < 0, \end{cases}
$$

so

$$
\begin{aligned}
u(x,t) &= u^+(x,t) - u^-(x,t) \\
&\equiv \int_{-\infty}^{+\infty} \Gamma_D(x - y, t)g^+(y)\, dy - \int_{-\infty}^{+\infty} \Gamma_D(x - y, t)g^-(y)\, dy.
\end{aligned}
$$

Therefore for every $x_0 > 0$, if $(x,t) \to (x_0, 0)$ we have

$$
u^+(x,t) \to g(x_0), \text{ and } u^-(x,t) \to 0,
$$

because g is continuous at x_0. Consequently u is continuous on the closed quadrant, *except possibly for the origin*, and in particular $u(x,0) = g(x)$, $x > 0$. Continuity at the origin holds if and only if $g(0) = 0$, because both g^+, g^- are continuous at $x = 0$. In addition, $\tilde{u} \in C^\infty$ on the half-plane $t > 0$ so $u \in C^\infty$ when $t > 0$, $x \geq 0$.

b) The strategy is completely similar to the previous one. Now we prolong g evenly for $x < 0$:

$$
\tilde{g}(x) = \begin{cases} g(x) & x \geq 0 \\ g(-x) & x \leq 0 \end{cases} \qquad \text{(even reflection)}
$$

---

[5] There are no problems in passing to the limit $x \to 0^+$ when $t > 0$.

and consider the global Cauchy problem with datum $\tilde{g}$. The solution is

$$
\begin{aligned}
\tilde{u}(x,t) &= \int_R \Gamma_D(x-y,t)\tilde{g}(y)\,dy = \\
&= \int_0^{+\infty} \Gamma_D(x-y,t)g(y)\,dy + \int_{-\infty}^0 \Gamma_D(x-y,t)g(-y)\,dy = \\
&= \int_0^{+\infty} \Gamma_D(x-y,t)g(y)\,dy + \int_0^{+\infty} \Gamma_D(x+y,t)g(y)\,dy.
\end{aligned}
$$

If $u(x,t)$ denotes $\tilde{u}$ restricted to the first quadrant, we have

$$
u(x,t) = \int_0^{+\infty} [\Gamma_D(x-y,t) + \Gamma_D(x+y,t)]\,g(y)\,dy. \tag{1.26}
$$

• *Analysis of (1.26)*. As before u is bounded and solves the heat equation on $x > 0, t > 0$. Note that $\tilde{g}$ is continuous also at $x = 0$, so u equals the Cauchy datum continuously on the half-line $x \geq 0$. To verify the Neumann condition we have to compute $u_x(0,t)$. Observe

$$
\partial_x \Gamma_D(x \pm y,t) = \partial_x \frac{1}{\sqrt{4\pi Dt}} \exp\left(-\frac{(x \pm y)^2}{4Dt}\right) = -\frac{x \pm y}{2Dt}\Gamma_D(x \pm y,t)
$$

so at $x = 0$

$$
\partial_x \Gamma_D(\pm y,t) = \mp \frac{y}{2Dt}\Gamma_D(y,t).
$$

For $t > 0$ we can differentiate the integrand, obtaining

$$
\begin{aligned}
u_x(0,t) &= \int_0^{+\infty} [\partial_x \Gamma_D(-y,t) + \partial_x \Gamma_D(y,t)]g(y)\,dy \\
&= \int_0^{+\infty} \frac{y}{2Dt}[\Gamma_D(y,t) - \Gamma_D(y,t)]g(y)\,dy \\
&= 0.
\end{aligned}
$$

In this case, too, the regularity of $\tilde{u}$ implies $u \in C^\infty$ when $t > 0, x \geq 0$.

c) If there existed distinct bounded and regular solutions for $t > 0, x \geq 0$, the reflection would generate bounded C^2 functions for $t > 0$, solving the same global Cauchy problem, thus contradicting the general theory.

Remark. The functions

$$
\Gamma_D^-(x,y,t) = \Gamma_D(x-y,t) - \Gamma_D(x+y,t)
$$

and

$$
\Gamma_D^+(x,y,t) = \Gamma_D(x-y,t) + \Gamma_D(x+y,t)
$$

are called *fundamental solutions for the Cauchy-Dirichlet and Cauchy-Neumann problems on the quadrant $t > 0, x > 0$*, respectively (see Figs. 1.7 and 1.8).

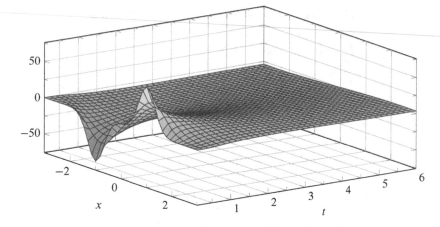

Fig. 1.7 $\Gamma_{1/4}^{-}(x,1,t) = \Gamma_{1/4}(x-1,t) - \Gamma_{1/4}(x+1,t)$

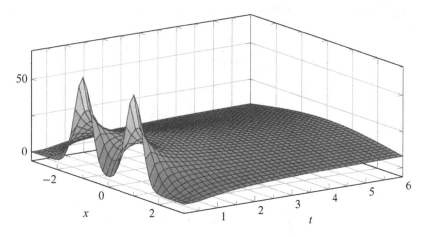

Fig. 1.8 $\Gamma_{1/4}^{+}(x,1,t) = \Gamma_{1/4}(x-1,t) + \Gamma_{1/4}(x+1,t)$

Problem 1.2.16 (Reflection method for a finite interval). *Adapting the reflection method of Problem 1.2.15, b), deduce a formula for solving*

$$\begin{cases} u_t - Du_{xx} = 0 & 0 < x < L,\, t > 0 \\ u_x(x,0) = g(x) & 0 \le x \le L \\ u_x(0,t) = u_x(1,t) = 0 & t > 0 \end{cases} \qquad (1.27)$$

where g is continuous on $[0,L]$.

Solution. With the given homogeneous Neumann conditions we first extend g to $(-L, L]$ in an even way:

$$\tilde{g}(x) = \begin{cases} g(x) & 0 \le x \le L \\ g(-x) & -L < x < 0 \end{cases} \qquad \text{(even reflection)},$$

and then prolong $\tilde{g}$ to $\mathbb{R}$ by setting it equal to zero outside $(-L, L]$; at last, we define

$$g^*(x) = \sum_{n=-\infty}^{+\infty} \tilde{g}(x - 2nL).$$

The function g^* is continuous, $2L$-periodic and it coincides with $\tilde{g}$ on $(-L, L]$. For any given x, at most one summand of the series is non-zero. Let us solve the global Cauchy problem, with datum g^*, using

$$u(x,t) = \int_{-\infty}^{+\infty} \Gamma_D(x - y, t)\, g^*(y)\, dy \qquad (1.28)$$

$$= \sum_{n=-\infty}^{+\infty} \int_{-\infty}^{+\infty} \Gamma_D(x - y, t)\, \tilde{g}(y - 2nL)\, dy.$$

Since $\tilde{g}(y - 2nL)$ vanishes outside $(2n-1)L < y \le (2n+1)L$, we can write

$$\begin{aligned}
u(x,t) \quad &= \quad \sum_{n=-\infty}^{+\infty} \int_{(2n-1)L}^{(2n+1)L} \Gamma_D(x - y, t)\, \tilde{g}(y - 2nL)\, dy \\[2mm]
&\underset{(y-2nL)\longrightarrow y}{=} \sum_{n=-\infty}^{+\infty} \int_{-L}^{L} \Gamma_D(x - y - 2nL, t)\, \tilde{g}(y)\, dy \\[2mm]
&= \quad \sum_{n=-\infty}^{+\infty} \int_{0}^{L} [\Gamma_D(x - y - 2nL, t) + \Gamma_D(x + y - 2nL, t)]g(y)\, dy \\[2mm]
&\equiv \quad \int_{0}^{L} N_D(x, y, t)\, g(y)\, dy,
\end{aligned}$$

where

$$N_D(x, y, t) = \sum_{n=-\infty}^{+\infty} [\Gamma_D(x - y - 2nL, t) + \Gamma_D(x + y - 2nL, t)].$$

The restriction of u to $[0, L]$ is the solution of (1.27), and we set out to check this fact.

Initial datum: by (1.28), u certainly solves the diffusion equation; since g^* is continuous, it is straightforward to see

$$u(x,t) \to g^*(x_0) = g(x_0), \text{ if } (x,t) \to (x_0, 0)$$

at $x_0 \in [0, L]$. Let us compute the flow at $x = 0$; for $t > 0$ we can differentiate the integrand, so it suffices to verify the conditions on the *kernel* N_D:

$$\partial_x N_D(0, y, t) = \frac{1}{2Dt} \sum_{n=-\infty}^{+\infty} [(-y - 2nL) \Gamma_D(y + 2nL, t)$$

$$+ (y - 2nL) \Gamma_D(y - 2nL, t)]$$

$$= \frac{1}{2Dt} \sum_{n=-\infty}^{+\infty} [-(y + 2nL) \Gamma_D(y + 2nL, t)$$

$$+ (y + 2nL) \Gamma_D(y + 2nL, t)]$$

$$= 0.$$

At $x = L$, changing the summation indices,

$$\partial_x N_D(L, y, t) = \frac{1}{2Dt} \sum_{n=-\infty}^{+\infty} [(L - y - 2nL) \Gamma_D(L - y - 2nL, t)$$

$$+ (L + y - 2nL) \Gamma_D(L + y - 2nL, t)]$$

$$= \frac{1}{2Dt} \sum_{n=-\infty}^{+\infty} [((2n + 1)L - y) \Gamma_D(L - y + 2nL, t)$$

$$+ (y - (2n + 1)L) \Gamma_D(L + y - 2nL, t)]$$

$$= 0.$$

Therefore the Neumann conditions hold.

Remark. The function

$$N_D = N_D(x, y, t)$$

is called *fundamental solution with Neumann conditions for the interval* $[0, L]$. Notice that, since the solution with datum $g(x) = 1$ is $u(x,t) = 1$, we have

$$1 = \int_0^L N_D(x, y, t)dy.$$

In particular, if $|g(x)| \le \varepsilon$, we deduce $|u(x,t)| \le \varepsilon$, which shows a continuous dependence on the initial datum.

Problem 1.2.17 (Duhamel's principle). *Consider the Neumann problem*

$$\begin{cases} u_t(x,t) - Du_{xx}(x,t) = f(x,t) & 0 < x < L, t > 0 \\ u(x,0) = 0 & 0 \le x \le L \\ u_x(0,t) = u_x(L,t) = 0 & t > 0. \end{cases} \qquad (1.29)$$

Let f be continuous for $0 \le x \le L, t \ge 0$. Show that if $v(x,t;\tau), t \ge \tau \ge 0$, solves

$$\begin{cases} v_t(x,t;\tau) - Dv_{xx}(x,t;\tau) = 0 & 0 < x < L, t > \tau \\ v(x,\tau;\tau) = f(x,\tau) & 0 \le x \le L \\ v_x(0,t;\tau) = v_x(L,t;\tau) = 0 & t > 0, \end{cases} \qquad (1.30)$$

then the solution to (1.29) is

$$u(x,t) = \int_0^t v(x,t;\tau)\, d\tau.$$

Find an explicit formula and examine the dependence of u on f.

Solution. As f is continuous for $0 \le x \le L, t \ge 0$, from the previous problem we know that the solution to (1.30) is for any $\tau, 0 \le \tau \le t$,

$$v(x,t;\tau) = \int_0^L N_D(x,y,t-\tau)\, f(y,\tau)\, dy$$

and that it is continuous on that set. Then for every $0 < x < L, t > 0$:

$$u_t(x,t) = v(x,t,t) + \int_0^t v_t(x,t;\tau)\, d\tau = f(x,t) + \int_0^t v_t(x,t;\tau)\, d\tau$$

$$u_{xx}(x,t) = \int_0^t v_{xx}(x,t;\tau)\, d\tau.$$

Therefore

$$u_t - Du_{xx} = f(x,t) \qquad 0 < x < L, t > 0$$

and moreover $u(x,0) = 0$. We conclude that u solves (1.29). An explicit expression for u is:

$$u(x,t) = \int_0^t v(x,t;\tau)\, d\tau = \int_0^t \int_0^L N_D(x,y,t-\tau)\, f(y,\tau)\, dy\, d\tau.$$

There is an alternative formula that arises from separation of variables. For $v(x,t;\tau)$ we find

$$v(x,t;\tau) = \frac{f_0(\tau)}{2} + \sum_{k=1}^{\infty} f_k(\tau)\, e^{-k^2\pi^2(t-\tau)/L} \cos\left(\frac{k\pi}{L}x\right)$$

where

$$f_0(\tau) = \frac{2}{L}\int_0^L f(y,\tau)\, dy,$$

$$f_k(\tau) = \frac{2}{L}\int_0^L f(y,\tau)\cos\left(\frac{k\pi}{L}x\right) dy,$$

and so

$$u\left(x,t\right)=\frac{1}{2}\int_{0}^{t}f_{0}\left(\tau\right)d\tau+\sum_{k=1}^{\infty}\cos\left(\frac{k\pi}{L}x\right)\int_{0}^{t}f_{k}\left(\tau\right)e^{-k^{2}\pi^{2}\left(t-\tau\right)/L}d\tau.$$

• *Continuous dependence.* Since

$$\int_{0}^{L}N_{D}\left(x,y,t-\tau\right)dy=1$$

for $t>\tau$, if $\left|f\left(y,\tau\right)\right|\leq\varepsilon$ when $0\leq\tau\leq T,0\leq y\leq L$, we have

$$\left|u\left(x,t\right)\right|\leq\int_{0}^{t}\int_{0}^{L}N_{D}\left(x,y,t-\tau\right)\left|f\left(y,\tau\right)\right|dyd\tau\leq T\varepsilon,$$

which shows continuous dependence on *finite* time intervals.

---

*Problem 1.2.18. Let g be bounded[a] ($\left|g\left(x\right)\right|\leq M$ for any $x\in\mathbb{R}$) and

$$u\left(x,t\right)=\int_{\mathbb{R}}\Gamma_{D}\left(x-y,t\right)g\left(y\right)dy.$$

Show that if g is continuous on x_{0}, then $u\left(x,t\right)\to g\left(x_{0}\right)$ for $\left(x,t\right)\to\left(x_{0},0\right)$.

[a] Actually it is enough to have constants C,A such that $\left|g\left(x\right)\right|\leq Ce^{Ax^{2}}$.

Solution. We recall $\int_{\mathbb{R}}\Gamma_{D}\left(x-y,t\right)dy=1$, for any $t>0,x\in\mathbb{R}$. Therefore we may write

$$u\left(x,t\right)-g\left(x_{0}\right)=\int_{\mathbb{R}}\Gamma_{D}\left(x-y,t\right)\left[g\left(y\right)-g\left(x_{0}\right)\right]dy. \qquad (1.31)$$

Given $\varepsilon>0$, let δ_{ε} be chosen so that if $\left|y-x_{0}\right|\leq 2\delta_{\varepsilon}$ then

$$\left|g\left(y\right)-g\left(x_{0}\right)\right|<\varepsilon.$$

Now write:

$$\int_{\mathbb{R}}\Gamma_{D}\left(x-y,t\right)\left[g\left(y\right)-g\left(x_{0}\right)\right]dy=\int_{\{|y-x_{0}|\leq2\delta_{\varepsilon}\}}\cdots dy+\int_{\{|y-x_{0}|>2\delta_{\varepsilon}\}}\cdots dy.$$

Then:

$$\left|\int_{\{|y-x_{0}|\leq2\delta_{\varepsilon}\}}\cdots dy\right|\leq\int_{\{|y-x_{0}|\leq2\delta_{\varepsilon}\}}\Gamma_{D}\left(x-y,t\right)\underbrace{\left|g\left(y\right)-g\left(x_{0}\right)\right|}_{\leq\varepsilon}dy\leq\varepsilon.$$

For the second integral, if $|x - x_0| \leq \delta_\varepsilon$ and $|y - x_0| > 2\delta_\varepsilon$, it follows $|x - y| > \delta_\varepsilon$; so, $|x - x_0| \leq \delta_\varepsilon$ implies:

$$\left| \int_{\{|y-x_0|>2\delta_\varepsilon\}} \cdots \, dy \right| \leq \int_{\{|y-x|>\delta_\varepsilon\}} \Gamma_D \, (x - y, t) \underbrace{|g \, (y) - g \, (x_0)|}_{\leq 2M} dy$$

$$\leq \frac{2M}{\sqrt{4\pi Dt}} \int_{\{|y-x|>\delta_\varepsilon\}} e^{-\frac{(x-y)^2}{4Dt}} \, dy = \left(y - x = z\sqrt{4\pi Dt} \right)$$

$$\leq \frac{2M}{\sqrt{\pi}} \int_{\frac{\delta_\varepsilon}{\sqrt{4\pi Dt}}}^{+\infty} e^{-z^2} dz \to 0 \quad \text{for } t \downarrow 0.$$

In conclusion, if $|x - x_0| \leq \delta_\varepsilon$ and $t > 0$ is small enough,

$$|u \, (x, t) - g \, (x_0)| \leq 2\varepsilon$$

as claimed.

1.2.4 Use of Fourier and Laplace transforms

Problem 1.2.19 (Fourier transform and fundamental solution). *Using the Fourier transform with respect to x recover the formula for the solution to the global Cauchy problem*

$$\begin{cases} u_t - Du_{xx} = f \, (x, t) & -\infty < x < \infty, \, t > 0 \\ u \, (x, 0) = g \, (x) & -\infty < x < \infty. \end{cases}$$

Solution. Define $\widehat{u} \, (\xi, t) = \int_{\mathbb{R}} u \, (x, t) \, e^{-ix\xi} d\xi$, the partial Fourier transform of u. Then $\widehat{u}$ solves the Cauchy problem (formally)

$$\begin{cases} \widehat{u}_t + D\xi^2 \widehat{u} = \widehat{f} \, (\xi, t) & -\infty < \xi < \infty, \, t > 0 \\ \widehat{u} \, (\xi, 0) = \widehat{g} \, (\xi) & -\infty < \xi < \infty, \end{cases}$$

where $\widehat{f}$ denotes the partial Fourier transform of f. We find

$$\widehat{u} \, (\xi, t) = \widehat{g} \, (\xi) \, e^{-D\xi^2 t} + \int_0^t e^{-D\xi^2 (t-s)} \widehat{f} \, (\xi, s) \, ds.$$

We remind that the inverse transform of the exponential $e^{-D\xi^2 t}$ is $\Gamma_D \, (x, t)$, and the inverse transform of a product is the convolution of the inverse transforms. This yields (at least formally)

$$u \, (x, t) = \int_{\mathbb{R}} \Gamma_D \, (x - y, t) g \, (y) \, dy + \int_0^t \int_{\mathbb{R}} \Gamma_D \, (x - y, t - s) f \, (y, s) \, ds.$$

Problem 1.2.20 (Dirichlet conditions on the half-line). *a) Using the Fourier sine transform find a formula for the bounded solution to*

$$\begin{cases} u_t(x,t) - u_{xx}(x,t) = 0 & x > 0,\, t > 0 \\ u(x,0) = 0 & x \geq 0 \\ u(0,t) = g\,(t) & t > 0, \end{cases} \qquad (1.32)$$

where g is continuous and bounded. Show that this is the only solution with the given properties.

b) Prove that without the condition that u is bounded, problem (1.32) does not have, in general, a unique solution. Hint. Use the functions $w_1\,(x,t) = e^x \cos(2t + x)$ and $w_2\,(x,t) = e^{-x} \cos(2t - x)$.

Solution. a) The Fourier *sine transform* in x is defined as:

$$S\,(u)\,(\xi,t) = U\,(\xi,t) = \frac{2}{\pi} \int_0^\infty u\,(x,t) \sin(\xi x)\,dx,$$

with inverse formula

$$u\,(x,t) = \int_0^\infty U\,(\xi,t) \sin(\xi x)\,d\xi.$$

Notice that U is an odd function in ξ. Assuming that both u and u_x vanish at infinity we have

$$S\,(u_{xx})\,(\xi,t) = \frac{2}{\pi}\xi u\,(0,t) - \xi^2 U\,(\xi,t),$$

and therefore U solves the problem

$$\begin{cases} U_t(\xi,t) + \xi^2 U\,(\xi,t) = \frac{2}{\pi}\xi g\,(t) & \xi > 0,\, t > 0 \\ U(\xi,0) = 0 & \xi \geq 0. \end{cases}$$

We have

$$U\,(\xi,t) = \frac{2}{\pi}\xi \int_0^t e^{-\xi^2(t-s)} g\,(s)\,ds.$$

Anti-transforming gives:

$$
\begin{aligned}
u\,(x,t) &= \int_0^\infty U\,(\xi,t) \sin(\xi x)\,d\xi = \frac{1}{\pi} \int_0^t g\,(s) \left[\int_0^\infty 2\xi e^{-\xi^2(t-s)} \sin(\xi x)\,d\xi \right] ds \\
&= -\int_0^t \frac{g\,(s)}{\pi(t-s)} \left\{ \left[\sin(\xi x) e^{-\xi^2(t-s)}\,d\xi \right]_0^\infty - x \int_0^\infty e^{-\xi^2(t-s)} \cos(\xi x)\,d\xi \right\} ds \\
&= \frac{x}{\pi} \int_0^t \frac{g\,(s)}{t-s} \left[\int_0^\infty e^{-\xi^2(t-s)} \cos(\xi x)\,d\xi \right] ds.
\end{aligned}
$$

Observe that

$$\int_0^\infty e^{-a\xi^2}\cos(\xi x)d\xi = \frac{1}{2}\int_{-\infty}^\infty e^{-a\xi^2 + i\xi x}d\xi = \sqrt{\frac{\pi}{4a}}e^{-\frac{x^2}{4a}}.$$

Substituting $a = t - s$ finally gives (at least formally)

$$u(x,t) = \frac{x}{2\sqrt{\pi}}\int_0^t \frac{g(s)}{(t-s)^{3/2}}e^{-\frac{x^2}{4(t-s)}}ds. \tag{1.33}$$

To show uniqueness, let u_1, u_2 be bounded solutions, so that also $v = u_1 - u_2$ is a bounded solution to (1.33) as well, with $g(t) \equiv 0$. But then its odd prolongation $(v(x,t) = -v(-x,t)$ on $x < 0)$ is regular, bounded and it solves the global Cauchy problem for the heat equation with null initial value. The general theory says v must vanish identically, i.e. u_1 and u_2 coincide.

b) The functions w_1, w_2 are solutions of $w_t - w_{xx} = 0$ (on the whole (x,t)-plane); additionally

$$w_1(x,0) = e^x\cos x, \quad w_1(0,t) = \cos 2t$$
$$w_2(x,0) = e^{-x}\cos x, \quad w_2(0,t) = \cos 2t.$$

We will modify these functions in order to have zero initial value and the same Dirichlet datum on $x = 0, t > 0$. To this end we recall from Problem 1.2.15 (page 29) that[6]

$$v_1(x,t) = \int_0^{+\infty} \Gamma^-(x,y,t)e^y\cos y\,dy \tag{1.34}$$

$$v_2(x,t) = \int_0^{+\infty} \Gamma^-(x,y,t)e^{-y}\cos y\,dy \tag{1.35}$$

solve $v_t - v_{xx} = 0$ on the quadrant $x > 0, t > 0$, with vanishing lateral datum and initial datum

$$v_1(x,0) = e^x\cos x, \quad v_2(x,0) = e^{-x}\cos x,$$

respectively. Then

$$u_1 = w_1 - v_1, \quad u_2 = w_2 - v_2$$

have zero initial value, Dirichlet value on $x = 0$ equal to $\cos 2t$ and it is not hard to check that they are different (for example at $\left(\frac{\pi}{2}, \frac{\pi}{2}\right)$). The problem therefore has no unique solution. Note that w_1 (hence u_1) is unbounded on the quadrant.

Problem 1.2.21 (Linear reaction coefficient). *Use Fourier transforms to solve*

$$\begin{cases} u_t = u_{xx} + xu & x \in \mathbb{R}, t > 0 \\ u(x,0) = g(x) & x \in \mathbb{R} \end{cases}$$

where g is continuous and L^2 on $\mathbb{R}$. Study the effect of the reaction term by choosing $g(x) = \delta(x)$. Does anything change if the reaction terms is $-xu$?

---

[6] See the remark at the end of Problem 1.2.15.

Solution. Let

$$\hat{u}(\xi,t) = \int_{\mathbb{R}} u(x,t) e^{-ix\xi} d\xi$$

be the partial Fourier transform of u. Since the transform of $xu(x,t)$ is $i\hat{u}_\xi(\xi,t)$, $\hat{u}$ satisfies (formally) the Cauchy problem

$$\begin{cases} \hat{u}_t - i\hat{u}_\xi = -\xi^2\hat{u} & -\infty < \xi < \infty, \; t > 0 \\ \hat{u}(\xi,0) = \hat{g}(\xi) & -\infty < \xi < \infty. \end{cases}$$

The equation is linear, non-homogeneous, of order one and can be solved by the method of characteristics, as described in Chap. 3. The characteristic curves have parametric equations

$$t = t(\tau), \xi = \xi(\tau), z = z(\tau)$$

and solve $(d/d\tau = \cdot)$

$$\begin{cases} \dot{t} = 1, & t(0) = 0 \\ \dot{\xi} = -i, & \xi(0) = s \\ \dot{z} = -\xi^2 z, & z(0) = \hat{g}(s). \end{cases}$$

The first two give $t = \tau$, $\xi = s - i\tau$, the third

$$z(\tau,s) = \hat{g}(s) e^{-s^2\tau - is\tau^2 + \frac{\tau^3}{3}}.$$

Eliminating the parameters s, τ we find

$$\hat{u}(\xi,t) = \hat{g}(\xi + it) e^{-\xi^2 t - 3i\xi t^2 + \frac{7}{3}t^3}$$

and then

$$u(x,t) = \frac{1}{2\pi} e^{\frac{7}{3}t^3} \int_{\mathbb{R}} e^{i\xi(x-3t^2)} e^{-\xi^2 t} \hat{g}(\xi + it) d\xi.$$

Now recall $\hat{g}(\xi + it)$ is the transform of

$$e^{xt} g(x),$$

while $e^{-\xi^2 t}$ that of

$$\Gamma_1(x,t) = \frac{1}{\sqrt{4\pi t}} e^{-\frac{x^2}{4t}}.$$

Hence

$$u(x,t) = \int_{\mathbb{R}} \Gamma_1(x - 3t^2 - y, t) e^{yt + \frac{7}{3}t^3} g(y) dy.$$

Choosing $g(x) = \delta(x)$ gives (Fig. 1.9)

$$u(x,t) = e^{\frac{7}{3}t^3} \Gamma_1(x - 3t^2, t) = \cdots = \frac{1}{\sqrt{4\pi t}} \exp\left\{\frac{1}{12}t^3 - \frac{(x - 6t^2)x}{4t}\right\}.$$

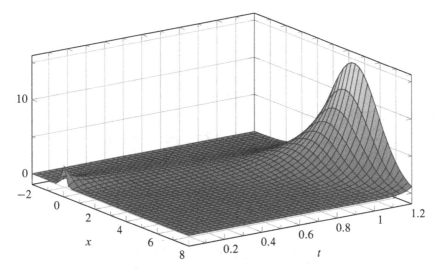

Fig. 1.9 Effect of the reaction term xu: the solution of Problem 1.2.21 for $t > 0.01$

What we see is that for $t \to \infty$ the solution diverges to $+\infty$ at any point x. The reaction term cancels out the damping effect of the diffusion. Had there been $-x$ instead of x nothing would have changed: setting $y = -x$ would reduce to the above case.

Problem 1.2.22. *The initial temperature of a homogeneous bar of length L and small cross-section is zero. Determine a formula for the temperature knowing that*

$$u\,(0+,t) = \delta\,(t)\,, \qquad u\,(L,t) = 0,\ t > 0,$$

where $\delta\,(t)$ is the Dirac distribution centred at the origin. Use the Laplace transform.

Solution. We assume that u admits Laplace transform with respect to t

$$\mathscr{L}\,(u)\,(x,\zeta) = U\,(x,\zeta) = \int_0^\infty e^{-\zeta t} u\,(x,t)\,dt$$

defined on the half-plane Re $\zeta > \alpha$, $\alpha > 0$ a suitable constant. Recalling that

$$\mathscr{L}\,(u_t)\,(x,\zeta) = \zeta U\,(x,\zeta) - u\,(x,0) = \zeta U\,(x,\zeta),$$

we deduce that U solves[7]

$$\begin{aligned} \zeta U - U_{xx} &= 0, & 0 < x < L \\ U\,(0,\zeta) &= 1, & U\,(L,\zeta) = 0. \end{aligned}$$

---

[7] Set $D = 1$ for simplicity.

Solving the ODE gives:

$$U(x, \zeta) = C_1 e^{\sqrt{\zeta}x} + C_2 e^{-\sqrt{\zeta}x}$$

with

$$\sqrt{\zeta} = \sqrt{|\zeta|} e^{i \frac{\arg \zeta}{2}}.$$

The boundary conditions force

$$C_1 + C_2 = 1, \qquad C_1 e^{\sqrt{\zeta}L} + C_2 e^{-\sqrt{\zeta}L} = 0,$$

whence

$$C_2 = \frac{e^{\sqrt{\zeta}L}}{2 \sinh\left(\sqrt{\zeta}L\right)} \quad \text{and} \quad C_1 = -\frac{e^{-\sqrt{\zeta}L}}{2 \sinh\left(\sqrt{\zeta}L\right)}.$$

Therefore

$$U(x, \zeta) = \frac{e^{\sqrt{\zeta}(L-x)} - e^{\sqrt{\zeta}(x-L)}}{e^{\sqrt{\zeta}L} - e^{-\sqrt{\zeta}L}}$$

$$= \frac{\sinh\left[\sqrt{\zeta}(L-x)\right]}{\sinh\left(\sqrt{\zeta}L\right)}.$$

To anti-transform U we recall[8] that the inverse transform of $\psi(a, \zeta) = e^{-a\sqrt{\zeta}}$ is

$$\Psi(a, t) = \frac{a}{2\sqrt{\pi}} t^{-3/2} e^{-a^2/4t}.$$

So let us try to write U as a sum of terms $e^{-a\sqrt{\zeta}}$:

$$\frac{e^{\sqrt{\zeta}(L-x)} - e^{\sqrt{\zeta}(x-L)}}{e^{\sqrt{\zeta}L} - e^{-\sqrt{\zeta}L}} = \frac{e^{-x\sqrt{\zeta}} - e^{-(2L-x)\sqrt{\zeta}}}{1 - e^{-2\sqrt{\zeta}L}}$$

$$= \left[e^{-x\sqrt{\zeta}} - e^{-(2L-x)\sqrt{\zeta}}\right] \sum_{n=0}^{\infty} e^{-2nL\sqrt{\zeta}}$$

$$= \sum_{n=0}^{\infty} e^{-(2nL+x)\sqrt{\zeta}} - \sum_{n=1}^{\infty} e^{-(2nL-x)\sqrt{\zeta}}$$

$$= \sum_{n=0}^{\infty} \psi(2nL + x, \zeta) - \sum_{n=1}^{\infty} \psi(2nL - x, \zeta).$$

Thus, noticing that Ψ is odd with respect to a:

$$u(x, t) = \sum_{n=-\infty}^{\infty} \Psi(2nL + x, t) = \frac{1}{2\sqrt{\pi} t^{3/2}} \sum_{n=-\infty}^{\infty} (2nL + x) e^{-(2nL+x)^2/4t}.$$

---

[8] Appendix B.

1.2.5 Problems in dimension higher than one

Problem 1.2.23 (Dirichlet on the rectangle). *The surfaces of a thin rectangular plate of length a and width b are thermally insulated while the four edges are kept at zero temperature. Determine the evolution of the temperature, knowing its initial value.*

Solution. The problem is essentially two-dimensional, so that we can suppose $u = u(x, y, t)$. Then (for simplicity we fix the diffusion coefficient $D = 1$) we have the problem

$$u_t - (u_{xx} + u_{yy}) = 0, \qquad 0 < x < a, 0 < y < b, t > 0$$

with initial condition

$$u(x, y, 0) = g(x, y)$$

and Dirichlet conditions

$$\begin{cases} u(0, y, t) = 0, \ u(a, y, t) = 0, & 0 < y < b, t > 0 \\ u(x, 0, t) = 0, \ u(x, b, t) = 0, & 0 < x < a, t > 0. \end{cases}$$

We can separate the variables. First we look for non-zero solutions of the form

$$u(x, y, t) = v(x, y) z(t)$$

satisfying the Dirichlet conditions. Substituting in the equation and separating variables gives:

$$\frac{v_{xx} + v_{yy}}{v} = \frac{z'}{z} = \lambda.$$

For z there are no problems; we have

$$z(t) = ce^{\lambda t}.$$

For v we obtain the *eigenvalue problem*

$$v_{xx} + v_{yy} = \lambda v \tag{1.36}$$

on the rectangle $(0, a) \times (0, b)$, with

$$\begin{cases} v(0, y) = 0, \ v(a, y) = 0, & 0 \le y \le b \\ v(x, 0) = 0, \ v(x, b) = 0, & 0 \le x \le a. \end{cases} \tag{1.37}$$

We separate again variables, setting

$$v(x, y) = X(x) Y(y).$$

Substituting in (1.36) and separating, we find

$$\frac{Y''(y)}{Y(y)} - \lambda = -\frac{X''(x)}{X(x)} = \mu$$

with μ constant. Set

$$\nu = -\lambda - \mu.$$

We have the following eigenvalue problems for X and Y:

$$\begin{cases} X'' + \mu X = 0 \text{ in } (0, a) \\ X(0) = X(a) = 0, \end{cases} \qquad \begin{cases} Y'' + \nu Y = 0 \text{ in } (0, b) \\ Y(0) = Y(b) = 0. \end{cases}$$

We have already solved these problems in previous exercises. The eigenvalues and the corresponding eigenfunctions are:

$$X_m(x) = A_m \sin\left(\frac{m\pi x}{a}\right), \qquad \mu_m = \frac{m^2\pi^2}{a^2}, \qquad m = 1, 2, \ldots$$
$$Y_n(x) = B_n \sin\left(\frac{n\pi y}{b}\right), \qquad \nu_n = \frac{n^2\pi^2}{b^2}, \qquad n = 1, 2, \ldots .$$

As

$$\lambda = -(\nu + \mu),$$

we conclude that the eigenvalues for problem (1.36), (1.37) are

$$\lambda_{mn} = -\pi^2\left(\frac{m^2}{a^2} + \frac{n^2}{b^2}\right), \qquad m, n = 1, 2, \ldots$$

with eigenfunctions

$$v_{mn}(x, y) = C_{mn} \sin\left(\frac{m\pi x}{a}\right) \sin\left(\frac{n\pi y}{b}\right), \qquad m, n = 1, 2, \ldots .$$

Summarising, we have the solutions

$$u_{mn}(x, y, t) = C_{mn} e^{-\pi^2\left(\frac{m^2}{a^2} + \frac{n^2}{b^2}\right)t} \sin\left(\frac{m\pi x}{a}\right) \sin\left(\frac{n\pi y}{b}\right),$$

which vanish on the boundary of the rectangle. To match the initial condition we superpose the functions u_{mn}:

$$u(x, y, t) = \sum_{m,n=1}^{\infty} C_{mn} e^{-\pi^2\left(\frac{m^2}{a^2} + \frac{n^2}{b^2}\right)t} \sin\left(\frac{m\pi x}{a}\right) \sin\left(\frac{n\pi y}{b}\right) \qquad (1.38)$$

and impose

$$\sum_{m,n=1}^{\infty} C_{mn} \sin\left(\frac{m\pi x}{a}\right) \sin\left(\frac{n\pi y}{b}\right) = g(x, y).$$

If we assume g can be expanded in double sines-Fourier series, it suffices that the C_{mn} equal the corresponding Fourier coefficients of g:

$$C_{mn} = \frac{4}{ab} \int_0^a \int_0^b \sin\left(\frac{m\pi x}{a}\right) \sin\left(\frac{n\pi y}{b}\right) g(x, y) \, dx dy.$$

As usual, if g is smooth enough, for instance of class C^1 on the closed rectangle, the series is uniformly convergent, and the fast convergence to zero of the exponentials ensures that (1.38) solve the problem.

Problem 1.2.24 (Fourier transform on the half-plane). *Let* $g = g(x, y) : \mathbb{R} \times [0, +\infty) \to \mathbb{R}$ *be continuous and bounded. Using Fourier transforms solve the following Dirichlet problem on* $S = \mathbb{R} \times (0, +\infty) \times (0, +\infty)$:

$$\begin{cases} u_t(x, y, t) - \Delta u(x, y, t) = 0 & x \in \mathbb{R}, \, y > 0, \, t > 0 \\ u(x, y, 0) = g(x, y) & x \in \mathbb{R}, \, y > 0 \\ u(x, 0, t) = 0 & x \in \mathbb{R}, \, t > 0. \end{cases}$$

Solution. Denote by $\widehat{u}(\xi, y, t) = \int_{\mathbb{R}} u(x, y, t) e^{-ix\xi} d\xi$ the partial Fourier transform of u in x. Since the transform of $u_{xx}(x, y, t)$ is $-\xi^2 \widehat{u}(\xi, y, t)$, $\widehat{u}(\xi, \cdot, \cdot)$ satisfies (formally) the Cauchy problem on the quadrant $y > 0$, $t > 0$,

$$\begin{cases} \widehat{u}_t - \widehat{u}_{yy} + \xi^2 \widehat{u} = 0 & y > 0, \, t > 0 \\ \widehat{u}(\xi, y, 0) = \widehat{g}(\xi, y) & y > 0 \\ \widehat{u}(\xi, 0, t) = 0 & t > 0, \end{cases}$$

where $\xi \in \mathbb{R}$. We eliminate the reaction term by setting $v(\xi, y, t) = e^{\xi^2 t} \widehat{u}(\xi, y, t)$; the function v solves $v_t - v_{yy} = 0$ with the same initial and boundary data. The reflection method used in Problem 1.2.15 on page 29 gives:

$$v(\xi, y, t) = \int_0^{+\infty} [\Gamma_1(y - z, t) - \Gamma_1(y + z, t)] \, \widehat{g}(\xi, z) \, dz$$

and then

$$u(\xi, y, t) = e^{-\xi^2 t} \int_0^{+\infty} [\Gamma_1(y - z, t) - \Gamma_1(y + z, t)] \, \widehat{g}(\xi, z) \, dz$$

where $\Gamma_1(y, t)$ is the fundamental solution for the operator $\partial_t - \partial_{yy}$. Note the inverse transform of $e^{-\xi^2 t} \widehat{g}(\xi, y)$ is

$$\int_{-\infty}^{+\infty} \Gamma_1(x - w, t) g(w, y) \, dw$$

and that

$$\Gamma_1(x, t) \Gamma_1(y, t) = \frac{1}{2\sqrt{\pi t}} e^{-\frac{x^2}{4t}} \frac{1}{2\sqrt{\pi t}} e^{-\frac{y^2}{4t}}$$

$$= \frac{1}{4\pi t} e^{-\frac{x^2 + y^2}{4t}} = \Gamma_1(x, y, t)$$

where $\Gamma_1(x, y, t)$ is the fundamental solution for the operator $\partial_t - (\partial_{xx} + \partial_{yy})$. The final formula reads:

$$u(x, y, t) = \int_{-\infty}^{+\infty} \int_0^{+\infty} [\Gamma_1(x - w, y - z, t) - \Gamma_1(x - w, y + z, t)] g(w, z) \, dz \, dw.$$

As g is continuous and bounded, the study of the solution goes as in Problem 1.2.15. In particular, for any $x_0 \in \mathbb{R}$ and $y_0 > 0$, if $(x, y, t) \to (x_0, y_0, 0)$

$$u(x, y, t) \to g(x_0, y_0),$$

so u is continuous on the closure of S *except* possibly for the half-plane $y = 0$. Continuity along $y = 0$ holds precisely if $g(x, 0) = 0$.

Problem 1.2.25 (Dirichlet in the ball). *Let B_R be a ball of radius R in $\mathbb{R}^3$, made of a homogeneous material and at temperature $U > 0$ (constant) at time $t = 0$. Describe how the temperature of B_R evolves at any of its points, in case the temperature on the surface is kept constant and equal 0. Verify that the temperature at the centre of the ball tends to zero exponentially as $t \to +\infty$.*

Solution. The problem is invariant under rotations, so the temperature depends only upon time and the distance from the centre, which we take to be the origin: $u = u(r, t)$ with $r = |\mathbf{x}|$. The equation for u is[9]:

$$u_t - \Delta u = u_t - (u_{rr} + \frac{2}{r} u_r) = 0, \qquad 0 < r < R, t > 0$$

with conditions

$$u(r, 0) = U, \qquad 0 \le r < R$$
$$u(R, t) = 0, \ |u(0, t)| < \infty, \qquad t > 0.$$

Using the identity

$$u_{rr} + \frac{2}{r} u_r = \frac{1}{r}(ru)_{rr} \tag{1.39}$$

we can set $v = ru$ and write for v a one-dimensional problem:

$$\begin{cases} v_t - v_{rr} = 0 & 0 < r < R, t > 0 \\ v(r, 0) = rU & 0 \le r < R \\ v(R, t) = v(0, t) = 0 & t > 0. \end{cases}$$

Recalling Problem 1.2.1 on page 3, we have

$$v(r, t) = \sum_{k=1}^{\infty} c_k \exp\left[-\frac{k^2 \pi^2}{R^2} t\right] \sin \frac{k \pi r}{R}$$

---

[9] For the expression of the Laplace operator Δ in spherical coordinates see Appendix B.

where

$$c_k = \frac{2U}{R} \int_0^R r \sin \frac{k\pi r}{R} dr = \frac{2RU}{\pi k} (-1)^{k+1}.$$

Finally:

$$u(r,t) = \frac{2RU}{\pi r} \sum_{k=1}^{\infty} \frac{(-1)^{k+1}}{k} \exp\left[-\frac{k^2\pi^2}{R^2}t\right] \sin \frac{k\pi r}{R}.$$

The analysis follows the lines of the one carried out for Problem 1.2.1. In particular, by the maximum principle, $u(r,t) \geq 0$; moreover, if $t > 0$, we can take the limit as $r \to 0$ and find the temperature in the middle:

$$0 \leq u(0,t) = 2U \sum_{k=1}^{\infty} (-1)^{k+1} \exp\left[-\frac{k^2\pi^2}{R^2}t\right] \leq 2U \exp\left[-\frac{\pi^2}{R^2}t\right],$$

since the series is alternating with decreasing terms. We deduce that $u(0,t) \to 0$ exponentially as $t \to +\infty$.

Problem 1.2.26 (Dirichlet in a cylinder). *Determine the temperature u inside the cylinder*

$$\mathcal{C} = \{(x,y,z) : r^2 \equiv x^2 + y^2 < R^2, 0 < z < b\},$$

knowing that the surface is kept at temperature $u = 0$, and the initial temperature is $g = g(r,z)$.

Solution. The problem to solve has axial symmetry and cylindrical coordinates are the most natural ones. Hence let $u = u(r,z,t)$; in these coordinates we have[10]

$$\Delta u = u_{rr} + \frac{1}{r}u_r + u_{zz}.$$

We have to find a *bounded* u such that:

$$\begin{cases} u_t - D(u_{rr} + \frac{1}{r}u_r + u_{zz}) = 0 & 0 < r < R, 0 < z < b, t > 0 \\ u(r,z,0) = g(r,z) & 0 < r < R, 0 < z < b, \\ u(r,z,t) = 0 & \text{if } r = R, z = 0, \text{ or } z = b, t > 0. \end{cases}$$

We seek solutions $u(r,z,t) = v(r,z) w(t)$, that satisfy the Dirichlet conditions. Substituting into the differential equation, with the usual procedure we are led to the following equation for w,

$$w'(t) = D\lambda w(t),$$

whence $w(t) = ce^{\lambda Dt}$, and to the eigenvalue problem

$$\begin{cases} v_{rr} + \frac{1}{r}v_r + v_{zz} = \lambda v & 0 < r < R, 0 < z < b, \\ v(r,z) = 0 & \text{if } r = R, z = 0 \text{ or } z = b, \end{cases} \tag{1.40}$$

---

[10] Appendix B.

where v must additionally be *bounded*. We solve also this problem by separation of variables, setting $v\,(r, z) = h\,(r)\,Z\,(z)$. Substituting gives:

$$\frac{\left[h''\,(r) + \frac{1}{r}h'\,(r) - \lambda h\,(r)\right]}{h\,(r)} = -\frac{Z''\,(z)}{Z\,(z)} = \mu \text{ (constant)}$$

from which the eigenvalue problem:

$$Z'' + \mu Z = 0, \qquad Z\,(0) = Z\,(b) = 0 \tag{1.41}$$

arises. Setting $v = -\lambda - \mu$, we find for h the equation

$$h''\,(r) + \frac{1}{r}h'\,(r) + vh\,(r) = 0, \qquad h\,(R) = 0, \; h \text{ bounded.} \tag{1.42}$$

Problem (1.41) is solved by

$$Z_m\,(z) = A_m \sin\left(\frac{m\pi z}{b}\right), \qquad \mu_m = \frac{m^2\pi^2}{b^2}, \qquad m = 1, 2, \dots \; .$$

Problem (1.42) has non-trivial solutions only if $v = \gamma^2 > 0$. This is clear by multiplying by rh and integrating by parts on $(0, R)$; this gives, in fact:

$$0 = \int_0^R (rhh'' + h'h)dr + v\int_0^R h^2 dr = [rhh']_0^R - \int_0^R r(h')^2 dr + v\int_0^R h^2 dr$$

and then

$$v = \int_0^R r(h')^2 dr \Big/ \int_0^R h^2 dr > 0.$$

The equation of problem (1.42) is thus a *Bessel equation of order zero*,[11] and its only bounded solutions are of the type

$$h\,(r) = J_0\,(\gamma r)$$

where $v = \gamma^2$, $\gamma > 0$, and

$$J_0\,(s) = \sum_{k=0}^{\infty} \frac{(-1)^k}{(2^k k!)^2} s^{2k} = 1 - \frac{s^2}{4} + \frac{s^4}{32} - \cdots$$

is the *Bessel function of order zero*.

For the condition $h\,(R) = 0$ to hold we must require

$$J_0\,(\gamma R) = 0.$$

Since J_0 has infinitely many simple zeroes $s_1 < s_2 < \cdots < s_n < \cdots$ in $(0, \infty)$, we obtain infinitely many eigenvalues

$$v_n = \gamma_n^2 = s_n^2/R^2,$$

---

[11] Appendix A.

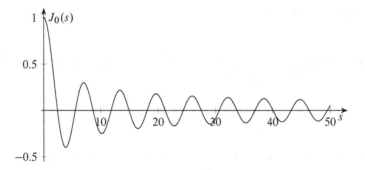

Fig. 1.10 Graph of the Bessel function $s \mapsto J_0(s)$

with eigenfunctions

$$h_n(r) = J_0\left(\frac{s_n r}{R}\right).$$

Recalling that $\lambda = -\nu - \mu$, we get ∞^2-many eigenvalues

$$\lambda_{nm} = -\frac{s_n^2}{R^2} - \frac{m^2 \pi^2}{b^2}$$

for problem (1.40), with eigenfunctions

$$v_{mn}(r,t) = c_{mn} \sin\left(\frac{m\pi z}{b}\right) J_0\left(\frac{s_n r}{R}\right) \qquad m, n = 1, 2, \ldots$$

Now since $w(t) = c e^{\lambda D t}$, we can construct a candidate solution as

$$u(r,z,t) = \sum_{m,n=1}^{\infty} c_{mn} \sin\left(\frac{m\pi z}{b}\right) J_0\left(\frac{s_n r}{R}\right) \exp\left(-D\left[\frac{s_n^2}{R^2} + \frac{m^2 \pi^2}{b^2}\right] t\right).$$

The coefficients c_{mn} should be chosen so that

$$\sum_{m,n=1}^{\infty} c_{mn} \sin\left(\frac{m\pi z}{b}\right) J_0\left(\frac{s_n r}{R}\right) = g(r,z). \tag{1.43}$$

To find them, we remind that the functions

$$\varphi_n(r) = J_0\left(\frac{s_n r}{R}\right)$$

have the following orthogonality properties[12]

$$\int_0^R r\varphi_n(r)\varphi_m(r)\,dr = \begin{cases} 0 & m \neq n \\ \frac{R^2}{2} J_1(s_n)^2 & m = n \end{cases}.$$

---

[12] Appendix A.

Let us multiply (1.43) by

$$rJ_0\left(s_k r/R\right)\sin\left(j\pi z/b\right)$$

and integrate over $(0, R)\times(0, b)$. Keeping in mind of the orthogonality of Bessel's functions and the trigonometric functions we obtain:

$$c_{jk}=\frac{4}{bR^2 J_1\left(s_k\right)^2}\int_0^R\int_0^b r\sin\left(\frac{j\pi z}{b}\right)J_0\left(\frac{s_k r}{R}\right)g\left(r,z\right)drdz.$$

The formal expression thus found is a solution provided g is smooth enough. Uniqueness can be proved using the energy method.

1.3 Further Exercises

1.3.1. *Let u denote the temperature of a homogeneous bar of density ρ and constant (small) cross-section, placed along the interval $0\le x\le L$. Set $u\left(x,0\right)=g\left(x\right)$. Suppose:*

i) *The lateral surface is not thermally insulated, and exchanges heat with the ambient. The latter is at temperature θ, and obeys Newton's law of cooling.*

ii) *The bar ends are insulated.*

iii) *The bar is heated by an electric current of intensity I.*

Write down the mathematical model describing the evolution of u for $t>0$.

1.3.2. *A pipe of constant cross-section S and length L is filled with a homogeneous substance of constant porosity α (the ratio of volume of the pores over the total volume). Inside the pipe is a gas, of concentration $u=u\left(x,t\right),0<x<L$, that diffuses under Nerst's law:*

$$Q=-Du_x$$

where $Q=Q\left(x,t\right)$ is the quantity of gas crossing from left to right the surface element S per unit of time, at the point x and at time t. The walls are leakproof. Write the mathematical model for u when $t>0$, in the following cases: $u\left(x,0\right)=g\left(x\right)$ and:

a) Beginning at $t=0$, at $x=0$ the gas concentration is $c=c\left(t\right)$, while at $x=L$ there is no flux.

b) A constant gaseous inflow c_0 is maintained at $x=0$, while a porous membrane diaphragm at $x=L$ lets the gas seep in accordance with Newton's law. There is no gas in the ambient.

1.3.3. *Let $D>0$ be a constant and $g\in C^1\left([0,\pi]\right)$. Solve by separation of variables the following mixed problem:*

$$\begin{cases}u_t(x,t)-Du_{xx}(x,t)=0 & 0<x<\pi,\ t>0\\ u(x,0)=2\sin\left(3x/2\right) & 0\le x\le\pi\\ u(0,t)=u_x(\pi,t)=0 & t>0.\end{cases}\tag{1.44}$$

Find a solution formula with a general initial profile $u\left(x,0\right)=g\left(x\right)$.

1.3.4. Let $D > 0, h > 0$ be constants, $g \in C^1([0, \pi])$. Solve by separation of variables the mixed problem:

$$\begin{cases} u_t(x,t) - Du_{xx}(x,t) = 0 & 0 < x < L, t > 0 \\ u(x,0) = U & 0 \leq x \leq L \\ u(0,t) = 0 & t > 0 \\ u_x(L,t) + hu(L,t) = 0 & t > 0. \end{cases}$$

1.3.5. Let u be a solution to
$$u_t = Du_{xx} + bu_x + cu.$$

a) Determine h and k so that the function

$$v(x,t) = u(x,t)e^{hx+kt}$$

solves $v_t - Dv_{xx} = 0$.

b) Write the formula for the global Cauchy problem with initial datum $u(x,0) = u_0(x)$.

1.3.6. Solve the following Dirichlet problem using separation of variables:

$$\begin{cases} u_t = u_{xx} + mu + \sin 2\pi x + 2\sin 3\pi x & 0 < x < 1, t > 0 \\ u(x,0) = 0 & 0 < x < 1 \\ u(0,t) = u(1,t) = 0 & t > 0. \end{cases}$$

1.3.7. (Periodicity at one end) Write the formal solution to the problem:

$$\begin{cases} u_t = Du_{xx} & 0 < x < 1, -\infty < t < -\infty \\ u(0,t) = 0, \quad u(1,t) = p(t) & -\infty < t < -\infty \end{cases}$$

where p is of class $C^1(\mathbb{R})$ and T-periodic: $p(t+T) = p(t)$. Study the case $D = 1$, $p(t) = \cos 2t$.

1.3.8. Let g be bounded $(|g(x)| \leq M$ for any $x \in \mathbb{R})$ and

$$u(x,t) = \int_{\mathbb{R}} \Gamma_D(x - y, t) g(y) \, dy.$$

Discuss

$$\lim_{(x,t) \to (x_0, 0)} u(x,t)$$

in the case g has a jump discontinuity at x_0.

1.3.9. Solve the global Cauchy problem with the characteristic function of the interval $(0, 1)$ as initial datum. Study the limit of the solution as (x,t) tends to $(x_0, 0)$.

1.3.10. Adapting the reflection method used in Problem 1.2.15 on page 29, find a formula for the solution to

$$\begin{cases} u_t - Du_{xx} = 0 & 0 < x < L, t > 0 \\ u(x,0) = g(x) & 0 \leq x \leq L \\ u(0,t) = u(L,t) = 0 & t > 0 \end{cases} \tag{1.45}$$

with g continuous on $[0, L]$, $g(0) = g(L) = 0$.

1.3.11. *Find the bounded solution of the problem*

$$u_t - u_{xx} = 0 \qquad \text{on the quadrant } x > 0,\ t > 0,$$

satisfying the following conditions:

a) $u(0,t) = 0,\ u(x,0) = U.$
b) $u(0,t) = U,\ u(x,0) = 0.$
c) $u(0,t) = 0,$

$$u(x,0) = \begin{cases} 0 & 0 < x < L \\ 1 & x > L. \end{cases}$$

1.3.12. *Referring to Problem 1.2.14 on page 27:*

a) compute the probability that the particle reaches L before the instant t, i.e.[13]

$$F(L,t) = \text{Prob}\{T_L \le t\}.$$

b) Determine the probability that the particle is absorbed at $x = L$ in the time interval $(t, t + dt)$.

1.3.13. (Reflecting barriers and Neumann conditions) *Refer to the random walk of Problem 1.2.14 (page 27) and consider the situation where a reflecting barrier is placed at $L = mh$: then the particle, at $L - \frac{h}{2}$ at time t and moving rightwards, is reflected and returns to $L - \frac{h}{2}$ at time $t + \tau$.*

a) Which problem does the transition probability $p = p(x,t)$ solve, when passing to the limit $h, \tau \to 0$, with $h^2/\tau = 2D$?
b) Find the explicit expression for p.

1.3.14. (Duhamel's principle) *Consider the Dirichlet problem*

$$\begin{cases} u_t(x,t) - Du_{xx}(x,t) = f(x,t) & 0 < x < L,\ t > 0 \\ u(x,0) = 0 & 0 \le x \le L \\ u(0,t) = u(L,t) = 0 & t > 0. \end{cases} \tag{1.46}$$

Let f be continuous for $0 \le x \le L,\ t \ge 0$. Show that if $v(x,t;\tau),\ t \ge \tau \ge 0$, is the solution to

$$\begin{cases} v_t(x,t;\tau) - Dv_{xx}(x,t;\tau) = 0 & 0 < x < L,\ t > \tau \\ v(x,\tau;\tau) = f(x,\tau) & 0 \le x \le L \\ v(0,t;\tau) = v(L,t;\tau) = 0 & t > 0 \end{cases} \tag{1.47}$$

then (1.29) is solved by

$$u(x,t) = \int_0^t v(x,t;\tau)\,d\tau.$$

Determine an explicit formula for u.

1.3.15. *A round wire with constant cross-section S and length L, (centred at the origin) with given initial temperature, is heated by a sine-wave current of intensity $I(t)$. Write the mathematical model for the temperature u and find the expression of u.*

---

[13] $F(L,t) = \text{Prob}\{T_L \le t\}$ is called (probability) distribution for the variable T_L. The derivative F_t is the (associated probability) density.

1.3.16. (Neumann condition on the half-line; Fourier cosine transform) **a)** *Taking into account Problem 1.2.20 on page 38, find a formula for a bounded solution to*

$$\begin{cases} u_t(x,t) - u_{xx}(x,t) = 0 & x > 0, \ t > 0 \\ u(x,0) = 0 & x \geq 0 \\ u_x(0,t) = g(t) & t > 0, \end{cases} \qquad (1.48)$$

where g is continuous and bounded. Prove that there is only one bounded solution.

b) *Prove that dropping the assumption on boundedness problem (1.48) does not, in general, guarantee uniqueness of the solution; you can use*

$$w_1(x,t) = e^x \sin(2t + x) \quad \text{and} \quad w_2(x,t) = -e^{-x} \sin(2t - x).$$

1.3.17. (Drift variable) *Using Fourier transforms solve:*

$$\begin{cases} u_t = u_{xx} + x u_x & x \in \mathbb{R}, \ t > 0 \\ u(x,0) = g(x) & x \in \mathbb{R}, \end{cases}$$

where g is continuous and L^2 on $\mathbb{R}$. Examine the effect of the transport term when $g(x) = \delta(x - x_0)$ (use the method of characteristics, Chap. 3).

1.3.18. *The initial temperature of a semi-infinite homogeneous bar with small cross-section is zero. Determine a formula for the temperature knowing that*

$$u(0+,t) = \delta(t), \qquad u(\infty,t) = 0,$$

where $\delta(t)$ denotes Dirac's distribution centred at the origin. Use the Laplace transform.

1.3.19. (Neumann on the ball) *Let B_R be a ball of radius R in $\mathbb{R}^3$ made of a homogeneous material. Describe how the temperature of B_R evolves at any of its points, in case the heat quantity q (constant) flows across the surface and the initial temperature is qr, with r being the distance of the point from the centre.*

1.3.20. (Fourier transform and fundamental solution) *Using Fourier transforms in **x**, recover the formula for the global Cauchy problem in dimension n:*

$$\begin{cases} u_t - D\Delta u = f(\mathbf{x},t) & \mathbf{x} \in \mathbb{R}^n, \ t > 0 \\ u(\mathbf{x},0) = g(\mathbf{x}) & \mathbf{x} \in \mathbb{R}^n. \end{cases}$$

Assuming g and f are in $L^2(\mathbb{R}^n)$ and $L^2(\mathbb{R}^{n+1})$ respectively, explain in which sense the initial condition is attained.

1.3.21. *Determine the temperature $u = u(x,y,z,t)$ in the region between the parallel planes $z = 0$ and $z = 1$, knowing that $u = 0$ on the planes, and initially*

$$u(x,y,z,0) = g(x,y,z).$$

1.3.22. (Maximum principle) *Consider in $\mathbb{R}^{n+1}$ the cylinder*

$$Q_T = \Omega \times (0,T),$$

where $\Omega \subset \mathbb{R}^n$ is a bounded domain with parabolic boundary

$$\partial_p Q_T = (\partial\Omega \times [0, T]) \cup (\Omega \times \{0\}).$$

Let

$$a = a(\mathbf{x}, t), \quad \mathbf{b} = \mathbf{b}(\mathbf{x}, t) \in \mathbb{R}^n, \quad c = c(\mathbf{x}, t)$$

be continuous on $\overline{Q}_T$ and such that

$$a(\mathbf{x}, t) \ge a_0 > 0.$$

Consider a function $u \in C^{2,1}(Q_T) \cap C(\overline{Q}_T)$ satisfying

$$\mathcal{L}u = u_t - a\Delta u + \mathbf{b}\cdot\nabla u + cu \le 0 \quad (\text{resp.} \ge 0) \qquad \text{on } Q_T.$$

a) Show that if $c(\mathbf{x}, t) \ge 0$ and u has a positive maximum (or negative minimum), then this value is taken on the parabolic boundary:

$$\max_{\overline{Q}_T} u = M > 0 \quad \Rightarrow \quad \max_{\partial_p Q_T} u = M.$$

(resp. $\min_{\overline{Q}_T} u = m < 0 \Longrightarrow \min_{\partial_p Q_T} u = m$.)

Deduce that if $u \le 0$ on $\partial_p Q_T$ (or $u \ge 0$) then $u \le 0$ in $\overline{Q}_T$ (resp. $u \ge 0$).

b) Deduce that if g is continuous on $\partial_p Q_T$, the Dirichlet problem

$$\begin{cases} \mathcal{L}u = 0 & \text{in } Q_T \\ u = g & \text{on } \partial_p Q_T \end{cases}$$

has a unique solution in $C^{2,1}(Q_T) \cap C(\overline{Q}_T)$.

c) Show that part b) holds even when the assumption $c(\mathbf{x}, t) \ge 0$ is replaced by the more general requirement $|c(x, t)| \le M$.

1.3.23. Provide an explicit formula for the solution to global Cauchy problem (in $\mathbb{R}^3$):

$$\begin{cases} u_t(\mathbf{x}, t) = a(t)\Delta u(\mathbf{x}, t) + \mathbf{b}(t)\cdot\nabla u(\mathbf{x}, t) + c(t)u(\mathbf{x}, t) & \mathbf{x} \in \mathbb{R}^3, t > 0 \\ u(\mathbf{x}, 0) = g(\mathbf{x}) & \mathbf{x} \in \mathbb{R}^3, \end{cases}$$

where a, b, c and f are continuous and $a(t) \ge a_0 > 0$.

1.3.24. Answer the following questions.

a) Let u solve $u_t - u_{xx} = -1$, in $0 < x < 1, t > 0$, with

$$u(x, 0) = 0 \quad \text{and} \quad u(0, t) = u(1, t) = \sin \pi t.$$

Can there exist a point $x_0, 0 < x_0 < 1$ such that $u(x_0, 1) = 1$?

b) Establish whether there exists a solution to

$$\begin{cases} u_t + u_{xx} = 0 & -1 < x < 1, 0 < t < T \\ u(x, 0) = |x| & -1 < x < 1 \\ u(0, t) = u(1, t) = 0 & 0 < t < T. \end{cases}$$

c) Verify that the function

$$u(x,t) = \partial_x \Gamma_1(x,t)$$

solves

$$\begin{cases} u_t - u_{xx} = 0 & x \in \mathbb{R}, t > 0 \\ u(x,0) = 0 & x \in \mathbb{R} \end{cases}$$

and that $u(x,t) \to 0$ if $t \to 0$, for any x given. Does this contradict the uniqueness of the solution to the global Cauchy problem?

d) Let $u = u(x,t)$ be a continuous solution to the Robin problem

$$\begin{cases} u_t - u_{xx} = 0 & 0 < x < 1, 0 < t < T \\ u(x,0) = \sin \pi x & 0 \le x \le 1 \\ -u_x(0,t) = u_x(1,t) = -hu, h > 0 & 0 \le t \le T. \end{cases}$$

Show that u cannot have a negative minimum. What is the maximum of u?

1.3.25. (Evolution of a chemical solution) *Consider a tube of length L and constant cross-section A, where x is the symmetry axis. The tube contains a saline solution of concentration c. Let A be small enough so that we can assume that the concentration c depends only on x and t, so that the diffusion of salt can be thought of as one-dimensional, along x. Let also the fluid speed be negligible.*

From the left end of the pipe, at $x = 0$, a solution of constant concentration C_0 enters at a rate of R_0 cm^3/s, while at the other end $x = L$ the solution is removed at the same speed.

Using Fick's law show that c solves a diffusion Neumann-Robin problem. Then find the explicit solution and verify that, for $t \to +\infty$, $c(x,t)$ tends to a steady state.

1.3.26. (Random particle subject to an elastic force) *The 1-dimensional motion of a random particle is subject to the following rules. Let N be a natural integer.*

1. *At each time step τ the particle takes one step of h units of length, starting from $x = 0$.*
2. *If the particle is at the point mh, $-N \le m \le N$, it moves to the right, or to the left, with probability*

$$p = \frac{1}{2}\left(1 - \frac{m}{N}\right) \quad \text{or} \quad q = \frac{1}{2}\left(1 + \frac{m}{N}\right)$$

independently of the previous step.

Prove that if $h^2/\tau = 2D$ and $N\tau = \gamma > 0$, when $h, \tau \to 0$ and $N \to \infty$, the limit transition probability $p(x,t)$ is a solution of the equation

$$p_t = Dp_{xx} + \frac{1}{\gamma}(xp)_x.$$

Give a formula for the solution with initial data $p(x,0) = \delta(x)$ when $D = \gamma = 1$.

1.3.27. *Solve the following initial-value/Neumann problem in the unit ball $B_1 = \{\mathbf{x} \in \mathbb{R}^3 : |\mathbf{x}| < 1\}$:*

$$\begin{cases} u_t = \Delta u & \mathbf{x} \in B_1, t > 0 \\ u(\mathbf{x},0) = |\mathbf{x}| & \mathbf{x} \in B_1 \\ u_\nu(\sigma,t) = 1 & \sigma \in \partial B_1, t > 0. \end{cases}$$

1.3.28. *Solve the following non-homogeneous initial-value/Dirichlet problem in B_1 ($u = u(r,t), r = |\mathbf{x}|$):*

$$\begin{cases} u_t - \left(u_{rr} + \dfrac{2}{r}u_r\right) = qe^{-t} & 0 < r < 1, t > 0 \\ u(r,0) = U & 0 \le r \le 1 \\ u(1,t) = 0 & t > 0. \end{cases}$$

1.3.29. *(An ...invasion problem) A population of density $P = P(x,y,t)$ and total mass $M(t)$ is initially ($t = 0$) concentrated at a single point on the plane (say, the origin $(0,0)$). It grows linearly at a rate $a > 0$ and spreads with diffusion constant D.*

a) Write the problem governing the evolution of P, then solve it.
b) Determine the evolution of the mass $M(t) = \int_{\mathbb{R}^2} P(x,y,t)\,dxdy$.
c) Let B_R be the circle centred at $(0,0)$ with radius R. Determine $R = R(t)$ so that

$$\int_{\mathbb{R}^2 \setminus B_{R(t)}} P(x,y,t)\,dxdy = M(0).$$

d) Call metropolitan area the region $B_{R(t)}$ and rural area the region $\mathbb{R}^2 \setminus B_{R(t)}$. Determine the velocity of the metropolitan advancing front.

1.3.1 Solutions

Solution 1.3.1. The temperature solves the following Neumann problem:

$$\begin{cases} u_t = \dfrac{\kappa}{\rho c_v}u_{xx} - \dfrac{\beta}{\rho c_v}(u - \theta) + \gamma\dfrac{I^2 R}{\rho c_v} & 0 < x < L, t > 0 \\ u_x(0,t) = u_x(L,t) = 0 & t > 0 \\ u(x,0) = g(x) & 0 \le x \le L, \end{cases}$$

where the term $\gamma\dfrac{I^2 R}{\rho c}$ is due to the heat produced by the current and $-\dfrac{\beta}{\rho c}(u - \theta)$ comes from the heat exchange with the surrounding ambient (Newton's law).

Solution 1.3.2. As the pipe is insulated, in either case the concentration of the gas satisfies the diffusion equation

$$u_t = au_{xx} \tag{1.49}$$

in $0 < x < L, t > 0$, with $a = D/\alpha$. In fact, the gas found at time t between x and $x + \Delta x$ is

$$\int_x^{x+\Delta x} \alpha u(x,t)\,dx,$$

so the conservation of mass gives:

$$\int_x^{x+\Delta x} \alpha u_t(x,t)\,dx = -D\left[u_x(x,t) - u_x(x + \Delta x, t)\right]$$

where we kept Nerst's law into account. Dividing by Δx and passing to the limit as $\Delta x \to 0$ gives (1.49). Let us consider the ends conditions.

a) We have a Dirichlet condition at $x = 0$:

$$u(0, t) = c(t)$$

and a homogeneous Neumann condition at $x = L$:

$$u_x(L, t) = 0.$$

b) Here we have a non-homogeneous Neumann condition at $x = 0$:

$$-Du_x(0, t) = c_0$$

and a Robin condition at $x = L$:

$$Du_x(L, t) = -hu(L, t).$$

Solution 1.3.3. The solution is (Fig. 1.11)

$$u(x, t) = 2e^{-\frac{9D^2}{4}t} \sin\left(\frac{3}{2}x\right).$$ (1.50)

In fact, writing $u(x, t) = v(x) w(t)$ leads to the equations

$$w'' - \lambda Dw = 0$$

solved by

$$w(t) = Ce^{\lambda Dt}, \ C \in \mathbb{R},$$

and the eigenvalue problem

$$v''(x) - \lambda v(x) = 0$$

with mixed conditions

$$v(0) = v'(\pi) = 0.$$

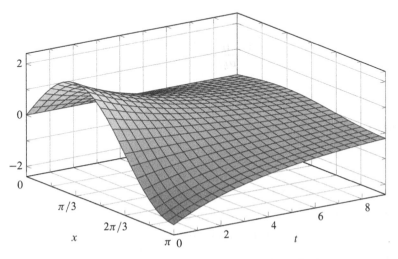

Fig. 1.11 $u(x, t) = 2e^{-t/4} \sin\left(\frac{3}{2}x\right)$

In case $\lambda = \mu^2 \geq 0$ we have only the zero solution. If $\lambda = -\mu^2 < 0$ the eigenvalues are

$$\lambda_k = -\left(\frac{(2k+1)}{2}\right)^2 \qquad k = 0, 1, \ldots$$

with eigenfunctions

$$v_k(x) = \sin\left(\frac{(2k+1)}{2}x\right).$$

Therefore we have infinitely many solutions

$$u_k(x,t) = c_k \sin\left(\frac{(2k+1)}{2}x\right) e^{-\left(\frac{(2k+1)}{2}\right)^2 Dt},$$

fulfilling the mixed conditions at the endpoints. For $c_1 = 2$, u_1 satisfies the initial condition as well. This gives (1.50).

Taking $u(x,0) = g(x)$ as initial datum produces as (formal) solution:

$$u(x,t) = \sum_{k=0}^{\infty} c_k \sin\left(\frac{(2k+1)}{2}x\right) e^{-\left(\frac{(2k+1)}{2}\right)^2 Dt}$$

where

$$c_k = \frac{2}{\pi} \int_0^{\pi} g(x) \sin\left(\frac{(2k+1)}{2}x\right) dx$$

are the Fourier coefficients of g with respect to the family $\{v_k\}$.

Solution 1.3.4. The solution is

$$u(x,t) = U \sum_{k=1}^{\infty} c_k e^{-\mu_k^2 Dt} \sin \mu_k x$$

where μ_k are the positive solutions to

$$h \tan \mu L = -\mu$$

and

$$c_k = \frac{\cos \mu_k L - 1}{\alpha_k \mu_k} \qquad (1.51)$$

with

$$\alpha_k = \frac{L}{2} - \frac{\sin(2\mu_k L)}{4\mu_k}. \qquad (1.52)$$

In fact, by setting $u(x,t) = v(x)w(t)$ we are led to

$$w'' - \lambda Dw = 0$$

solved by

$$w(t) = Ce^{\lambda Dt}, \quad C \in \mathbb{R},$$

and to the Sturm-Liouville problem

$$v''(x) - \lambda v(x) = 0$$

with mixed conditions

$$v(0) = v'(L) + hv(L) = 0.$$

If $\lambda = \mu^2 \geq 0$ the only solution is the trivial one, If $\lambda = -\mu^2 < 0$ we obtain

$$v(x) = C_1 \cos \mu x + C_2 \sin \mu x,$$

and since

$$v'(x) = -\mu C_1 \sin \mu x + \mu C_2 \cos \mu x,$$

the mixed conditions read

$$\begin{cases} C_1 = 0 \\ (\mu \cos \mu L + h \sin \mu L)C_2 = 0. \end{cases}$$

The eigenvalues are therefore the positive solutions μ_k, $k \geq 1$, to the equation

$$h \tan \mu L = -\mu$$

while the eigenfunctions are

$$v_k(x) = \sin \mu_k x.$$

Equations (1.51) and (1.52) follow from the fact that

$$\int_0^L \sin \mu_k x \, dx = \frac{\cos \mu_k L - 1}{\mu_k} \quad \text{and} \quad \int_0^L \sin^2 \mu_k x \, dx = \alpha_k.$$

Solution 1.3.5. a) We have

$$v_t = [u_t + ku]e^{hx+kt}$$
$$v_x = [u_x + hu]e^{hx+kt} \qquad v_{xx} = [u_{xx} + 2hu_x + h^2 u]e^{hx+kt}$$

and hence, by $u_t = Du_{xx} + bu_x + cu$,

$$v_t - Dv_{xx} = e^{hx+kt}[u_t - Du_{xx} - 2Dhu_x + (k - Dh^2)u] =$$
$$= e^{hx+kt}[(b - 2Dh)u_x + (k - Dh^2 + c)u].$$

If we choose

$$h = \frac{b}{2D} \qquad k = \frac{b^2}{4D} - c$$

the function v solves the heat equation $v_t - Dv_{xx} = 0$.

b) The formula is (Fig. 1.12)

$$u(x,t) = e^{\left(c - \frac{b^2}{4D}\right)t} \int_{\mathbb{R}} e^{\frac{b}{2D}(y-x)} \Gamma_D(y - x, t) u_0(y) \, dy.$$

Solution 1.3.6. Using the previous exercise we may simplify, and set

$$w(x,t) = u(x,t)e^{-mt}.$$

The function w satisfies

$$w_t - w_{xx} = e^{-mt}[\sin 2\pi x + 2 \sin 3\pi x]$$

with homogeneous Dirichlet conditions. Define $w(x,t) = v(x)z(t)$. The eigenfunctions associated to the Dirichlet problem are

$$v_k(x) = \sin k\pi x, \qquad k = 1, 2, \ldots$$

with eigenvalues $\lambda_k = -k^2\pi^2$. The right-hand side of the differential equation has the form

$$e^{-mt}[v_2(x) + v_3(x)],$$

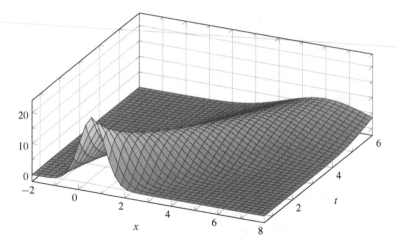

Fig. 1.12 Exercise 1.3.5: the function $(x,t) \mapsto 30e^{2x-t}\Gamma_{1/4}(x,t)$ $(t > 0.6)$

whence the candidate solution will have the form

$$w(x,t) = c_2(t)\sin 2\pi x + 2c_3(t)\sin 3\pi x,$$

due to the homogeneous Dirichlet conditions. The $c_j(t)$, $j = 1,2$, are determined imposing $c_j(0) = 0$ and

$$w_t - w_{xx} = [c_j'(t) + j^2\pi^2 c_j(t)]\sin j\pi x = e^{-mt}\sin j\pi x,$$

i.e.

$$c_j'(t) + j^2\pi^2 c_j(t) = e^{-mt}, \qquad c_j(0) = 0.$$

Then, if $m \neq j^2\pi^2$, we find

$$c_j(t) = \frac{1}{j^2\pi^2 - m}\left(e^{-mt} - e^{-j^2\pi^2 t}\right),$$

while, if $m = j^2\pi^2$,

$$c_j(t) = te^{-j^2\pi^2 t}.$$

The solution is therefore (Fig. 1.13)

$$u(x,t) = e^{mt}[c_2(t)\sin 2\pi x + 2c_3(t)\sin 3\pi x].$$

Solution 1.3.7. Let us expand p in Fourier series; it is convenient to use the complex form:

$$p(t) = \sum_{n=-\infty}^{+\infty} p_n \exp\left(\frac{2\pi int}{T}\right)$$

where

$$p_n = \frac{1}{T}\int_0^T p(t)\exp\left(\frac{2\pi int}{T}\right)dt.$$

We recall that since p is real-valued, we have

$$p_{-n} = \overline{p_n}.$$

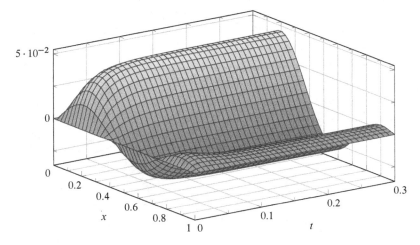

Fig. 1.13 The solution to Exercise 1.3.6; $m = 0.5, 0 < t < 0.3$. Reaction, source and diffusion balance each other

It is reasonable to assume u is T-periodic in t, so we set

$$u(x,t) = \sum_{n=-\infty}^{+\infty} p_n u_n(x) \exp\left(\frac{2\pi i n t}{T}\right)$$

and we set out to find the coefficients u_n from:

$$u_t - D u_{xx} = \sum_{n=-\infty}^{+\infty} \left[\left(\frac{2\pi i n}{T}\right) u_n(x) - D u_n''(x)\right] \exp\left(\frac{2\pi i n t}{T}\right) = 0,$$

whence

$$u_n''(x) - \frac{2\pi i n}{DT} u_n(x) = 0, \qquad n = 0, \pm 1, \pm 2, \ldots \tag{1.53}$$

and

$$u_n(0) = 0, \; u_n(1) = 1. \tag{1.54}$$

As u is real, too, we have

$$u_{-n}(x) = \overline{u_n(x)}$$

so it suffices to determine u_n for $n = 0, 1, 2, \ldots$ With $n = 0$, we have

$$u_0(x) = x.$$

With $n > 0$, the general solution of (1.53) is:

$$u_n(x) = a_n \exp\{c_n(1+i)x\} + b_n \exp\{-c_n(1+i)x\}, \qquad c_n = \sqrt{\frac{\pi n}{DT}}.$$

Equations (1.54) hold if:

$$a_n + b_n = 0, \qquad a_n \exp\{c_n(1+i)\} + b_n \exp\{-c_n(1+i)\} = 1$$

which give, after elementary manipulations:

$$a_n = -b_n = \frac{1}{2i \sin[c_n(1+i)]}.$$

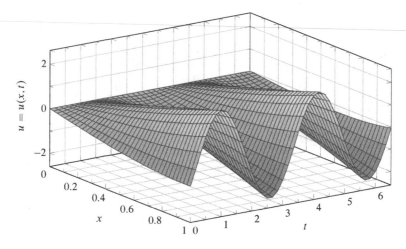

Fig. 1.14 The solution to Exercise 1.3.7

Finally, returning to real functions:

$$u(x,t) = p_0 x + \sum_{n=1}^{\infty} \mathrm{Re} \left\{ p_n \frac{\sin[c_n(1+i)]x}{\sin[c_n(1+i)]} \exp\left(-\frac{2\pi i n t}{T}\right)\right\}.$$

In case

$$p(t) = \cos 2t = \frac{e^{it} + e^{-it}}{2}$$

we have $T = \pi$, $c_n = \sqrt{n}$,

$$p_1 = p_{-1} = \frac{1}{2}, \quad p_n = 0 \text{ if } n \neq \pm 1$$

so

$$a_1 = -b_1 = \frac{1}{4i \sin(1+i)}, \quad a_n = b_n = 0 \text{ if } n \neq 1.$$

Then:

$$u_1(x) = \frac{1}{4i \sin(1+i)} \left[e^{(1+i)x} - e^{-(1+i)x} \right] = \frac{\sin[(1+i)x]}{2\sin(1+i)}$$

and (see Fig. 1.14)

$$u(x,t) = u_1(x) e^{-2it} + \overline{u_1(x)} e^{2it} = 2\,\mathrm{Re}\left[u_1(x)e^{-2it}\right] = \mathrm{Re}\left[\frac{\sin[(1+i)x]}{\sin(1+i)} \exp(-2it)\right].$$

Solution 1.3.8.   Up to translating the x-axis we may assume $x_0 = 0$. Let us set $l^+ = g(0+)$ and $l^- = g(0-)$, and introduce the extensions

$$g^+(x) = \begin{cases} g(x) & x > 0 \\ l^+ & x \le 0, \end{cases} \qquad g^-(x) = \begin{cases} l^- & x > 0 \\ g(x) & x \le 0. \end{cases}$$

The maps g^+ and g^- are continuous on $\mathbb{R}$. Observe that we can write

$$u(x,t) = \int_{-\infty}^{0} \Gamma_D(x-y,t) g(y)\, dx + \int_{0}^{+\infty} \Gamma_D(x-y,t) g(y)\, dy \equiv u^+(x,t) + u^-(x,t)$$

and

$$u^+ (x,t) = \int_{\mathbb{R}} \Gamma_D (x - y, t) g^+ (y) \, dy - l^+ \int_0^{+\infty} \Gamma_D (x - y, t) \, dy$$

$$u^- (x,t) = \int_{\mathbb{R}} \Gamma_D (x - y, t) g^- (y) \, dy - l^- \int_{-\infty}^0 \Gamma_D (x - y, t) \, dy.$$

From Problem 1.2.18 (page 36) we know that if $(x, t) \to (0, 0)$ then:

$$\int_{\mathbb{R}} \Gamma_D (x - y, t) g^+ (y) \, dy \to l^+, \qquad \int_{\mathbb{R}} \Gamma_D (x - y, t) g^- (y) \, dy \to l^-.$$

Thus, it is enough to examine the limits of

$$\int_0^{+\infty} \Gamma_D (x - y, t) \, dy \quad \text{and} \quad \int_{-\infty}^0 \Gamma_D (x - y, t) \, dy.$$

We have

$$\int_0^{+\infty} \Gamma_D (x - y, t) \, dy = \frac{1}{\sqrt{4 D \pi t}} \int_0^{+\infty} e^{-\frac{(x-y)^2}{4Dt}} \, dy = \frac{1}{\sqrt{\pi}} \int_{-\frac{x}{\sqrt{4Dt}}}^{+\infty} e^{-z^2} \, dz$$

$$\int_{-\infty}^0 \Gamma_D (x - y, t) \, dy = \frac{1}{\sqrt{\pi}} \int_{-\infty}^{-\frac{x}{\sqrt{4Dt}}} e^{-z^2} \, dz$$

from which we deduce the following facts:

a) As the limit of $x/\sqrt{t}$ for $(x, t) \to (0, 0)$ does not exist, neither does the limit of u^+. Similarly, the limits of u^- and u do not exist.

b) If $x = o(t)$ (this implies, in particular, that $(x, t) \to (0, 0)$ tangentially along the t-axis) then $u^+ \to l^+/2$, $u^- \to l^-/2$ and $u \to (l^+ + l^-)/2$.

c) Take $t = o(|x|)$ (implying, in particular, $(x, t) \to (0, 0)$ tangentially to the x-axis); if $x > 0$, then $u^+ \to l^+$ and $u^- \to 0$, whereas if $x < 0$, then $u^+ \to 0$ and $u^- \to l^-$. In the former case $u \to l^+$, in the latter $u \to l^-$.

Solution 1.3.9. The solution is

$$u (x,t) = \int_0^1 \Gamma_D (x - y, t) \, dy.$$

Set $(x, t) \to (x_0, 0)$. From the previous exercise we infer:

$$|x_0| > 1 \quad \Longrightarrow \quad u (x,t) \to 0,$$
$$0 < x_0 < 1 \quad \Longrightarrow \quad u (x,t) \to 1,$$

and if $x_0 = 0$ or $x_0 = 1$, the limit does not exist.

Solution 1.3.10. With the given homogeneous Dirichlet conditions we extend g to $[-L, L]$ to an odd map

$$\tilde{g}(x) = \begin{cases} g(x) & 0 \leq x \leq L \\ -g(-x) & -L \leq x < 0 \end{cases} \qquad \text{(odd reflection).}$$

Then we extend $\tilde{g}$ to $\mathbb{R}$ by setting it equal zero outside $[-L, L]$, and define

$$g^* (x) = \sum_{n=-\infty}^{+\infty} \tilde{g} (x - 2nL).$$

The map g^* is continuous (for $g(0) = g(L) = 0$), periodic of period $2L$ and it coincides with $\widetilde{g}$ on $[-L, L]$. Note that, for any x, only one summand survives. Let us solve the global Cauchy problem using the formula

$$u(x,t) = \int_{-\infty}^{+\infty} \Gamma_D (x - y, t) g^* (y) \, dy \tag{1.55}$$

$$= \sum_{n=-\infty}^{+\infty} \int_{-\infty}^{+\infty} \Gamma_D (x - y, t) \widetilde{g} (y - 2nL) \, dy.$$

We remind that $\widetilde{g}(y - 2nL)$ is zero outside the interval

$$(2n - 1)L \leq y \leq (2n + 1)L,$$

so we can write

$$u(x,t) = \sum_{n=-\infty}^{+\infty} \int_{(2n-1)L}^{(2n+1)L} \Gamma_D (x - y, t) \widetilde{g} (y - 2nL) \, dy$$

$$\underset{(y-2nL) \longrightarrow y}{=} \sum_{n=-\infty}^{+\infty} \int_{-L}^{L} \Gamma_D (x - y - 2nL, t) \widetilde{g} (y) \, dy$$

$$= \sum_{n=-\infty}^{+\infty} \int_{0}^{L} [\Gamma_D (x - y - 2nL, t) - \Gamma_D (x + y - 2nL, t)] g(y) \, dy$$

$$= \int_{0}^{L} G_D (x, y, t) \, g(y) \, dy$$

where

$$G_D (x, y, t) = \sum_{n=-\infty}^{+\infty} [\Gamma_D (x - y - 2nL, t) - \Gamma_D (x + y - 2nL, t)].$$

The restriction of u to $[0, L]$ is a solution of (1.27). Now let us examine the initial condition. From (1.28), u is certainly a solution of the diffusion equation; since g^* is continuous, we have

$$u(x,t) \to g^* (x_0) = g(x_0), \text{ if } (x, t) \to (x_0, 0)$$

if $x_0 \in [0, L]$. Let us verify the Dirichlet data at the endpoints. When $t > 0$ we can take the limits as $x \to 0^+, L^-$ inside the integral, so it will be enough to consider the Dirichlet conditions on the *kernel* G_D; after an easy change of index in the sum, we find:

$$G_D (0, y, t) = \sum_{n=-\infty}^{+\infty} \Gamma_D (y + 2nL, t) - \Gamma_D (y - 2nL, t)$$

$$= \sum_{n=-\infty}^{+\infty} [\Gamma_D (y + 2nL, t) - \Gamma_D (y + 2nL, t)]$$

$$= 0.$$

At $x = L$

$$= \sum_{n=-\infty}^{+\infty} [\Gamma_D (L - y - 2nL, t) - \Gamma_D (L + y - 2nL, t)]$$

$$= \sum_{n=-\infty}^{+\infty} [\Gamma_D (L - y - 2nL, t) - \Gamma_D (-L + y + 2nL, t)]$$

$$= 0.$$

This shows that the Dirichlet conditions hold and, by maximum principle, u is the unique solution of problem (1.27).

Remark. The function $G_D = G_D (x, y, t)$ is called *fundamental solution of the heat equation with Dirichlet conditions for the interval* $[0, L]$.

Solution 1.3.11. The problems of parts a) and b) are invariant under parabolic dilations, so it is reasonable to seek solutions of the form given in Problem 1.2.13 (page 26):

$$u(x, t) = C_1 + C_2 \, \mathrm{erf} \left(\frac{x}{2\sqrt{t}} \right).$$

Since $\mathrm{erf}(0) = 0$, $\mathrm{erf}(+\infty) = 1$, it is easy to check that the solution (the only bounded one) to problem a) is

$$u(x, t) = U \, \mathrm{erf} \left(\frac{x}{2\sqrt{t}} \right).$$

Analogously, the only bounded solution to problem b) reads[14]:

$$u(x, t) = U \left[1 - \mathrm{erf} \left(\frac{x}{2\sqrt{t}} \right) \right].$$

For the solution to c), we use (1.25) in Problem 1.2.15 (page 29), which tells

$$u(x, t) = \int_{L}^{+\infty} [\Gamma_1(x - y, t) - \Gamma_1(x + y, t)] \, dy$$

$$= \frac{1}{\sqrt{\pi}} \left[\int_{\frac{(L-x)}{2\sqrt{t}}}^{+\infty} e^{-z^2} dz - \int_{\frac{(L+x)}{2\sqrt{t}}}^{+\infty} e^{-z^2} dy \right] =$$

$$= \frac{1}{2} \left[\mathrm{erf} \, \frac{(L + x)}{2\sqrt{t}} - \mathrm{erf} \, \frac{(L - x)}{2\sqrt{t}} \right].$$

The graph of u is shown in Fig. 1.15.

Solution 1.3.12. a) From Problem 1.2.14 on page 27 the probability equals

$$F(L, t) = 1 - \int_{-\infty}^{L} p^A (x, t) \, dx = 1 - \frac{1}{\sqrt{2\pi t}} \int_{-L}^{L} e^{-\frac{x^2}{2t}} dx$$

$$= \frac{2}{\sqrt{2\pi t}} \int_{L}^{+\infty} e^{-\frac{x^2}{2t}} dx \quad \underset{x=\sqrt{2t}y}{=} \quad \frac{2}{\sqrt{\pi}} \int_{\frac{L}{\sqrt{2t}}}^{+\infty} e^{-\frac{y^2}{2}} dy.$$

b) The required probability coincides with the probability that the particle first reaches $x = L$ at a time between t and $t + dt$. The latter event coincides with $\{t < T_L \leq t + dt\}$, so if $dt \sim 0$,

$$P\{t < T_L \leq t + dt\} = F(L, t + dt) - F(L, t) \simeq F_t(L, t)dt.$$

---

[14] One can also use the formula of Problem 1.2.20 (page 38).

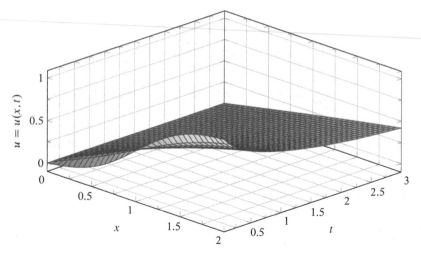

Fig. 1.15 The function $\frac{1}{2}\left(\operatorname{erf}\frac{1+x}{2\sqrt{t}} - \operatorname{erf}\frac{1-x}{2\sqrt{t}}\right), 0 < x < 2, 0.05 < t < 3$

Consequently

$$P\{t < T_L \le t + dt\} = \frac{L}{\sqrt{2\pi}\,t^{3/2}}e^{-\frac{L^2}{2t}}\,dt.$$

Remark. It might be instructive to see another method as well. Let us interpret F as the rate at which a unit quantity of heat, initially concentrated at the origin, will propagate in presence of a heat sink placed at $x = L$. In this case the interpretation is simple: F is nothing but the outgoing heat flow per unit of time at the point L. Fourier's law prescribes

$$F(L,t) = -\frac{1}{2}p_x^A(L,t) = \frac{L}{\sqrt{2\pi}\,t^{3/2}}e^{-\frac{L^2}{2t}},$$

in agreement with the result previously found.

Solution 1.3.13.   a) The transition probability $p = p(x,t)$ solves

$$p_t - \frac{1}{2}p_{xx} = 0$$

in $(-\infty, L) \times (0, \infty)$, and $p(x,0) = \delta$ in $(-\infty, L)$. Moreover, we must have

$$\int_{-\infty}^{L} p(x,t)\,dx = 1. \tag{1.56}$$

To see what happens at $x = L$, let us use the reflection principle. By the latter, a particle has the same probability of being at $L - h/2$, at time $t + \tau$, and at $L - 3h/2$. By the theorem of total probability we can write:

$$p\left(L - \frac{1}{2}h, t + \tau\right) = \frac{1}{2}p\left(L - \frac{3}{2}h, t\right) + \frac{1}{2}p\left(L - \frac{1}{2}h, t\right). \tag{1.57}$$

Now, since

$$p\left(L - \frac{1}{2}h, t + \tau\right) = p\left(L - \frac{1}{2}h, t\right) + p_t\left(L - \frac{1}{2}h, t\right)\tau + o(\tau)$$

and

$$p\left(L - \frac{3}{2}h, t\right) = p\left(L - \frac{1}{2}h, t\right) - px\left(L - \frac{1}{2}h, t\right)h + o(h),$$

substituting into (1.57) gives, after a few simplifications,

$$p_t\left(L - \frac{1}{2}h, t\right)\tau + o(\tau) = -px\left(L - \frac{1}{2}h, t\right)h + o(h).$$

Let us divide by h and take the limit $h \to 0$. Since $h^2/\tau = 1$ we have $\frac{\tau}{h} \to 0$, and then

$$p_x(L, t) = 0 \qquad t > 0. \tag{1.58}$$

This is a homogeneous Neumann condition. To sum up p solves, besides (1.56),

$$\begin{cases} p_t - \frac{1}{2}p_{xx} = 0 & \text{on } -\infty < x < L, \ t > 0 \\ p(x,0) = \delta(x) & -\infty < x < L \\ p_x(L,t) = 0 & t > 0. \end{cases}$$

b) We will use the method of images and place a second walk starting at $2L$, the symmetric point to the origin with respect to L. By virtue of the heat equation's linearity, we consider a linear combination of the two fundamental solutions

$$p^R(x,t) = \Gamma_D(x,t) + \Gamma_D(x - 2L, t) = \Gamma_D(x,t) + \Gamma_D(2L - x, t). \tag{1.59}$$

The function p^R thus defined is precisely the solution needed, because for $-\infty < x < L$

$$p^R(x,0) = \Gamma_D(x,0) + \Gamma_D(2L - x, 0) = \delta(x)$$

and

$$p_x^R(x,t) = \frac{1}{\sqrt{4\pi Dt}}\left\{-\frac{x}{2Dt}e^{-\frac{x^2}{4Dt}} + \frac{2L-x}{2Dt}e^{-\frac{(2L-x)^2}{4Dt}}\right\},$$

so

$$p_x(L,t) = 0$$

(which could have been inferred from the symmetry of p^R, without computations). Finally,

$$\int_{-\infty}^{L} p^R(x,t)\,dx = \int_{-\infty}^{L} \{\Gamma_D(x,t) + \Gamma_D(2L - x, t)\}\,dx =$$

letting $2L - x = z$ in the last integrand,

$$= \int_{-\infty}^{L} \Gamma_D(x,t)\,dx + \int_{L}^{+\infty} \Gamma_D(z,t)\,dz = 1.$$

Hence also (1.56) holds.

Solution 1.3.14. As f is continuous for $0 \le x \le L$, $t \ge 0$, from Exercise 1.3.10 we deduce that the solution to problem (1.30) is, for any τ, $0 \le \tau \le t$,

$$v(x,t;\tau) = \int_0^L G_D(x,y,t-\tau)\,f(y,\tau)\,dy.$$

The latter is continuous on the same set. We also have, for any $0 < x < L$, $t > 0$:

$$u_t(x,t) = v(x,t,t) + \int_0^t v_t(x,t;\tau)\,d\tau = f(x,t) + \int_0^t v_t(x,t;\tau)\,d\tau$$

$$u_{xx}(x,t) = \int_0^t v_{xx}(x,t;\tau)\,d\tau.$$

Hence

$$u_t - Du_{xx} = f(x,t) \qquad 0 < x < L, t > 0$$

and

$$u(x,0) = 0.$$

We conclude that u solves (1.29). An explicit formula is:

$$u(x,t) = \int_0^t v(x,t;\tau)\,d\tau = \int_0^t \int_0^L G_D(x,y,t-\tau)\,f(y,\tau)\,dy\,d\tau.$$

An alternative expression comes from separating the variables. In fact, for $v(x,t;\tau)$ we find

$$v(x,t;\tau) = \sum_{k=1}^{\infty} f_k(\tau)\,e^{-k^2\pi^2(t-\tau)/L} \sin\left(\frac{k\pi}{L}x\right)$$

where

$$f_k(\tau) = \frac{2}{L}\int_0^L f(y,\tau)\sin\left(\frac{k\pi}{L}x\right)dy$$

and then

$$u(x,t) = \sum_{k=1}^{\infty} \sin\left(\frac{k\pi}{L}x\right)\int_0^t f_k(\tau)\,e^{-k^2\pi^2(t-\tau)/L}d\tau.$$

Solution 1.3.15. Let $R = L/2\pi$ be the radius and use coordinates θ and t, then set $u = u(\theta,t)$. The temperature u varies with continuity, whence $u(\theta,t)$ is periodic in θ, of period 2π. We isolate a (cylindrical) portion of the wire V, between θ and $\theta + d\theta$ of length $ds = R d\theta$. A current $I = I(t)$ generates a source of intensity γI^2, where γ depends on the physical characteristics of the wire (electric resistance, density, diffusivity). More precisely, $\gamma I^2 R d\theta$ represents the quantity of heat generated per unit of time in the infinitesimal line element between θ and $\theta + d\theta$. The conservation of energy tells

$$\frac{d}{dt}\int_V c_v\rho u\,dv = \int_{\partial V} -\kappa\frac{du}{ds}\boldsymbol{\tau}\cdot\boldsymbol{v}\,d\sigma + \gamma I^2 R\,d\theta \tag{1.60}$$

where $\boldsymbol{\tau}$ is the unit tangent vector to the wire and $\boldsymbol{v}$ the unit normal to the boundary of the cylinder. Then:

$$\frac{d}{dt}\int_V c_v\rho u\,dv = c_v\rho SR\int_\theta^{\theta+d\theta} u_t(\theta',t)\,d\theta'. \tag{1.61}$$

Moreover, rrom $\boldsymbol{\tau}\cdot\boldsymbol{v} = 0$ on the boundary of the wire $d\sigma = S$ at θ, $\theta + d\theta$, and

$$du/ds = u_\theta d\theta/ds = u_\theta/R,$$

it follows that

$$\int_{\partial V} -\frac{\kappa}{R}u_\theta\boldsymbol{\tau}\cdot\boldsymbol{v}\,d\sigma = -\frac{\kappa S}{R}[u_\theta(\theta+d\theta,t) - u_\theta(\theta,t)]. \tag{1.62}$$

Comparing eqs. (1.60), (1.61) and (1.62), dividing by $d\theta$ and passing to the limit as $d\theta \to 0$ produces the equation

$$u_t = Ku_{\theta\theta} + f(t), \qquad \text{in } 0 < \theta < 2\pi, t > 0$$

with

$$K = \frac{\kappa}{c_v \rho R^2}, \quad f(t) = \frac{\gamma I^2}{c_v \rho S}.$$

Moreover,

$$\begin{cases} u(\theta,0) = g(\theta) & 0 \le \theta \le 2\pi \\ u(0,t) = u(2\pi,t), \; u_\theta(0,t) = u_\theta(2\pi,t) & t > 0, \end{cases}$$

where the initial datum g has period 2π.

When searching for solutions of the type $v(\theta) w(t)$ we are lead to the following eigenvalue problem:

$$v'' + \lambda v = 0, \; v(0) = v(2\pi), \; v'(0) = v'(2\pi),$$

whose solutions are

$$v_n(\theta) = A \cos n\theta + B \sin n\theta, \quad n = 0,1,2\ldots.$$

Therefore we look for a solution like:

$$u(\theta,t) = \sum_{n=0}^{\infty} [A_n(t) \cos n\theta + B_n(t) \sin n\theta]$$

and we impose

$$\sum_{k=0}^{\infty} \left[A_n'(t) + n^2 K A_n(t) \right] \cos n\theta + \left[B_n'(t) + n^2 B_n(t) \right] \sin n\theta = f(t)$$

with

$$A_n(0) = a_n, \quad B_n(0) = b_n, \quad n = 0,1,2\ldots$$

where a_n and b_n are the Fourier coefficients of g. Necessarily

$$A_0'(t) = f(t), A_0(0) = \frac{a_0}{2}$$

and when $n > 0$,

$$A_n'(t) + n^2 K A_n(t) = 0, \quad A_n(0) = a_n$$
$$B_n'(t) + n^2 K B_n(t) = 0, \quad B_n(0) = b_n.$$

This gives:

$$A_0(t) = \frac{a_0}{2} + \int_0^t f(s)\,ds, \; A_n(t) = a_n e^{-n^2 Kt}, \; B_n(t) = b_n e^{-n^2 Kt}.$$

Solution 1.3.16. **a)** We use the Fourier *cosine transform* in x, defined by

$$C(u)(\xi,t) = U(\xi,t) = \frac{2}{\pi} \int_0^{\infty} u(x,t) \cos(\xi x)\,dx,$$

whose inverse formula is given by

$$u(x,t) = \int_0^{\infty} U(\xi,t) \cos(\xi x)\,d\xi.$$

Notice that U is even in ξ. For functions vanishing with their first x-derivatives, as x goes to infinity, we have

$$C(u_{xx})(\xi,t) = -\frac{2}{\pi} u_x(0,t) - \xi^2 U(\xi,t).$$

Then we look for U solving

$$\begin{cases} U_t(\xi,t) + \xi^2 U(\xi,t) = -\frac{2}{\pi} g(t) & \xi > 0, t > 0 \\ U(\xi,0) = 0 & \xi \ge 0. \end{cases}$$

Hence

$$U(\xi,t) = -\frac{2}{\pi} \int_0^t e^{-\xi^2(t-s)} g(s)\, ds.$$

Transforming back:

$$u(x,t) = \int_0^\infty U(\xi,t)\cos(\xi x)\, d\xi = -\frac{2}{\pi} \int_0^t g(s) \left[\int_0^\infty e^{-\xi^2(t-s)} \cos(\xi x)\, d\xi \right] ds$$

$$= -\frac{1}{\sqrt{\pi}} \int_0^t \frac{g(s)}{\sqrt{t-s}} e^{-\frac{x^2}{4(t-s)}}\, ds$$

$$= -2 \int_0^t \Gamma(x,t-s)\, g(s)\, ds$$

where

$$\Gamma(x,t) = \frac{1}{\sqrt{4\pi t}} \exp\left(-\frac{x^2}{4t}\right)$$

is the fundamental solution for the operator $\partial_t - \partial_{xx}$.

b) The proof of non-uniqueness is carried out, with minimal variations, as in Problem 1.2.20 (page 38). The details are left to the reader.

Solution 1.3.17. Let us denote with

$$\widehat{u}(\xi,t) = \int_{\mathbb{R}} u(x,t) e^{-ix\xi}\, dx$$

the partial Fourier transform of u. Recalling that the transform of xu_x is $-\xi\widehat{u}_\xi - \widehat{u}$, then $\widehat{u}$ (formally) satisfies the Cauchy problem

$$\begin{cases} \widehat{u}_t + \xi\widehat{u}_\xi = -(\xi^2+1)\widehat{u} & -\infty < \xi < \infty,\ t > 0 \\ \widehat{u}(\xi,0) = \widehat{g}(\xi) & -\infty < \xi < \infty. \end{cases}$$

The differential equation is linear, non-homogeneous and of order one. The characteristic curves (see Chap. 3) have parametric equations

$$t = t(\tau),\ \xi = \xi(\tau),\ z = z(\tau)$$

and solve $(d/d\tau = \ ^{\cdot})$

$$\begin{cases} \dot{t} = 1, & t(0) = 0 \\ \dot{\xi} = \xi, & \xi(0) = s \\ \dot{z} = -(\xi^2+1)z, & z(0) = \widehat{g}(s). \end{cases}$$

From the first two we get $t = \tau,\ \xi = se^\tau$, while the third one gives

$$z(\tau,s) = \widehat{g}(s) \exp\left(-\frac{1}{2}s^2 e^{2\tau} + \frac{1}{2}s^2 - \tau\right).$$

Eliminating the parameters $s,\ \tau$, we find

$$\widehat{u}(\xi,t) = \widehat{g}\left(\xi e^{-t}\right) \exp\left(-\frac{\xi^2}{2} + \frac{\xi^2}{2}e^{-2t} - t\right),$$

and so

$$u(x,t) = \frac{1}{2\pi} e^{-t} \int_{\mathbb{R}} \widehat{g}\left(\xi e^{-t}\right) \exp\left\{-\frac{1-e^{-2t}}{2}\xi^2 + i\xi x\right\} d\xi.$$

We observe that $\widehat{g}\left(\xi e^{-t}\right)$ is the transform of $e^t g\left(xe^t\right)$, while if we set

$$a(t) = \frac{1-e^{-2t}}{2},$$

then $e^{-\xi^2 a(t)}$ is the transform[15] of

$$\Gamma_1(x,a(t)) = \frac{1}{\sqrt{4\pi a(t)}} e^{-\frac{x^2}{4a(t)}}.$$

Therefore

$$u(x,t) = \int_{\mathbb{R}} \Gamma_1(y,a(t)) g\left(e^t(x-y)\right)dy = e^{-t}\int_{\mathbb{R}} \Gamma_1\left(x - e^{-t}y, a(t)\right) g(y)\, dy.$$

In case

$$g(x) = \delta(x-x_0),$$

we find (Fig. 1.16)

$$u(x,t) = \frac{e^{-t}}{\sqrt{2\pi(1-e^{-2t})}} \exp\left\{-\frac{(x-e^{-t}x_0)^2}{2(1-e^{-2t})}\right\}. \tag{1.63}$$

When $x_0 \neq 0$ the Gaussian curve *shifts* (to the right if $x_0 > 0$, to the left if negative) with speed decreasing exponentially in time. When $x_0 = 0$ there is no shift; the damping effect remains exponential by virtue of the term e^{-t}.

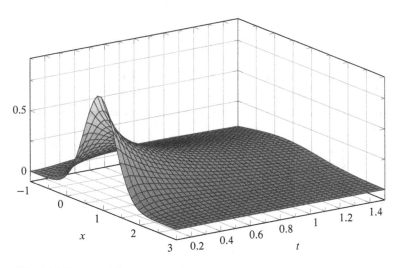

Fig. 1.16 Exponential transport and damping for the Gaussian curve in (1.63)

---

[15] Appendix B.

Solution 1.3.18. Supposing u admits a Laplace transform in t, define

$$\mathcal{L}(u)(x,\zeta) = U(x,\zeta) = \int_0^\infty e^{-\zeta t} u(x,t)\,dt.$$

Then recalling that

$$\mathcal{L}(u_t)(x,\zeta) = \zeta U(x,\zeta) - u(x,0) = \zeta U(x,\zeta),$$

we see that U solves[16]

$$\zeta U - U_{xx} = 0, \qquad x > 0$$
$$U(0,\zeta) = 1, \qquad U(\infty,\zeta) = 0.$$

Solving the ODE we find

$$U(x,\zeta) = C_1 e^{\sqrt{\zeta}x} + C_2 e^{-\sqrt{\zeta}x},$$

where

$$\sqrt{\zeta} = \sqrt{|\zeta|} e^{i\frac{\arg\zeta}{2}}.$$

Now, imposing the boundary conditions gives

$$C_1 = 0,\ C_2 = 1,$$

hence

$$U(x,\zeta) = e^{-\sqrt{\zeta}x}.$$

Finally, anti-transforming[17] produces

$$u(x,t) = \frac{x}{2\sqrt{\pi}t^{3/2}} e^{-\frac{x^2}{4t}}.$$

Solution 1.3.19. The problem is rotation-invariant and thus the temperature depends only on time and the distance from the centre, which we take to be the origin. Thus $u = u(r,t)$, with $r = |\mathbf{x}|$. The equation for u reads[18]:

$$u_t - \Delta u = u_t - \frac{1}{r}(ru)_{rr} = 0, \qquad 0 < r < R, t > 0$$

with

$$\begin{cases} u(r,0) = qr & 0 \le r < R \\ u_r(R,t) = q,\ |u(0,t)| < \infty & t > 0. \end{cases}$$

As in Problem 1.2.25 (page 46) we may set $v = ru$ and write the one-dimensional mixed Dirichlet-Robin problem:

$$\begin{cases} v_t - v_{rr} = 0 & 0 < r < R, t > 0 \\ v(r,0) = qr^2 & 0 \le r < R \\ v(0,t) = 0,\ Rv_r(R,t) = v(R,t) + R^2 q & t > 0. \end{cases}$$

Let us also define

$$w(r,t) = v(r,t) - qr^2,$$

---

[16] We set $D = 1$ for simplicity.
[17] Appendix B.
[18] Recall identity (1.39).

so that

$$\begin{cases} w_t - w_{rr} = 2q & 0 < r < R, t > 0 \\ w(r,0) = 0 & 0 \le r < R \\ w(0,t) = 0, \ Rw_r(R,t) = w(R,t) & t > 0. \end{cases}$$

Remembering the solution to Exercise 1.3.4, the problem has as eigenvalues the numbers μ_k/R, $k \ge 1$, i.e. the positive solutions of $\tan\mu = \mu$. The eigenfunctions are

$$w_k(r) = \sin\frac{\mu_k r}{R},$$

so the solution is

$$w(r,t) = \sum_{k=1}^{\infty} c_k(t)\sin\frac{\mu_k r}{R}$$

where the c_k are chosen so that (recall that $w_k'' + \frac{\mu_k^2}{R^2}w_k = 0$)

$$w_t - w_{rr} = \sum_{k=1}^{\infty}\left[c_k'(t) + \frac{\mu_k^2}{R^2}c_k(t)\right]\sin\frac{\mu_k r}{R} = 2q.$$

Since

$$2q = \sum_{k=1}^{\infty} q_k \sin\frac{\mu_k r}{R}$$

where

$$q_k = 2q\frac{\cos\mu_k}{\alpha_k\mu_k}, \quad \alpha_k = \frac{1}{2} - \frac{\sin\mu_k}{4\mu_k},$$

we must have, also by the zero initial condition:

$$c_k'(t) + \frac{\mu_k^2}{R^2}c_k(t) = q_k, \qquad c_k(0) = 0, \qquad k = 1,2,\ldots.$$

This gives

$$c_k(t) = \frac{q_k R^2}{\mu_k^2}\left(1 - e^{-\frac{\mu_k^2}{R^2}t}\right).$$

and finally

$$u(r,t) = qr + \frac{2qR^2}{r}\sum_{k=1}^{\infty}\frac{q_k}{\mu_k^2}\left(1 - e^{-\frac{\mu_k^2}{R^2}t}\right)\sin\frac{\mu_k^2 r}{R^2}.$$

Solution 1.3.20.   Define

$$\widehat{u}(\xi,t) = \int_{\mathbb{R}^n} u(x,t)e^{-ix\cdot\xi}\,d\xi,$$

the partial Fourier transform of u. Thus $\widehat{u}$ (formally) satisfies the Cauchy problem

$$\begin{cases} \widehat{u}_t + D\,|\xi|^2\,\widehat{u} = \widehat{f}(\xi,t) & \xi \in \mathbb{R}^n, t > 0 \\ \widehat{u}(\xi,0) = \widehat{g}(\xi) & \xi \in \mathbb{R}^n \end{cases}$$

where $\widehat{f}$ is the partial Fourier transform of f. Hence

$$\widehat{u}(\xi,t) = \widehat{g}(\xi)e^{-D|\xi|^2 t} + \int_0^t e^{-D|\xi|^2(t-s)}\widehat{f}(\xi,s)\,ds.$$

We remind that the inverse transform of $e^{-D|\boldsymbol{\xi}|^2 t}$ is[19]

$$\Gamma_D(\mathbf{x}, t) = \frac{1}{(4\pi Dt)^{n/2}} \exp\left(-\frac{|\mathbf{x}|^2}{4Dt}\right)$$

and the inverse transform of a product is the convolution of the respective anti-transforms; then we get

$$u(\mathbf{x}, t) = \int_{\mathbb{R}^n} \Gamma_D(\mathbf{x} - \mathbf{y}, t) g(\mathbf{y}) \, d\mathbf{y} + \int_0^t \int_{\mathbb{R}^n} \Gamma_D(\mathbf{x} - \mathbf{y}, t - s) f(\mathbf{y}, s) \, ds$$

$$\equiv u_1(\mathbf{x}, t) + u_2(\mathbf{x}, t).$$

The *analysis of the procedure* goes exactly as in Problem 1.2.19 (page 37); the conclusion is that

$$u(\mathbf{x}, t) \to f(\mathbf{x}, t)$$

in $L^2(\mathbb{R}^n)$ as $t \to 0$.

Solution 1.3.21. We have to solve the following Dirichlet problem (setting $D = 1$):

$$B \begin{cases} u_t(x, y, z, t) = \Delta u(x, y, z, t) & (x, y) \in \mathbb{R}^2, \, 0 < z < 1, \, t > 0 \\ u(x, y, z, 0) = g(x, y, z) & (x, y) \in \mathbb{R}^2, \, 0 < z < 1 \\ u(x, y, 0, t) = u(x, y, 1, t) = 0 & (x, y) \in \mathbb{R}^2, \, t > 0 \end{cases}$$

(The Laplacian is intended with respect to the spatial coordinates). In order to solve it we use a two-dimensional Fourier transform in x and y

$$\widehat{u}(\xi, \eta, z, t) = \int_{\mathbb{R}^2} u(x, y, z, t) e^{-i(x\xi + y\eta)} d\xi d\eta.$$

Then $\widehat{u}(\xi, \eta, \cdot, \cdot)$ is the (formal) solution to:

$$\begin{cases} \widehat{u}_t - \widehat{u}_{zz} + (\xi^2 + \eta^2)\widehat{u} = 0 & 0 < z < 1, \, t > 0 \\ \widehat{u}(\xi, \eta, z, 0) = \widehat{g}(\xi, \eta, z) & 0 \le z \le 1 \\ \widehat{u}(\xi, \eta, 0, t) = \widehat{u}(\xi, \eta, 1, t) = 0 & t > 0. \end{cases}$$

If we think of ξ, η as fixed, we can separate the variables and look for solutions of the form

$$\widehat{u}(z, t) = v(z) w(t).$$

In the usual way we find

$$w_n(t) = \alpha_n(\xi, \eta) e^{-(\xi^2 + \eta^2)t} e^{-n^2 \pi^2 t}, \qquad n = 1, 2, \ldots$$

and

$$v_n(z) = \beta_n(\xi, \eta) \sin(n\pi z), \qquad n = 1, 2, \ldots$$

By superposing the products $w_n(t) v_n(z)$ and adding the initial condition, we construct the (formal) solution

$$\widehat{u}(\xi, \eta, z, t) = \sum_{n=1}^{\infty} \gamma_n(\xi, \eta) e^{-(\xi^2 + \eta^2)t} e^{-n^2 \pi^2 t} \sin(n\pi z)$$

where

$$\gamma_n(\xi, \eta) = 2 \int_0^1 \widehat{g}(\xi, \eta, z) \sin(n\pi z) \, dz.$$

---

[19] Appendix B.

Now we transform back, and call $c_n(x, y)$ the inverse transform of $\gamma_n(\xi, \eta)$ and also recall that $e^{-(\xi^2+\eta^2)t}$ is the transform of $\Gamma_1(x, y, t)$. We find:

$$u(x, y, z, t) = \sum_{n=1}^{\infty} \left(\int_{\mathbb{R}} \Gamma_1(x - x_1, y - y_1, t) c_n(x_1, y_1) \, dx_1 dy_1 \right) e^{-n^2\pi^2 t} \sin(n\pi z).$$

Solution 1.3.22. **a)** We will mimic the proof of the case $\mathcal{L}u = u_t - D\Delta u$[20].
  1. Given $\varepsilon > 0$, $T - \varepsilon > 0$ we set $v = u - \varepsilon t$. By contradiction, if

$$(\mathbf{x_0}, t_0) \in \overline{Q}_{T-\varepsilon} \setminus \partial_p Q_{T-\varepsilon} = \Omega \times (0, T - \varepsilon]$$

were a positive maximum point for v on $\overline{Q}_{T-\varepsilon}$, we would have:

$$v_t(\mathbf{x_0}, t_0) \geq 0, \ \Delta v(\mathbf{x_0}, t_0) \leq 0, \ \nabla u(\mathbf{x_0}, t_0) = \mathbf{0}, \ c(\mathbf{x_0}, t_0) u(\mathbf{x_0}, t_0) \geq 0,$$

contradicting $\mathcal{L}v \leq -\varepsilon$. Therefore

$$\max_{\overline{Q}_{T-\varepsilon}} u \leq \max_{\overline{Q}_{T-\varepsilon}} v + \varepsilon T \leq \max_{\partial_p Q_{T-\varepsilon}} v + \varepsilon T \leq \max_{\partial_p Q_T} u + \varepsilon T.$$

  2. By taking the limit $\varepsilon \to 0$ in the previous item we see that if $\max_{\overline{Q}_T} u = M > 0$, then $\max_{\partial_p Q_T} u = M$. If now $u \leq 0$ on $\partial_p Q_T$, then u cannot be positive at some other point.

b) If u, v solve the same problem then $w = u - v$ vanishes on the parabolic boundary, whence it must be zero everywhere on Q_T.

c) If $|c(\mathbf{x}, t)| \leq K$, without assumptions on the sign, we can reduce to the previous situation by setting

$$z(\mathbf{x}, t) = e^{-Kt} w(\mathbf{x}, t).$$

In fact,

$$\mathcal{L}z = e^{-Kt} [w_t - a\Delta w + \mathbf{b} \cdot \nabla w + (c - K)w] = -Kz$$

so

$$\mathcal{L}z + Kz = z_t - a\Delta z + \mathbf{b} \cdot \nabla z + (c + K)z = 0$$

and the coefficient of z, i.e. $c + K$, is ≥ 0. Hence $z = 0$, implying

$$w = u - v = 0.$$

Solution 1.3.23. The idea is to reduce to the diffusion equation $U_t = \Delta U$ by repeated change of variables. Let us proceed step by step.

Step 1. We eliminate the reaction term by writing

$$C(t) = \int_0^t c(s) \, ds$$

and defining

$$w(\mathbf{x}, t) = e^{-C(t)} u(\mathbf{x}, t).$$

The function w solves

$$w_t = a(t)\Delta w + \mathbf{b}(t) \cdot \nabla w, \qquad w(\mathbf{x}, 0) = g(\mathbf{x}).$$

---

[20] [18, Chap. 2, Sect. 2].

Step 2. Now we get rid of the transport term, by noting that if

$$\mathbf{B}(t) = \int_0^t \mathbf{b}(s)\,ds,$$

it follows

$$\frac{\partial}{\partial t} w\left(\mathbf{z} - \mathbf{B}(t), t\right) = w_t\left(\mathbf{z} - \mathbf{B}(t), t\right) - \mathbf{b}(t) \cdot \nabla w\left(\mathbf{z} - \mathbf{B}(t), t\right).$$

So we set

$$\mathbf{x} = \mathbf{z} - \mathbf{B}(t)$$

and

$$h(\mathbf{z}, t) = w\left(\mathbf{z} - \mathbf{B}(t), t\right).$$

Then h solves

$$h_t = a(t)\,\Delta w, \qquad h(\mathbf{z}, 0) = g(\mathbf{z}).$$

Step 3. We eliminate the coefficient $a(t)$ by rescaling in time. Set

$$A(t) = \int_0^t a(s)\,ds.$$

As

$$a(s) \geq a_0 > 0,$$

A is invertible and we may put

$$U(\mathbf{z}, \tau) = h\left(\mathbf{z}, A^{-1}(\tau)\right).$$

Then

$$U_\tau = h_t\frac{1}{a},$$

and therefore U solves

$$U_\tau = \Delta U, \qquad U(\mathbf{z}, 0) = g(\mathbf{z}).$$

We can write

$$U(\mathbf{z}, \tau) = \frac{1}{(4\pi\tau)^{3/2}} \int_{\mathbb{R}^3} \exp\left\{-\frac{(\mathbf{z} - \mathbf{y})^2}{4\tau}\right\} g(\mathbf{y})\,d\mathbf{y}.$$

Finally, going back to the original variables, we obtain

$$u(\mathbf{x}, t) = \frac{1}{(4\pi A(t))^{3/2}} \int_{\mathbb{R}^3} \exp\left\{C(t) - \frac{(\mathbf{x} + \mathbf{B}(t) - \mathbf{y})^2}{4A(t)}\right\} g(\mathbf{y})\,d\mathbf{y}.$$

Solution 1.3.24. **a)** No, for otherwise $(x_0, 1)$ would be an internal positive maximum and

$$u_t(x_0, 1) = 0, u_{xx}(x_0, 1) \leq 0,$$

violating the equation

$$u_t - u_{xx} = -1.$$

b) Such a solution cannot exist, because $|x|$ is not regular enough to be taken as initial datum for the *backward* equation (see Problem 1.2.7 on page 19).

c) There is no contradiction. The example shows only that, in order to have uniqueness, $u(x, t) \to 0$ for any given x is not sufficient. Note, by the way, that $u(x, t) \to \infty$ as $(x, t) \to (0, 0)$ along the parabola $x^2 = t$.

d) If

$$\min u = u\,(x_0, t_0) = m < 0$$

then $t_0 > 0$ and $x_0 = 0$ or $x_0 = 1$. We may suppose $u > m$ for $0 \leq t < t_0$, i.e. that t_0 is the first instant at which u assumes the value m. By Hopf's principle (Problem 1.2.11 on page 24)

$$u_x\,(0, t_0) > 0 \quad \text{or} \quad u_x\,(1, t_0) < 0.$$

Both violate the Robin condition, and therefore $m \geq 0$. A similar argument shows that the maximum of u, which is positive, cannot be reached along either of the half-lines $x = 0$, $x = 1$. Hence it must coincide with the maximum of the initial datum

$$\max u = \max \sin \pi x = 1.$$

Solution 1.3.25. The concentration c satisfies the equation

$$c_t = D c_{xx} \quad 0 < x < L, t > 0.$$

If we denote by $\mathbf{i}$ the unit vector along the x-axis, according to *Fick's law* the flux entering at $x = 0$ is given by

$$\int_A \mathbf{q}\,(c\,(0, t)) \cdot \mathbf{i}\,dxdy = \int_A -D c_x\,(0, t)\,dxdy = -D A c_x\,(0, t) = C_0 R_0$$

while the outgoing flow at $x = L$ is

$$\int_A \mathbf{q}\,(c\,(L, t)) \cdot \mathbf{i}\,dxdy = \int_A -D c_x\,(L, t)\,dxdy = -D A c_x\,(L, t) = c\,(L, t)\,R_0.$$

Therefore we deduce the following Neumann-Robin conditions

$$c_x\,(0, t) = -B \quad \text{and} \quad c_x\,(L, t) + E c\,(L, t) = 0,$$

where we denoted

$$B = \frac{C_0 R_0}{DA} \quad \text{and} \quad E = \frac{R_0}{DA};$$

the problem is also associated to the initial condition $c\,(x, 0) = c_0\,(x)$.

First we determine the stationary solution c^{St}, which satisfies the conditions

$$\begin{cases} c^{St}_{xx} = 0 & 0 < x < L, t > 0 \\ c^{St}_x\,(0, t) = -B, \; c^{St}_x\,(L, t) + E c^{St}\,(L, t) = 0 & t > 0. \end{cases}$$

We find

$$c^{St}\,(x) = B\,(L - x) + \frac{B}{E}.$$

Now we analyse the transient function $u\,(x, t) = c\,(x, t) - c^{St}\,(x)$, which solves the following problem

$$\begin{cases} u_t = D u_{xx} & 0 < x < L, t > 0 \\ u_x\,(0, t) = 0, \; u_x\,(L, t) + E u\,(L, t) = 0 & t > 0 \\ u\,(x, 0) = c_0\,(x) - c^{St}\,(x) & 0 < x < L. \end{cases}$$

In this way, we are brought back to homogeneous boundary conditions, and we can use the method of separation of variables as in Problem 1.2.6 (page 16); we set $u(x, t) = y(x)w(t)$ and deduce

$$\frac{w'(t)}{D w'(t)} = \frac{y''(x)}{y(x)} = \lambda.$$

In particular, w satisfies the equation $w'(t) = \lambda D w(t)$, whose solution is $V(t) = e^{\lambda D t}$, while y is a solution of the eigenvalue problem

$$\begin{cases} y''(x) - \lambda y(x) = 0 \\ y'(0,t) = 0, \; y'(L,t) + Ey(L,t) = 0. \end{cases}$$

If $\lambda > 0$, the general solution is $y(x) = c_1 e^{-\sqrt{\lambda}x} + c_1 e^{\sqrt{\lambda}x}$. The boundary conditions give

$$\begin{cases} -c_1 + c_2 = 0 \\ c_1(E - \sqrt{\lambda})e^{-\sqrt{\lambda}L} + c_2(E + \sqrt{\lambda})e^{\sqrt{\lambda}L} = 0. \end{cases} \tag{1.64}$$

Now we have that

$$\det \begin{pmatrix} -1 & 1 \\ (E - \sqrt{\lambda})e^{-\sqrt{\lambda}L} & (E + \sqrt{\lambda})e^{\sqrt{\lambda}L} \end{pmatrix} = -(E + \sqrt{\lambda})e^{\sqrt{\lambda}L} - (E - \sqrt{\lambda})e^{-\sqrt{\lambda}L}$$

$$= (E + \sqrt{\lambda})e^{-\sqrt{\lambda}L} \left(\frac{\sqrt{\lambda} - E}{\sqrt{\lambda} + E} - e^{2\sqrt{\lambda}L} \right) < 0,$$

since $e^{2\sqrt{\lambda}L} > 1$ and $(\sqrt{\lambda} - E)/(\sqrt{\lambda} + E) < 1$. System (1.64) has the only solution $c_1 = c_2 = 0$. We can reach the same conclusion even in the case $\lambda = 0$. If $\lambda < 0$, we find the conditions

$$\begin{cases} U'(0) = \sqrt{-\lambda}c_2 = 0 \\ U'(L,t) + EU(L,t) = c_1 \left[\cos\left(\sqrt{-\lambda}L\right) - \sqrt{-\lambda}\sin\left(\sqrt{-\lambda}L\right) \right] = 0. \end{cases}$$

So, λ satisfies the equation

$$\cot\left(\sqrt{-\lambda}L\right) = \sqrt{-\lambda}.$$

Therefore the solution is

$$u(x,t) = \sum_{m=1}^{\infty} u_m e^{-Dk_m^2 t} \cos(k_m x)$$

where the eigenfunctions $\cos(k_m x)$ and the corresponding eigenvalues $\lambda_m = k_m^2$ are related to the points k_m, with $0 < k_m < m\pi/L$ and $m > 0$, where the two functions $f_1(\xi) = \cot L\xi$, $f_2(\xi) = \xi$ intersect. Furthermore, each u_m is the coefficient of the Fourier series[21] of $u(x,0)$ with respect to the eigenfunction $\cos(k_m x)$, namely

$$u_m = \frac{1}{\alpha_m} \int_0^L u(x,0)\cos(k_m x)\,dx \qquad \alpha_m = \int_0^L \cos^2(k_m x)\,dx.$$

Regarding the concentration c, we finally deduce the formula

$$c(x,t) = \frac{B}{E} + B(L - x) + \sum_{m=1}^{\infty} u_m e^{-Dk_m^2 t} \cos(k_m x).$$

Since $k_m > 0$ for every m, as t goes to $+\infty$ every term of the series converges to zero exponentially, and therefore c settles to the steady solution c^{St}

$$c(x,t) \to C_0 + \frac{C_0 R_0}{DA}(L - x).$$

---

[21] Appendix B.

Solution 1.3.26. By the total-probability formula we can write the following difference equation for the transition probability $p = p(x,t)$, where $x = mh, t = N\tau$:

$$p(x, t + \tau) = \frac{1}{2}\left(1 + \frac{m+1}{N}\right) p(x + h, t) + \frac{1}{2}\left(1 - \frac{m-1}{N}\right) p(x - h, t)$$

or

$$p(x, t + \tau) = \frac{1}{2}[p(x + h, t) + p(x - h, t)] + \frac{1}{2}\frac{1}{N}\frac{(x+h)\,p(x+h,t) - (x-h)\,p(x-h,t)}{h}.$$

Using Taylor formulas we get, after simple calculations,

$$p_t \tau + o(\tau) = \frac{1}{2}p_{xx}h^2 + o\left(h^2\right) + \frac{1}{N}\left[(xp)_x + o(h)\right]$$

where p and its derivatives are evaluated at (x, t).

Dividing by τ and letting $h, \tau \to 0$, since $h^2/\tau = 2D$ and $N\tau = \gamma$, we obtain

$$p_t = Dp_{xx} + \frac{1}{\gamma}(xp)_x. \tag{1.65}$$

To find the required formula for $D = \gamma = 1$, write the equation in the form $p_t = p_{xx} + xp_x + p$. Thus the function $u(x,t) = p(x,t)e^{-t}$ satisfies the differential equation in Exercise 1.3.17. Recalling formula (63) with $x_0 = 0$, we end up with the solution (Fig. 1.17)

$$p(x, t) = \frac{1}{\sqrt{2\pi\left(1 - e^{-2t}\right)}} \exp\left\{-\frac{x^2}{2\left(1 - e^{-2t}\right)}\right\}. \tag{1.66}$$

Note that $p > 0, \int_{\mathbb{R}} p(x,t)\,dx = 1$ and $p(x,t) \to 0$ as $t \to 0^+, x \neq 0$. Therefore the function in (1.66) is the fundamental solution for the equation (1.65). Also observe the absence of decay in time; indeed, as $t \to \infty$, (1.66) exponentially approaches the standard Gaussian.

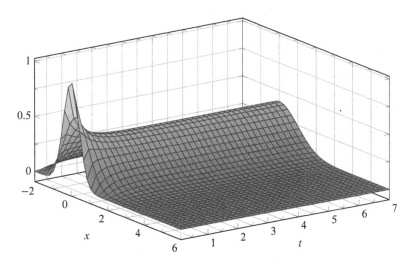

Fig. 1.17 The fundamental solution of eq. (1.65) (compare with Fig. 1.16)

Solution 1.3.27. The data are radially symmetric, so the solution depends only on the radius $r = |\mathbf{x}|$ and on time (prove this fact using uniqueness of the solution). We can separate the variables, as in Problems 1.2.25 and 1.2.5 (pages 46 and 14, respectively). We find:

$$u(r,t) = \frac{r^2}{2} + 3t + \frac{3}{8}\sum_{n=1}^{\infty}\frac{(-1)^n}{n^2\pi^2}\left\{1 + \frac{6}{n^2\pi^2}[1-(-1)^n]\right\}e^{-\lambda_n^2 t}\frac{\cos(\lambda_n r)}{r}, \qquad \text{with } \lambda_n = n\pi.$$

Solution 1.3.28. By separation of variables the solution is

$$u(r,t) = \frac{2}{r}\sum_{n=1}^{\infty}\frac{(-1)^n}{\lambda_n}\sin(\lambda_n r)\left\{\frac{q}{1-\lambda_n^2}\left(e^{-t} - e^{-\lambda_n^2 t}\right) - Ue^{-\lambda_n^2 t}\right\}, \qquad \text{with } \lambda_n = n\pi.$$

Solution 1.3.29. **a)** We are considering a global Cauchy problem:

$$\begin{cases} P_t - D\Delta P = aP & (x,y)\in\mathbb{R}^2,\ t>0 \\ P(x,y,0) = \delta_2(x,y)M(0) & (x,y)\in\mathbb{R}^2, \end{cases}$$

where Dirac's delta is centred at the origin of $\mathbb{R}^2$. We argue as in Exercise 1.3.5 and reduce to a similar problem without reaction term. More precisely, the function $u(x,y,t) = e^{-at}P(x,y,t)$ solves

$$\begin{cases} u_t - D\Delta u = 0 & (x,y)\in\mathbb{R}^2,\ t>0 \\ u(x,y,0) = \delta_2(x,y)M(0) & (x,y)\in\mathbb{R}^2. \end{cases}$$

Hence $u(x,y,t) = M(0)\Gamma_D(x,y,t)$, and then

$$P(x,y,t) = M(0)e^{at}\Gamma_D(x,y,t) = \frac{M(0)}{4\pi Dt}\exp\left(at - \frac{|\mathbf{x}|^2}{4Dt}\right).$$

b) We have

$$M(t) = M(0)e^{at}\int_{\mathbb{R}^2}\Gamma_D(x,y,t)\,dx\,dy = M(0)e^{at}.$$

c) Passing to polar coordinates yields:

$$\int_{\mathbb{R}^2\setminus B_R}P(x,y,t)\,dxdy = \int_R^{+\infty}\frac{M(0)}{4\pi Dt}\exp\left(at - \frac{r^2}{4Dt}\right)\cdot 2\pi r\,dr = M(0)\exp\left(at - \frac{R^2}{4Dt}\right),$$

for any R. Comparing with $M(0)$ gives $R(t) = 2t\sqrt{aD}$.

d) The metropolitan front advances with constant velocity $R'(t) = 2\sqrt{aD}$.

2

The Laplace Equation

2.1 Backgrounds

Denote by Ω a domain (an open connected subset) of $\mathbb{R}^n$ and by $B_r(\mathbf{x})$ the open n-dimensional ball of radius r and centre $\mathbf{x}$. A C^2 function u is *harmonic on* Ω if $\Delta u = 0$ on Ω.

- *Mean-value property.* The function u is harmonic on Ω if and only if the following averaging property holds: for any $B_r(\mathbf{x})$ with $\overline{B}_r(\mathbf{x}) \subset \Omega$,

$$u(\mathbf{x}) = \frac{1}{|B_r(\mathbf{x})|} \int_{B_r(x)} u(\mathbf{y}) \, d\mathbf{y} \quad \text{and} \quad u(\mathbf{x}) = \frac{1}{|\partial B_r(\mathbf{x})|} \int_{\partial B_R(x)} u(\sigma) \, d\sigma.$$

- *Maximum principle.* If Ω is bounded, u is harmonic on Ω and continuous on $\overline{\Omega}$, then either u is constant or

$$\min_{\partial \Omega} u < u(\mathbf{x}) < \max_{\partial \Omega} u \qquad \text{for every } \mathbf{x} \in \Omega.$$

A frequently-used consequence: let u, v be harmonic on Ω (bounded), and continuous on $\overline{\Omega}$. If $u \geq v$ on $\partial \Omega$ then $u \geq v$ in Ω.

- *Subharmonic/superharmonic functions.* A function $u \in C(\Omega)$, $\Omega \subset \mathbb{R}^n$ is called *subharmonic* if for any ball $\overline{B}_R(\mathbf{x}) \subset \Omega$

$$u(\mathbf{x}) \leq \frac{1}{|B_R(\mathbf{x})|} \int_{B_R(x)} u(\mathbf{y}) \, d\mathbf{y}, \qquad u(\mathbf{x}) \leq \frac{1}{|\partial B_R(\mathbf{x})|} \int_{\partial B_R(x)} u(\mathbf{y}) \, d\mathbf{y}.$$

It is called *superharmonic* if the above inequalities are reversed.

Equivalently, u is subharmonic if for any harmonic v on $B_R(\mathbf{x})$ that is greater or equal to u on the boundary, then $u \leq v$ in the entire ball. If additionally Ω is connected, and $u \in C(\overline{\Omega})$ is subharmonic and assumes its maximum at an interior point of Ω, then u must be constant. Furthermore, the maximum of two subharmonics is subharmonic too.

If $u \in C^2(\Omega)$, then u is subharmonic if and only if $-\Delta u \leq 0$ in Ω.

© Springer International Publishing Switzerland 2015
S. Salsa, G. Verzini, *Partial Differential Equations in Action. Complements and Exercises*,
UNITEXT – La Matematica per il 3+2 87, DOI 10.1007/978-3-319-15416-9_2

• *Poisson formula.* Let u be harmonic on $B_r(\mathbf{p}) \subset \mathbb{R}^n$ and continuous on $\overline{B}_r(\mathbf{p})$. Then

$$u(\mathbf{x}) = \frac{r^2 - |\mathbf{x} - \mathbf{p}|^2}{\omega_n r} \int_{\partial B_r(\mathbf{p})} \frac{u(\sigma)}{|\sigma - \mathbf{x}|^n} d\sigma,$$

where $\omega_n = |\partial B_1(\mathbf{p})|$. Therefore $u \in C^\infty(B_r(\mathbf{p}))$.

• *Harnack's inequality.* Let u be harmonic and non-negative in $B_r(\mathbf{p}) \subset \mathbb{R}^n$. For any $\mathbf{x} \in B_r(\mathbf{p})$ we have

$$\frac{r^{n-2}(r - |\mathbf{x}|)}{(r + |\mathbf{x}|)^{n-1}} u(\mathbf{p}) \le u(\mathbf{x}) \le \frac{r^{n-2}(r + |\mathbf{x}|)}{(r - |\mathbf{x}|)^{n-1}} u(\mathbf{p}).$$

In particular

$$\max_{B_{r/2}(\mathbf{p})} u \le 3^n \min_{B_{r/2}(\mathbf{p})} u.$$

• *Liouville's theorem.* A harmonic function u on $\mathbb{R}^n$, with $u(\mathbf{x}) \ge 0$ for every $\mathbf{x} \in \mathbb{R}^n$, is constant.

Therefore the only harmonic functions on $\mathbb{R}^n$ that are bounded either from above or from below are the constant functions.

• *Fundamental solution and potentials.* The function

$$\Gamma(\mathbf{x}) = \begin{cases} -\dfrac{1}{2\pi} \log |\mathbf{x}| & n = 2, \\ \dfrac{1}{(n-2)\omega_n} \dfrac{1}{|\mathbf{x}|^{n-2}} & n \ge 3 \end{cases}$$

is the solution of

$$-\Delta\Gamma(\mathbf{x}) = \delta_n(\mathbf{x}) \text{ in } \mathbb{R}^n \quad (\delta_n \text{ is the } n\text{-dimensional } Dirac's \ delta \ function \text{ at the origin})$$

and is called the *fundamental solution of the Laplace operator*.

Using Γ one can reconstruct an arbitrary $u \in C^2(\overline{\Omega})$, Ω bounded with regular boundary, as the sum of three terms:

$$u(\mathbf{x}) = \int_{\partial\Omega} \Gamma(\mathbf{x} - \sigma) \partial_{\nu_\sigma} u \, d\sigma - \int_{\partial\Omega} u \, \partial_{\nu_\sigma} \Gamma(\mathbf{x} - \sigma) \, d\sigma - \int_\Omega \Gamma(\mathbf{x} - \mathbf{y}) \Delta u \, d\mathbf{y},$$

called *simple-layer potential*, *double-layer potential*, and *Newtonian potential* respectively. Above, $\partial_{\nu_\sigma} = \nabla \cdot \nu_\sigma$ and ν_σ is the outer unit normal at $\sigma \in \partial\Omega$.

An integral of the form

$$\mathcal{N}(\mathbf{x}; f) = \int_\Omega \Gamma(\mathbf{x} - \mathbf{y}) f(\mathbf{y}) \, d\mathbf{y}$$

is called a *Newtonian potential*. In three dimensions it represents the electrostatic potential generated by a charge distribution of density f in Ω. If $f \in C(\overline{\Omega})$ then $\mathcal{N} \in C^1(\mathbb{R}^n)$. If $f \in C^1(\overline{\Omega})$ then $\mathcal{N} \in C^2(\Omega)$, and $-\Delta\mathcal{N} = f$ in Ω.

An integral of the form

$$\mathcal{D}\,(\mathbf{x};\mu) = \int_{\partial\Omega} \mu\,(\boldsymbol{\sigma})\,\partial_{\nu_\sigma}\Gamma\,(\mathbf{x}-\boldsymbol{\sigma})\,d\sigma$$

is called *double-layer potential of* μ. In three dimensions it represents the electrostatic potential generated by a *dipole* distribution of momentum μ on $\partial\Omega$. Let $\Omega \subset \mathbb{R}^n$ be a bounded, C^2 domain. Then

$$\mathcal{D}\,(\mathbf{x};1) = \int_{\partial\Omega} \partial_{\nu_\sigma}\Gamma\,(\mathbf{x}-\boldsymbol{\sigma})\,d\sigma = \begin{cases} -1 & \mathbf{x}\in\Omega \\ -\dfrac{1}{2} & \mathbf{x}\in\partial\Omega \\ 0 & \mathbf{x}\in\mathbb{R}^n\setminus\overline{\Omega}. \end{cases}$$

More generally, let μ be a continuous function on $\partial\Omega$. Then $D\,(\mathbf{x};\mu)$ is harmonic on $\mathbb{R}^n\setminus\partial\Omega$ and the following *jump relations* hold for every $\mathbf{x}\in\partial\Omega$:

$$\lim_{\mathbf{z}\to\mathbf{x},\,\mathbf{z}\in\mathbb{R}^n\setminus\overline{\Omega}} \mathcal{D}\,(\mathbf{z};\mu) = \mathcal{D}\,(\mathbf{x};\mu) + \frac{1}{2}\mu\,(\mathbf{x}) \tag{2.1}$$

and

$$\lim_{\mathbf{z}\to\mathbf{x},\,\mathbf{z}\in\Omega} \mathcal{D}\,(\mathbf{z};\mu) = \mathcal{D}\,(\mathbf{x};\mu) - \frac{1}{2}\mu\,(\mathbf{x}). \tag{2.2}$$

On the contrary, *normal derivatives across $\partial\Omega$ are continuous*.

An integral of the form

$$\mathcal{S}\,(\mathbf{x},\psi) = \int_{\partial\Omega} \Gamma\,(\mathbf{x}-\boldsymbol{\sigma})\,\psi\,(\boldsymbol{\sigma})\,d\sigma$$

is called *single-layer potential* of ψ. In three dimensions it represents the electrostatic potential generated by a charge distribution of density ψ on $\partial\Omega$. If Ω is a C^2 domain and ψ is continuous on $\partial\Omega$, then $\mathcal{S}$ is *continuous across $\partial\Omega$* and

$$\Delta\mathcal{S} = 0 \qquad \text{in } \mathbb{R}^n\setminus\partial\Omega,$$

because one can differentiate inside the integral. On the other hand, the normal derivative of $\mathcal{S}$ undergoes a jump when crossing $\partial\Omega$. More precisely, for every $\mathbf{x}\in\partial\Omega$, setting $\mathbf{z}_t = \mathbf{x}+t\boldsymbol{\nu}_\mathbf{x}$, we have

$$\lim_{t\to 0^+} \partial_{\nu_\mathbf{x}}\mathcal{S}\,(\mathbf{z}_t,\psi) = \int_{\partial\Omega} \partial_{\nu_\mathbf{x}}\Gamma\,(\mathbf{x}-\boldsymbol{\sigma})\,\psi\,(\boldsymbol{\sigma})\,d\sigma - \frac{1}{2}\psi\,(\mathbf{x}) \tag{2.3}$$

and

$$\lim_{t\to 0^-} \partial_{\nu_\mathbf{x}}\mathcal{S}\,(\mathbf{z}_t,\psi) = \int_{\partial\Omega} \partial_{\nu_\mathbf{x}}\Gamma\,(\mathbf{x}-\boldsymbol{\sigma})\,\psi\,(\boldsymbol{\sigma})\,d\sigma + \frac{1}{2}\psi\,(\mathbf{x}). \tag{2.4}$$

• *Green function.* Γ is the fundamental solution for the operator $-\Delta$ on the entire space. We can define a fundamental solution for $-\Delta$ in Ω, when the latter has, for example, Lipschitz boundary; the idea is that the fundamental solution should represent the potential generated by a unit charge at some point $\mathbf{y}$ inside a conductor occupying Ω and *connected to ground* at the boundary. We denote by $G(\mathbf{x}, \mathbf{y})$ this function, called the *Green function in Ω*. For any given $\mathbf{y}$ it holds

$$-\Delta G(\cdot, \mathbf{y}) = \delta_n(\cdot - \mathbf{y}) \qquad \text{in } \Omega,$$

and since the conductor is grounded, we also know that

$$G(\cdot, \mathbf{y}) = 0, \qquad \text{on } \partial\Omega.$$

Therefore

$$G(\mathbf{x}, \mathbf{y}) = \Gamma(\mathbf{x} - \mathbf{y}) - g(\mathbf{x}, \mathbf{y})$$

where g, as function of $\mathbf{x}$, solves, for any fixed $\mathbf{y}$, the Dirichlet problem

$$\begin{cases} \Delta g = 0 & \text{in } \Omega \\ g(\cdot, \mathbf{y}) = \Gamma(\cdot - \mathbf{y}) & \text{on } \partial\Omega. \end{cases}$$

2.2 Solved Problems

- **2.2.1 – 2.2.12** : General properties of harmonic functions.
- **2.2.13 – 2.2.24** : Boundary-value problems. Solution methods.
- **2.2.25 – 2.2.31** : Potentials and Green functions.

2.2.1 General properties of harmonic functions

**Problem 2.2.1 (Analyticity of harmonic functions). *Consider a harmonic function u in a domain $\Omega \subset \mathbb{R}^n$. Prove the following properties:*

a) If $\overline{B}_R(\mathbf{p}) \subset \Omega$, for any multi-index $\alpha = (\alpha_1, \ldots, \alpha_n)$

$$|D^\alpha u(\mathbf{p})| \le \frac{(n|\alpha|)^{|\alpha|}}{R^{|\alpha|}} \max_{\partial B_R(\mathbf{p})} |u| \qquad (|\alpha| = \alpha_1 + \cdots + \alpha_n) \qquad (2.5)$$

(use induction).

b) Deduce the Taylor expansion

$$u(\mathbf{x}) = \sum_{|\alpha|=0}^{\infty} \frac{D^\alpha u(\mathbf{p})}{\alpha!}(\mathbf{x} - \mathbf{p})^\alpha \qquad (\alpha! = \alpha_1! \cdots \alpha_n!)$$

for $\mathbf{x}$ close enough to $\mathbf{p}$; therefore, u is (real) analytic on Ω.

c) Deduce that if u has a local maximum or minimum at $\mathbf{p} \in \Omega$, u is constant in Ω.

Solution. a) Let u be harmonic in $\Omega \subset \mathbb{R}^n$. Then u belongs to $C^\infty(\Omega)$ and therefore, for any multi-index $\alpha = (\alpha_1, \ldots, \alpha_n)$ the derivative of order α exists:

$$v = D^\alpha u = \frac{\partial^{\alpha_1}}{\partial x_1^{\alpha_1}} \cdots \frac{\partial^{\alpha_n}}{\partial x_n^{\alpha_n}} u.$$

In particular we can compute the Laplacian of v:

$$\Delta v = \Delta(D^\alpha u) = \sum_{i=1}^{n} \frac{\partial^2}{\partial x_i^2}(D^\alpha u) = \sum_{i=1}^{n} D^\alpha \left(\frac{\partial^2}{\partial x_i^2} u \right) = D^\alpha(\Delta u) = 0$$

(as u is C^∞, Schwarz's theorem allows to swap derivatives).

Take w harmonic on Ω, $\overline{B}_r(\mathbf{q}) \subset \Omega$. Since w_{x_j} is harmonic in Ω, by the mean-value property and the Gauss formula we may write:

$$\left| w_{x_j}(\mathbf{q}) \right| = \frac{n}{\omega_n r^n} \left| \int_{B_r(\mathbf{q})} w_{x_j}(\mathbf{y}) \, d\mathbf{y} \right| =$$

$$= \frac{n}{r} \cdot \frac{1}{\omega_n r^{n-1}} \left| \int_{\partial B_r(\mathbf{q})} w(\sigma) v_j \, d\sigma \right| \leq \frac{n}{r} \max_{\partial B_r(\mathbf{q})} |w|. \quad (2.6)$$

Now we use induction. When $|\alpha| = 1$ it suffices to rewrite (2.6) with $w = u$, $r = R$, $\mathbf{q} = \mathbf{p}$. So suppose (2.5) holds for any $|\alpha| = k$ (and any $\mathbf{p}$, R such that $\overline{B}_R(\mathbf{p}) \subset \Omega$). Using (2.6) with $w = D^\alpha u$, $r = R/(k+1)$, $\mathbf{q} = \mathbf{p}$ we obtain

$$\left| (D^\alpha u)_{x_j}(\mathbf{p}) \right| \leq \frac{n(k+1)}{R} \max_{\partial B_{R/(k+1)}(\mathbf{p})} |D^\alpha u| = \frac{n(k+1)}{R} |D^\alpha u(\mathbf{q}_{max})|, \quad (2.7)$$

where $\mathbf{q}_{max} \in \partial B_{R/(k+1)}$ is the maximum point. Since $B_{kR/(k+1)}(\mathbf{q}_{max}) \subset B_R(\mathbf{p})$ (see Fig. 2.1), we can use equation (2.5), with $kR/(k+1)$ replacing R and $\mathbf{q}_{max}$ instead of $\mathbf{p}$

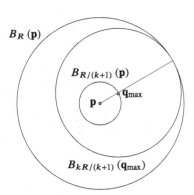

Fig. 2.1 Use of the inductive assumption (Problem 2.2.1)

(by inductive hypothesis). We find

$$|D^\alpha u(\mathbf{q}_{\max})| \le \frac{(nk)^k}{R^k} \cdot \frac{(k+1)^k}{k^k} \max_{\partial B_{kR/(k+1)}(\mathbf{q}_{\max})} |u| . \qquad (2.8)$$

From (2.7) and (2.8), together with the fact that

$$\max_{\partial B_{kR/(k+1)}(\mathbf{q}_{\max})} |u| \le \max_{\partial B_R(\mathbf{p})} |u|$$

by the maximum principle, we deduce (2.5) for $|\alpha| = k + 1$, and hence the claim.

b) Let $\overline{B}_{2R}(\mathbf{p}) \subset \Omega$. Then u is smooth in $B_R(\mathbf{p})$. If $0 < r \le R$ we can use the Taylor expansion with Lagrange remainder and write, for any $\mathbf{x} \in B_r(\mathbf{p})$,

$$u(\mathbf{x}) = \sum_{|\alpha|=0}^{k-1} \frac{D^\alpha u(\mathbf{p})}{\alpha!}(\mathbf{x}-\mathbf{p})^\alpha + R_k(\mathbf{x}),$$

where

$$R_k(\mathbf{x}) = \sum_{|\alpha|=k} \frac{D^\alpha u(\mathbf{p})}{\alpha!}(\boldsymbol{\xi}_\mathbf{x}-\mathbf{p})^\alpha$$

for some $\boldsymbol{\xi}_\mathbf{x} \in B_r(\mathbf{p})$. The known formula[1]

$$n^k = (1+\cdots+1)^k = k! \sum_{|\alpha|=k} \frac{1}{\alpha!}$$

and part a) give

$$|\boldsymbol{\xi}_\mathbf{x}| \le r \quad \Longrightarrow \quad |R_k(\mathbf{x})| \le \frac{(nk)^k}{R^k} \max_{\partial B_R(\boldsymbol{\xi}_\mathbf{x})} |u| \cdot r^k \cdot \frac{n^k}{k!}$$

$$= \underbrace{\frac{k^k}{k!} \left(\frac{n^2 r}{R}\right)^k}_{a_k} \max_{\partial B_{2R}(\mathbf{p})} |u| ,$$

so there remains to prove that $a_k \to 0$ as k goes to infinity. This actually follows by observing that

$$\frac{a_{k+1}}{a_k} = \left(1+\frac{1}{k}\right)^k \frac{n^2 r}{R} < e\frac{n^2 r}{R} \le \frac{e}{3},$$

and so

$$a_{k+1} \le a_1 \left(\frac{e}{3}\right)^k,$$

provided we choose r small enough (it suffices to take $r = R/(3n^2)$).

---

[1] This can be proved by induction on n, using Newton's binomial formula.

c) By definition if $\mathbf{p}$ is a local minimum for u, then it is also an absolute minimum for u restricted on $B_R(\mathbf{p})$, if R is small enough. The maximum principle implies that u is constant on $B_R(\mathbf{p})$, and by the uniqueness of analytic prolongations u is constant on Ω.

Problem 2.2.2 (Harmonic functions with polynomial growth). *Let u be harmonic in $\mathbb{R}^n$, and $\gamma > 0$, $C > 0$ such that*

$$|u(\mathbf{x})| \le C(1 + |\mathbf{x}|)^\gamma \qquad \text{for every } \mathbf{x} \in \mathbb{R}^n. \tag{2.9}$$

Using Problem 2.2.1 a), prove that u is a polynomial of degree less than or equal to γ.

Solution. Let $k > \gamma$ be an integer. As u is harmonic on $\mathbb{R}^n$, from (2.5) and (2.9) we have, for any $\mathbf{p} \in \mathbb{R}^n$, $R > 0$ and $|\alpha| = k$,

$$|D^\alpha u (\mathbf{p})| \le \frac{(nk)^k}{R^k} \max_{\partial B_R(\mathbf{p})} |u| \le C (nk)^k \frac{1 + R^\gamma}{R^k}.$$

Letting R go to infinity we see $D^\alpha u(\mathbf{p}) = 0$, for any $\mathbf{p}$. But this holds for derivatives of any order greater than γ, so u is a polynomial of degree less than or equal to γ.

Problem 2.2.3 (Harmonic polynomials). *Find all harmonic polynomials of degree n in two variables.*

Solution. Suppose

$$P_n(x, y) = \sum_{k=0}^{n} c_k x^{n-k} y^k,$$

indicates the generic harmonic polynomial of degree n in the variables x, y, with non-zero coefficients c_k. Differentiating

$$\Delta P_n(x, y) = \sum_{k=0}^{n-2} c_k(n - k)(n - k - 1)x^{n-k-2} y^k + \sum_{h=2}^{n} c_h h(h - 1)x^{n-h} y^{h-2}$$

$$= \sum_{k=0}^{n-2} c_k(n - k)(n - k - 1)x^{n-k-2} y^k + \sum_{k=0}^{n-2} c_{k+2}(k + 2)(k + 1)x^{n-k-2} y^k$$

$$= \sum_{k=0}^{n-2} [c_k (n - k)(n - k - 1) + c_{k+2}(k + 2)(k + 1)] x^{n-k-2} y^k.$$

For P_n to be harmonic, each summand must necessarily have zero coefficient, so for any k

$$c_k (n - k)(n - k - 1) + c_{k+2}(k + 2)(k + 1) = 0,$$

and then

$$c_{k+2} = -\frac{(n - k)(n - k - 1)}{(k + 2)(k + 1)} c_k.$$

Therefore the even coefficients depend on the choice of c_0, while the odd ones depend on c_1. Let us consider the former ones:

$$c_2 = -\frac{n(n-1)}{2 \cdot 1} c_0 = -\frac{n(n-1)(n-2)!}{2! \cdot (n-2)!} c_0 = -\binom{n}{2} c_0,$$

$$c_4 = -\frac{(n-2)(n-3)}{4 \cdot 3} c_2 = \frac{n(n-1)(n-2)(n-3)}{4 \cdot 3 \cdot 2 \cdot 1} c_0 =$$

$$= \frac{n(n-1)(n-2)(n-3)(n-4)!}{4! \cdot (n-4)!} c_0 = \binom{n}{4} c_0.$$

By induction

$$c_{2h} = (-1)^h \binom{n}{2h} c_0, \quad \text{and, analogously,} \quad c_{2h+1} = (-1)^h \binom{n}{2h+1} \frac{c_1}{n}.$$

In conclusion

$$P_n(x, y) = \sum_{k=0}^{n} \tilde{c}_k \binom{n}{k} x^{n-k} y^k, \quad \text{with} \quad \tilde{c}_k = \begin{cases} (-1)^h c_0 & \text{for } k = 2h \\ (-1)^h \frac{c_1}{n} & \text{for } k = 2h+1 \end{cases}$$

and c_0, c_1 arbitrary.

Problem 2.2.4 (Subharmonic functions and a variant of Liouville's theorem). *Let u be harmonic on $\Omega \subset \mathbb{R}^n$. Prove:*

a) If $F \in C^2(\mathbb{R})$ is convex then $w = F(u)$ is subharmonic on Ω.

b) If $\Omega = \mathbb{R}^n$ and

$$\int_{\mathbb{R}^n} u^2(x)\, dx = M < \infty$$

then $u \equiv 0$.

Solution. a) Since

$$w_{x_j} = F'(u) u_{x_j}, \quad w_{x_j x_j} = F''(u) u_{x_j}^2 + F'(u) u_{x_j x_j}$$

it follows

$$\Delta w = \sum_{j=1}^{n} \left[F''(u) u_{x_j}^2 + F'(u) u_{x_j x_j} \right] = F''(u) |\nabla u|^2 + F'(u) \Delta u$$

$$= F''(u) |\nabla u|^2 \geq 0$$

as F is convex.

b) Take $x \in \mathbb{R}^n$ and $R > 0$. Because u is harmonic and $F(s) = s^2$ is convex, u^2 is subharmonic and we may write

$$u^2(x) \leq \frac{1}{|B_R(x)|} \int_{B_R(x)} u^2(y) dy \leq \frac{M}{|B_R(x)|}.$$

But $\frac{M}{|B_R(\mathbf{x})|} \to 0$ for $R \to \infty$, hence $u^2(\mathbf{x}) = 0$ and, since $\mathbf{x}$ is arbitrary, $u \equiv 0$ in $\mathbb{R}^n$.

---

*Problem 2.2.5 (Liouville theorem for subharmonic functions in the plane). *Let u be a (continuous) subharmonic function, bounded from above in $\mathbb{R}^2$.*

a) *Verify that for any $\varepsilon > 0$ the function $w_\varepsilon(\mathbf{x}) = u(\mathbf{x}) - \varepsilon \log |\mathbf{x}|$ satisfies*

$$\max_{\overline{B}_e} w_\varepsilon = \max_{\partial B_e} w_\varepsilon = \max_{\partial B_e} u,$$

where $B_e = \{\mathbf{x} \in \mathbb{R}^2 : |\mathbf{x}| > 1\}$.

b) *Deduce that u is constant.*

Solution. a) Since $u(\mathbf{x}) \leq M$ for any $\mathbf{x}$, for any given $\varepsilon > 0$ there exists $R_\varepsilon > 0$ such that

$$|\mathbf{x}| \geq R_\varepsilon \quad \Longrightarrow \quad w_\varepsilon(\mathbf{x}) \leq M - \varepsilon \log R_\varepsilon \leq \max_{\partial B_e} w_\varepsilon \qquad (2.10)$$

(take $R_\varepsilon = \exp[(M - \max_{\partial B_e} w_\varepsilon)/\varepsilon]$). On the other hand $v(\mathbf{x}) = \log |\mathbf{x}|$ is harmonic on $\mathbb{R}^2 \setminus \{\mathbf{0}\}$, so w_ε is subharmonic on the annulus $\{1 < |\mathbf{x}| < R_\varepsilon\}$, and assumes its maximum on the closure, at points $\mathbf{x}$ with $|\mathbf{x}| = 1$ or $|\mathbf{x}| = R_\varepsilon$. By (2.10) we deduce

$$w_\varepsilon(\mathbf{x}) \leq \max_{\partial B_e} w_\varepsilon \qquad \text{also when } 1 < |\mathbf{x}| < R_\varepsilon.$$

b) Letting $\varepsilon \to 0$ we have $w_\varepsilon \to u$ uniformly on compact subsets in $\overline{B}_e$. The previous result forces

$$\max_{\overline{B}_e} u = \max_{\partial B_e} u.$$

Therefore, for any compact set $K \supset B_1$

$$\max_K u = \max_{B_1} u = \max_{\partial B_1} u.$$

So the subharmonic function u has a maximum in the interior, and thus it is constant on K, for any K.

---

*Problem 2.2.6 (Harnack inequality on compact sets). *Consider a domain Ω in $\mathbb{R}^n$ and a compact subset $K \subset \mathbb{R}^n$. Prove that there exists a constant $\gamma > 0$, depending only on K and Ω, such that for any harmonic, non-negative function on Ω one has*

$$\max_K u \leq \gamma \min_K u.$$

Solution. K is compact, so we can find a finite number of balls $B_{R_j} = B_{R_j}(\mathbf{p}_j)$, $j = 1, \ldots, k$ inside Ω such that

i) K is contained in the union of the $B_{R_j/2}$.

ii) $\emptyset \neq B_{R_j/2} \cap B_{R_{j+1}/2} \ni \mathbf{z}_j$.

In particular $\min_j R_j$ decreases (and k increases) as the distance between K and $\partial\Omega$ gets smaller. From Harnack's inequality we have

$$\max_{B_{R_j/2}} u \le 3^n \min_{B_{R_j/2}} u \le 3^n u(\mathbf{z}_j) \le 3^n \max_{B_{R_{j+1}/2}} u \le 3^{2n} \min_{B_{R_{j+1}/2}} u.$$

Take $\mathbf{x}, \mathbf{y} \in K$, and to fix ideas suppose $\mathbf{x} \in B_{R_{j_1}/2}, \mathbf{y} \in B_{R_{j_2}/2}$, with $j_1 \le j_2$. Iterating the previous inequality we get

$$u(\mathbf{x}) \le 3^{j_2-j_1+1} u(\mathbf{y}) \le 3^{kn} u(\mathbf{y}).$$

But $\mathbf{x}, \mathbf{y}$ were arbitrary, so (with $\gamma = 3^{kn}$) the claim follows.

---

*Problem 2.2.7 (Series of harmonic functions). *Consider harmonic, non-negative functions u_i, $i \in \mathbb{N}$, defined on a domain Ω in $\mathbb{R}^n$. Using Harnack's inequality show that if $\sum_{i=0}^\infty u_i$ converges at some $\mathbf{x}_0 \in \Omega$, then it converges uniformly on any compact set $K \subset \Omega$. Deduce that the sum U of the series is non-negative and harmonic everywhere on Ω.*

Solution. We wish to prove uniform convergence on any given compact $K \subset \Omega$. Since Ω is connected, we may assume K is connected and that it contains $\mathbf{x}_0$ (if not, just choose another compact set containing the original one). By Harnack's inequality (Problem 2.2.6)

$$\max_K u_i \le \gamma \min_K u_i \le \gamma u_i(\mathbf{x}_0).$$

Then

$$\sum_{i=0}^\infty \max_K u_i(\mathbf{x}) \le \gamma \sum_{i=0}^\infty u_i(\mathbf{x}_0) < \infty$$

and by the Weierstrass criterion the series $\sum_{i=0}^\infty u_i(\mathbf{x})$ converges uniformly on K.

But $K \subset \Omega$ is arbitrary, so the series converges at any point in Ω, implying that the sum U is defined on Ω, and is non-negative as sum of non-negative terms. Also the u_i are continuous on Ω, hence the uniform convergence on compact subsets of Ω guarantees that U is continuous on Ω.

To show U is harmonic, then, it suffices to show it satisfies the mean-value property[2]. For any $B_r(\mathbf{x}) \subset\subset \Omega$,

$$\frac{1}{|B_r(\mathbf{x})|} \int_{B_r(\mathbf{x})} U(\mathbf{y})\, dy = \frac{1}{|B_r(\mathbf{x})|} \int_{B_r(\mathbf{x})} \left(\sum_{i=0}^\infty u_i(\mathbf{y})\right) dy$$

$$= \sum_{i=0}^\infty \left(\frac{1}{|B_r(\mathbf{x})|} \int_{B_r(\mathbf{x})} u_i(\mathbf{y})\, dy\right)$$

$$= \sum_{i=0}^\infty u_i(\mathbf{x}) = U(\mathbf{x}),$$

and the claim follows (we rely on uniform convergence in order to swap sum and integral).

---

[2] With a little extra effort one could prove $\Delta U = \sum \Delta u_i = 0$.

Problem 2.2.8. *Let u be a positive harmonic function in $B_1(0) \setminus \{0\} \subset \mathbb{R}^n$. Show that there exists a constant $\gamma > 0$, depending only on n, such that*

$$u(\mathbf{x}) \geq \gamma u(\mathbf{y})$$

whenever $0 < |\mathbf{x}| = |\mathbf{y}| \leq \frac{1}{2}$.

Solution. Let $|\mathbf{x}| = |\mathbf{y}| = \frac{1}{2}$. Since $\partial B_{1/2}(0)$ is a compact subset of $B_1(0) \setminus \{0\}$, Harnack's inequality (Problem 2.2.6 on page 89) gives

$$u(\mathbf{x}) \geq \gamma u(\mathbf{y}) \qquad (2.11)$$

with $\gamma > 0$, dependent only on n. Now, let $0 < R \leq 1$ and set $U(\mathbf{z}) = u(R\mathbf{z})$. Then U is harmonic on $B_{1/R}(0) \setminus \{0\}$ and in particular in $B_1(0) \setminus \{0\}$. From (2.11) we infer

$$u(R\mathbf{z}) = U(\mathbf{z}) \geq \gamma U(\mathbf{w}) = \gamma u(R\mathbf{w})$$

whenever $|\mathbf{z}| = |\mathbf{w}| = \frac{1}{2}$. Letting $\mathbf{x} = R\mathbf{z}, \mathbf{w} = R\mathbf{y}$, we get $u(\mathbf{x}) \geq \gamma u(\mathbf{y})$ whenever $0 < |\mathbf{z}| = |\mathbf{w}| = \frac{R}{2} \leq \frac{1}{2}$.

**Problem 2.2.9 (Hopf principle). *Fix $\Omega \subset \mathbb{R}^n$ and $u \in C^2(\Omega) \cap C^1(\overline{\Omega})$, harmonic and positive on Ω. Let $\mathbf{x}_0 \in \partial\Omega$ be a zero of u. If there exists a ball (Fig. 2.2)*

$$B_R(\mathbf{p}) \subset \Omega \qquad \text{such that} \qquad \partial\Omega \cap \overline{B_R(\mathbf{p})} = \{\mathbf{x}_0\},$$

then $\partial_\nu u(\mathbf{x}_0) > 0$, where $\nu = (\mathbf{p} - \mathbf{x}_0)/R$.
In particular, if $\partial\Omega$ is C^1 at $\mathbf{x}_0$ then ν is the inward unit normal to $\partial\Omega$ at the point $\mathbf{x}_0$.

Solution. We set $r = |\mathbf{x} - \mathbf{p}|$ and consider the annulus

$$C_R = \left\{ \frac{R}{2} < r < R \right\},$$

contained in Ω and touching $\partial\Omega$ at $\mathbf{x}_0$. Let w be the harmonic function on C_R that vanishes on ∂B_R and equals the minimum of u on $\partial B_{R/2}$. By the maximum principle $u \geq v$

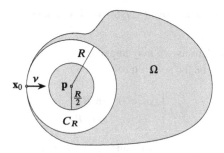

Fig. 2.2 Touching ball at $\mathbf{x}_0$

on C_R. But since $u(\mathbf{x}_0) = w(\mathbf{x}_0) = 0$, we also have

$$\partial_\nu u(\mathbf{x}_0) = \lim_{h \to 0+} \frac{u(\mathbf{x}_0 + h\nu) - u(\mathbf{x}_0)}{h} \geq \lim_{h \to 0+} \frac{w(\mathbf{x}_0 + h\nu) - w(\mathbf{x}_0)}{h} = \partial_\nu w(\mathbf{x}_0).$$

Now it suffices to prove $\partial_\nu w(\mathbf{x}_0) > 0$. The function w is radial, that is

$$w = w(|\mathbf{x} - \mathbf{p}|) = w(r),$$

and hence it has the form (we consider only $n > 2$, and leave the case $n = 2$ to the reader)

$$w(r) = \frac{C_1}{r^{n-2}} + C_2.$$

Define

$$m = \inf_{\partial B_{R/2}} u.$$

Imposing $w(R) = 0$ and $w(R/2) = m$, we find

$$w(r) = \frac{m}{2^{n-2} - 1}\left[\left(\frac{R}{r}\right)^{n-2} - 1\right],$$

and then

$$\partial_\nu w(\mathbf{x}_0) = -w'(R) = \frac{m(n-2)}{R(2^{n-2} - 1)} > 0,$$

whence the claim.

**Problem 2.2.10 (Removable singularities). *Let $\Omega \subset \mathbb{R}^2$ be an open bounded domain, $\mathbf{x}_0 \in \Omega$ and $u \in C^2(\Omega \setminus \{\mathbf{x}_0\})$ such that, for some $M > 0$,*

$$\Delta u(\mathbf{x}) = 0, \quad |u(\mathbf{x})| \leq M \qquad \text{for } x \in \Omega \setminus \{\mathbf{x}_0\}.$$

Show that u can be extended to a harmonic function at $\mathbf{x}_0$, that is, there exists $\tilde{u} \in C^2(\Omega)$ with

$$\tilde{u}(\mathbf{x}) = u(\mathbf{x}) \text{ for } \mathbf{x} \neq \mathbf{x}_0, \qquad \Delta\tilde{u} = 0 \text{ in } \Omega. \qquad (2.12)$$

(see Exercise 2.3.20 for an improved version of this result).

Solution. Using translations we may suppose $\mathbf{x}_0 = \mathbf{0}$. Let $R > 0$ be so small that $\overline{B}_R = \overline{B}_R(\mathbf{0}) \subset \Omega$. Let v be the solution to

$$\begin{cases} \Delta v = 0 & \text{in } B_R \\ v = u & \text{on } \partial B_R. \end{cases}$$

The maximum principle ensures that the range of v lies between the maximum and minimum of u on ∂B_R, so $|v(\mathbf{x})| \leq M$.

Now $w = u - v$ is harmonic on $B_R \setminus \{0\}$, it vanishes on ∂B_R and $|w(x)| \le 2M$. If we show that $w \equiv 0$ on $B_R \setminus \{0\}$, then

$$\tilde{u}(x) = \begin{cases} u(x) & x \in \Omega \setminus B_R \\ v(x) & x \in B_R \end{cases}$$

satisfies (2.12). Consider $0 < r < R$, $B_r = B_r(0)$; the function

$$h(x) = 2M \frac{\log(|x|/R)}{\log(r/R)}$$

solves

$$\begin{cases} \Delta h = 0 & \text{in } B_R \setminus \overline{B_r} \\ h = 0 & \text{on } \partial B_R \\ h = 2M & \text{on } \partial B_r. \end{cases} \tag{2.13}$$

The maximum principle implies that $-h \le w \le h$ in $\overline{B_R} \setminus B_r$, i.e.

$$|w(x)| \le 2M \frac{\log(|x|/R)}{\log(r/R)} \qquad \text{for } r \le |x| \le R.$$

Fix $x \ne 0$. The previous inequality holds for any $r \le |x|$. In particular, then, we let r tend to 0, obtaining $|w(x)| = 0$. But x was arbitrary, so w is identically zero away from the origin; as we already remarked, this proves the claim.

Problem 2.2.11 (Radial replacement). *Let u be harmonic in $B_1(0) \setminus (0)$, $B_1(0) \subset \mathbb{R}^n$, and suppose $u(x) \to 0$ as $|x| \to 1$. Define the radial replacement of u by the formula*

$$u^R(x) = \frac{1}{n\omega_n |x|^{n-1}} \int_{\partial B_{|x|}(0)} u(\sigma) \, d\sigma.$$

Show that, for some constant C,

$$u^R(x) = \begin{cases} C \log |x| & n = 2 \\ C\left(|x|^{2-n} - 1\right) & n \ge 3 \end{cases} \tag{2.14}$$

and therefore u^R is harmonic on $B_1(0) \setminus (0)^a$.

---

^{a} *Hint.* Set $g(|x|) = u^R(x)$ and show that $r^{n-1}g'(r)$ is constant on $(0, 1)$.

Solution. Let

$$g(r) = \frac{1}{n\omega_n r^{n-1}} \int_{\partial B_r(0)} u(\sigma) \, d\sigma = \frac{1}{n\omega_n} \int_{\partial B_1(0)} u(r\omega) \, d\omega.$$

Then, $u^R(\mathbf{x}) = g(|\mathbf{x}|)$ and $g(1) = 0$. We want to show that $r^{n-1}g(r)$ is constant on $(0, 1)$. In fact,

$$g'(r) = \frac{1}{n\omega_n} \int_{\partial B_1(0)} \nabla u(r\omega) \cdot \omega \, d\omega = \frac{1}{n\omega_n r^{n-1}} \int_{\partial B_r(0)} \partial_\nu u(\sigma) \, d\sigma$$

where $\nu = \sigma/r$ is the outward unit normal at the point σ. Since $\Delta u = 0$ in $B_1(0) \setminus (0)$, by applying Green's formula to the annulus $A_{r,s}, 0 < r, s < 1$ we deduce that

$$0 = \int_{A_{r,s}} \Delta u = \int_{\partial B_s(0)} \partial_\nu u(\sigma) \, d\sigma - \int_{\partial B_r(0)} \partial_\nu u(\sigma) \, d\sigma.$$

Therefore

$$r^{n-1} g'(r) = \frac{1}{n\omega_n} \int_{\partial B_r(0)} \partial_\nu u(\sigma) \, d\sigma = C \quad (C \in \mathbb{R})$$

or

$$g'(r) = \frac{C}{r^{n-1}}$$

from which (2.14) follows by integration, taking into account that $g(1) = 0$.

---

*Problem 2.2.12 (Radial replacement of positive harmonic functions). Let u be positive and harmonic on $B_1(0) \setminus (0) \subset \mathbb{R}^n$, and continuously vanishing on $\partial B_1(0)$. Denote by u^R its radial replacement (see Problem 2.2.11).

a) Show that if $u \geq u^R$ in $B_1(0) \setminus (0)$ then $u = u^R$.

b) Show that u coincides with its radial replacement u^R:

$$u(\mathbf{x}) = u^R(\mathbf{x}) = \begin{cases} C \log |\mathbf{x}| & n = 2 \\ C\left(|\mathbf{x}|^{2-n} - 1\right) & n \geq 3 \end{cases} \quad (2.15)$$

where C is some constant[a].

---

[a] *Hint.* Show that the set of all $t \in [0, 1]$ such that $u \geq tu^R$ in $\overline{B_1}(0) \setminus (0)$ is non-empty, closed and open in $[0, 1]$.

Solution. a) We claim that $u \geq u^R$ implies $u = u^R$. In fact, if $u \geq u^R$ and there is a point $\mathbf{x}$ such that $u(\mathbf{x}) > u^R(\mathbf{x})$, then $u(\mathbf{y}) > u^R(\mathbf{y})$ in a neighbourhood of $\mathbf{x}$ on $\partial B_{|\mathbf{x}|}(0)$. Thus, integrating on $\partial B_{|\mathbf{x}|}(0)$, we get the contradiction

$$u^R(\mathbf{x}) > \left(u^R\right)^R(\mathbf{x}) = u^R(\mathbf{x})$$

since u^R is radial.

b) To show that $u \geq u^R$ in $B_1(0) \setminus (0)$ define the set

$$E = \left\{t \in [0, 1] : u \geq tu^R \text{ in } \overline{B_1}(0) \setminus (0)\right\}.$$

1. E is not empty because it contains 0.
2. E is closed: indeed, if a sequence $\{t_m\} \subset E$ converges to t_0, in fact, then

$$u(\mathbf{x}) \geq t_m u^R(\mathbf{x})$$

for all $\mathbf{x} \in \overline{B_1}(\mathbf{0}) \setminus (\mathbf{0})$, and passing to the limit we get $u(\mathbf{x}) \geq t_0 u^R(\mathbf{x})$ for all $\mathbf{x} \in \overline{B_1}(\mathbf{0}) \setminus (\mathbf{0})$. Thus $t_0 \in E$ and E is closed.

3. Finally, E is open in $[0, 1]$. Suppose that $u \geq t^* u^R$ in $\overline{B_1}(\mathbf{0}) \setminus (\mathbf{0})$. Then clearly $u \geq t u^R$ in $\overline{B_1}(\mathbf{0}) \setminus (\mathbf{0})$ for any $0 \leq t < t^*$. Set $w = u - t^* u^R \geq 0$. Then w is harmonic, and $w = 0$ on $\partial B_1(\mathbf{0})$. If $w \equiv 0$ then $u = t^* u^R$; as in part a), by integrating and recalling that $(u^R)^R = u^R$ we obtain that $t^* = 1$, and the proof is completed. Otherwise, $w > 0$ on $B_1(\mathbf{0}) \setminus \{\mathbf{0}\}$, and by Problem 2.2.8 (page 91),

$$w(\mathbf{x}) \geq \gamma w^R(\mathbf{x})$$

for all $0 < |\mathbf{x}| \leq 1/2$. By the maximum principle $w(\mathbf{x}) > \gamma w^R(\mathbf{x})$ on all of $B_1(\mathbf{0}) \setminus (\mathbf{0})$ since $w(\mathbf{x}) - \gamma w^R(\mathbf{x}) = 0$ on $\partial B_1(\mathbf{0})$. Thus

$$u - t^* u^R \geq \gamma \left(u^R - t^* u^R \right) \text{ in } \overline{B_1}(\mathbf{0}) \setminus (\mathbf{0})$$

or, rearranging the terms,

$$u - \left[t^* + \gamma \left(1 - t^* \right) \right] u^R \geq 0.$$

Therefore $[0, t^* + \gamma(1 - t^*)) \subset E$ and E is open.

We deduce that $E = [0, 1]$. Setting $t = 1$, we get $u \geq u^R$, and the conclusion follows by part a).

2.2.2 Boundary-value problems. Solution methods

Problem 2.2.13 (Mixed problem on a rectangle, separation of variables). *On the rectangle*

$$\{Q = (x, y) : 0 < x < a, \ 0 < y < b\}$$

solve the mixed problem:

$$\begin{cases} \Delta u = 0 & \text{in } Q \\ u(x, 0) = 0, \ u(x, b) = g(x) & 0 \leq x \leq a \\ u(0, y) = u_x(a, y) = 0 & 0 \leq y \leq b \end{cases}$$

where $g \in C^2(\mathbb{R})$, $g(0) = g'(a) = 0$.

Solution. We seek (nontrivial) harmonic functions of the form $u(x, y) = v(x) w(y)$, with $v(0) = v'(a) = 0$, $w(0) = 0$. Substituting in $\Delta u = 0$ we find $v''(x) w(y) + v(x) w''(y) = 0$, thus

$$\frac{v''(x)}{v(x)} = -\frac{w''(y)}{w(y)} = \lambda$$

and λ constant. The eigenvalue problem for v reads

$$\begin{cases} v''(x) - \lambda v(x) = 0 \\ v(0) = v'(a) = 0, \end{cases}$$

hence $\lambda_k = -\frac{(2k+1)^2\pi^2}{4a^2}$, $k \geq 0$, and

$$v_k(x) = \sin \frac{(2k+1)\pi x}{2a}.$$

With these λ_k, we get for w the equation

$$w''(y) + \lambda_k w(y) = 0,$$

with $w(0) = 0$. Writing the general integral of the ODE as

$$w(y) = c_1 \sinh \frac{(2k+1)\pi y}{2a} + c_2 \cosh \frac{(2k+1)\pi y}{2a},$$

we see that $w(0) = 0$ implies $c_2 = 0$.

So we may write the candidate solution in the form

$$u(x,y) = \sum_{k=0}^{\infty} a_k \sin \frac{(2k+1)\pi x}{2a} \sinh \frac{(2k+1)\pi y}{2a}. \tag{2.16}$$

Now we need to find the coefficients a_k from the condition $u(x,b) = g(x)$. By the assumptions made on g we have

$$g(x) = \sum_{k=0}^{\infty} g_k \sin \frac{(2k+1)\pi x}{2a}$$

where, after two integrations by parts

$$g_k = \frac{2}{a} \int_0^a g(x) \sin \frac{(2k+1)\pi x}{2a} dx = -\frac{8a}{(2k+1)^2\pi^2} \int_0^a g''(x) \sin \frac{(2k+1)\pi x}{2a} dx,$$

and $\sum_0^{\infty} |g_k| < \infty$. It will suffice to choose

$$a_k = g_k \left[\sinh \frac{(2k+1)\pi b}{2a} \right]^{-1}$$

to get $u(x,b) = g(x)$.

• *Analysis of (2.16)*. In $\overline{Q}$ we have

$$\left| a_k \sin \frac{(2k+1)\pi x}{2a} \right| \sinh \frac{(2k+1)\pi y}{2a} \leq |g_k|$$

so the series defining u is uniformly convergent in $\overline{Q}$, and hence $u \in C(\overline{Q})$. Inside the domain we can differentiate without problem the single terms of the series as many times as we want, because the coefficients a_k go to zero fast enough. This guarantees that u is harmonic on Q. For the same reason, and since $g'(a) = 0$, the Neumann datum along $x = a$ is also continuously achieved.

Problem 2.2.14 (Poisson-Dirichlet in the disc, separation of variables). *Solve the non-homogeneous Dirichlet problem:*

$$\begin{cases} \Delta u(x, y) = y & \text{in } B_1 \\ u = 1 & \text{on } \partial B_1. \end{cases}$$

Solution. First, we break up the problem, which is non-homogeneous both in the equation and also in the boundary condition, in two sub-problems each with only one non-homogeneity. Set $u = v + w$, where

$$\begin{cases} \Delta v(x, y) = y & \text{in } B_1 \\ v = 0 & \text{on } \partial B_1, \end{cases} \qquad \begin{cases} \Delta w(x, y) = 0 & \text{in } B_1 \\ w = 1 & \text{on } \partial B_1. \end{cases}$$

The latter system immediately tells that $w(x, y) \equiv 1$ is the only solution (by the maximum principle). As for the former, we pass to polar coordinates, $V(r, \theta) = v(r \cos \theta, r \sin \theta)$. Then V is 2π-periodic in θ, and using the polar expression[3] for Δ we see that V satisfies

$$V_{rr} + \frac{1}{r} V_r + \frac{1}{r^2} V_{\theta\theta} = r \sin \theta,$$

with $V(1, \theta) = 0$ and V bounded. The right-hand side $y = r \sin \theta$ suggests to seek solutions of the type

$$V(r, \theta) = b_1(r) \sin \theta, \quad \text{with } b_1(1) = 0 \text{ and } b_1 \text{ bounded.}$$

Substituting, we obtain for b_1 the equation

$$b_1''(r) \sin \theta + \frac{1}{r} b_1'(r) \sin \theta - \frac{1}{r^2} b_1(r) \sin \theta = r \sin \theta,$$

i.e. the ODE

$$r^2 b_1'' + r b_1' - b_1 = r^3. \tag{2.17}$$

The associated homogeneous equation is of Euler type, and may be reduced to constant coefficients by setting $s = \log r$. Alternatively we could seek solutions of the form r^a, $a \in \mathbb{R}$. Substituting, we have

$$a(a-1)r^a + ar^a - r^a = (a+1)(a-1)r^a = 0,$$

whence $a = \pm 1$. The general integral for the homogeneous equation thus reads

$$C_1 r + C_2 r^{-1} \qquad (C_1, C_2 \in \mathbb{R}).$$

---

[3] Appendix B.

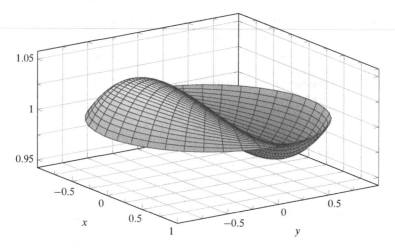

Fig. 2.3 The function $u(x, y) = 1 + y(x^2 + y^2 - 1)/8$

Let us now look for a particular non-homogeneous solution to (2.17) of the form Ar^3. We find

$$6Ar^3 + 3Ar^3 - Ar^3 = r^3,$$

so $A = 1/8$. Overall, the general integral of (2.17) is given by

$$b_1(r) = \frac{1}{8}r^3 + C_1 r + C_2 r^{-1},$$

with arbitrary C_1, C_2. The boundedness of b_1 implies $C_2 = 0$, while $b_1(1) = 0$ forces $C_1 = -1/8$. Thus

$$V(r, \theta) = \frac{1}{8}r(r^2 - 1)\sin\theta,$$

or in Cartesian coordinates (Fig. 2.3)

$$u(x, y) = 1 + \frac{1}{8}(x^2 + y^2 - 1)y.$$

Problem 2.2.15 (Dirichlet on an annulus, separation of variables). *Consider the annulus*

$$C_{1,R} = \{(r, \theta) : 1 < r < R\}.$$

a) *Given g, h in $C^1(\mathbb{R})$ and 2π-periodic, solve the Dirichlet problem*

$$\begin{cases} \Delta u = 0 & \text{in } C_{1,R} \\ u(1, \theta) = g(\theta) & 0 \le \theta \le 2\pi \\ u(R, \theta) = h(\theta) & 0 \le \theta \le 2\pi. \end{cases}$$

b) *Study the special case $g(\theta) = \sin\theta$ and $h(\theta) = 1$.*

Solution. a) As g, h are regular and periodic we can expand them in Fourier series

$$g(\theta) = \frac{a_0}{2} + \sum_{n=1}^{+\infty} (a_n \cos n\theta + b_n \sin n\theta),$$

$$h(\theta) = \frac{A_0}{2} + \sum_{n=1}^{+\infty} (A_n \cos n\theta + B_n \sin n\theta)$$

with convergent series

$$\sum_{n=1}^{+\infty} (|a_n| + |b_n|), \qquad \sum_{n=1}^{+\infty} (|A_n| + |B_n|). \tag{2.18}$$

The domain symmetry suggests to look for solutions of the type

$$u(r, \theta) = \alpha_0(r) + \sum_{n=1}^{+\infty} [\alpha_n(r) \cos n\theta + \beta_n(r) \sin n\theta]. \tag{2.19}$$

The boundary conditions require that at $r = 1$ and $r = R$ the solution coefficients must coincide with the coefficients of g and h respectively. On the other hand, in polar coordinates

$$\Delta u = u_{rr} + \frac{1}{r} u_r + \frac{1}{r^2} u_{\theta\theta}. \tag{2.20}$$

We have

$$u_r(r, \theta) = \alpha_0'(r) + \sum_{n=1}^{+\infty} [\alpha_n'(r) \cos n\theta + \beta_n'(r) \sin n\theta],$$

$$u_{rr}(r, \theta) = \alpha_0''(r) + \sum_{n=1}^{+\infty} [\alpha_n''(r) \cos n\theta + \beta_n''(r) \sin n\theta],$$

$$u_{\theta\theta}(r, \theta) = -\sum_{n=1}^{+\infty} n^2 [\alpha_n(r) \cos n\theta + \beta_n(r) \sin n\theta].$$

Substituting into equation (2.20) we get

$$\alpha_0''(r) + \frac{\alpha_0'(r)}{r} + \sum_{n=1}^{+\infty} \left\{ \left[\alpha_n''(r) + \frac{1}{r}\alpha_n'(r) - \frac{n^2}{r^2}\alpha_n(r) \right] \cos n\theta + \right.$$

$$\left. + \left[\beta_n''(r) + \frac{1}{r}\beta_n'(r) - \frac{n^2}{r^2}\beta_n(r) \right] \sin n\theta \right\} = 0$$

from which we obtain the family of problems, for $n \geq 0$:

$$\begin{cases} \alpha_0''(r) + \dfrac{1}{r}\alpha_0'(r) = 0 \\ \alpha_0(1) = \dfrac{a_0}{2}, \quad \alpha_0(R) = \dfrac{A_0}{2}, \end{cases} \tag{2.21}$$

$$\begin{cases} \alpha_n''(r) + \dfrac{1}{r}\alpha_n'(r) - \dfrac{n^2}{r^2}\alpha_n(r) = 0 \\ \alpha_n(1) = a_n, \quad \alpha_n(R) = A_n, \end{cases} \qquad \begin{cases} \beta_n''(r) + \dfrac{1}{r}\beta_n'(r) - \dfrac{n^2}{r^2}\beta_n(r) = 0 \\ \beta_n(1) = b_n, \quad \beta_n(R) = B_n. \end{cases}$$

$$(2.22)$$

The equation in (2.21) has order one in α_0', and is solved by $C_1 + C_2 \log r$. Using the constraints we find

$$\alpha_0(r) = \frac{a_0}{2} + \frac{A_0 - a_0}{2\log R}\log r.$$

To find the general solution of the (Euler) equations in (2.22) we first seek special solutions of the form r^γ, with γ to be determined. Substituting into (2.22) we obtain

$$(\gamma(\gamma - 1) + \gamma - n^2)r^{\gamma-2} = 0$$

so $\gamma = \pm n$. The general integral is $C_1 r^n + C_2 r^{-n}$. The boundary conditions force[4]:

$$\alpha_n(r) = a_n K_n(r) r^{-n} + A_n H_n(r)\left(\frac{r}{R}\right)^n$$

and

$$\beta_n(r) = b_n K_n(r) r^{-n} + B_n H_n(r)\left(\frac{r}{R}\right)^n,$$

where

$$H_n(r) = \frac{1 - r^{-2n}}{1 - R^{-2n}} \quad \text{and} \quad K_n(r) = \frac{1 - R^{-2n}r^{2n}}{1 - R^{-2n}}.$$

Substituting into (2.19) we finally get the solution. Since

$$H_n(r) \le \frac{1}{1 - R^{-1}}, \qquad K_n(r) \le \frac{1}{1 - R^{-1}} \quad \text{and} \quad \frac{r}{R} < 1, \frac{1}{r} < 1,$$

by (2.18), we see that the series in (2.19) is absolutely and uniformly convergent for $1 \le r \le R$. Moreover, we can differentiate term by term, and therefore (2.19) is the unique solution. Note that the solution has the form

$$c_0 + c_1 \frac{\log r}{\log R} + \sum_{n=1}^{\infty}(c_n r^n + d_n r^{-n})$$

reminiscent of the Laurent series of an analytic function on the plane.

b) The functions g and h are already finite Fourier sums. In particular, using the notation of part a), the only non-zero coefficients are $b_1 = 1$ and $A_0 = 2$. Hence

$$\alpha_0(r) = \frac{\log r}{\log R}, \qquad \beta_1(r) = \frac{1 - R^{-2}r^2}{1 - R^{-2}}r^{-1} = \frac{R^2 - r^2}{(R^2 - 1)r}$$

and $\alpha_n(r) \equiv \beta_m(r) \equiv 0$ for $n \ge 1$, $m \ge 2$. To sum up, the solution is (Fig. 2.4)

$$u(r, \theta) = \frac{\log r}{\log R} + \frac{R^2 - r^2}{(R^2 - 1)r}\sin\theta,$$

as a direct computation shows.

---

[4] Rearranging terms, with a little patience.

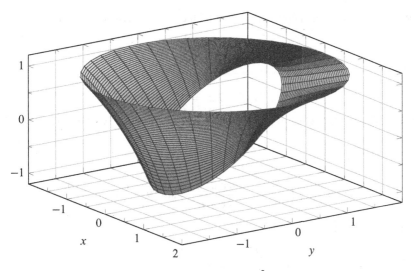

Fig. 2.4 $u\,(r,\theta) = \frac{\log r}{\log 2} + \frac{4-r^2}{3r}\sin\theta$

Problem 2.2.16 (Dirichlet problem on the ball). *Determine, at least formally, the so-lution of the Dirichlet problem (in spherical coordinates)*

$$\begin{cases} \Delta u = 0 & \text{in } B_1 \\ u\,(1,\varphi,\theta) = g\,(\theta) & \text{on } \partial B_1, \end{cases}$$

where g is $(2\pi$-periodic and) continuous on the ball

$$B_1 = \{(r,\varphi,\theta) = 0 \le r < 1, 0 \le \varphi \le 2\pi, 0 < \theta < \pi\}.$$

Solution. Since the problem is invariant under rotations about the z-axis, we look for functions $u = u\,(r,\theta)$. The Laplace equation for u becomes[5]

$$\frac{\partial}{\partial r}\left(r^2\frac{\partial u}{\partial r}\right) + \frac{1}{\sin\theta}\frac{\partial}{\partial\theta}\left(\sin\theta\frac{\partial u}{\partial\theta}\right) = 0$$

with $u(1,\theta) = g\,(\theta)$ and u bounded. Let us separate variables, and seek for solutions of the form $u\,(r,\theta) = v\,(\theta)\,w\,(r)$. Inserting into the equation and rearranging, we find

$$\frac{1}{w}\left(r^2 w'\right)' = -\frac{1}{v\sin\theta}(\sin\theta\,v')'.$$

---

[5] Appendix B.

Setting either side equal a constant λ, we find the two eigenvalue problems:

$$v'' + \frac{\cos\theta}{\sin\theta}v' + \lambda v = 0, \quad v \text{ bounded} \tag{2.23}$$

and

$$r^2 w'' + 2rw' - \lambda w = 0, \quad w \text{ bounded}. \tag{2.24}$$

Let us consider (2.23). Changing variable from θ to $x = \cos\theta$, we have x in $[-1, 1]$ and

$$v' = \frac{dv}{d\theta} = \frac{dv}{dx}\frac{dx}{d\theta} = (-\sin\theta)\frac{dv}{dx}$$

$$v'' = \frac{d^2v}{dx^2}\left(\frac{dx}{d\theta}\right)^2 + \frac{dv}{dx}\frac{d^2x}{d\theta^2} = \sin^2\theta\frac{d^2v}{dx^2} - \cos\theta\frac{dv}{dx}.$$

Substituting into (2.23) we find

$$\sin^2\theta\frac{d^2v}{dx^2} - 2\cos\theta\frac{dv}{dx} + \lambda v = 0$$

i.e.

$$\left(1 - x^2\right)\frac{d^2v}{dx^2} - 2x\frac{dv}{dx} + \lambda v = 0$$

which is a Legendre equation[6], with

$$v(-1) < \infty, \quad v(1) < \infty.$$

The eigenvalues are the integers $\lambda_n = n(n+1)$, $n \geq 0$, and the eigenfunctions, recursively defined by the Rodrigues formula

$$L_n(x) = \frac{1}{2^n n!}\frac{d^n}{dx^n}\left(x^2 - 1\right)^n,$$

are called *Legendre polynomials*. The normalised polynomials

$$L_n^*(x) = \sqrt{\frac{2n+1}{2}}L_n(x)$$

form an orthonormal basis of $L^2(-1, 1)$. With $\lambda = \lambda_n$ equation (2.24) reads

$$r^2 w'' + 2rw' - n(n+1)w = 0, \quad w \text{ bounded},$$

solved by $w_n(r) = cr^n$. Thus we have found the following bounded solutions with separated variables

$$u_n(r, \theta) = r^n L_n(\cos\theta), \quad n \geq 0.$$

---

[6] Appendix A.

Superposing the functions u_n we can write

$$u(r, \theta) = \sum_{n=0}^{\infty} a_n r^n L_n (\cos \theta), \tag{2.25}$$

so the Dirichlet condition now requires

$$\sum_{n=0}^{\infty} a_n L_n (\cos \theta) = g(\theta), \quad 0 \leq \theta \leq \pi$$

or, equivalently,

$$\sum_{n=0}^{\infty} a_n L_n(x) = g\left(\cos^{-1} x\right), \quad -1 \leq x \leq 1. \tag{2.26}$$

As $f(x) = g\left(\cos^{-1} x\right)$ is continuous on $[-1, 1]$, $F \in L^2(-1, 1)$ and (2.26) holds (at least) in $L^2(-1, 1)$, with

$$a_n = \frac{2n + 1}{2} \int_{-1}^{1} g\left(\cos^{-1} x\right) L_n(x) \, dx.$$

Hence (2.25) is, at least formally, the solution we wanted.

Problem 2.2.17 (Planar reflection). *Let*

$$B_1^+ = \{(x, y) \in \mathbb{R}^2 : x^2 + y^2 < 1, y > 0\}$$

denote the upper half-disc and take $u \in C^2(B_1^+) \cap C(\overline{B_1^+})$ *harmonic on* B_1^+ *with* $u(x, 0) = 0$. *Prove that the (continuous) function*

$$U(x, y) = \begin{cases} u(x, y) & y \geq 0 \\ -u(x, -y) & y < 0 \end{cases}$$

obtained through an odd reflection of u *with respect to the* x*-axis, is harmonic on the entire disc* B_1.

Solution. Call v the solution to

$$\begin{cases} \Delta v(x, y) = 0 & \text{in } B_1 \\ v = U & \text{on } \partial B_1 \end{cases}$$

which exists, is unique and is defined by Poisson's formula. The function $v(x, -y)$ is harmonic in B_1, and has on the boundary the same values of v with opposite signs. Set $w(x, y) = v(x, y) + v(x, -y)$. We have:

$$\begin{cases} \Delta w(x, y) = 0 & \text{in } B_1 \\ w = 0 & \text{on } \partial B_1. \end{cases}$$

By uniqueness, it must be $w \equiv 0$. Therefore $v(x, y) = -v(x, -y)$, i.e. v is odd with respect to the horizontal axis, and in particular $v(x, 0) = 0$. So, v solves

$$\begin{cases} \Delta v\,(x, y) = 0 & \text{in } B_1^+ \\ v = u & \text{on } \partial B_1^+. \end{cases}$$

As u solves this problem as well, by uniqueness $v \equiv u \equiv U$ on B_1^+. But both v and U are odd in y, so $v \equiv U$ on B_1 and then $\Delta U = 0$ on B_1.

Problem 2.2.18 (Poisson kernel on the upper half-plane). *Let $g \in L^1(\mathbb{R}) \cap C(\mathbb{R})$ be bounded. Show that there is a unique bounded and continuous solution on $\{y \geq 0\}$ to*

$$\begin{cases} \Delta u(x, y) = 0 & x \in \mathbb{R}, \ y > 0 \\ u(x, 0) = g(x) & x \in \mathbb{R}. \end{cases}$$

Using the partial Fourier transform write a representation formula for u.

Solution. Uniqueness follows from the reflection principle (Problem 2.2.17) and Liouville's theorem. In fact, if u_1, u_2 are bounded solutions with the same datum g, the difference $w = u_1 - u_2$ is harmonic on $y > 0$ and has null datum. Extending w to an odd function on $y < 0$ gives a harmonic, bounded function on the plane, so constant. The constant is zero as w vanishes on $y = 0$; therefore $u_1 = u_2$.

For a representation formula, let us set

$$\widehat{u}\,(\xi, y) = \int_{\mathbb{R}} e^{-i\xi x} u\,(x, y)\,dx.$$

Then

$$\widehat{u_x}\,(\xi, y) = i\xi \widehat{u}\,(\xi, y), \qquad \widehat{u_{xx}}\,(\xi, y) = -\xi^2 \widehat{u}\,(\xi, y),$$

and $\widehat{u}$ solves

$$\begin{cases} \widehat{u}_{yy}\,(\xi, y) - \xi^2 \widehat{u}\,(\xi, y) = 0 \\ \widehat{u}(\xi, 0) = \widehat{g}\,(\xi). \end{cases}$$

The general integral of the ODE is:

$$\widehat{u}\,(\xi, y) = c_1\,(\xi)\,e^{|\xi| y} + c_2\,(\xi)\,e^{-|\xi| y}.$$

The fact that u is bounded says that[7] $c_1\,(\xi) = 0$, while $y = 0$ forces

$$c_2\,(\xi) = \widehat{g}\,(\xi).$$

Hence

$$\widehat{u}\,(\xi, y) = \widehat{g}\,(\xi)\,e^{-|\xi| y}.$$

---

[7] $e^{|\xi| y}$ admits no anti-transform.

The anti-transform[8] of $e^{-|\xi|y}$ is the harmonic function (*Poisson kernel on the half-plane*)

$$\frac{1}{\pi} \frac{y}{x^2 + y^2}$$

so we obtain

$$u(x, y) = \frac{1}{\pi} \int_{\mathbb{R}} \frac{y}{(x - s)^2 + y^2} g(s)\, ds. \tag{2.27}$$

Remark 1. Without assuming boundedness there would be infinitely many solutions, for instance $u(x, y) + cy$, $u(x, y) + cxy$, $u(x, y) + ce^x \sin y$, ...

• *Analysis of* (2.27)**. We will show that (2.27) is effectively the required solution, i.e. that it is harmonic on $y > 0$, bounded, and that it agrees with g continuously on $y = 0$. For $y > 0$ we can differentiate to any order the integrand, and since the Poisson kernel is harmonic, so is u. Now for $y > 0$,

$$\frac{1}{\pi} \int_{\mathbb{R}} \frac{y}{(x - s)^2 + y^2}\, ds = \frac{1}{\pi} \int_{\mathbb{R}} \frac{y}{x^2 + y^2}\, dx = \frac{1}{\pi} \left[\arctan\left(\frac{x}{y}\right) \right]_{-\infty}^{+\infty} = 1. \tag{2.28}$$

Thus

$$|u(x, y)| \leq \frac{1}{\pi} \int_{\mathbb{R}} \frac{y}{(x - s)^2 + y^2} |g(x)|\, ds \leq \sup_{\mathbb{R}} |g|$$

and therefore u is bounded. Let now $(x, y) \to (x_0, 0)$; given $\varepsilon > 0$, we take

$$|g(s) - g(x_0)| < \varepsilon$$

for $|s - x_0| < \delta_\varepsilon$. Then

$$|u(x, y) - g(x_0)| = \frac{1}{\pi} \int_{\mathbb{R}} \frac{y}{(s - x)^2 + y^2} |g(s) - g(x_0)|\, ds$$

$$= \underbrace{\frac{1}{\pi} \int_{\{|s - x_0| < \delta_\varepsilon\}} \cdots\, ds}_{I_1} + \underbrace{\frac{1}{\pi} \int_{\{|s - x_0| \geq \delta_\varepsilon\}} \cdots\, ds}_{I_2}.$$

From (2.28) we infer $I_1 < \varepsilon$. As for I_2, notice that if $|x - x_0| < \delta_\varepsilon/2$, the triangle inequality implies

$$\{s : |s - x_0| \geq \delta_\varepsilon\} \subset \left\{s : |s - x| \geq \frac{\delta_\varepsilon}{2}\right\},$$

so

$$I_2 \leq \frac{y \cdot 2 \sup_{\mathbb{R}} |g|}{\pi} \int_{\{|s - x| \geq \delta_\varepsilon/2\}} \frac{1}{(s - x)^2}\, d = \frac{8 \sup_{\mathbb{R}} |g|}{\pi \delta_\varepsilon} y,$$

and thence $I_2 < \varepsilon$ for y small enough.

---

[8] Appendix B.

In summary, if (x, y) is close enough to $(x_0, 0)$,

$$|u(x, y) - g(x_0)| < 2\varepsilon$$

or, in other words $u(x, y) \to g(x_0)$ for $(x, y) \to (x_0, 0)$.

Remark 2. The above discussion shows that everything carries through if g is just bounded and continuous, without assuming integrability.

Problem 2.2.19 (Using the reflection principle). *Solve in*

$$B_1^+ = \{x^2 + y^2 < 1, y > 0\}$$

the problem

$$\begin{cases} \Delta u(x, y) = 0 & \text{in } B_1^+ \\ u = g & \text{on } \partial B_1^+, \end{cases}$$

with g continuous on ∂B_1.

Solution. To use reflections we modify u in order to have vanishing datum on the diameter $y = 0$. Set

$$h(x) = \begin{cases} g(1, 0) & \text{for } x > 1 \\ g(x, 0) & \text{for } -1 \le x \le 1 \\ g(-1, 0) & \text{for } x < -1. \end{cases}$$

The function thus defined is continuous, bounded on $\mathbb{R}$ and coincides with $g(x, 0)$ along $y = 0$, $|x| \le 1$. By Problem 2.2.18 (in particular Remark 2) the function

$$v(x, y) = \frac{1}{\pi} \int_{\mathbb{R}} \frac{y}{(x - s)^2 + y^2} h(s) \, ds$$

is harmonic on $y > 0$ and

$$v(x, 0) = g(x, 0)$$

for $-1 \le x \le 1$. Then

$$w = u - v$$

is harmonic on B_1^+ and moreover,

$$w(x, 0) = 0 \qquad \text{for } -1 \le x \le 1,$$

$$w = g - v \equiv \widetilde{g} \qquad \text{on } \partial B_1^+ \cap \{y > 0\}.$$

We extend $\widetilde{g}$ in an odd way on $\partial B_1^+ \cap \{y < 0\}$. From the reflection principle and the Poisson formula we deduce

$$w(x, y) = \frac{1 - x^2 - y^2}{2\pi} \int_0^{2\pi} \frac{\widetilde{g}(\cos\theta, \sin\theta)}{(x - \cos\theta)^2 + (y - \sin\theta)^2} \, d\theta.$$

The solution is then $u = w + v$.

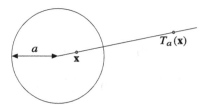

Fig. 2.5 Kelvin transform

Problem 2.2.20 (Kelvin transform in $\mathbb{R}^2$). *Fix $a > 0$ and $\mathbf{x} \in \mathbb{R}^2 \setminus \{\mathbf{0}\}$. We call Kelvin transform the map sending $\mathbf{x}$ to the point*

$$\mathbf{y} = T_a(\mathbf{x}) = \frac{a^2}{|\mathbf{x}|^2}\mathbf{x}. \tag{2.29}$$

a) Verify that the map T_a is regular on $\mathbb{R}^2 \setminus \{\mathbf{0}\}$, invertible with inverse $T_a^{-1} = T_a$.

b) Prove that if u is harmonic on $\Omega \subset \mathbb{R}^2 \setminus \{\mathbf{0}\}$ then

$$v(\mathbf{x}) = u(T_a(\mathbf{x}))$$

is harmonic on $T_a^{-1}(\Omega)$.

Solution. a) In $\mathbb{R}^2 \setminus \{\mathbf{0}\}$, T_a is C^∞ (as composition of smooth functions) and $\mathbf{y} = T_a(\mathbf{x}) \neq \mathbf{0}$ if $\mathbf{x} \neq \mathbf{0}$ (Fig. 2.5). Let us find the inverse. Computing the modulus of (2.29) gives

$$|\mathbf{y}| = \frac{a^2}{|\mathbf{x}|^2}|\mathbf{x}| = \frac{a^2}{|\mathbf{x}|}, \qquad \text{i.e.} \quad |\mathbf{x}| = \frac{a^2}{|\mathbf{y}|}.$$

Substituting into (2.29) we obtain $\mathbf{y} = \frac{a^2|\mathbf{y}|^2}{a^4}\mathbf{x} = \frac{|\mathbf{y}|^2}{a^2}\mathbf{x}$. So whenever $\mathbf{y} \neq \mathbf{0}$, T_a is invertible and

$$T_a^{-1}(\mathbf{y}) = \mathbf{x} = \frac{a^2}{|\mathbf{y}|^2}\mathbf{y} = T_a(\mathbf{y}).$$

b) As $\mathbf{x}$ and $T_a(\mathbf{x})$ are parallel vectors, it is convenient to use polar coordinates: write $\mathbf{x} = (r, \theta)$, so that

$$T_a(\mathbf{x}) = (\rho, \theta),$$

with $r\rho = a^2$. Setting $u = u(\rho, \theta)$, we have

$$v(r, \theta) = u\left(\frac{a^2}{r}, \theta\right) = u(\rho, \theta), \qquad v_{\theta\theta}(r, \theta) = u_{\theta\theta}\left(\frac{a^2}{r}, \theta\right) = u_{\theta\theta}(\rho, \theta),$$

$$v_r(r, \theta) = -\frac{a^2}{r^2}u_\rho\left(\frac{a^2}{r}, \theta\right) = -\frac{\rho^2}{a^2}u_\rho(\rho, \theta),$$

$$v_{rr}(r, \theta) = \frac{2a^2}{r^3}u_\rho\left(\frac{a^2}{r}, \theta\right) + \frac{a^4}{r^4}u_{\rho\rho}\left(\frac{a^2}{r}, \theta\right) = \frac{2\rho^3}{a^4}u_\rho(\rho, \theta) + \frac{\rho^4}{a^4}u_{\rho\rho}(\rho, \theta),$$

and so

$$\Delta v = v_{rr}(r, \theta) + \frac{1}{r} v_r(r, \theta) + \frac{1}{r^2} v_{\theta\theta}(r, \theta)$$

$$= \frac{2\rho^3}{a^4} u_\rho(\rho, \theta) + \frac{\rho^4}{a^4} u_{\rho\rho}(\rho, \theta) - \frac{\rho}{a^2}\frac{\rho^2}{a^2} u_\rho(\rho, \theta) + \frac{\rho^2}{a^4} u_{\theta\theta}(r, \theta)$$

$$= \frac{\rho^4}{a^4} \left(u_{\rho\rho}(\rho, \theta) + \frac{1}{\rho} u_\rho(\rho, \theta) + \frac{1}{\rho^2} u_{\theta\theta}(\rho, \theta) \right)$$

$$= \frac{\rho^4}{a^4} \Delta u(\rho, \theta).$$

Hence, if u is harmonic on its domain then also v is.

Problem 2.2.21 (Using Kelvin transforms). *Let T_a denote the Kelvin transform introduced in Problem 2.2.20. We write $\mathbf{x} = (x_1, x_2)$ and $\mathbf{y} = (y_1, y_2)$.*

a) Show that the straight line $x_1 = a$ is mapped by T_a onto the circle

$$\left(y_1 - \frac{a}{2} \right)^2 + y_2^2 = \frac{a^2}{4} \tag{2.30}$$

and the image of the half-plane $x_1 > a$ is the corresponding enclosed disc. Similarly, check that the line $x_2 = a$ is sent to the circle

$$y_1^2 + \left(y_2 - \frac{a}{2} \right)^2 = \frac{a^2}{4}. \tag{2.31}$$

b) On the quadrant $Q = \{\mathbf{x} \in \mathbb{R}^2 : x_1 > 1, x_2 > 1\}$ solve the Dirichlet problem

$$\begin{cases} \Delta u(x) = 0 & x \in Q \\ u(x_1, 1) = 0 & x_1 > 1 \\ u(1, x_2) = 1 & x_2 > 1 \\ u \text{ bounded} & \text{in } Q. \end{cases}$$

Check that the solution is unique[a].

c) On

$$\Omega = \{\mathbf{x} \in \mathbb{R}^2 : x_1^2 + x_2^2 - x_1 < 0, \ x_1^2 + x_2^2 - x_2 < 0\},$$

intersection of two discs (Fig. 2.7), solve the Dirichlet problem:

$$\begin{cases} \Delta v = 0 & \text{in } \Omega \\ v = 1 & \text{on } \{\mathbf{x} : x_1^2 + x_2^2 - x_1 = 0, \ x_1^2 + x_2^2 - x_2 < 0\} \\ v = 0 & \text{on } \{\mathbf{x} : x_1^2 + x_2^2 - x_1 < 0, \ x_1^2 + x_2^2 - x_2 = 0\}. \end{cases}$$

[a] Use the reflection principle of Problem 2.2.17 (page 103) and the result of Problem 2.2.10 (page 92).

Solution. a) The Kelvin transform $\mathbf{y} = T_a(\mathbf{x})$ reads, explicitly:

$$(y_1, y_2) = \left(\frac{a^2}{x_1^2 + x_2^2} x_1, \frac{a^2}{x_1^2 + x_2^2} x_2 \right),$$

with inverse (as seen in the previous problem)

$$(x_1, x_2) = \left(\frac{a^2}{y_1^2 + y_2^2} y_1, \frac{a^2}{y_1^2 + y_2^2} y_2 \right).$$

Using the latter, the vertical line $x_1 = a$ is mapped to:

$$\frac{a^2}{y_1^2 + y_2^2} y_1 = a$$

i.e. (2.30). Analogously, the right half-plane $x_1 > a$ is transformed into

$$\frac{a^2}{y_1^2 + y_2^2} y_1 > a,$$

which is the interior of the circle. In general, all vertical lines $x_1 = b$, $b \neq 0$, are mapped onto circles centred on the horizontal axis and passing through the origin. The claim about horizontal lines $x_2 = a$ is identical, so we leave the details to the reader.

b) To find a solution, we start by observing that the domain is a "cone", i.e. the union of rays emanating from $(1,1)$. The boundary is made of two rays, on each of which the Dirichlet datum is constant. So it is natural to seek solutions that are constant (and bounded) on half-straight lines from $(1, 1)$; in other words, solutions of the form $U(r, \theta) = \psi(\theta)$, where (r, θ) are polar coordinates with pole at $(1, 1)$. If the solution is of this kind, then

$$\begin{cases} \Delta u = U_{rr}(r, \theta) + \dfrac{1}{r} U_r(r, \theta) + \dfrac{1}{r^2} U_{\theta\theta}(r, \theta) = \dfrac{1}{r^2} \psi''(\theta) = 0 \quad 0 < \theta < \dfrac{\pi}{2} \\ \psi(0) = 0 \\ \psi\left(\frac{\pi}{2}\right) = 1, \end{cases}$$

i.e. $\psi(\theta) = 2\theta / \pi$. Going back to the Cartesian frame system, we find

$$u(x_1, x_2) = \frac{2}{\pi} \arctan \left(\frac{x_2 - 1}{x_1 - 1} \right).$$

• The same conclusion is reached by noting that the complex function

$$w = \log(z - 1) = \log|z - 1| + i \arg(z - 1)$$

is holomorphic on Q. Its imaginary part

$$\arg(z - 1) = \arctan \left(\frac{x_2 - 1}{x_1 - 1} \right)$$

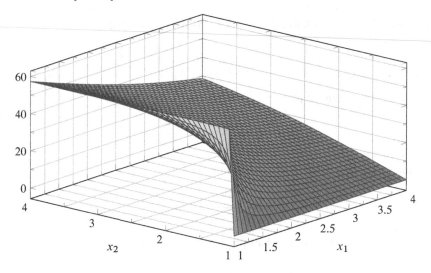

Fig. 2.6 $u\,(x,y) = \frac{2}{\pi}\arctan\frac{x_2-1}{x_1-1}$

is therefore harmonic on Q, and 0 on $x_2 = 1$, $x_1 > 1$, while it equals $\pi/2$ on $x_2 = 1$, $x_1 > 1$. The function (Fig. 2.6)

$$u(x_1,x_2) = \frac{2}{\pi}\arg(z-1) = \frac{2}{\pi}\arctan\left(\frac{x_2-1}{x_1-1}\right)$$

is therefore the solution we wanted.

• *Uniqueness.* The difficulties hide in the discontinuity of the datum at $(1,1)$, besides the fact that the domain is unbounded. Assume that there exist two solutions u and v and set $w = u - v$. Since the datum is discontinuous, we may assume that u, v, and hence also w, are continuous only on $\overline{Q} \setminus \{(1,1)\}$. The function w solves

$$\begin{cases} \Delta w(\mathbf{x}) = 0 & \mathbf{x} \in Q \\ w(\mathbf{x}) = 0 & \mathbf{x} \in \partial Q \setminus \{(1,1)\} \\ w \text{ bounded} & \text{on } Q \end{cases}$$

(as difference of bounded functions, w is bounded). Let us extend w to $\mathbb{R}^2$ by reflection, as follows:

$$w\,(x, 1-k) = -w\,(x, 1+k), \quad x > 1, k > 0 \quad \text{(odd w.r.t. } y = 1)$$

and

$$w\,(1-h, y) = -w\,(1+h, y), \quad h > 0, y \in \mathbb{R} \quad \text{(odd w.r.t. } x = 1).$$

The reflection principle (Problem 2.2.17 on page 103) guarantees that the extended w is harmonic on the whole plane, except at $(1,1)$ at most. On the other hand w is bounded,

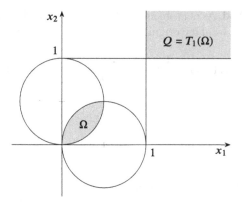

Fig. 2.7 Domain of Problem 2.2.21 c)

so we may use Problem 2.2.10 (page 92), whereby w has a removable singularity at $(1, 1)$: setting $w(1, 1) = 0$ gives a harmonic and bounded function on the plane.

By Liouville's theorem w is constant, and zero on the axes, hence identically zero; this means that u coincides with v.

c) By part a) the Kelvin map T_1 transforms the half-plane $x_1 > 1$ in the disc $y_1^2 + y_2^2 - y_1 < 0$, and the half-plane $x_2 > 1$ into $y_1^2 + y_2^2 - y_2 < 0$, and vice versa (recall that $T_1^{-1} = T_1$). Therefore the intersection of the two discs (see Fig. 2.7), i.e. Ω, is mapped to the quadrant $\{x_1 > 1\} \cap \{x_2 > 1\}$:

$$Q = T_1(\Omega).$$

Set $v(\mathbf{x}) = u(T_1(\mathbf{x}))$, with u harmonic as determined above:

$$v(x_1, x_2) = \frac{2}{\pi} \arctan\left(\frac{x_1^2 + x_2^2 - x_2}{x_1^2 + x_2^2 - x_1}\right).$$

The function v is defined on Ω, and it is not hard to check that it fulfils the boundary conditions. By Problem 2.2.20 (page 107), moreover, since u is harmonic also v is harmonic, and must be the required solution.

Problem 2.2.22 (Dirichlet problem on the half-plane, discontinuous data). *Solve on the half-plane:*

$$\begin{cases} \Delta u(x, y) = 0 & x > 0, \, y \in \mathbb{R} \\ u(0, y) = 1 & y > 0 \\ u(0, y) = -1 & y < 0 \\ u \text{ bounded on the half-plane.} \end{cases}$$

Verify that the solution is unique, using the reflection principle and Problem 2.2.10 (page 92).

Solution. 1st method. As in Problem 2.2.21.b), we try to find a solution that is constant on half-lines emanating from the origin. Set $u = u(r, \theta) = \psi(\theta)$, so that

$$\Delta u = u_{rr}(r, \theta) + \frac{1}{r}u_r(r, \theta) + \frac{1}{r^2}u_{\theta\theta}(r, \theta) = \frac{1}{r^2}\psi''(\theta).$$

Then ψ solves

$$\begin{cases} \psi''(\theta) = 0 & -\frac{\pi}{2} < \theta < \frac{\pi}{2} \\ \psi(\frac{\pi}{2}) = 1 \\ \psi(-\frac{\pi}{2}) = -1 \end{cases}$$

so $\psi(\theta) = \frac{2}{\pi}\theta$ or, in Cartesian coordinates,

$$u(x, y) = \frac{2}{\pi}\arctan\frac{y}{x}.$$

2nd method. The complex function

$$w = \log z = \log|z| + i \arg z$$

is holomorphic on $x > 0$. Its imaginary part

$$\arg z = \arctan\frac{y}{x}$$

is harmonic on $x > 0$, and is equal to $\pi/2$ and $-\pi/2$ along $x = 0$, $y > 0$ and $x = 0$, $y < 0$, respectively. Hence

$$u(x, y) = \frac{2}{\pi}\arg z = \frac{2}{\pi}\arctan\frac{y}{x}.$$

3rd method. We use the Poisson kernel for the half-plane $x > 0$ (Problem 2.2.18 on page 104):

$$u(x, y) = \frac{1}{\pi}\int_0^\infty \frac{x}{x^2 + (y-s)^2}ds - \frac{1}{\pi}\int_{-\infty}^0 \frac{x}{x^2 + (y-s)^2}ds =$$

$$= \frac{x}{\pi}\int_0^\infty \left(\frac{1}{x^2 + (y-s)^2} - \frac{1}{x^2 + (y+s)^2}\right)ds =$$

$$= \frac{2}{\pi}\arctan\frac{y}{x}.$$

• Now we check that the problem admits at most one solution. We may argue in analogy to Problem 2.2.21,b), where the details can be found. So let u and v be solutions, and set $w = u - v$, where

$$\begin{cases} \Delta w(x, y) = 0 & x > 0, \ y \in \mathbb{R} \\ w(0, y) = 0 & y \in \mathbb{R} \setminus \{0\} \\ w \text{ bounded.} \end{cases}$$

Extend w to $\mathbb{R}^2$ in an odd way:

$$\tilde{w}(x, y) = \begin{cases} w(x, y) & x \geq 0 \\ -w(-x, y) & x \leq 0. \end{cases}$$

By the reflection principle (Problem 2.2.17 on page 103) $\tilde{w}$ is harmonic on $\mathbb{R}^2 \setminus \{(0,0)\}$, and can be extended harmonically to the origin as well, by Problem 2.2.10 (page 92). As $\tilde{w}$ is harmonic and bounded everywhere, by Liouville's theorem it is constant. Now, since it vanishes on the x-axis, the constant is necessarily zero: $\tilde{w} \equiv 0$, and u coincides with v.

Problem 2.2.23 (Uniqueness for the exterior Dirichlet problem). *Let $\Omega \subset \mathbb{R}^2$ be a bounded domain containing the origin and $\Omega_e = \mathbb{R}^2 \setminus \overline{\Omega}$. Prove that the following problem*

$$\begin{cases} \Delta u = 0 & \text{in } \Omega_e \\ u = g & \text{on } \partial\Omega_e, \text{ with } g \text{ continuous} \\ u \text{ bounded} & \text{in } \Omega_e \end{cases} \qquad (2.32)$$

has at most one solution of class $C^2(\Omega_e) \cap C(\overline{\Omega}_e)$, by answering the following questions:

a) Solve the Dirichlet problem

$$\begin{cases} \Delta v = 0 & \text{in } C_{r,R} \\ v = 0 & \text{on } \partial B_r \\ v = 1 & \text{on } \partial B_R, \end{cases}$$

on the annulus $C_{r,R} = B_R \setminus \overline{B_r}$, where

$$B_r \subset \Omega \subset B_R$$

and B_r, B_R are centred at the origin.

b) Show that given two solutions u_1, u_2 of (2.32),

$$w = \begin{cases} u_1 - u_2 & \text{in } \Omega_e \\ 0 & \text{in } \Omega \end{cases} \qquad (2.33)$$

satisfies

$$|w| \leq Mv \qquad \text{in } C_{r,R}$$

where v is as in part a) and M is a suitable constant.

c) Let R tend to infinity in the previous inequality, and deduce $w \equiv 0$. Conclude that (2.32) admits at most one solution.

Solution. a) The Dirichlet problem is solved on the annulus in Problem 2.2.15 (page 98). Here the problem is rotationally invariant, thus we seek *radially symmetric* solutions (also called *radial solutions*), which on the plane are of the form

$$v(\rho) = C_1 + C_2 \log \rho.$$

The initial conditions force

$$v(\rho) = \frac{\log (\rho/r)}{\log (R/r)}.$$

Note v is non-negative on $C_{r,R}$.

b) The function w in (2.33) is continuous on $\mathbb{R}^2$, identically zero on Ω and harmonic and bounded on Ω_e.

Let M be a positive number such that $|w| \le M$ in Ω_e. Then

$$W_+ = Mv - w$$

solves

$$\begin{cases} \Delta W_+ = 0 & \text{in } C_{r,R} \cap \Omega_e \\ W_+|_{\partial\Omega} = Mv|_{\partial\Omega} \ge 0 \\ W_+|_{\partial B_R} = M - w|_{\partial B_R} \ge 0. \end{cases}$$

By the maximum principle, $W_+ \ge 0$, i.e.

$$w \le Mv$$

on $C_{r,R} \cap \Omega_e$. Given the definition of w, the inequality extends easily to $C_{r,R}$. Similarly, let

$$W_- = -Mv - w.$$

The same argument shows $W_- \le 0$, so

$$w \ge -Mv$$

on $C_{r,R}$. In conclusion:

$$-Mv \le w \le Mv \quad \text{in } C_{r,R},$$

which was our claim.

c) In b) we showed that

$$|w(\rho, \theta)| \le M \frac{\log (\rho/r)}{\log (R/r)}, \tag{2.34}$$

with $r < \rho < R$ and r, R such that

$$B_r \subset \Omega \subset B_R.$$

In particular, Ω being bounded, the inequality holds for R arbitrarily large. So fix r, and let (ρ, θ) be a given point in Ω_e. If $R > \rho$, then $(\rho, \theta) \in C_{r,R}$. Since

$$\frac{\log (\rho/r)}{\log (R/r)} \to 0 \qquad \text{as } R \to +\infty,$$

from (2.34) it follows that $w(\rho, \theta) = 0$, and therefore $u_1 \equiv u_2$. Thus, problem (2.32) admits, at most, one solution.

Problem 2.2.24 (Uniqueness for the exterior Neumann problem). *Let $\Omega \subset \mathbb{R}^2$ be a bounded domain containing the origin and $\Omega_e = \mathbb{R}^2 \setminus \overline{\Omega}$. Prove that the problem*

$$
\begin{cases}
\Delta u = 0 & \text{in } \Omega_e \\
\partial_\nu u = g & \text{on } \partial\Omega_e, \text{ with } g \text{ continuous} \\
u \text{ bounded} & \text{in } \Omega_e
\end{cases}
\tag{2.35}
$$

has at most one solution of class $C^2(\Omega_e) \cap C^1(\overline{\Omega}_e)$ up to additive constants, following the steps below.

a) Using the Kelvin transform (Problem 2.2.20 on page 107), prove that a solution u of problem (2.35) satisfies

$$
\left| \frac{\partial u}{\partial x_i}(\mathbf{x}) \right| \leq \frac{C}{|\mathbf{x}|^2} \qquad (i = 1, 2)
$$

whenever $|\mathbf{x}|$ is large enough, for a suitable constant C.

b) Let u_1, u_2 be solutions of (2.35) and $w = u_1 - u_2$ in Ω_e. Suitably integrating by parts and using part a), prove that for large R

$$
\int_{\Omega_e \cap B_R} |\nabla w(\mathbf{x})|^2 \, d\mathbf{x} \leq \frac{C'}{R}.
$$

c) Let R go to infinity and deduce w is constant, whence (2.35) has, up to additive constants, at most one solution.

Solution. a) Take $a > 0$ so large that $B = B_a(0) \supset \Omega$, and hence $B_e = \{\mathbf{x} \in \mathbb{R}^2 : |\mathbf{x}| > a\} \subset \Omega_e$. Consider the Kelvin transform

$$
\mathbf{y} = T_a(\mathbf{x}) = \frac{a^2}{|\mathbf{x}|^2}\mathbf{x}, \qquad \mathbf{x} = T_a^{-1}(\mathbf{y}) = T_a(\mathbf{y}) = \frac{a^2}{|\mathbf{y}|^2}\mathbf{y}.
$$

From Problem 2.2.20 (page 107), $T_a(B_e) = B \setminus \{0\}$ and

$$
v(\mathbf{y}) = u(T_a(\mathbf{y}))
$$

is harmonic on that set. As u is bounded, so is v. The result of Problem 2.2.10 (page 92) says that v is extendable to B harmonically. Consequently $v \in C^2(B)$, so it has bounded first (and second) derivatives on any compact subset of B. In particular there exists a constant c_0 such that, for $j = 1, 2$,

$$
\left| \frac{\partial v}{\partial y_j}(\mathbf{y}) \right| \leq c_0 \quad \text{for } |\mathbf{y}| \leq \frac{a}{2}.
$$

But

$$\frac{\partial u}{\partial x_i}(\mathbf{x}) = \sum_{j=1}^{2} \frac{\partial v}{\partial y_j}(\mathbf{y}) \frac{\partial y_j}{\partial x_i}(\mathbf{x}) = \sum_{j=1}^{2} \frac{\partial v}{\partial y_j}(\mathbf{y}) \left(\frac{a^2}{|\mathbf{x}|^2} \delta_{ij} - \frac{a^2 x_i x_j}{|\mathbf{x}|^4} \right),$$

where δ_{ij} is the usual Kronecker symbol ($\delta_{ij} = 1$ if $i = j$, 0 otherwise). At this point it is enough to notice that $|x_i x_j| \leq |\mathbf{x}|^2$ (for any i, j) to have

$$\left| \frac{\partial u}{\partial x_i}(\mathbf{x}) \right| \leq \frac{2a^2}{|\mathbf{x}|^2} \sum_{j=1}^{2} \left| \frac{\partial v}{\partial y_j}(\mathbf{y}) \right| \leq \frac{4a^2 c_0}{|\mathbf{x}|^2}, \qquad \text{for } |\mathbf{x}| \geq 2a. \tag{2.36}$$

It might be useful to notice that the estimates obtained only depend on the fact u is harmonic and bounded (the Neumann condition was not used).

b) Integrating by parts, we have

$$\int_{\Omega_e \cap B_R} \varphi(\mathbf{x}) \Delta w(\mathbf{x}) \, d\mathbf{x} = - \int_{\Omega_e \cap B_R} \nabla w(\mathbf{x}) \cdot \nabla \varphi(\mathbf{x}) \, d\mathbf{x} + \int_{\partial(\Omega_e \cap B_R)} \varphi(\mathbf{x}) \partial_\nu w(\mathbf{x}) \, d\sigma,$$

where φ is an arbitrary smooth function (the formula holds since the domain of integration is bounded). To obtain the L^2 norm of ∇w we must choose $\varphi = w$. Then

$$\int_{\Omega_e \cap B_R} w(\mathbf{x}) \Delta w(\mathbf{x}) \, d\mathbf{x} = \int_{\Omega_e \cap B_R} |\nabla w(\mathbf{x})|^2 \, d\mathbf{x} - \int_{\partial(\Omega_e \cap B_R)} w \partial_\nu w \, d\sigma.$$

Recalling that $\Delta w = 0$ in Ω and $\partial_\nu w = 0$ on $\partial\Omega$ we find

$$\int_{\Omega_e \cap B_R} |\nabla w(\mathbf{x})|^2 \, d\mathbf{x} = \int_{\partial B_R} w \partial_\nu w \, d\sigma \leq M \int_{\partial B_R} |\nabla w| \, d\sigma,$$

where M is some constant such that $|w| \leq M$ on Ω_e (w is bounded as difference of bounded functions). If $R \geq 2a$, from (2.36) we obtain (on ∂B_R we have $|\mathbf{x}| = R$)

$$\int_{\Omega \cap B_R} |\nabla w(\mathbf{x})|^2 \, d\mathbf{x} \leq 8a^2 c_0 M \int_{\partial B_R} \frac{1}{|\mathbf{x}|^2} \, d\sigma = \frac{16\pi a^2 c_0 M}{R}.$$

c) The inequality holds for any R large enough, so we may take the limit to obtain

$$\int_{\Omega} |\nabla w(\mathbf{x})|^2 \, d\mathbf{x} \leq 0,$$

i.e. $|\nabla w(\mathbf{x})|^2 = 0$ almost everywhere on Ω_e. By regularity the claim follows.

2.2.3 Potentials and Green functions

Problem 2.2.25 (Newtonian potential). *Compute the Newtonian potential of a homogeneous mass distribution (of density $\mu = 1$) on the disc B_1 centred at the origin.*

Solution. The required potential (defined up to additive constants) is

$$u(x, y) = -\frac{1}{4\pi} \int_{B_1} \log \left[(x - \xi)^2 + (y - \eta)^2\right] d\xi d\eta$$

and it is C^1 on the entire $\mathbb{R}^2$. For an explicit expression, first we observe that the problem is invariant under rotations and therefore u has radial symmetry. Thus $u = u(r)$, $r^2 = x^2 + y^2$, and

$$\Delta u = \begin{cases} -1 & 0 \le r < 1 \\ 0 & r > 1 \end{cases}$$

together with

$$u(1-) = u(1+), \qquad u_r(1-) = u_r(1+). \tag{2.37}$$

For $r > 1$, u is a radial harmonic functions, thus

$$u(r) = a \log r + b, \qquad r > 1.$$

On the other hand, from

$$u_{rr} + \frac{1}{r} u_r = -1,$$

we easily find

$$u(r) = c + d \log r - \frac{1}{4} r^2, \qquad r < 1.$$

Since u is bounded for $r < 1$, it follows $d = 0$. By (2.37) we obtain

$$c - \frac{1}{4} = b \qquad \text{and} \qquad a = -\frac{1}{2}.$$

Choosing $c = 1/4$, then, we find

$$u(r) = \begin{cases} \frac{1}{4}(1 - r^2) & r \le 1 \\ -\frac{1}{2} \log r & r > 1. \end{cases}$$

Problem 2.2.26. Let $u = u(\mathbf{x})$ be the double-layer potential with density μ on the circle $C = \{\mathbf{x} = (x_1, x_2) : \rho = 1\}$, $\rho^2 = x_1^2 + x_2^2$. Outside the circle u is given by the harmonic function

$$u(\mathbf{x}) = -\frac{x_1 x_2}{\rho^4}.$$

Compute μ.

Solution. Note that since the double-layer potential of a constant is zero outside the disk bounded by C, μ is determined up to an additive constant.

The relation between u and μ is given, at any point $\mathbf{z} = (z_1, z_2)$ on the circle, by the formula

$$u^E (\mathbf{z}) - u^I (\mathbf{z}) = \mu (\mathbf{z})$$

where

$$u^E (\mathbf{z}) = \lim_{\mathbf{x} \to \mathbf{z}, \rho > 1} u (\mathbf{x}), \quad u^I (\mathbf{z}) = \lim_{\mathbf{x} \to \mathbf{z}, \rho < 1} u (\mathbf{x}).$$

We know that $u^E (\mathbf{z}) = -z_1 z_2$. To compute u^I we use the continuity of the normal derivative $\frac{\partial u}{\partial \rho}$ across C. We find

$$\frac{\partial u^E}{\partial \rho} = \frac{\partial u^I}{\partial \rho} (\mathbf{z}) = 2z_1 z_2. \tag{2.38}$$

We deduce that u is harmonic inside C and satisfies the Neumann condition (2.38). Note that $\int_C z_1 z_2 = 0$, so that the Neumann problem has solutions, differing by additive constants. Therefore inside C we find $u (\mathbf{x}) = x_1 x_2 + k, k \in \mathbb{R}$, and

$$\mu (\mathbf{z}) = u^E (\mathbf{z}) - u^I (\mathbf{z}) = -2z_1 z_2 + k \quad k \in \mathbb{R}.$$

Problem 2.2.27. Let $f \in C(\mathbb{R}^2)$ be a function with compact support K and define

$$u (\mathbf{x}) = -\frac{1}{2\pi} \int_{\mathbb{R}^2} \log |\mathbf{x} - \mathbf{y}| \ f (\mathbf{y}) d\mathbf{y}.$$

Show that

$$u (\mathbf{x}) = -\frac{M}{2\pi} \log |\mathbf{x}| + O(|\mathbf{x}|^{-1}), \quad as |\mathbf{x}| \to +\infty$$

where

$$M = \int_K f (\mathbf{y}) d\mathbf{y}.$$

Consequence: $u (\mathbf{x}) \to 0$ at infinity if and only if $M = 0$.

Solution. We have

$$\left| u (\mathbf{x}) + \frac{M}{2\pi} \log |\mathbf{x}| \right| = \frac{1}{2\pi} \left| \int_{\mathbb{R}^2} [\log |\mathbf{x} - \mathbf{y}| - \log |\mathbf{x}|] \ f (\mathbf{y}) d\mathbf{y} \right|$$

$$\leq \frac{1}{2\pi} \int_K \left| \log \left(\frac{|\mathbf{x} - \mathbf{y}|}{|\mathbf{x}|} \right) \right| \ |f (\mathbf{y})| d\mathbf{y}.$$

Since K is compact we may assume that $K \subset B_R (0), R \geq 1$. If $\mathbf{y} \in K$ then $|\mathbf{y}| \leq R$. Let $|\mathbf{x}| \geq 2R$. Then

$$|\mathbf{x} - \mathbf{y}| \geq |\mathbf{x}| - |\mathbf{y}| \geq |\mathbf{x}| - R \geq \frac{|\mathbf{x}|}{2} \geq 1,$$

so the last integral is finite.

Assume $|\mathbf{x} - \mathbf{y}| \ge |\mathbf{x}|$. Since $\log(1 + t) \le t, t > -1$,

$$0 < \log\left(\frac{|\mathbf{x} - \mathbf{y}|}{|\mathbf{x}|}\right) \le \log\left(\frac{|\mathbf{x}| + |\mathbf{y}|}{|\mathbf{x}|}\right) \le \log\left(1 + \frac{R}{|\mathbf{x}|}\right) \le \frac{R}{|\mathbf{x}|}.$$

Now let $|\mathbf{x} - \mathbf{y}| \le |\mathbf{x}|$. We have, since $|\mathbf{y}| \le |\mathbf{x}|/2$,

$$\frac{|\mathbf{x}|}{|\mathbf{x} - \mathbf{y}|} \le 1 + \frac{|\mathbf{y}|}{|\mathbf{x}| - |\mathbf{y}|} \le 1 + \frac{2|\mathbf{y}|}{|\mathbf{x}|},$$

so that

$$\left|\log\left(\frac{|\mathbf{x} - \mathbf{y}|}{|\mathbf{x}|}\right)\right| = \log\left(\frac{|\mathbf{x}|}{|\mathbf{x} - \mathbf{y}|}\right) \le \log\left(1 + \frac{2|\mathbf{y}|}{|\mathbf{x}|}\right) \le \frac{2R}{|\mathbf{x}|}.$$

Thus

$$\left|u(\mathbf{x}) + \frac{M}{2\pi}\log|\mathbf{x}|\right| \le \frac{2R}{\pi |\mathbf{x}|}\int_K |f(\mathbf{y})|\, d\mathbf{y},$$

which yields the desired result.

Problem 2.2.28 (Computing Green functions). *Determine the Green functions of the Laplace operator for the following sets.*

a) The half-plane $P^+ = \{\mathbf{x} = (x_1, x_2) : x_2 > 0\}$.

b) The disc $B_1 = \{\mathbf{x} = (x_1, x_2) : x^2 + y^2 < r\}$.

c) The half-disc $B_1^+ = \{\mathbf{x} = (x_1, x_2) : x^2 + y^2 < r, x_2 > 0\}$.

Solution. a) For any given $\mathbf{y} \in P^+$, the Green function $G = G(\mathbf{x}, \mathbf{y})$ is harmonic on the half-plane, $G(\mathbf{x}, \mathbf{y}) = 0$ on $x_2 = 0$, and

$$\Delta_{\mathbf{x}} G(\mathbf{x}, \mathbf{y}) = -\delta_2(\mathbf{x} - \mathbf{y}), \qquad (2.39)$$

where $\delta_2(\mathbf{x} - \mathbf{y})$ is the Dirac distribution at $\mathbf{y}$. Let us use the *method of images*. We know that the fundamental solution

$$\Gamma(\mathbf{x} - \mathbf{y}) = -\frac{1}{2\pi}\log|\mathbf{x} - \mathbf{y}|$$

satisfies (2.39). If $\mathbf{y} = (y_1, y_2)$, we define $\widetilde{\mathbf{y}} = (y_1, -y_2)$, the mirror image of $\mathbf{y}$ with respect to the y_1-axis. The function $\Gamma(\mathbf{x} - \widetilde{\mathbf{y}})$ is harmonic on P^+ and coincides with $\Gamma(\mathbf{x} - \mathbf{y})$ on $x_2 = 0$. Hence

$$G(\mathbf{x}, \mathbf{y}) = \Gamma(\mathbf{x} - \mathbf{y}) - \Gamma(\mathbf{x} - \widetilde{\mathbf{y}}).$$

b) Using again the method of images, define

$$\mathbf{y}^* = T_1(\mathbf{y}) = \frac{\mathbf{y}}{|\mathbf{y}|^2}$$

to be the Kelvin image of $\mathbf{y} \neq \mathbf{0}$. For $|\mathbf{x}| = 1$,

$$
\begin{aligned}
|\mathbf{x} - \mathbf{y}^*|^2 &= 1 - \frac{2\mathbf{x} \cdot \mathbf{y}}{|\mathbf{y}|^2} + \frac{1}{|\mathbf{y}|^2} \\
&= \frac{1}{|\mathbf{y}|^2} \left(1 - 2\mathbf{x} \cdot \mathbf{y} + |\mathbf{y}|^2 \right) \\
&= \frac{1}{|\mathbf{y}|^2} |\mathbf{x} - \mathbf{y}|^2 \, .
\end{aligned}
$$

If $\mathbf{y} \neq \mathbf{0}$, set

$$
G(\mathbf{x}, \mathbf{y}) = -\frac{1}{2\pi} \{ \log |\mathbf{x} - \mathbf{y}| - \log(|\mathbf{y}| \, |\mathbf{x} - \mathbf{y}^*|) \} \, .
$$

This gives

$$
G(\mathbf{x}, \mathbf{y}) = 0
$$

for $|\mathbf{x}| = 1, \mathbf{y} \neq \mathbf{0}$ and

$$
\Delta_{\mathbf{x}} G(\mathbf{x}, \mathbf{y}) = -\delta(\mathbf{x} - \mathbf{y}) \quad \text{in } B_1 \, .
$$

When $\mathbf{y} = \mathbf{0}$, we simply define

$$
G(\mathbf{x}, \mathbf{0}) = -\frac{1}{2\pi} \log |\mathbf{x}| \, .
$$

When $\mathbf{x} \neq \mathbf{0}$ and $\mathbf{y} \to \mathbf{0}$, note that

$$
G(\mathbf{x}, \mathbf{y}) \to G(\mathbf{x}, \mathbf{0}) \, .
$$

   c) Call G_{B_1} the Green function for the disc B_1 of the previous part, and define $\widetilde{\mathbf{y}} = (y_1, -y_2)$. Then

$$
G_{B_1}^+ (\mathbf{x}, \mathbf{y}) = G_{B_1}(\mathbf{x}, \mathbf{y}) - G_{B_1}(\mathbf{x}, \widetilde{\mathbf{y}}) \, .
$$

---

*Problem 2.2.29 (Symmetry of Green functions). *Let $G(\mathbf{x}, \mathbf{y})$ be the Green function associated to the Laplace operator on a bounded, smooth domain $\Omega \subset \mathbb{R}^3$. Prove that*

$$
G(\mathbf{x}_1, \mathbf{x}_2) = G(\mathbf{x}_2, \mathbf{x}_1)
$$

for any pair $\mathbf{x}_1, \mathbf{x}_2 \in \Omega$.

   Solution. First recall that (in dimension three) the Green function G can be written as

$$
G(\mathbf{x}, \mathbf{y}) = \Gamma(\mathbf{x} - \mathbf{y}) - g(\mathbf{x}, \mathbf{y}) = \frac{1}{4\pi |\mathbf{x} - \mathbf{y}|} - g(\mathbf{x}, \mathbf{y}),
$$

where $g(\mathbf{x}, \cdot)$ is harmonic on Ω for fixed $\mathbf{x}$, continuous on $\overline{\Omega}$ and satisfies

$$
g(\mathbf{x}, \cdot) = \Gamma(\mathbf{x} - \cdot) \quad \text{on } \partial\Omega \, .
$$

In particular, $G(\mathbf{x}, \cdot)$ is non-negative on Ω, null on $\partial\Omega$ and

$$
G(\mathbf{x}, \mathbf{y}) \leq \frac{1}{4\pi |\mathbf{x} - \mathbf{y}|} \qquad \text{in } \Omega \times \Omega. \tag{2.40}
$$

If $\mathbf{x}_1 = \mathbf{x}_2$ the claim is trivial, so we take $\mathbf{x}_1 \neq \mathbf{x}_2$. We carve out of Ω two balls $B_r(\mathbf{x}_1)$, $B_r(\mathbf{x}_2)$ with radius r small enough so to be disjoint. On the resulting domain

$$\Omega_r = \Omega \setminus (B_r(\mathbf{x}_1) \cup B_r(\mathbf{x}_2)),$$

the functions $u(\mathbf{y}) = G(\mathbf{x}_1, \mathbf{y})$, $v(\mathbf{y}) = G(\mathbf{x}_2, \mathbf{y})$ are harmonic and vanish on $\partial\Omega$. Hence we can invoke Green's identity:

$$\int_{\Omega_r} (v\Delta u - u\Delta v) \, d\mathbf{x} = \int_{\partial\Omega_r} (v\partial_\nu u - u\partial_\nu v) \, d\sigma.$$

Because of our choice for u and v the latter reduces to

$$\int_{\partial B_r(\mathbf{x}_1) \cup \partial B_r(\mathbf{x}_2)} (v\partial_\nu u - u\partial_\nu v) \, d\sigma = 0$$

i.e.

$$\int_{\partial B_r(\mathbf{x}_1)} (v\partial_\nu u - u\partial_\nu v) \, d\sigma = \int_{\partial B_r(\mathbf{x}_2)} (u\partial_\nu v - v\partial_\nu u) \, d\sigma. \tag{2.41}$$

Let us compute the limit of the left-hand side when $r \to 0$. Since v is smooth near $\mathbf{x}_1$, we have $|\nabla v| \leq M$ on $\partial B_r(\mathbf{x}_1)$ provided r is small enough. From (2.40) we also have

$$0 \leq u \leq \frac{1}{4\pi r} \text{ on } \partial B_r(\mathbf{x}_1).$$

Then

$$\left| \int_{\partial B_r(\mathbf{x}_1)} u\partial_\nu v \, d\sigma \right| \leq \int_{\partial B_r(\mathbf{x}_1)} u \, |\partial_\nu v| \, d\sigma \leq \frac{M}{4\pi r} 4\pi r^2 = Mr \to 0 \qquad \text{for } r \to 0.$$

On the other hand

$$\int_{\partial B_r(\mathbf{x}_1)} v\partial_\nu u \, d\sigma = \frac{1}{4\pi} \int_{\partial B_r(\mathbf{x}_1)} v\partial_\nu \left(\frac{1}{|\sigma - \mathbf{x}_1|} \right) d\sigma + \int_{\partial B_r(\mathbf{x}_1)} v\partial_\nu g(\mathbf{x}_1, \sigma) \, d\sigma.$$

The last integrand is a smooth function in a neighbourhood of $\mathbf{x}_1$, so the integral tends to 0 as $r \to 0$. Moreover

$$\partial_\nu \left(\frac{1}{|\sigma - \mathbf{x}_1|} \right) = \nabla \left(\frac{1}{|\sigma - \mathbf{x}_1|} \right) \cdot \nu = \frac{\sigma - \mathbf{x}_1}{|\sigma - \mathbf{x}_1|^3} \cdot \frac{\sigma - \mathbf{x}_1}{|\sigma - \mathbf{x}_1|} = \frac{1}{|\sigma - \mathbf{x}_1|^2},$$

and then

$$\frac{1}{4\pi} \int_{\partial B_r(\mathbf{x}_1)} v\partial_\nu \left(\frac{1}{|\sigma - \mathbf{x}_1|} \right) d\sigma = \frac{1}{4\pi} \int_{\partial B_r(\mathbf{x}_1)} \frac{1}{|\sigma - \mathbf{x}_1|^2} v \, d\sigma =$$

$$= \frac{1}{|\partial B_r(\mathbf{x}_1)|} \int_{\partial B_r(\mathbf{x}_1)} v \, d\sigma \to v(\mathbf{x}_1) \text{ for } r \to 0.$$

Overall, we have found

$$\int_{\partial B_r(\mathbf{x}_1)} (v\partial_\nu u - u\partial_\nu v) \, d\sigma \to v(\mathbf{x}_1) \text{ for } r \to 0. \tag{2.42}$$

Similar computations show

$$\int_{\partial B_r(\mathbf{x}_2)} (u\partial_\nu v - v\partial_\nu u)\, d\sigma \to u(\mathbf{x}_2) \text{ for } r \to 0. \qquad (2.43)$$

Taking the limit in (2.41), and with the aid of (2.42) and (2.43), we finally get

$$v(\mathbf{x}_1) = u(\mathbf{x}_2),$$

that is, the desired symmetry for G:

$$G(\mathbf{x}_2, \mathbf{x}_1) = G(\mathbf{x}_1, \mathbf{x}_2).$$

Remark. The Green function for the Laplace operator is always symmetric, in any dimension. The proof is identical to the above one.

---

*Problem 2.2.30 (Poisson formula, double-layer potential). *Recover the Poisson formula in the plane by representing the solution of*

$$\begin{cases} \Delta u = 0 & \text{in } B_R \\ u = g & \text{on } \partial B_R \end{cases}$$

as double-layer potential.

Solution. We have to find a function $\mu : \partial B_R \to \mathbb{R}$ such that the solution u of the given Dirichlet problem reads

$$u(\mathbf{x}) = \int_{\partial B_R} \frac{\partial}{\partial \nu}\left(-\frac{1}{2\pi}\log|\mathbf{x} - \boldsymbol{\sigma}|\right)\mu(\boldsymbol{\sigma})\, d\sigma = \frac{1}{2\pi}\int_{\partial B_R} \frac{(\mathbf{x} - \boldsymbol{\sigma})\cdot \mathbf{v}(\boldsymbol{\sigma})}{|\mathbf{x} - \boldsymbol{\sigma}|^2}\mu(\boldsymbol{\sigma})\, d\sigma.$$

Let us briefly recall the properties of double-layer potentials. First of all, if μ is continuous, the above u is harmonic on B_R. In fact, if $\mathbf{x} \notin \partial B_R$ the denominator of the integrand never vanishes, so we can differentiate inside the integral, obtaining that u is harmonic. The unknown density μ is determined so to fulfill the boundary condition. We remind that if $\mathbf{x} \in B_R, \mathbf{z} \in \partial B_R$ and $\mathbf{x} \to \mathbf{z}$, then

$$u(\mathbf{x}) \to \frac{1}{2\pi}\int_{\partial B_R} \frac{(\mathbf{z} - \boldsymbol{\sigma})\cdot \mathbf{v}(\boldsymbol{\sigma})}{|\mathbf{z} - \boldsymbol{\sigma}|^2}\mu(\boldsymbol{\sigma})\, d\sigma - \frac{1}{2}\mu(\mathbf{z}).$$

From $u(\mathbf{x}) \to g(\mathbf{z})$, we obtain the integral equation

$$\frac{1}{2\pi}\int_{\partial B_R} \frac{(\mathbf{z} - \boldsymbol{\sigma})\cdot \mathbf{v}(\boldsymbol{\sigma})}{|\mathbf{z} - \boldsymbol{\sigma}|^2}\mu(\boldsymbol{\sigma})\, d\sigma - \frac{1}{2}\mu(\mathbf{z}) = g(\mathbf{z}).$$

Now observe $v(\sigma) = \sigma/R$ (note that $\sigma \in \partial B_R$, so $|\sigma| = R$). Substituting we get

$$
\begin{aligned}
g(\mathbf{z}) + \frac{1}{2}\mu(\mathbf{z}) &= \frac{1}{2\pi R}\int_{\partial B_R}\frac{\mathbf{z}\cdot\sigma - |\sigma|^2}{|\mathbf{z}|^2 - 2\mathbf{z}\cdot\sigma + |\sigma|^2}\mu(\sigma)\,d\sigma = \\
&= \frac{1}{2\pi R}\int_{\partial B_R}\frac{\mathbf{z}\cdot\sigma - R^2}{2(R^2 - \mathbf{z}\cdot\sigma)}\mu(\sigma)\,d\sigma = -\frac{1}{4\pi R}\int_{\partial B_R}\mu(\sigma)\,d\sigma.
\end{aligned}
\tag{2.44}
$$

We have to compute the integral in the last term to finally get μ. To this end we integrate (2.44) on ∂B_R:

$$
\int_{\partial B_R}g(\sigma)\,d\sigma = 2\pi R\cdot\left(-\frac{1}{4\pi R}\int_{\partial B_R}\mu(\sigma)\,d\sigma\right) - \frac{1}{2}\int_{\partial B_R}\mu(\sigma)\,d\sigma,
$$

whence

$$
\int_{\partial B_R}\mu(\sigma)\,d\sigma = -\int_{\partial B_R}g(\sigma)\,d\sigma.
$$

Substituting in (2.44) gives

$$
\mu(\mathbf{z}) = -2g(\mathbf{z}) + \frac{1}{2\pi R}\int_{\partial B_R}g(\sigma)\,d\sigma.
$$

Now that we have μ, we revert to the initial definition of u, and recall that if $\mathbf{x} \in B_R$,

$$
\frac{1}{2\pi}\int_{\partial B_R}\frac{(\mathbf{x} - \sigma)\cdot v(\sigma)}{|\mathbf{x} - \sigma|^2}\,d\sigma = 1.
$$

Then

$$
\begin{aligned}
u(\mathbf{x}) &= \frac{1}{2\pi}\int_{\partial B_R}\frac{(\mathbf{x} - \sigma)\cdot v(\sigma)}{|\mathbf{x} - \sigma|^2}\left[-2g(\sigma) + \frac{1}{2\pi R}\int_{\partial B_R}g(\sigma)\,d\sigma\right]d\sigma = \\
&= \frac{1}{2\pi}\int_{\partial B_R}\frac{-2(\mathbf{x} - \sigma)\cdot v(\sigma)}{|\mathbf{x} - \sigma|^2}g(\sigma)\,d\sigma - \frac{1}{2\pi R}\int_{\partial B_R}g(\sigma)\,d\sigma.
\end{aligned}
$$

Substituting $v(\sigma) = \sigma/R$, we obtain

$$
\begin{aligned}
&= \frac{1}{2\pi R}\int_{\partial B_R}\frac{-2\mathbf{x}\cdot\sigma + 2R^2}{|\mathbf{x} - \sigma|^2}g(\sigma)\,d\sigma - \frac{1}{2\pi R}\int_{\partial B_R}g(\sigma)\,d\sigma = \\
&= \frac{1}{2\pi R}\int_{\partial B_R}\frac{R^2 - |\mathbf{x}|^2 + |\mathbf{x} - \sigma|^2}{|\mathbf{x} - \sigma|^2}g(\sigma)\,d\sigma - \frac{1}{2\pi R}\int_{\partial B_R}g(\sigma)\,d\sigma = \\
&= \frac{R^2 - |\mathbf{x}|^2}{2\pi R}\int_{\partial B_R}\frac{g(\sigma)}{|\mathbf{x} - \sigma|^2}\,d\sigma,
\end{aligned}
$$

which is precisely the Poisson formula on the plane.

*Problem 2.2.31 (Non-homogeneous Poisson-Dirichlet problem). *Prove the represen-tation formula*

$$u(\mathbf{x}) = -\int_{\partial\Omega} h(\sigma)\, \partial_\nu G(\mathbf{x}, \sigma)\, d\sigma - \int_\Omega f(\mathbf{y})\, G(\mathbf{x}, \mathbf{y})\, d\mathbf{y},$$

for the solution of the problem

$$\begin{cases} \Delta u = f & \text{in } \Omega \\ u = h & \text{on } \partial\Omega \end{cases}$$

where G is the Green function of $\Omega \subset \mathbb{R}^2$.

Solution. Recall that in dimension two

$$G(\mathbf{x}, \mathbf{y}) = \Gamma(\mathbf{x} - \mathbf{y}) - g(\mathbf{x}, \mathbf{y})$$
$$= -\frac{1}{2\pi} \log |\mathbf{x} - \mathbf{y}| - g(\mathbf{x}, \mathbf{y}),$$

where $g(\mathbf{x}, \cdot)$ solves

$$\begin{cases} \Delta_\mathbf{y} g(\mathbf{x}, \mathbf{y}) = 0 & \mathbf{y} \in \Omega \\ g(\mathbf{x}, \mathbf{y}) = -\frac{1}{2\pi} \log |\mathbf{x} - \mathbf{y}| & \mathbf{y} \in \partial\Omega. \end{cases}$$

The function u can be written as sum of three potentials (Newtonian, double- and single layer)

$$u(\mathbf{x}) = \int_{\partial\Omega} \partial_\nu u(\sigma)\, \Gamma|\mathbf{x} - \sigma|\, d\sigma - \int_{\partial\Omega} h(\sigma)\, \partial_\nu \Gamma|\mathbf{x} - \sigma|\, d\sigma - \int_\Omega f(\mathbf{y})\, \Gamma|\mathbf{x} - \mathbf{y}|\, d\mathbf{y}.$$

At the same time, applying

$$\int_\Omega (\psi \Delta\varphi - \varphi\Delta\psi)\, d\mathbf{x} = \int_{\partial\Omega} (\psi \partial_\nu \varphi - \varphi\partial_\nu \psi)\, d\sigma \tag{2.45}$$

to

$$\varphi = u \quad \text{and} \quad \psi = g(\mathbf{x}, \cdot),$$

gives

$$0 = -\int_{\partial-} g(\mathbf{x}, \sigma)\, \partial_\nu u(\sigma)\, d\sigma + \int_{\partial\Omega} h(\sigma)\, \partial_\nu g(\sigma)\, d\sigma + \int_\Omega g(\mathbf{y})\, f(\mathbf{y})\, d\mathbf{y}. \tag{2.46}$$

Adding up (2.46) and (2.45) furnishes straightaway the claim.

2.3 Further Exercises

2.3.1. (Mixed problem on the square, separation of variables) *Solve, on the square*

$$\{Q = (x, y): 0 < x < 1, 0 < y < 1\},$$

the problem

$$\begin{cases} \Delta u = 0 & \text{in } Q \\ u_y\,(x,0) = \sin\frac{\pi x}{2},\, u\,(x,1) = 0 & 0 \le x \le 1 \\ u\,(0,y) = u_x\,(1,y) = 0 & 0 \le y \le 1. \end{cases}$$

2.3.2. *Determine all harmonic functions on the annulus*

$$a < r < b \qquad (r^2 = x^2 + y^2)$$

that satisfy the following boundary conditions:

a) $u\,(a,\theta) = 0,\, u\,(b,\theta) = \cos\theta.$
b) $u\,(a,\theta) = \cos\theta,\, u\,(b,\theta) = U\sin 2\theta.$

2.3.3. *Let $B_{1,2} = \{(r,\theta) \in \mathbb{R}^2 : 1 < r < 2\}$. Discuss whether the Neumann problem*

$$\begin{cases} \Delta u = -1 & \text{in } B_{1,2} \\ u_\nu = \cos\theta & \text{on } r = 1 \\ u_\nu = \lambda(\cos\theta)^2 & \text{on } r = 2, \end{cases}$$

has solutions, depending on the parameter $\lambda \in \mathbb{R}$.

2.3.4. (Neumann-Robin problem on a half-strip) *Solve, on the half-strip*

$$S = \{(x,y) : 0 < x < \infty,\, 0 < y < 1\}$$

the problem

$$\begin{cases} \Delta u = 0 & \text{in } Q \\ u_y\,(x,0) = u_y\,(x,1) + hu\,(x,1) = 0 & 0 \le x \le \infty \\ u\,(0,y) = g\,(y),\, u\,(\infty, y) = 0 & 0 \le y \le 1 \end{cases}$$

with $h > 0$.

2.3.5. *Show that for any integer $n \ge 1$*

$$u(x,y) = \frac{1}{n^2}\sinh ny \cos nx$$

solves the Cauchy problem

$$\begin{cases} \Delta u = 0 & \text{in } x \in \mathbb{R},\, y > 0 \\ u(x,0) = 0 & \text{for } x \in \mathbb{R} \\ u_y(x,0) = \frac{1}{n}\cos nx & \text{for } x \in \mathbb{R}. \end{cases}$$

Deduce that the Cauchy problem for the Laplace operator is not well posed, because the solution does not depend continuously on the data.

2.3.6. *Let $u = u\,(\mathbf{x})$ be harmonic on $\mathbb{R}^n$. Using the mean-value property show that the function*

$$v\,(\mathbf{x}) = u\,(M\mathbf{x})$$

is harmonic on $\mathbb{R}^n$, where M is a rotation $\mathbb{R}^n$ (represented by an orthogonal matrix).

2.3.7. *Let Q denote the square*

$$\{(x, y) : -1 \le x \le 1, \, -1 \le y \le 1\}$$

and L_i, $i = 0, \ldots 3$, its sides, labelled counter-clockwise from the bottom one

$$L_0 = \{(x, -1) : -1 < x < 1\}.$$

The solution u of

$$\begin{cases} \Delta u = 0 & \text{in } Q \\ u = 1 & \text{on } L_0 \\ u = 0 & \text{on } L_i, \, i = 1, 2, 3, \end{cases} \tag{2.47}$$

is continuous and bounded on $\overline{Q}$, except at the corners $\mathbf{p} = (-1, 0)$ and $\mathbf{q} = (1, 0)$. Compute $u(0, 0)$.

2.3.8. *Verify that*

$$u(x, y) = \frac{1 - x^2 - y^2}{1 - 2x + x^2 + y^2}$$

is harmonic on $B_1(0, 0)$. Since the numerator vanishes on $\partial B_1(0, 0)$, should we not have $u \equiv 0$, by uniqueness? Explain this (apparent) contradiction.

2.3.9. *Let Ω be a bounded domain in $\mathbb{R}^n$ and suppose $u \in C^2(\Omega) \cap C(\overline{\Omega})$ is a solution of the equation $\Delta u = u^3 - u$ with $u = 0$ on $\partial \Omega$. Show that $|u| \le 1$.*

2.3.10. *Let $u \ge 0$ be harmonic on $B_4(0, 0) \subset \mathbb{R}^2$ with $u(1, 0) = 1$. Using Harnack's inequality find upper and lower bounds for $u(-1, 0)$.*

2.3.11. *Determine for which $\alpha \in \mathbb{R}$ the function*

$$u(\mathbf{x}) = |\mathbf{x}|^\alpha$$

is subharmonic on $\mathbb{R}^n \setminus \{0\}$.

2.3.12. *Let u be a harmonic function on the open set $\Omega \subset \mathbb{R}^2$, continuous on $\overline{\Omega}$. Call (x_0, y_0) a point in Ω where $u(x_0, y_0) = 2$ and let E_1 be the set*

$$E_1 = \{(x, y) \in \Omega : u \ge 1\}.$$

Show that ∂E_1 cannot consist only of a (smooth) closed curve contained in Ω.

2.3.13. *Solve, on*

$$B_1^+ = B_1 \cap \{y > 0\} = \{(x, y) \in \mathbb{R}^2 : x^2 + y^2 < 1, \, y > 0\}$$

the Dirichlet problem:

$$\begin{cases} \Delta u(x, y) = 0 & \text{in } \Omega \\ u(x, 0) = 0 & -1 \le x \le 1 \\ u(x, y) = y^3 & x^2 + y^2 = 1, \, y \ge 0. \end{cases}$$

2.3.14. *(Neumann problem and reflection principle) Consider*

$$B_1^+ = \left\{(x, y) \in \mathbb{R}^2 : x^2 + y^2 < 1, \, y > 0\right\}$$

and $u \in C^2(B_1^+) \cap C(\overline{B_1^+})$, harmonic on B_1^+, with $u_y(x, 0) = 0$.

a) Prove that

$$U(x, y) = \begin{cases} u(x, y) & y \geq 0 \\ u(x, -y) & y < 0, \end{cases}$$

obtained by evenly reflecting u about the x-axis, is harmonic on the whole B_1.

b) Let u be the solution to the mixed problem

$$\begin{cases} \Delta u(x, y) = 0 & \text{in } B_1^+ \\ u(x, y) = x^2 & \text{on } \partial B_1^+, y > 0 \\ u_y(x, 0) = 0 & -1 \leq x \leq 1 \end{cases}$$

on $B_1^+ = \{x^2 + y^2 < 1, y > 0\}$. Compute $u(0, 0)$.

** 2.3.15.   *Consider the Neumann problem*

$$\begin{cases} \Delta u(x, y) = 0 & \text{in } P^+ \\ u_y(x, 0) = g(x) & x \in \mathbb{R} \end{cases} \tag{2.48}$$

on the half-plane $P^+ = \{(x, y) : y > 0\}$*, where g is smooth, zero outside an interval $[a, b]$, and $\int_a^b g = 0$. Prove that the problem possesses a bounded solution $u \in C^1(\overline{P^+})$, unique up to an additive constant, and find it explicitly.*

2.3.16.   (Dirichlet problem on a circular sector. Counterexample to Hopf's principle) **a)** *In the sector*

$$S_\alpha = \{(r, \theta) : r < 1, 0 < \theta < \alpha < 2\pi\},$$

solve the Dirichlet problem

$$\begin{cases} \Delta u = 0 & r < 1, 0 < \theta < \alpha \\ u(r, 0) = u(r, \alpha) = 0 & r \leq 1 \\ u(1, \theta) = g(\theta) & 0 \leq \theta \leq \alpha \end{cases}$$

where g is regular and $g(0) = g(\alpha) = 0$.

b) *If $\alpha < \pi$, show that there exist positive harmonic functions on S_α that vanish, together with their gradients, at the origin.*

2.3.17.   *Using the known Poisson formula for the disc and the Kelvin transform, find the Poisson formula for the problem (written in polar coordinates):*

$$\begin{cases} \Delta u = 0 & \text{in } B_e \\ u(1, \theta) = g(\theta) & 0 \leq \theta \leq 2\pi, \text{ with } g \in C^1 \text{ and } 2\pi\text{-periodic} \\ |u| \leq M & \text{in } B_e \end{cases}$$

on the exterior of the circle $B_e = \{x \in \mathbb{R}^2 : |x| > 1\}$*.*

2.3.18. Let B_1 be the unit ball centred at the origin in $\mathbb{R}^3$, and u be the solution to Dirichlet problem

$$\begin{cases} \Delta u(x, y, z) = 0 & \text{in } B_1 \\ u(x, y, z) = x^4 + y^4 + z^4 & \text{on } \partial B_1. \end{cases}$$

Compute the maximum and minimum values of u on $\overline{B}_1$.

2.3.19. In relation to Problem 2.2.17 (page 103), state and prove the reflection principle in dimension three. Deduce that the Dirichlet problem on the half-space has at most one bounded solution.

2.3.20. (Removable singularities) Given the ball $B = B_1(0) \subset \mathbb{R}^n$, let u be harmonic on $B \setminus \{0\}$ and such that, as $|x| \to 0$

$$\begin{cases} \dfrac{u(x)}{\log |x|} \to 0 & n = 2 \\ |x|^{n-2} u(x) \to 0 & n \geq 3. \end{cases}$$

Prove that the possible discontinuity at 0 is removable and, therefore, that u can be defined at 0 so to become harmonic on the whole B (see Problem 2.2.10 on page 92).

2.3.21. Referring to Problem 2.2.5 (page 89), give examples of:

a) Non-constant subharmonic functions on $\mathbb{R}^2$ that are bounded from below.

b) Non-constant subharmonic functions on $\mathbb{R}^3$ that are bounded.

2.3.22. Suppose that the functions u_i, $i \in \mathbb{N}$, are harmonic on the domain $\Omega \subset \mathbb{R}^n$, and that they form a pointwise-monotone increasing sequence. Show that if the sequence converges at some point $x_0 \in \Omega$, then it converges uniformly on every compact set $K \subset \Omega$. Deduce that the limit U is harmonic on Ω.

2.3.23. (Kelvin transform in $\mathbb{R}^3$) Take $x \in \mathbb{R}^3 \setminus \{0\}$, $a > 0$, and set

$$T_a(x) = \frac{a^2}{|x|^2} x.$$

Verify that T_a is a smooth map of $\mathbb{R}^3 \setminus \{0\}$ to itself, with smooth inverse, and that if u is harmonic on $\Omega \subset \mathbb{R}^3$ then

$$v(x) = \frac{a}{|x|} u(T_a(x))$$

is harmonic on $T_a^{-1}(\Omega)$.

2.3.24. (Exterior Dirichlet problem) Define $B_e = \{x \in \mathbb{R}^3 : |x| > 1\}$. Using the Kelvin transform in three dimensions, solve the problem

$$\begin{cases} \Delta u = 0 & \text{in } B_e \\ u = g & \text{on } \partial B_e \\ u(x) \to A & \text{for } |x| \to \infty, \end{cases}$$

where g is continuous. Examine in detail the case $g(x) = x_1$.

2.3.25. (Uniqueness for the exterior Robin problem) *Let* $\Omega \subset \mathbb{R}^2$ *be a smooth, bounded domain, and* $\Omega_e = \mathbb{R}^2 \setminus \Omega$. *Prove that the Robin problem*

$$\begin{cases} \Delta u = 0 & \text{in } \Omega_e \\ \partial_\nu u + \alpha u = g & \text{on } \partial\Omega_e, \text{ with } g \text{ continuous and } \alpha \geq 0 \\ u \text{ bounded} & \text{in } \Omega_e \end{cases}$$

has, at most, one solution in $C^2(\Omega_e) \cap C^1(\overline{\Omega}_e)$.

2.3.26. *Consider the Neumann problem on the exterior domain* $\Omega_e \subset \mathbb{R}^3$

$$\begin{cases} \Delta u = 0 & \text{in } \Omega_e \\ \partial_\nu u = g & \text{on } \partial\Omega_e \\ u \to 0 & \text{for } |\mathbf{x}| \to \infty. \end{cases} \tag{2.49}$$

Show that if $|u(\mathbf{x})| \leq M |\mathbf{x}|^{-1-\varepsilon}$, *with* $\varepsilon > 0$, *as* $|\mathbf{x}| \to \infty$, *then*

$$\int_{\partial\Omega} g \, d\sigma = 0$$

is a necessary condition for solving (2.49).

2.3.27. (A variant of Liouville's theorem) *Prove that a harmonic function* u *on* $\mathbb{R}^n$ *such that*

$$\int_{\mathbb{R}^n} |\nabla u(\mathbf{x})|^2 \, d\mathbf{x} < +\infty \tag{2.50}$$

is constant.

2.3.28. (Sublinearity of stress functions) *Denote by* Ω *the cross-section of a cylinder parallel to the* z*-axis. An external force produces a shear stress on each section. If* σ_1 *and* σ_2 *denote the scalar components of the stress on the planes* (x, z) *and* (y, z), *there is a function* $v = v(x, y; z)$ *(called stress function) such that*

$$v_x = \sigma_1, \qquad v_y = \sigma_2.$$

In suitable units v *solves the problem*

$$\begin{cases} v_{xx} + v_{yy} = -2 & \text{in } \Omega \\ v = 0 & \text{on } \partial\Omega. \end{cases}$$

Assuming $v \in C^2(\Omega) \cap C^1(\overline{\Omega})$, *prove that* $|\nabla v|^2$ *assumes its maximum on* $\partial\Omega$.

2.3.29. (Bôcher's theorem). *Let* u *be harmonic and positive on* $B_1(0) \setminus (0) \subset \mathbb{R}^n$, *and continuous up to the boundary* $\partial B_1(0)$. *Show that:*

$$u(\mathbf{x}) = \begin{cases} a \log |\mathbf{x}| + v(\mathbf{x}) & n = 2 \\ a |\mathbf{x}|^{2-n} + v(\mathbf{x}) & n \geq 3 \end{cases} \tag{2.51}$$

for some constant $a \geq 0$, *where* v *is harmonic and continuous on* $\overline{B}_1(0)$.

2.3.30. *The potential* $u = u(x, y)$ *of a distribution of mass inside the disc* $r^2 = x^2 + y^2 < 1$ *is given by*

$$u(r) = \frac{\pi}{8} \left(1 - r^4\right).$$

Determine the density μ and the Newtonian potential for $r \geq 1$.

2.3.31. *Compute the Newtonian potential of a homogeneous distribution of mass (density $\mu = 1$) in the ball $B_1 \subset \mathbb{R}^3$, at the origin.*

2.3.32. *Let $u = u(\mathbf{x})$ be the single-layer potential with density μ on the unit circle $C = \{\mathbf{x} : |\mathbf{x}| = 1\}$. Outside the circle u is given by the harmonic function*

$$u(\mathbf{x}) = \frac{x_1}{\rho^2}\left(1 + \frac{x_2}{\rho^2}\right),$$

where $\mathbf{x} = (x_1, x_2)$ and $\rho = |\mathbf{x}|$. Compute μ.

2.3.33. *Consider the harmonic function*

$$u(x_1, x_2) = \frac{x_1}{x_1^2 + x_2^2}$$

outside a bounded smooth domain Ω in $\mathbb{R}^2$. Can u be represented as a double-layer potential with bounded density μ on $\partial\Omega$?

2.3.34. *Determine Green functions for these sets:*

a) The half-space $P^+ = \{\mathbf{x} = (x_1, x_2, x_3) : x_3 > 0\}$.
b) The ball $B_1 = \{\mathbf{x} = (x_1, x_2, x_3) : x^2 + y^2 + z^2 < r\}$.
c) The half-ball $B_1^+ = \{\mathbf{x} = (x_1, x_2, x_3) : x^2 + y^2 + z^2 < r, x_3 > 0\}$.

2.3.1 Solutions

Solution 2.3.1. The solution is

$$u(x, y) = \frac{2}{\pi}\left\{\sinh\frac{\pi y}{2} - \tanh\frac{\pi}{2}\cosh\frac{\pi y}{2}\right\}\sin\frac{\pi x}{2}.$$

Solution 2.3.2. The solutions are (Fig. 2.8):

a) $u(r, \theta) = \dfrac{b}{b^2 - a^2}\left(r - \dfrac{a^2}{r}\right)\cos\theta.$

b) $u(r, \theta) = \dfrac{a^2}{b^2 + a^2}\left(r - \dfrac{b^2}{r}\right)\cos\theta + \dfrac{b^2 U}{b^2 + a^2}\left(r^2 + \dfrac{a^4}{r^2}\right)\sin 2\theta.$

Solution 2.3.3. We have to check the compatibility condition

$$\int_{B_{1,2}} \Delta u(\mathbf{x})\, d\mathbf{x} = \int_{\partial B_{1,2}} \partial_\nu u(\mathbf{s})\, ds,$$

that is

$$-|B_{1,2}| = -\int_{\partial B_1} \cos\theta\, ds + \int_{\partial B_2} \lambda \cos^2\theta\, ds,$$

or

$$-3\pi = 2\lambda\pi.$$

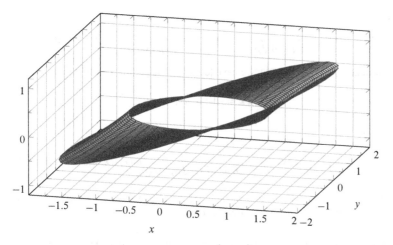

Fig. 2.8 $u(r,\theta) = \frac{2}{3}\left(r - \frac{1}{r}\right)\cos\theta$

Thus the problem can not be solved unless $\lambda = -3/2$. Separating the variables we find that for $\lambda = -3/2$ the problem is solved by

$$u(r,\theta) = a + \frac{1}{2}\log r - \frac{1}{4}r^2 + \frac{1}{3}\left(r + \frac{4}{r}\right)\cos\theta - \frac{1}{5}\left(r^2 + \frac{1}{r^2}\right)\cos 2\theta,$$

with $a \in \mathbb{R}$ arbitrary.

Solution 2.3.4. The solution is

$$u(x,y) = \sum_{n=1}^{\infty}\left\{\frac{2\left(h^2 + \lambda_n^2\right)}{h + \left(h^2 + \lambda_n^2\right)}\int_0^1 g(z)\cos\lambda_n z\,dz\right\}e^{-\lambda_n x}\cos\lambda_n y$$

where λ_n are the positive solutions of $\lambda\tan\lambda = h$.

Solution 2.3.5. As

$$\left|\frac{1}{n}\cos nx\right| \leq \frac{1}{n},$$

the datum $u_y(x,0) = \frac{1}{n}\cos nx$ tends to 0 uniformly on $\mathbb{R}$. On the other hand for large n the solution may become arbitrarily large, even at points (x,y) with $|y|$ extremely small. In fact, given a small $\delta > 0$ we have

$$\lim_{n\to\infty} u(0,\delta) = \lim_{n\to\infty}\frac{1}{n^2}\sinh n\delta = +\infty$$

and thus the solution does not depend continuously on the data.

Solution 2.3.6. As composition of continuous functions, v is continuous on $\mathbb{R}^n$. Hence it suffices to prove that v fulfils one of the averaging formulas on $B_R(x)$, for any $x \in \mathbb{R}^n$ and any $R > 0$; for instance:

$$u(\mathbf{x}) = \frac{1}{|B_R(\mathbf{x})|} \int_{B_R(\mathbf{x})} v(\mathbf{y})\, d\mathbf{y} = \frac{1}{|B_R(\mathbf{x})|} \int_{B_R(\mathbf{x})} u(M\mathbf{y})\, d\mathbf{y}.$$

As M is orthogonal ($M^T = M^{-1}$), we have $|\det M| = 1$. Setting $\mathbf{z} = M\mathbf{y}$ then,

$$d\mathbf{z} = |\det M|\, d\mathbf{y} = d\mathbf{y},$$

and we can rewrite

$$\frac{1}{|B_R(\mathbf{x})|} \int_{B_R(\mathbf{x})} v(\mathbf{y})\, d\mathbf{y} = \frac{1}{|B_R(M\mathbf{x})|} \int_{B_R(M\mathbf{x})} u(\mathbf{z})\, d\mathbf{z} = u(M\mathbf{x}) = v(\mathbf{x}),$$

where the mean-value property of u was employed (as $|B_R(\mathbf{x})| = |B_R(M\mathbf{x})|$). So, v satisfies the mean-value property and is therefore harmonic on $\mathbb{R}^n$.

Solution 2.3.7. Suppose for the moment that the solution of problem (2.47) is unique. The value $u(0,0)$ could be computed directly from the analytic expression of u obtained from variable separation. A better way to proceed exploits the domain's symmetry, and avoids explicit computations, as follows. Let $M : \mathbb{R}^2 \to \mathbb{R}^2$ be the *clockwise* $\pi/2$-rotation. By Exercise 2.3.6 (page 125) the function

$$u_1(\mathbf{x}) = u(M\mathbf{x})$$

is harmonic on the square, it equals 1 on L_1 and 0 on the other sides. Analogously,

$$u_2(\mathbf{x}) = u(M^2\mathbf{x}),$$

is 1 along L_2 and 0 on the rest, while

$$u_3(\mathbf{x}) = u(M^3\mathbf{x})$$

is 1 on L_3 and 0 elsewhere. But then $v = u + u_1 + u_2 + u_3$ is a solution to

$$\begin{cases} \Delta v = 0 & \text{in } Q \\ v = 1 & \text{on } \partial Q, \end{cases}$$

and additionally bounded and continuous on $\overline{Q}$ without the vertices. Since we are assuming uniqueness, we immediately get $v(\mathbf{x}) \equiv 1$. At the same time $M\mathbf{0} = \mathbf{0}$, so $v(\mathbf{0}) = 4u(\mathbf{0})$, and then

$$u(0,0) = \frac{1}{4}.$$

We are left to prove uniqueness. The problem is that the Dirichlet datum is discontinuous at the corners $\mathbf{p} = (-1,0)$ and $\mathbf{q} = (0,1)$, so, *a priori*, we cannot use the maximum principle. Yet we can invoke the reflection principle as follows. Let U_1, U_2 be harmonic, continuous on $\overline{Q}$ except at the corners $\mathbf{p}, \mathbf{q}$. Then $w = U_2 - U_1$ is harmonic on Q, continuous on $\overline{Q} \setminus \{\mathbf{p}, \mathbf{q}\}$ and equal 0 on the boundary minus the two vertices. Moreover, applying Problem 2.2.10 (page 92) to the symmetric extension of w, depicted in Fig. 2.9, we know w is extendable with continuity to the corners as well. To sum up, $w \in C(\overline{Q})$ is harmonic and null on the boundary, hence null overall, and therefore $U_1 = U_2$ on Q.

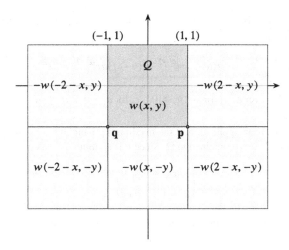

Fig. 2.9 Symmetric extension of w (Exercise 2.3.7)

Solution 2.3.8. The function u is indeed harmonic on the disc, as a computation shows, but it is unbounded, hence not continuous, at $(1,0)$, where the denominator vanishes. The maximum principle cannot be used.

Solution 2.3.9. If the maximum of u is attained at a point x_0 and $u(x_0) > 1$ we have

$$u^3(x_0) - u(x_0) > 0 \quad \text{and} \quad \Delta u(x_0) \leq 0,$$

a contradiction. Therefore $u \leq 1$; analogously we could prove that $u \geq -1$.

Solution 2.3.10. Let $\mathbf{x} = (1,0)$, $\mathbf{y} = (-1,0)$. As u is harmonic and non-negative, it is possibile to use Harnack's inequality. In this case $R = 4$ and $|\mathbf{x}| = |\mathbf{y}| = 1$. Then

$$\frac{3}{5}u(0,0) \leq u(\pm 1, 0) \leq \frac{5}{3}u(0,0),$$

from which

$$u(-1,0) \leq \frac{5}{3}u(0,0) \leq \frac{25}{9}u(1,0) = \frac{25}{9}$$

and

$$u(-1,0) \geq \frac{3}{5}u(0,0) \leq \frac{9}{25}u(1,0) = \frac{9}{25}.$$

It would have been possible to apply Harnack's inequality directly, by considering a disc centred at $(1,0)$. If so, the larger disc where u satisfies the assumptions is $B_3(1,0)$, $|\mathbf{x} - \mathbf{y}| = 2$ and Harnack tells, since $u(1,0) = 1$,

$$\frac{1}{5} \leq u(-1,0) \leq 5.$$

The estimate thus obtained, however, is worse than the previous one (no surprise, since we used the harmonicity of u only on a subset of $B_4(0,0)$).

Solution 2.3.11. Note u is C^2 on $\mathbb{R}^n \setminus \{0\}$, for any α. Since u is radial, $u(r) = r^\alpha$, we can compute the Laplacian as[9]:

$$\Delta u = u_{rr} + \frac{n-1}{r} u_r = \alpha(\alpha-1)r^{\alpha-2} + (n-1)\alpha r^{\alpha-2} = (\alpha^2 + (n-2)\alpha)r^{\alpha-2}.$$

Therefore $\Delta u \geq 0$ whenever $\alpha^2 + (n-2)\alpha \geq 0$, i.e. for $\alpha \geq 0$ or $\alpha \leq -n + 2$.

Solution 2.3.12. This is an application of the maximum principle. Let us suppose, by contradiction, that ∂E_1 is a closed curve contained in Ω. Since u is continuous and equals 1 on ∂E_1, it solves

$$\begin{cases} \Delta u = 0 & \text{in } E_1 \\ u = 1 & \text{on } \partial E_1. \end{cases}$$

By the maximum principle, $u \equiv 1$ on E_1. But by definition $\mathbf{x}_0 \in E_1$ and $u(\mathbf{x}_0) = 2$ by assumption, a contradiction.

Solution 2.3.13. The solution is unique, because B_1^+ is a Lipschitz domain and the boundary datum is continuous and so the maximum principle applies. To find it we use the reflection principle (Problem 2.2.17 on page 103) and solve

$$\begin{cases} \Delta u(x, y) = 0 & \text{in } B_1 \\ u(x, y) = y^3 & \text{on } \partial B_1. \end{cases}$$

We have chosen the datum on $\partial B_1 \cap \{y < 0\}$ in order to have an odd function in y. In polar coordinates, we have

$$\begin{cases} u_{rr} + \dfrac{1}{r}u_r + \dfrac{1}{r^2}u_{\theta\theta} = 0 & r < 1, 0 \leq \theta \leq 2\pi \\ u(1, \theta) = \sin^3 \theta = \dfrac{3}{4}\sin\theta - \dfrac{1}{4}\sin 3\theta & 0 \leq \theta \leq 2\pi. \end{cases}$$

The boundary datum is a finite trigonometric sum, thus we seek solutions of the type

$$u(r, \theta) = b_1(r)\sin\theta + b_3(r)\sin 3\theta.$$

Substituting we find $b_1(r) = \beta_1 r$ and $b_3(r) = \beta_3 r^3$, and the boundary condition produces $\beta_1 = 3/4$, $\beta_3 = -1/4$, so

$$v(r, \theta) = \frac{3}{4}r\sin\theta - \frac{1}{4}r^3\sin 3\theta.$$

But $v(r, 0) = v(r, \pi) = 0$, hence the restriction of v to $\overline{B}_1 \cap \{y \geq 0\}$ is the solution to the original problem. Going back to Cartesian coordinates, the solution reads

$$u(x, y) = \frac{1}{4}y(3 - 3x^2 + y^2).$$

---

[9] Appendix B.

• As an alternative approach notice that the datum is a degree-three polynomial. From Problem 2.2.3 (page 87) we know that a homogeneous harmonic polynomial of degree three looks like

$$P(x, y) = ax^3 + 3bx^2 y - 3axy^2 - by^3.$$

On $\partial B_1 = \{x^2 + y^2 = 1\}$

$$P(x, y) = ax\left(1 - y^2\right) + 3by\left(1 - y^2\right) - by^3 = ax\left(1 - y^2\right) + 3by - 4by^3.$$

The solution is found by fixing $a = 0, b = -1/4$, and setting

$$u(x, y) = P(x, y) + \frac{3}{4}y = -\frac{3}{4}x^2 y + \frac{1}{4}y^3 + \frac{3}{4}y$$

is precisely the solution found earlier.

Solution 2.3.14.   **a)** Let v denote the solution to

$$\begin{cases} \Delta v(x, y) = 0 & \text{in } B_1 \\ v = U & \text{on } \partial B_1 \end{cases}$$

which we know exists, is unique and is given by the Poisson formula. The function $v(x, -y)$ is harmonic on B_1 and agrees with v on the boundary, so

$$v(x, y) = v(x, -y),$$

implying

$$v_y(x, y) = -v_y(x, -y)$$

and in particular

$$v_y(x, 0) = -v_y(x, 0).$$

Therefore $v_y(x, 0) = 0$, making v the solution on B_1^+ to the same mixed problem that u solves. Then $u = v$ on B_1^+, and because it is even in y, it coincides with U on B_1. In conclusion, U is harmonic on B_1.

b) Define

$$U(x, y) = \begin{cases} u(x, y) & y \geq 0 \\ u(x, -y) & y < 0 \end{cases}.$$

Then U is harmonic on B_1 and coincides with u on $\overline{B_1^+}$. In polar coordinates:

$$u(r, \theta) = \frac{1 - r^2}{2\pi} \int_{-\pi}^{\pi} \frac{\cos^2 \varphi}{r^2 + 1 - 2r \cos(\varphi - \theta)} \, d\varphi$$

and

$$u(0, \theta) = U(0, \theta) = \frac{1}{2\pi} \int_{-\pi}^{\pi} \cos^2 \varphi \, d\varphi = \frac{1}{2}.$$

Solution 2.3.15.   Let u and v be bounded solutions of (2.48) in $C^1(\overline{P^+})$, and set $w = u - v$. Extend w evenly on $y < 0$. As $w = 0$ on $y = 0$, the new function is bounded and harmonic on $\mathbb{R}^2$ by the reflection principle. Liouville's theorem ensures w is constant, and so u, v differ by a constant.

To find a solution explicitly we may proceed in two ways.

1st method. We use the Fourier transform. Set

$$\widehat{u}(\xi, y) = \int_{\mathbb{R}} e^{-i\xi x} u(x, y)\, dx.$$

Then

$$\widehat{u_x}(\xi, y) = i\xi \widehat{u}(\xi, y), \quad \widehat{u_{xx}}(\xi, y) = -\xi^2 \widehat{u}(\xi, y)$$

and $\widehat{u}$ solves

$$\begin{cases} \widehat{u}_{yy}(\xi, y) - \xi^2 \widehat{u}(\xi, y) = 0, & y > 0 \\ \widehat{u}_y(\xi, 0) = \widehat{g}(\xi). \end{cases}$$

The general integral of the ODE is

$$\widehat{u}(\xi, y) = c_1(\xi) e^{|\xi| y} + c_2(\xi) e^{-|\xi| y}.$$

The boundedness[10] of u forces $c_1(\xi) = 0$, while the constraint on $y = 0$ requires

$$-c_2(\xi) = \widehat{g}(\xi)$$

whence

$$\widehat{u}(\xi, y) = -\widehat{g}(\xi)\, \frac{e^{-|\xi| y}}{|\xi|}. \tag{2.52}$$

Now, $|\widehat{g}|$ is integrable on $\mathbb{R}$ and has null average on (a, b), so

$$\widehat{g}(0) = \int_a^b g(x)\, dx = 0.$$

Moreover g is smooth, so when $\xi \to 0$

$$\widehat{g}(\xi) \sim \widehat{g}'(0)\, \xi.$$

Formula (2.52) defines

$$u(x, y) = -\frac{1}{2\pi} \int_{\mathbb{R}} \widehat{g}(\xi)\, \frac{e^{-|\xi| y}}{|\xi|} e^{i\xi x}\, d\xi,$$

as a bounded function for $y \geq 0$:

$$|u(x, y)| \leq \frac{1}{2\pi} \int_{\mathbb{R}} \frac{|\widehat{g}(\xi)|}{|\xi|}\, d\xi < \infty.$$

We will show below that the anti-transform of $\dfrac{e^{-|\xi| y}}{|\xi|}$ is

$$-\frac{1}{2\pi} \log\left(x^2 + y^2\right).$$

---

[10] $e^{|\xi| y}$ has no inverse Fourier transform.

Given this, the Neumann problem is solved by

$$u(x, y) = \frac{1}{2\pi} \int_{\mathbb{R}} \log[(x - z)^2 + y^2] \, g(z) \, dz.$$

Remark. To compute the anti-transform of $\frac{e^{-|\xi|y}}{|\xi|}$ we observe the following:

1. $e^{-|\xi|y}$ is the transform (with respect to x) of[11]

$$\frac{1}{\pi} \frac{y}{x^2 + y^2}.$$

2. $xu(x)$ transforms to

$$i \frac{d}{d\xi} \widehat{u}(\xi)$$

and the transform (in x) of

$$\frac{1}{\pi} \frac{2xy}{x^2 + y^2}$$

is

$$2i \frac{d}{d\xi} e^{-|\xi|y} = -2iy \, \text{sign}(\xi) \, e^{-|\xi|y}.$$

3. If $h(x, y) = \log(x^2 + y^2)$,

$$\frac{y}{\pi} h_x(x, y) = \frac{1}{\pi} \frac{2xy}{x^2 + y^2}$$

and

$$\widehat{h_x} = i\xi \widehat{h}.$$

Putting all this together yields

$$\frac{i\xi y}{\pi} \widehat{h}(\xi) = -2iy \, \text{sign}(\xi) \, e^{-|\xi|y}$$

i.e.

$$\frac{e^{-|\xi|y}}{|\xi|} = -\frac{1}{2\pi} \widehat{h}(\xi, y).$$

We deduce that the anti-transform of $\frac{e^{-|\xi|y}}{|\xi|}$ must be

$$-\frac{1}{2\pi} \log(x^2 + y^2).$$

2nd method. Let v be the harmonic conjugate to u, so that

$$f(z) = u(x, y) + iv(x, y), \qquad (z = x + iy)$$

is an analytic function on $\mathbb{C}$. The Cauchy-Riemann equations $u_x = v_y$, $u_y = -v_x$ imply that v_x is harmonic and $v_x(x, 0) = -g(x)$. Now, the formula for the solution of the Dirichlet problem (see

---

[11] Appendix B.

Problem 2.2.18 on page 104) gives

$$v_x(x, y) = \frac{1}{\pi} \int_{\mathbb{R}} \frac{y}{x^2 + y^2} g(x - s)\, ds = -u_y(x, y).$$

A primitive (in y) of $\frac{y}{x^2+y^2}$ is given by $\frac{1}{2}\log(x^2 + y^2)$, and therefore

$$u(x, y) = -\frac{1}{2\pi} \int_{\mathbb{R}} \log(x^2 + y^2) g(x - s)\, ds,$$

which is the unique solution up to additive constant.

Solution 2.3.16. We seek solutions with separated variables:

$$u_n = w(r) v(\theta).$$

Then v solves the eigenvalue problem

$$v''(\theta) + \lambda v(\theta) = 0,\ v(0) = 0,\ v(\alpha) = 0.$$

We get $\lambda = n^2\pi^2/\alpha^2$, $n > 0$, with eigenfunctions

$$v_n(\theta) = \sin\left(\frac{n\pi}{\alpha}\theta\right).$$

For w, instead, we have

$$r^2 w''(r) + r w'(r) - \frac{n^2\pi^2}{\alpha^2} w(r) = 0$$

with w bounded, and then

$$w_n(r) = c r^{n\pi/\alpha}.$$

To comply with the Dirichlet datum on $r = 1, 0 < \theta < \alpha$, we take linear combinations

$$u(r, \theta) = \sum_{n=1}^{+\infty} c_n r^{n\pi/\alpha} \sin\left(\frac{n\pi}{\alpha}\theta\right).$$

Necessarily

$$u(1, \theta) = \sum_{n=1}^{+\infty} c_n \sin\left(\frac{n\pi}{\alpha}\theta\right) = g(\theta)$$

and therefore $c_n = g_n$, where

$$g_n = \frac{2}{\alpha} \int_0^\alpha g(\theta) \sin\left(\frac{n\pi}{\alpha}\theta\right) d\theta$$

are the coefficients of the sine expansion of g.

b) We choose the special datum

$$g(\theta) = \sin\left(\frac{\pi}{\alpha}\theta\right).$$

The corresponding solution

$$u(r, \theta) = r^{\pi/\alpha} \sin\left(\frac{\pi}{\alpha}\theta\right)$$

is positive on S_α. Note

$$u_r(r, \theta) = \frac{\pi}{\alpha} r^{(\pi-\alpha)/\alpha} \sin\left(\frac{\pi}{\alpha}\theta\right),$$

$$u_\theta(r, \theta) = \frac{\pi}{\alpha} r^{\pi/\alpha} \cos\left(\frac{\pi}{\alpha}\theta\right),$$

and so

$$|\nabla u|^2 = (u_r)^2 + \frac{1}{r^2}(u_\theta)^2 = \frac{\pi^2}{\alpha^2} r^{2(\pi-\alpha)/\alpha}.$$

Hence we see that

$$\lim_{r\to 0} |\nabla u|^2 = \begin{cases} 0 & \text{for } \alpha < \pi \\ 1 & \text{for } \alpha = \pi \\ +\infty & \text{for } \alpha > \pi. \end{cases}$$

As a consequence, for $\alpha < \pi$ the gradient of u at the origin is zero, and hence no version of Hopf's principle holds (irrespective of the orientation given to the outward normal at that point; as a matter of fact, this domain contains no ball in the sense of Problem 2.2.9 on page 91). Note, furthermore, that for $\alpha > \pi$, u is an example of a harmonic function, defined on an open set, that is continuous but not C^2 on the closure (it is not even differentiable at (0,0)!).

Solution 2.3.17. The Kelvin transform (Problem 2.2.20 on page 107)

$$\mathbf{y} = T_1(\mathbf{x}) = \mathbf{x}/|\mathbf{x}|^2$$

gives $T_1(\Omega) = B_1 \setminus \{0\}$. Set $v(\mathbf{y}) = u(\mathbf{y}/|\mathbf{y}|^2)$. Then v is harmonic and bounded on $B_1 \setminus \{0\}$ (u is bounded). By Problem 2.2.10 (page 92) v admits a harmonic extension to the entire ball B_1, which means that

$$\begin{cases} \Delta v = 0 & \text{in } B_1 \\ v = g & \text{on } B_1. \end{cases}$$

By Poisson's formula we have

$$v(\mathbf{y}) = \frac{1 - |\mathbf{y}|^2}{2\pi} \int_{\partial B_1} \frac{g(\sigma)}{|\mathbf{y} - \sigma|^2} \, d\sigma.$$

Anti-transforming,

$$\mathbf{y} = \frac{\mathbf{x}}{|\mathbf{x}|^2}, \quad u(\mathbf{x}) = \frac{1}{|\mathbf{x}|} v\left(\frac{\mathbf{x}}{|\mathbf{x}|^2}\right).$$

Since

$$|\mathbf{x}| \left| \frac{\mathbf{x}}{|\mathbf{x}|^2} - \sigma \right| = |\mathbf{x} - \sigma|,$$

we finally get

$$u(\mathbf{x}) = \frac{|\mathbf{x}|^2 - 1}{2\pi} \int_{\partial B_1} \frac{g(\sigma)}{|\mathbf{x} - \sigma|^2} \, d\sigma.$$

Solution 2.3.18. The function u belongs to $C^2(B_1) \cap C(\overline{B_1})$ and is harmonic, thus by the maximum principle it attains extreme values on ∂B_1. As $u|_{\partial B_1}$ is known, what we have is a constrained optimisation problem. The constraint

$$g(x, y, z) = x^2 + y^2 + z^2 = 1$$

is regular (the gradient ∇g has no zeroes), so we can use Lagrange multipliers and find the stationary points of

$$\Phi(x, y, z, \lambda) = x^4 + y^4 + z^4 - 2\lambda(x^2 + y^2 + z^2 - 1).$$

The system to be solved is

$$\begin{cases} x(x^2 - \lambda) = 0 \\ y(y^2 - \lambda) = 0 \\ z(z^2 - \lambda) = 0 \\ x^2 + y^2 + z^2 = 1, \end{cases}$$

and it is algebraic. For symmetry reasons we can consider only the solutions

$$\left(\frac{\sqrt{2}}{2}, \frac{\sqrt{2}}{2}, 0\right), \qquad \left(\frac{\sqrt{3}}{3}, \frac{\sqrt{3}}{3}, \frac{\sqrt{3}}{3}\right), \qquad \text{and} \quad (1, 0, 0),$$

corresponding to the critical values $1/2$, $1/3$, 1. The maximum of u is therefore 1, the minimum $1/3$.

Solution 2.3.19. The reflection principle in $\mathbb{R}^3$ can be formulated in the following way:
Define

$$B_1^+ = \left\{(x, y, z) \in \mathbb{R}^3 : x^2 + y^2 + z^2 < 1, z > 0\right\}$$

and let $u \in C^2(B_1^+) \cap C(\overline{B_1^+})$ be harmonic on B_1^+ and such that $u(x, y, 0) = 0$. Then the function

$$U(x, y, z) = \begin{cases} u(x, y, z) & z \geq 0 \\ -u(x, y, -z) & z < 0, \end{cases}$$

generated by an odd reflection in the variable z, is harmonic on the whole ball B_1.
  The proof is exactly the same as in Problem 2.2.17 (page 103), so we will recall only the major steps. Call v the solution to

$$\begin{cases} \Delta v(x, y, z) = 0 & \text{in } B_1 \\ v = U & \text{on } \partial B_1 \end{cases}$$

and set

$$w(x, y, z) = v(x, y, z) + v(x, y, -z).$$

Then w is harmonic on B_1 and vanishes on the boundary, and by uniqueness $w \equiv 0$. Hence $v(x, y, z) = -v(x, y, -z)$, and $v(x, y, 0) = 0$. Consequently v and u solve the same Dirichlet problem on B_1^+, so again $v \equiv u \equiv U$ on B_1^+. Now both v and U are odd with respect to z, whence $v \equiv U$ on B_1. In particular U is harmonic on B_1.
  If the Dirichlet problem on $z > 0$ had two bounded solutions u and v, the difference $w = u - v$ would vanish on the plane $z = 0$. Extending w to $z > 0$ in an odd way would produce a *bounded* harmonic function in $\mathbb{R}^3$. By Liouville's theorem w would be constant, hence zero (for it would vanish on $z = 0$).

Solution 2.3.20. We shall follow the ideas used in Problem 2.2.10 (page 92), and provide the proof details of the proof for the case $n \geq 3$, leaving it to the reader to address dimension two. Call v the solution (smooth and bounded) of

$$\begin{cases} \Delta v = 0 & \text{in } B \\ v = u & \text{on } \partial B \end{cases}$$

and set $w = u - v$. Then w is harmonic on $B \setminus \{0\}$, it vanishes on ∂B and

$$|\mathbf{x}|^{n-2} w(\mathbf{x}) \to 0 \quad \text{for } |\mathbf{x}| \to 0. \tag{2.53}$$

If we show that $w(\mathbf{x}) = 0$ for any $\mathbf{x} \neq \mathbf{0}$ the claim is immediate. So let $0 < r < 1$ and

$$M_r = \max_{|\mathbf{x}|=r} |w(\mathbf{x})|.$$

Then (2.53) reads

$$\lim_{r \to 0+} r^{n-2} M_r = 0. \tag{2.54}$$

Call h the solution to

$$\begin{cases} \Delta h = 0 & \text{in } B \setminus \overline{B_r} \\ h = 0 & \text{on } \partial B \\ h = M_r & \text{on } \partial B_r. \end{cases}$$

It is not hard to check that

$$h(\mathbf{x}) = M_r \frac{-|\mathbf{x}|^3 + 2|\mathbf{x}|^2 - 2}{r^{2-n} - 1} = \frac{|\mathbf{x}|^{2-n} - 1}{1 - r^{n-2}} r^{n-2} M_r.$$

As $w = 0$ on ∂B, the maximum principle implies

$$-h(\mathbf{x}) \leq w(\mathbf{x}) \leq h(\mathbf{x})$$

on $\overline{B} \setminus B_r$. Fix $\mathbf{x} \neq \mathbf{0}$. The previous inequality holds for any $r \leq |\mathbf{x}|$. Let r tend to 0 and exploit (2.54), so that

$$|w(\mathbf{x})| \leq \frac{|\mathbf{x}|^{2-n} - 1}{1 - r^{n-2}} (r^{n-2} M_r) \to 0 \quad \text{for } r \to 0,$$

i.e. $w(\mathbf{x}) = 0$.

Solution 2.3.21. **a)** For example, $u(\mathbf{x}) = |\mathbf{x}|^2$ satisfies the requirements.

b) We saw in Problem 2.2.5 (page 89) that such a function cannot exist in dimension two, because the fundamental solution in $\mathbb{R}^2$ is *unbounded* at infinity. This fact suggests using the fundamental solution outside a bounded set to construct an example. An example is the C^2 function

$$u(\mathbf{x}) = \begin{cases} -|\mathbf{x}|^3 + 2|\mathbf{x}|^2 - 2 & |\mathbf{x}| \leq 1 \\ -|\mathbf{x}|^{-1} & |\mathbf{x}| \geq 1. \end{cases}$$

Another example is given by

$$u(\mathbf{x}) = \begin{cases} -1 & |\mathbf{x}| \leq 1 \\ -|\mathbf{x}|^{-1} & |\mathbf{x}| \geq 1, \end{cases}$$

which coincides with $\max\{-|\mathbf{x}|^{-1}, -1\}$.

Solution 2.3.22. We can use Problem 2.2.7 (page 90): in fact, if we set $v_i = u_{i+1} - u_i$, then v_i is harmonic and non-negative on Ω, and the convergence of the sequence reduces to that of the (telescopic) series $\sum_{i=0}^{\infty} v_i$.

Solution 2.3.23. Checking the smoothness of T_a and T_a^{-1} goes exactly as in the two-dimensional case (Problem 2.2.20 on page 107), and so we leave it to the reader. To verify that

$$v(\mathbf{x}) = \frac{a}{|\mathbf{x}|} u\left(T_a(\mathbf{x})\right) = \frac{a}{|\mathbf{x}|} u\left(\frac{a^2}{|\mathbf{x}|^2}\mathbf{x}\right)$$

is harmonic, it is convenient to pass to spherical coordinates. Write $\mathbf{x} = (r, \theta, \psi)$, $0 \le \theta \le 2\pi$, $0 \le \psi \le \pi$, so $T_a(\mathbf{x}) = (\rho, \theta, \psi)$ with $r\rho = a^2$. Recall that

$$\Delta v = v_{rr} + \frac{2}{r}v_r + \frac{1}{r^2}\left[\frac{1}{\sin^2\psi}v_{\theta\theta} + v_{\psi\psi} + \frac{\cos\psi}{\sin\psi}v_\psi\right] \equiv v_{rr} + \frac{2}{r}v_r + \frac{1}{r^2}\Delta_S v.$$

Since $\mathbf{x}$ and $T_a(\mathbf{x})$ are collinear vectors, the Kelvin transform leaves the 'spherical part' Δ_S of the operator Δ invariant (Δ_S is called *Laplace-Beltrami operator*). For $u = u(\rho, \theta, \psi)$ we have

$$v(r, \theta, \psi) = \frac{a}{r}u\left(\frac{a^2}{r}, \theta, \psi\right)$$

$$v_r(r, \theta, \psi) = -\frac{a}{r^2}u\left(\frac{a^2}{r}, \theta, \psi\right) - \frac{a^3}{r^3}u_\rho\left(\frac{a^2}{r}, \theta, \psi\right)$$

$$v_{rr}(r, \theta, \psi) = \frac{2a}{r^3}u\left(\frac{a^2}{r}, \theta, \psi\right) + \frac{4a^3}{r^4}u_\rho\left(\frac{a^2}{r}, \theta, \psi\right) + \frac{a^5}{r^5}u_{\rho\rho}\left(\frac{a^2}{r}, \theta, \psi\right)$$

$$\Delta_S v(r, \theta, \psi) = \frac{a}{r}\Delta_S u\left(\frac{a^2}{r}, \theta, \psi\right).$$

Substituting

$$\begin{aligned}
\Delta v &= \frac{a^5}{r^5}u_{\rho\rho}\left(\frac{a^2}{r}, \theta, \psi\right) + \frac{2a^3}{r^4}u_\rho\left(\frac{a^2}{r}, \theta, \psi\right) + \frac{a}{r^3}\Delta_S u\left(\frac{a^2}{r}, \theta, \psi\right) \\
&= \frac{\rho^5}{a^5}\left(u_{\rho\rho}(\rho, \theta, \psi) + \frac{2}{\rho}u_\rho(\rho, \theta, \psi) + \frac{1}{\rho^2}\Delta_S u(\rho, \theta, \psi)\right) = \\
&= \frac{\rho^5}{a^5}\Delta u.
\end{aligned}$$

In conclusion, if u is harmonic on its domain, so is v.

Solution 2.3.24. Since $u \to A$ at infinity, the problem has a unique solution. First note that

$$\widetilde{u}(|\mathbf{x}|) = a\left(1 - \frac{1}{|\mathbf{x}|}\right)$$

is harmonic on B_e, becomes 0 when $|\mathbf{x}| = 1$ and $\widetilde{u}(|\mathbf{x}|) \to A$ for $|\mathbf{x}| \to \infty$. Set $w = u - \widetilde{u}$, so $w \to 0$ at infinity. The Kelvin transform T_1 maps B_e bijectively onto the unit ball minus the origin. Moreover, every point of ∂B_e is fixed. Setting

$$v(\mathbf{y}) = \frac{1}{|\mathbf{y}|}w\left(\frac{\mathbf{y}}{|\mathbf{y}|^2}\right),$$

we obtain that v is harmonic on the punctured ball. Since $w(|\mathbf{x}|) \to 0$ at infinity, we have

$$|\mathbf{y}|\, v(\mathbf{y}) \to 0 \quad \text{as } |\mathbf{y}| \to 0,$$

and then Exercise 2.3.18 implies that v extends harmonically to the whole ball B_1, with datum g on ∂B_1 (we still call v the extension). By Poisson's formula

$$v(\mathbf{y}) = \frac{1 - |\mathbf{y}|^2}{4\pi} \int_{\partial B_1} \frac{g(\boldsymbol{\sigma})}{|\mathbf{y} - \boldsymbol{\sigma}|^3} \, d\sigma.$$

Transforming back gives

$$\mathbf{y} = \frac{\mathbf{x}}{|\mathbf{x}|^2}, \qquad w(\mathbf{x}) = \frac{1}{|\mathbf{x}|} v\left(\frac{\mathbf{x}}{|\mathbf{x}|^2}\right).$$

Since

$$|\mathbf{x}| \left| \frac{\mathbf{x}}{|\mathbf{x}|^2} - \boldsymbol{\sigma} \right| = |\mathbf{x} - \boldsymbol{\sigma}|,$$

we then obtain

$$w(\mathbf{x}) = \frac{|\mathbf{x}|^2 - 1}{4\pi} \int_{\partial B_1} \frac{g(\boldsymbol{\sigma})}{|\mathbf{x} - \boldsymbol{\sigma}|^3} \, d\sigma.$$

In conclusion

$$u(|\mathbf{x}|) = a\left(1 - \frac{1}{|\mathbf{x}|}\right) + \frac{|\mathbf{x}|^2 - 1}{4\pi} \int_{\partial B_1} \frac{g(\boldsymbol{\sigma})}{|\mathbf{x} - \boldsymbol{\sigma}|^3} \, d\sigma.$$

In case $g(\mathbf{x}) = x_1$, $v(\mathbf{y}) = y_1$ is the harmonic extension of v to $\{0 \le |\mathbf{y}| < 1\}$. The anti-transformation produces

$$u(\mathbf{x}) = a\left(1 - \frac{1}{|\mathbf{x}|}\right) + \frac{x_1}{|\mathbf{x}|^3}.$$

Solution 2.3.25. One can proceed exactly as in Problem 2.2.24 (page 115). Let u_1, u_2 be two solutions. Then $w = u_1 - u_2$ is harmonic on Ω_e and has zero Robin datum on $\partial\Omega_e$. As in 2.2.24.a),

$$\left| \frac{\partial w}{\partial x_i}(\mathbf{x}) \right| \le \frac{C}{|\mathbf{x}|^2}$$

for $|\mathbf{x}|$ large enough; in fact, the estimate does not depend on the condition on $\partial\Omega_e$. We multiply $\Delta w = 0$ by w, integrate on $\Omega_e \cap B_R$ by parts to obtain

$$0 = \int_{\Omega_e \cap B_R} w(\mathbf{x}) \Delta w(\mathbf{x}) \, d\mathbf{x} = \int_{\Omega_e \cap B_R} |\nabla w(\mathbf{x})|^2 \, d\mathbf{x} - \int_{\partial(\Omega_e \cap B_R)} w(\mathbf{x}) \partial_{\boldsymbol{\nu}} w(\mathbf{x}) \, d\sigma$$

just like in 2.2.24.b). Since w is bounded, i.e. $|w| \le M$, we infer

$$\int_{\Omega_e \cap B_R} |\nabla w(\mathbf{x})|^2 \, d\mathbf{x} = \int_{\partial\Omega_e} w(\mathbf{x}) \partial_{\boldsymbol{\nu}} w(\mathbf{x}) \, d\sigma + \int_{\partial B_R} w(\mathbf{x}) \partial_{\boldsymbol{\nu}} w(\mathbf{x}) \, d\sigma \le$$

$$\le -\int_{\partial\Omega_e} \alpha w^2(\mathbf{x}) \, d\sigma + M \int_{\partial B_R} \left| \nabla w(\mathbf{x}) \cdot \frac{\mathbf{x}}{|\mathbf{x}|} \right| \, d\sigma \le$$

$$[\text{as } \alpha \ge 0] \le M \left(2\pi R \int_{\partial B_R} |\nabla w(\mathbf{x})|^2 \, d\sigma \right)^{1/2} \le$$

$$\le M \left(2\pi R \int_{\partial B_R} \frac{C}{|\mathbf{x}|^4} \, d\sigma \right)^{1/2} = C_0 R^{-3/2}.$$

Taking the limit for $R \to +\infty$, we find:

$$\int_{\Omega_e} |\nabla w(\mathbf{x})|^2 \, d\mathbf{x} = 0.$$

As w was assumed C^1 up to the boundary, $\nabla w(\mathbf{x}) = 0$ on $\overline{\Omega}_e$, i.e. w is constant. Since $\partial_\nu w + \alpha w = 0$ on $\partial \Omega_e$, it follows that the constant is zero.

Solution 2.3.26. Take $R > 0$, so that $\Omega \subset B_R(0)$. Integrate the equation over $\Omega_e \cap B_R(0)$ to get

$$\int_{\partial \Omega} g \, d\sigma = \int_{\{|\mathbf{x}|=R\}} \partial_\nu u \, d\sigma.$$

Suppose R is large, so that $B_{R/2}(\mathbf{x}) \subset \Omega_e$ for any $|\mathbf{x}| = R$. Problem 2.2.1 a) (page 84) gives us the estimate

$$|\nabla u(\mathbf{x})| \le \sqrt{3} \frac{2n}{R} \max_{\partial B_{R/2}(\mathbf{x})} |u| \le \sqrt{3} \frac{2n}{R} \max_{R/2 \le |\mathbf{y}| \le 3R/2} |u|.$$

Using the decay estimate for u, we can write

$$\left| \int_{\partial \Omega} g \, d\sigma \right| \le \int_{\{|\mathbf{x}|=R\}} |\nabla u| \, d\sigma \le 4\pi R^2 \cdot \sqrt{3} \frac{2n}{R} \frac{M \cdot 2^{1+\varepsilon}}{R^{1+\varepsilon}} \to 0 \qquad \text{for } R \to +\infty.$$

Solution 2.3.27. Observe that

$$v(\mathbf{x}) = |\nabla u(\mathbf{x})|^2$$

is subharmonic, as sum of squares of harmonic functions (Problem 2.2.4 on page 88). Then, for any $\mathbf{x} \in \mathbb{R}^n$,

$$v(\mathbf{x}) \le \frac{1}{|B_R(\mathbf{x})|} \int_{B_R(\mathbf{x})} v(\mathbf{y}) \, d\mathbf{y},$$

i.e., using (2.50),

$$|\nabla u(\mathbf{x})|^2 \le \frac{1}{\omega_n R^n} \int_{B_R(\mathbf{x})} |\nabla u(\mathbf{y})|^2 d\mathbf{y} \to 0 \qquad \text{for } R \to +\infty.$$

Therefore $\nabla u(\mathbf{x}) \equiv 0$ in $\mathbb{R}^n$, and the claim follows.

Solution 2.3.28. Observe first that

$$u(x, y) = v(x, y) + (x^2 + y^2)/2$$

solves $\Delta u = 0$. Hence u is in $C^\infty(\Omega)$, whence also

$$v = u - (x^2 + y^2)/2$$

is in $C^\infty(\Omega)$. The derivatives v_x, v_y are continuous on $\overline{\Omega}$, and harmonic, so

$$|\nabla v|^2 = v_x^2 + v_y^2$$

is subharmonic, being a sum of squares of harmonic functions. It follows that $|\nabla v|^2$ is maximised on $\partial \Omega$.

Solution 2.3.29. We present the solution for $n = 2$; the proof for $n > 2$ is analogous. Let v be the harmonic extension of u to $B_1(0)$, that is $\Delta v = 0$ in B_1, $v = u$ on ∂B_1. Since u is continuous on ∂B_1, then v is bounded. Then $w(\mathbf{x}) = u(\mathbf{x}) - v(\mathbf{x}) - \log|\mathbf{x}|$ is harmonic on $B_1(0) \setminus (0)$, $w = 0$ on $\partial B_1(0)$, and it is positive by the maximum principle, since $w \to +\infty$ as $|\mathbf{x}| \to 0$.

Therefore w satisfies the hypotheses in Problem 2.2.12 (page 94), and

$$w(\mathbf{x}) = u(\mathbf{x}) - v(\mathbf{x}) - \log|\mathbf{x}| = C \log|\mathbf{x}|,$$

or

$$u(\mathbf{x}) = (C+1)\log|\mathbf{x}| + v(\mathbf{x}).$$

Solution 2.3.30. The relationship between the Newtonian potential in the plane and the generating density is

$$\Delta u = u_{rr} + \frac{1}{r}u_r = -\mu.$$

Since $\Delta u = -2\pi r^2$, we deduce

$$\mu(r) = 2\pi r^2.$$

The potential (defined up to additive constants) is C^1 in $\mathbb{R}^2$, harmonic for $r > 1$ and radially symmetric, so it must be of the form

$$a \log r + b.$$

Furthermore,

$$u(1-) = u(1+), \qquad u_r(1-) = u_r(1+). \tag{2.55}$$

Then

$$b = 0 \quad \text{and} \quad a = -\frac{\pi}{2}$$

and the potential for $r > 1$ is $u(r) = -\frac{\pi}{2}\log r$.

Solution 2.3.31. Up to additive constants the potential is

$$u(x,y,z) = \frac{1}{4\pi}\int_{B_1} \frac{1}{\left[(x-\xi)^2 + (y-\eta)^2 + (z-\zeta)^2\right]^{1/2}}\, d\xi\, d\eta\, d\zeta.$$

This is C^1 on $\mathbb{R}^3$. For an explicit expression note u is radially symmetric, $u = u(r)$, $r^2 = x^2 + y^2 + z^2$, (the problem is invariant under rotations) and solves

$$\Delta u = \begin{cases} -1 & 0 \le r < 1 \\ 0 & r > 1 \end{cases}$$

with

$$u(1-) = u(1+), \qquad u_r(1-) = u_r(1+). \tag{2.56}$$

Radial harmonic functions in $\mathbb{R}^3$ have the form

$$\frac{a}{r} + b,$$

while radial solutions of $\Delta u = -1$ satisfy

$$u_{rr} + \frac{2}{r}u_r = -4\pi$$

which have general integral

$$v(r) = c_1 + \frac{c_2}{r} - \frac{1}{6}r^2.$$

But u is bounded for $r < 1$, so $c_2 = 0$. From (2.56)

$$c_1 - \frac{1}{6} = a + b \qquad \text{and} \qquad -a = -\frac{1}{3}.$$

Choosing $b = 0$ (zero potential at infinity), we have

$$u(r) = \begin{cases} \frac{1}{2}\left(1 - \frac{r^2}{3}\right) & r \le 1 \\ \frac{1}{3r} & r > 1. \end{cases}$$

Solution 2.3.32. The relationship between u and μ is given, at any point $\mathbf{z} = (z_1, z_2)$ on the circle, by the formula

$$\frac{\partial u^I}{\partial \rho}(\mathbf{z}) - \frac{\partial u^E}{\partial \rho}(\mathbf{z}) = \mu(\mathbf{z})$$

where (the limits are taken radially)

$$\frac{\partial u^E}{\partial \rho}(\mathbf{z}) = \lim_{\mathbf{x}\to\mathbf{z}, \rho>1} \frac{\partial u^E}{\partial \rho}(\mathbf{x}), \quad \frac{\partial u^I}{\partial \rho}(\mathbf{z}) = \lim_{\mathbf{x}\to\mathbf{z}, \rho<1} \frac{\partial u^I}{\partial \rho}(\mathbf{x}).$$

We know that $\frac{\partial u^E}{\partial \rho}(\mathbf{z}) = -z_1 - 2z_1z_2$. To compute $\frac{\partial u^I}{\partial \rho}$ we first use the continuity of u across C. We find

$$u^E(\mathbf{z}) = u^I(\mathbf{z}) = z_1 + z_1z_2. \tag{2.57}$$

We deduce that u is harmonic inside C and satisfies the Dirichlet condition (2.57). Therefore inside C we find $u(\mathbf{x}) = x_1 + x_1x_2$, and so

$$\frac{\partial u^I}{\partial \rho}(\mathbf{z}) = z_1 + 2z_1z_2.$$

Finally,

$$\mu(\mathbf{z}) = \frac{\partial u^I}{\partial \rho}(\mathbf{z}) - \frac{\partial u^E}{\partial \rho}(\mathbf{z}) = 2z_1 + 4z_1z_2.$$

Solution 2.3.33. The answer is no. In fact, let

$$u(\mathbf{x}) = \int_{\partial\Omega} \frac{\langle \mathbf{x} - \boldsymbol{\sigma}, \boldsymbol{\nu}_\sigma \rangle}{|\mathbf{x} - \boldsymbol{\sigma}|^3} \mu(\boldsymbol{\sigma})\, d\sigma \qquad \mathbf{x} = (x_1, x_2)$$

with $|\mu(\boldsymbol{\sigma})| \le M$. Assume $\Omega \subset B_R(0)$. Then $|\boldsymbol{\sigma}| \le R$ for $\boldsymbol{\sigma} \in \partial\Omega$ and, if $|\mathbf{x}| \ge 2R$, we have

$$|\mathbf{x} - \boldsymbol{\sigma}| \ge |\mathbf{x}| - |\boldsymbol{\sigma}| \ge \frac{|\mathbf{x}|}{2}.$$

Using $|\langle \mathbf{x} - \boldsymbol{\sigma}, \boldsymbol{\nu}_\sigma \rangle| \le |\mathbf{x} - \boldsymbol{\sigma}|$, we infer

$$|u(\mathbf{x})| \le \int_{\partial\Omega} \left| \frac{\langle \mathbf{x} - \boldsymbol{\sigma}, \boldsymbol{\nu}_\sigma \rangle}{|\mathbf{x} - \boldsymbol{\sigma}|^3} \mu(\boldsymbol{\sigma}) \right| d\sigma \le \int_{\partial\Omega} \frac{\mu(\boldsymbol{\sigma})}{|\mathbf{x} - \boldsymbol{\sigma}|^2} d\sigma \le \frac{4M\,|\partial\Omega|}{|\mathbf{x}|^2}.$$

In particular, we should have

$$|u(x_1, 0)| = \frac{1}{|x_1|} \le \frac{4M \, |\partial\Omega|}{|x_1|^2}$$

which is a contradiction for x_1 large.

Solution 2.3.34. **a)** The Green function $G = G(\mathbf{x}, \mathbf{y})$ is harmonic on the half-space, $G(\mathbf{x}, \mathbf{y}) = 0$ on $x_3 = 0$ for any given $\mathbf{y} \in P^+$, and

$$\Delta_{\mathbf{x}} G(\mathbf{x}, \mathbf{y}) = -\delta_3(\mathbf{x} - \mathbf{y}) \qquad (2.58)$$

where $\delta_3(\mathbf{x} - \mathbf{y})$ is the Dirac distribution at $\mathbf{y}$. Let us use the *methods of images*. The fundamental solution

$$\Gamma(\mathbf{x} - \mathbf{y}) = -\frac{1}{4\pi \, |\mathbf{x} - \mathbf{y}|}$$

satisfies (2.39). Set $\widetilde{\mathbf{y}} = (y_1, y_2, -y_3)$ to be the mirror image of $\mathbf{y} = (y_1, y_2, y_3)$ with respect to the plane $y_3 = 0$. The function $\Gamma(\mathbf{x} - \widetilde{\mathbf{y}})$ is harmonic on P^+ and coincides with $\Gamma(\mathbf{x} - \mathbf{y})$ on $x_3 = 0$. Hence

$$G(\mathbf{x}, \mathbf{y}) = \Gamma(\mathbf{x} - \mathbf{y}) - \Gamma(\mathbf{x} - \widetilde{\mathbf{y}}).$$

b) We invoke the method of images once more. This time let, for $\mathbf{y} \ne \mathbf{0}$,

$$\mathbf{y}^* = T_1(\mathbf{y}) = \frac{\mathbf{y}}{|\mathbf{y}|^2}$$

be the mirror of $\mathbf{y}$ under the Kelvin transform. As in the case $n = 2$, for $|\mathbf{x}| = 1$ we have

$$|\mathbf{x} - \mathbf{y}^*|^2 = 1 - \frac{2\mathbf{x} \cdot \mathbf{y}}{|\mathbf{y}|^2} + \frac{1}{|\mathbf{y}|^2} = \frac{1}{|\mathbf{y}|^2}\left(1 - 2\mathbf{x} \cdot \mathbf{y} + |\mathbf{y}|^2\right) = \frac{1}{|\mathbf{y}|^2}|\mathbf{x} - \mathbf{y}|^2.$$

If $\mathbf{y} \ne \mathbf{0}$ define

$$G(\mathbf{x}, \mathbf{y}) = \frac{1}{4\pi}\left\{\frac{1}{|\mathbf{x} - \mathbf{y}|} - \frac{1}{|\mathbf{y}| \, |\mathbf{x} - \mathbf{y}^*|}\right\}.$$

Then $G(\mathbf{x}, \mathbf{y}) = 0$ for $|\mathbf{x}| = 1$, $\mathbf{y} \ne \mathbf{0}$, and

$$\Delta_{\mathbf{x}} G(\mathbf{x}, \mathbf{y}) = -\delta_3(\mathbf{x} - \mathbf{y}) \qquad \text{in } B_1.$$

When $\mathbf{y} = \mathbf{0}$, put

$$G(\mathbf{x}, \mathbf{0}) = \frac{1}{4\pi}\left\{\frac{1}{|\mathbf{x}|} - 1\right\}.$$

Note how, for $\mathbf{x} \ne \mathbf{y}$ and $\mathbf{y} \to \mathbf{0}$,

$$G(\mathbf{x}, \mathbf{y}) \to G(\mathbf{x}, \mathbf{0}).$$

c) Denote by G_{B_1} the Green function for the ball B_1 constructed in part b). Set $\widetilde{\mathbf{y}} = (y_1, y_2, -y_3)$. Then

$$G_{B_1}^+(\mathbf{x}, \mathbf{y}) = G_{B_1}(\mathbf{x}, \mathbf{y}) - G_{B_1}(\mathbf{x}, \widetilde{\mathbf{y}}).$$

3

First Order Equations

3.1 Backgrounds

The first part of the present chapter is devoted to *first order scalar conservation laws*, of the type

$$u_t + q(u)_x = 0, \tag{3.1}$$

where q is a smooth funciton. One seeks solutions $u = u(x,t)$ defined on the half-plane $\{t \geq 0\}$, and typically subject to a *(Cauchy or) initial condition*

$$u(x,0) = g(x), \qquad x \in \mathbb{R}.$$

The equation represents a *convection* or *trasport* model. The *velocity* v is related to the *flux function* q by the relationship

$$q(u) = v(u)u.$$

• *Characteristics and local solution.* The straight lines

$$x(t) = q'(g(\xi))t + \xi, \tag{3.2}$$

along which u is constant, are the *characteristic lines* for the equation (3.1). If only one characteristic passes through the point (x,t), and this "starts " from $(\xi,0)$, then

$$u(x,t) = g(\xi).$$

One says that "the characteristic carries the datum $g(\xi)$".

Since $u(x(t),t) = g(\xi)$ along the characteristic (3.2), for any t small enough the solution u is implicitly defined by the equation

$$u = g\left(x - q'(u)t\right). \tag{3.3}$$

© Springer International Publishing Switzerland 2015
S. Salsa, G. Verzini, *Partial Differential Equations in Action. Complements and Exercises*,
UNITEXT – La Matematica per il 3+2 87, DOI 10.1007/978-3-319-15416-9_3

• *Rankine-Hugoniot conditions.* When two characteristics, carrying distinct data, intersect, they cause a jump discontinuity in the solution, and (3.3) is no longer valid. The solution should be interpreted suitably, in *weak sense*. The discontinuity curve Γ is called a *shock wave*. If Γ is regular and has equation $x = s(t)$, the Rankine-Hugoniot conditions hold: if we call u^+, u^- the limit values of u as it approaches Γ from the right and left respectively, one has

$$s'(t) = \frac{q(u^+(s(t),t)) - q(u^-(s(t),t))}{u^+(s(t),t) - u^-(s(t),t)}.$$

Let q be strictly concave or convex.

• *Rarefaction waves.* In the areas of the upper half-plane $\{t > 0\}$ not reached by the characteristics carrying the initial datum it is usually possible to construct the solution (defined coherently with all other areas where it is defined) as a *rarefaction wave*. The wave centred at (x_0, t_0), given by

$$u(x,t) = R\left(\frac{x - x_0}{t - t_0}\right), \qquad \text{where } R = (q')^{-1} \text{ (the inverse of } q'),$$

is a rarefaction wave centred at (x_0, t_0).

• An *entropy condition* along the shock wave is:

$$q(u^+(s,t)) < s'(t) < q(u^-(s,t)).$$

Its purpose is to select among the weak solutions the physically significant one.

The second part of the chapter is devoted to *first order quasilinear equations* of the type

$$a(x, y, u)u_x + b(x, y, u)u_y = c(x, y, u), \tag{3.4}$$

with a, b, c being C^1 functions on their domain.

• If $u = u(x, y)$ is a C^1 solution and $P_0 = (x_0, y_0, z_0)$ is a point on the graph $(z_0 = u(x_0, y_0))$, then the vector $(a(P_0), b(P_0), c(P_0))$ is tangent to the graph of u at P_0. The solutions graphs are therefore *integral surfaces for the vector field* (a, b, c), i.e. unions of *integral curves (characteristic curves)* solving the *characteristic system*

$$\frac{dx}{dt} = a(x, y, z), \qquad \frac{dy}{dt} = b(x, y, z), \qquad \frac{dz}{dt} = c(x, y, z), \tag{3.5}$$

where $z(t) = u(x(t), y(t))$.

If $c \equiv 0$ then z must be constant on the plane curves solutions of the *reduced characteristic system*

$$\frac{dx}{dt} = a(x, y, z), \qquad \frac{dy}{dt} = b(x, y, z);$$

if, moreover, $a = a(u)$ and $b \equiv 1$ the curves are straight lines, in agreement with conservation laws (which correspond to this situation, with t replacing y).

• A *Cauchy problem* for (3.5) is given by assigning the value of u along a curve of the xy-plane or, equivalently, by requiring to find an integral surface for the vector field (a, b, c) that contains a given space curve Γ_0. If Γ_0 has parametric equations

$$x(s) = f(s), \quad y(s) = g(s), \quad z(s) = u(x(s), y(s)) = h(s), \quad s \in I, \qquad (3.6)$$

one has to solve the characteristic system (3.5) with the family of initial conditions (3.6), depending on the parameter $s \in I$. Under our assumptions, for any given s, the Cauchy problem (3.5), (3.6) has exactly one solution

$$x = X(s, t), \quad y = Y(s, t), \quad z = Z(s, t), \qquad (3.7)$$

for t in a neighbourhood of 0. If the first two equations can be solved for suitable variables $s = S(x, y), t = T(x, y)$, the solution of equation (3.4) reads

$$z = u(x, y) = Z(S(x, y), T(x, y)).$$

Let

$$P_0 = (x_0, y_0, z_0) = (X(s_0, 0), Y(s_0, 0), Z(s_0, 0)) = (f(s_0), g(s_0), h(s_0)),$$

and consider the Jacobian

$$J(s_0, 0) = \det \begin{pmatrix} \partial_s X(s_0, 0) & \partial_s Y(s_0, 0) \\ \partial_t X(s_0, 0) & \partial_t Y(s_0, 0) \end{pmatrix} = \det \begin{pmatrix} f'(s_0) & g'(s_0) \\ a(P_0) & b(P_0) \end{pmatrix}.$$

Based on the inverse function theorem, the following possibilities can occur:

a) $J(s, 0)$ *is different from zero in* I. Then in a neighbourhood of Γ_0 there is a unique solution $u = u(x, y)$ for the Cauchy problem, defined parametrically by (3.7).

b) $J(s_0, 0) = 0$ *for some* $s_0 \in I$. A C^1 solution in a neighbourhood of P_0 may exist only if Γ_0 is *characteristic at* P_0, i.e.

$$\text{rank} \begin{pmatrix} f'(0) & g'(0) & h'(0) \\ a(P_0) & b(P_0) & c(P_0) \end{pmatrix} = 1.$$

If this is not the case, then, there are no C^1 solutions in a neighbourhood of P_0 (there might exist less regular solutions).

In particular, if Γ_0 is characteristic for any $s \in I$ there are infinitely many C^1 solutions in a neighbourhood of Γ_0.

• A function $\varphi = \varphi(x, y, z)$ of class C^1 is a *first integral* for the characteristic system (3.5) if it is constant along characteristic curves, in other words if

$$\nabla\varphi \cdot (a, b, c) \equiv 0.$$

If φ and ψ are two *independent* first integrals for (3.5) (the vectors $\nabla\varphi$ and $\nabla\psi$ are everywhere linearly independent), the general solution $z = u(x, y)$ of the quasilinear equation is defined implicitly by

$$F(\varphi(x,y,z),\psi(x,y,z)) = 0,$$

where $F = F(h,k)$ is an arbitrary C^1 function such that $F_h\varphi_z + F_k\psi_z \neq 0$.

3.2 Solved Problems

- **3.2.1 – 3.2.11** : Conservation laws and applications.
- **3.2.12 – 3.2.21** : Characteristics for linear and quasilinear equations.

3.2.1 Conservation laws and applications

Problem 3.2.1 (Burgers equation, shock waves). *Study the global Cauchy problem for Burgers equation*

$$\begin{cases} u_t + uu_x = 0 & x \in \mathbb{R}, t > 0 \\ u(x,0) = g(x) & x \in \mathbb{R} \end{cases}$$

where

a) $g(x) = \begin{cases} 1 & x < -1 \\ -1/2 & -1 < x < 1 \\ -1 & x > 1, \end{cases}$ **b)** $g(x) = \begin{cases} 0 & x \le 0, x > 1 \\ 2x & 0 \le x < 1. \end{cases}$

Solution. a) The Burgers equation is a conservation law of the type

$$u_t + q(u)_x = 0$$

with $q(u) = u^2/2$ and $q'(u) = u$.

The characteristic emanating from the point $(\xi, 0)$ on the xt-plane, along which the solution is constant and equals $g(\xi)$, has equation

$$x = q'(g(\xi))t + \xi = g(\xi)t + \xi = \begin{cases} t + \xi & \xi < -1 \\ -\frac{1}{2}t + \xi & -1 < \xi < 1 \\ -t + \xi & \xi > 1. \end{cases}$$

As q' is increasing (q is convex) and g has decreasing discontinuities, the characteristic slopes decrease when crossing the datum discontinuities. Then the characteristics then intersect, for small times, near $x = -1$ and also $x = 1$ (Fig. 3.1).

Therefore from both points we have shock waves $x = s(t)$, which can be determined using the Rankine-Hugoniot condition

$$s'(t) = \frac{q(u^+(s(t),t)) - q(u^-(s(t),t))}{u^+(s(t),t) - u^-(s(t),t)} = \frac{1}{2}[u^+(s(t),t) + u^-(s(t),t)].$$

Fig. 3.1 Characteristics for Problem 3.2.1 a) (small times)

Near $(x, t) = (-1, 0)$ we have $u^- \equiv 1$, $u^+ \equiv -1/2$, so

$$\begin{cases} s_1'(t) = \frac{1}{4} \\ s_1(0) = -1 \end{cases} \qquad \text{whence } x = s_1(t) = \frac{1}{4}t - 1.$$

Similarly, near $(x, t) = (1, 0)$, $u^- \equiv -1/2$, $u^+ \equiv -1$ and

$$\begin{cases} s_2'(t) = -\frac{3}{4} \\ s_2(0) = 1 \end{cases} \qquad \text{whence } x = s_2(t) = -\frac{3}{4}t + 1.$$

Consequently, for small times, the solution $u(x, t)$ equals $-1/2$ for

$$\frac{1}{4}t - 1 < x < -\frac{3}{4}t + 1.$$

As t increases, this interval gets smaller, until it disappears for $t = 2$ (and $x = -1/2$). At this point the two shock waves collide, and the surviving characteristics carry the datum $u^- \equiv 1$ (left) and $u^+ \equiv -1$ (right); this generates a third shock curve $x = s_3(t)$, where

$$\begin{cases} s_3'(t) = 0 \\ s_3(2) = -\frac{1}{2} \end{cases} \qquad \text{thus } \quad x = s_3(t) = -\frac{1}{2}.$$

Overall, the only entropic solution is (Fig. 3.2)

$$u(x, t) = \begin{cases} 1 & x < \min\left(\frac{1}{4}t - 1, -\frac{1}{2}\right) \\ -\frac{1}{2} & \frac{1}{4}t - 1 < x < -\frac{3}{4}t + 1 \\ -1 & x < \max\left(-\frac{3}{4}t + 1, -\frac{1}{2}\right). \end{cases}$$

b) In this case the characteristics are

$$x = \begin{cases} \xi & \xi \le 0,\ \xi > 1 \\ 2\xi t + \xi & 0 \le \xi < 1. \end{cases}$$

In particular, $u(x, t) \equiv 0$ as $x \le 0$, $t \ge 0$. When $0 < \xi < 1$, if t is small, the implicit solution is given in implicit form by

$$u = g\left(x - q'(u)t\right) = 2(x - ut),$$

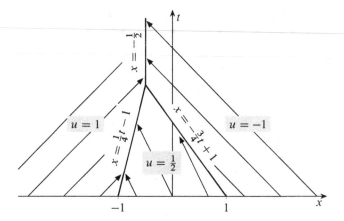

Fig. 3.2 Characteristic lines for Problem 3.2.1 a)

whence

$$u(x,t) = \frac{2x}{2t+1}.$$

Alternatively, from the characteristic $x = 2\xi t + \xi$ we find

$$\xi = \frac{x}{2t+1}, \quad \text{and hence } u(x,t) = g(\xi) = \frac{2x}{2t+1}.$$

As before, the decreasing discontinuity of g at $x = 1$, plus the convexity of q, cause the formation of a shock wave $x = s(t)$ satisfying the Rankine-Hugoniot condition. Since here $u^-(x,t) = 2x/(2t+1)$, $u^+ \equiv 0$, we have

$$\begin{cases} s_1'(t) = \dfrac{s(t)}{2t+1} \\ s_1(0) = 1. \end{cases}$$

The (ordinary) equation is linear and with separate variables. Integrating and imposing the initial condition gives $s(t) = \sqrt{2t+1}$. The required solution is thus (Fig. 3.3)

$$u(x,t) = \begin{cases} 0 & x \le 0,\ x > \sqrt{2t+1} \\ \dfrac{2x}{2t+1} & 0 \le x < \sqrt{2t+1}. \end{cases}$$

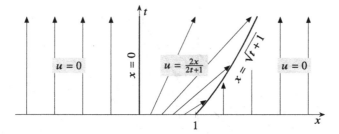

Fig. 3.3 Characteristic lines for Problem 3.2.1 b)

Problem 3.2.2 (Burgers equation, rarefaction vs. shock). *Solve the problem*

$$\begin{cases} u_t + u u_x = 0 & x \in \mathbb{R},\, t > 0 \\ u(x,0) = g(x) & x \in \mathbb{R} \end{cases}$$

where

a) $\quad g(x) = \begin{cases} 1 & x \le -1 \\ -x & -1 \le x < 0 \\ 1 & x > 1, \end{cases}$
    b) $\quad g(x) = \begin{cases} 0 & x < 0 \\ 1 & 0 < x < 1 \\ 0 & x > 1. \end{cases}$

Solution. a) As in the previous problem the characteristics are

$$x = q'(g(\xi))t + \xi = g(\xi)t + \xi = \begin{cases} t + \xi & \xi \le -1 \text{ or } \xi > 0 \\ -\xi t + \xi & -1 \le \xi < 0. \end{cases}$$

This time, though, g has an increasing discontinuity at $x = 0$; since $q(u) = u^2/2$ is convex (and hence q' is increasing), the slope of the characteristic has an increasing jump when ξ crosses 0 from left to right. Hence we expect that a region of the xt-plane will not be met by any characteristic. In this case the only entropic solution in this region is a rarefaction wave. On the other hand the characteristics corresponding to $-1 \le \xi < 0$ form a family of straight lines through the point $(x,t) = (0,1)$. Consequently, for $t < 1$, the solution is constructed by taking

$$\xi = \frac{x}{1-t} \qquad \text{and consequently } u(x,t) = g(\xi) = -\frac{x}{1-t}.$$

So for $t < 1$, no other characteristic enters the sector between the characteristics $x = 0$ and $x = t$, and the solution is given by a rarefaction wave. In general, a rarefaction wave centred at (x_0, t_0) has equation

$$u(x,t) = R\left(\frac{x - x_0}{t - t_0}\right) \qquad \text{where } R(y) = (q')^{-1}(y).$$

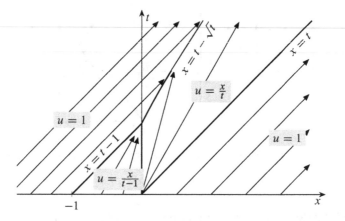

Fig. 3.4 Characteristic lines for Problem 3.2.2 a)

Since here $R(y) = y$, we find

$$u(x,t) = \frac{x}{t}, \qquad 0 \le x \le t, \ t < 1.$$

Alternatively we may put $\xi = 0$ and $g(\xi) = u(x,t)$ in the characteristics equation to get

$$x = u(x,t)t$$

and hence $u = x/t$. Note that a rarefaction wave is constant along the straight lines through the origin, also called characteristics.

When $t > 1$ the characteristics carrying $u^- \equiv 1$ hit the rarefaction characteristics, along which $u^+(x,t) = x/t$, and generate a shock curve Γ satisfying

$$\begin{cases} s_1'(t) = \dfrac{s(t)}{2t} \\ s_1(1) = 0. \end{cases}$$

This gives $s(t) = t - \sqrt{t}$. Note that Γ does not meet the characteristic $x = t$. Finally, we have (Figs. 3.4 and 3.5)

$$u(x,t) = \begin{cases} 1 & x \le t - 1 \text{ for } t < 1 \\ 1 & x < t - \sqrt{t} \text{ for } t \ge 1 \\ x/(t-1) & t - 1 \le x \le 0 \text{ for } t < 1 \\ x/t & \max(0, t - \sqrt{t}) < x \le t \\ 1 & x \ge t. \end{cases}$$

b) The function g has an increasing jump at $x = 0$ and a decreasing one at $x = 1$. Since q is convex, we expect a rarefaction wave at $(0,0)$; after some time, the latter inter-

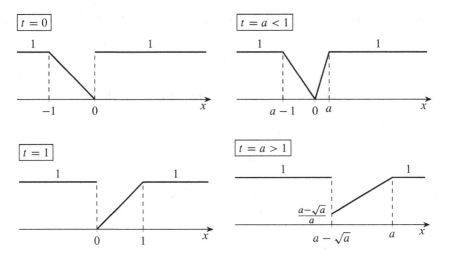

Fig. 3.5 Solution to Problem 3.2.2 a), at various times

acts with a shock wave emanating from $(1, 0)$. The characteristic line from the point $(\xi, 0)$ is

$$x = \xi + q'(g(\xi))t = \xi + g(\xi)t = \begin{cases} \xi & \xi < 0 \text{ or } \xi > 1 \\ t + \xi & 0 < \xi < 1. \end{cases}$$

By varying ξ we deduce immediately the following properties for the solution (Fig. 3.6):

- $u(x, t)$ equals 0 when $x < 0$ (vertical characteristics).
- The characteristics $x = 0$ and $x = t$ bound the region occupied by a rarefaction wave centred at the origin, at least until some instant time t_0 to be determined.
- From $(1, 0)$ starts a shock wave; on the right $u(x, t)$ is 0, while on the left, at least until t_0, $u(x, t)$ equals 1.
- For times larger than t_0 the shock interacts on the left with the rarefaction wave.

If we argue as in the previous situation, the rarefaction wave is

$$u(x, t) = \frac{x}{t}, \qquad 0 \le x \le t.$$

Concerning the shock wave, for small t we have $u^+ = 0$ and $u^- = 1$, so

$$\begin{cases} s'(t) = \frac{1}{2} \\ s(0) = 1, \end{cases} \quad \text{and thus } x = s(t) = \frac{1}{2}t + 1.$$

What we have said holds until the characteristic $x = t$ intersects the shock curve, that is up to $t_0 = 2$. For later times we still have a shock wave with $u^+ = 0$, but now

$$u^-(s,t) = \frac{s}{t},$$

corresponding to the value of u carried by the rarefaction wave. Therefore

$$\begin{cases} s'(t) = \dfrac{s(t)}{2t} \\ s(2) = 2. \end{cases}$$

The ODE is linear, and with separated variable, and has one solution

$$s(t) = \sqrt{2t}.$$

Summarising,

$$u(x,t) = \begin{cases} 0 & x \leq 0 \\ \frac{x}{t} & 0 \leq x < \min\left(t, \sqrt{2t}\right) \\ 1 & t \leq x < \frac{1}{2}t + 1, \text{ with } t < 2 \\ 0 & x > \max\left(\frac{1}{2}t + 1, \sqrt{2t}\right). \end{cases}$$

The shock speed is $1/2$ until $t = 2$ and then becomes negative, $-1/2t^{3/2}$. The strength equals the jump value of u across the shock, i.e. 1 until $t = 2$, and then fades to zero as $t \to \infty$ (Fig. 3.6).

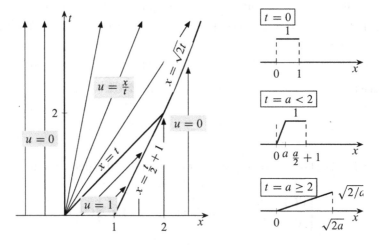

Fig. 3.6 Problem 3.2.2 b): characteristics and shock wave (*left*); solution at various times (*right*)

Problem 3.2.3 (Non-extendability). *Consider the Cauchy problem:*

$$\begin{cases} u_t + u^2 u_x = 0 & x \in \mathbb{R}, t > 0 \\ u(x,0) = x & x \in \mathbb{R}. \end{cases}$$

a) Check whether the family of characteristics admits an envelope.

b) Find an explicit formula for the solution and discuss whether it may be extended to the whole half plane $\{t > 0\}$.

Solution. a) The PDE is written as conservation law with $q(u) = u^3/3$, $q'(u) = u^2$. Note how the initial datum $g(x) = x$ is *unbounded* when $x \to \pm\infty$. The characteristic from $(\xi, 0)$ has equation:

$$x = \xi + q'(g(\xi))t = \xi + \xi^2 t.$$

To establish whether this family, depending on ξ, admits an envelope, we must solve for x and t the system

$$\begin{cases} x = \xi + \xi^2 t \\ 0 = 1 + 2\xi t. \end{cases}$$

The second equation is just the first one differentiated with respect to ξ. The parameter ξ can be eliminated and we find that the envelope lies in the quadrant $x < 0, t > 0$ and coincides with the hyperbola $4xt = -1$ (Fig. 3.7).

b) The solution $u = u(x,t)$ is defined implicitly by

$$u = g\left(x - q'(u)t\right)$$

at least for small times. In our case, since $g(x) = x$, we find

$$u = x - u^2 t.$$

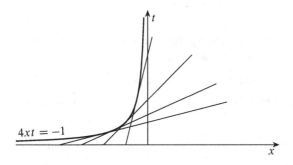

Fig. 3.7 Envelope of characteristics for Problem 3.2.3

Solving for u, we get

$$u^{\pm}(x,t) = \frac{-1 \pm \sqrt{1 + 4xt}}{2t}, \quad x \geq -\frac{1}{4t}.$$

Let us determine $\lim_{t \to 0^+} u^{\pm}(x,t)$. For given x:

$$\lim_{t \to 0^+} u^-(x,t) = \lim_{t \to 0^+} \frac{-1 - \sqrt{1 + 4xt}}{2t} = -\infty,$$

while

$$\lim_{t \to 0^+} u^+(x,t) = \lim_{t \to 0^+} \frac{-1 + \sqrt{1 + 4xt}}{2t} = \lim_{t \to 0^+} \frac{4xt}{2t\left(1 + \sqrt{1 + 4xt}\right)} = x.$$

Only u^+ satisfies the initial condition, and is therefore the unique solution, defined in the region $\{x \geq -1/4t\}$ and regular inside. This region is bounded above by the envelope of the characteristics, which becomes a barrier beyond which the characteristics do not carry initial data. Moreover, since the initial datum tends to $-\infty$ as $\xi \to -\infty$, and the characteristics tend to flatten horizontally, there is no coherent way to extend the definition of u beyond the envelope, in the quadrant $x < 0, t > 0$.

On the contrary, the formula

$$u(x,t) = \frac{-1 + \sqrt{1 + 4xt}}{2t}$$

defines the solution on $x \geq 0, t \geq 0$ as well.

Problem 3.2.4 (A traffic model, vehicle path). *The following problem models what happens at a traffic light:*

$$\begin{cases} \rho_t + v_m \left(1 - \frac{2\rho}{\rho_m}\right)\rho_x = 0 & x \in \mathbb{R}, t > 0 \\ \rho(x,0) = \begin{cases} \rho_m & x < 0 \\ 0 & x > 0, \end{cases} \end{cases}$$

where ρ is the density of cars, ρ_m the maximum density, v_m the maximum speed allowed. Determine the solution and calculate:

a) The density of cars at the light for any $t > 0$.

b) The time taken by a car placed at $x_0 < 0$ at time $t = 0$ to get past the light.

Solution. a) The equation is written as conservation law with

$$q(\rho) = \rho v(\rho) = v_m \rho \left(1 - \frac{\rho}{\rho_m}\right)$$

where $v(\rho)$ is the speed when the cars are in an area of density ρ. The characteristic through $(\xi, 0)$ is

$$x = v_m \left(1 - \frac{2\rho(\xi, 0)}{\rho_m} \right) t + \xi.$$

When $\xi < 0$ we find

$$x = -v_m t + \xi.$$

Thus in the region $x < -v_m t$ we have $\rho(x, t) = \rho_m$. When $\xi > 0$

$$x = v_m t + \xi$$

and if $x > v_m t$ we have $\rho(x, t) = 0$. In the sector $-v_m t \leq x \leq v_m t$ we can join the values ρ_m and 0 with a rarefaction wave centred at the origin. Setting

$$q'(\rho) = v_m \left(1 - \frac{2\rho}{\rho_m} \right) = y$$

we can find the inverse function

$$R(y) = (q')^{-1}(y) = \frac{\rho_m}{2} \left(1 - \frac{y}{v_m} \right),$$

and the rarefaction wave is

$$\rho(x, t) = R\left(\frac{x}{t} \right) = \frac{\rho_m}{2} \left(1 - \frac{x}{v_m t} \right).$$

To sum up, the solution is

$$\rho(x, t) = \begin{cases} \rho_m & x < -v_m t \\ \frac{\rho_m}{2} \left(1 - \frac{x}{v_m t} \right) & -v_m t \leq x \leq v_m t \\ 0 & x > v_m t. \end{cases}$$

Therefore the vehicle density at the traffic light is

$$\rho(0, t) = \frac{\rho_m}{2},$$

constant in time.

b) In the present model the speed of a vehicle at x at time t depends only on the density:

$$v(\rho) = v_m \left(1 - \frac{\rho}{\rho_m} \right).$$

Denote by $x = x(t)$ the law of motion of the car, with $x(0) = x_0 < 0$. Initially the car does not move, until time t_0, with $x_0 = -v_m t_0$; after t_0 the car moves within the region of the rarefaction wave as long as $x(t) < v_m t$, in particular before it reaches the traffic light; after that it moves with constant speed v_m. Therefore, after t_0 and before reaching

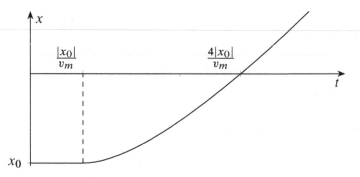

Fig. 3.8 Problem 3.2.4: path of the car starting from $x = x_0 < 0$ at time $t = 0$. The traffic light is reached at time $t = 4|x_0|/v_m$

the light, x solves the Cauchy problem

$$\begin{cases} x'(t) = v(\rho(x(t), t)) = \frac{v_m}{2}\left(1 + \frac{x(t)}{v_m t}\right) \\ x(t_0) = -v_m t_0. \end{cases}$$

Integrating the (linear) equation gives

$$x(t) = v_m(t - 2\sqrt{t_0 t}),$$

and hence $x(t) = 0$ for $t = 4t_0 = 4|x_0|/v_m$ (Fig. 3.8).

Problem 3.2.5 (Traffic model; normalised density). *Let ρ be the vehicle density in the model of Problem 3.2.4. Normalise the density by setting $u(x,t) = \rho(x,t)/\rho_m$, so that $0 \leq u \leq 1$. Check that u solves*

$$u_t + v_m(1 - 2u)u_x = 0, \qquad x \in \mathbb{R},\ t > 0. \tag{3.8}$$

Determine the solution to (3.8) with initial condition

$$u(x,0) = g(x) = \begin{cases} 1/3 & x \leq 0 \\ 1/3 + 5x/12 & 0 \leq x \leq 1 \\ 3/4 & x \geq 1. \end{cases}$$

Solution. The initial datum $g = g(x)$ is shown in Fig. 3.9, left.

Elementary computations show that u satisfies eq. (3.8). We have $q'(u) = v_m(1-2u)$ and hence $q(u) = v_m(u - u^2)$. The characteristic issued from $(\xi, 0)$ is

$$x = \xi + v_m(1 - 2g(\xi))t, \tag{3.9}$$

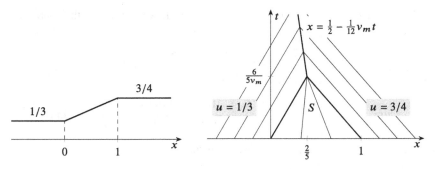

Fig. 3.9 Initial datum and characteristics for Problem 3.2.5

i.e.

$$x = \xi + \frac{1}{3}v_m t \quad \text{for } \xi \le 0$$

$$x = \xi + \left(\frac{1}{3} - \frac{5}{6}\xi\right)v_m t \quad \text{for } 0 \le \xi \le 1$$

$$x = \xi - \frac{1}{2}v_m t \quad \text{for } \xi \ge 1.$$

We can see that the characteristics meet, creating a shock wave. The starting point of it is the point with smallest time coordinate, at which the characteristics intersect for $0 \le \xi \le 1$. In this case the characteristics form a pencil depending on the parameter ξ, and the pencil base point, where all characteristics with $0 \le \xi \le 1$ meet, is

$$(x_0, t_0) = \left(\frac{2}{5}, \frac{6}{5v_m}\right).$$

This is shown in Fig. 3.9, right.

The shock curve Γ, of equation, say, $x = s(t)$, is thus emanating from $(2/5, 6/(5v_m))$. On the right of Γ $u^+ = 3/4$, while on the left $u^- = 1/3$. The Rankine-Hugoniot condition gives

$$s'(t) = \frac{q(u^+) - q(u^-)}{u^+ - u^-} = -\frac{1}{12}v_m.$$

Since $s\left(6/(5v_m)\right) = 2/5$ we get the straight line

$$s(t) = \frac{1}{2} - \frac{1}{12}v_m t.$$

Thus we have found the (entropic) solution, for $t > t_0 = 6/(5v_m)$.

Suppose now $t < t_0$. To compute the solution in the region

$$S = \left\{(x, t) : 0 \le t < \frac{6}{5v_m}, \frac{1}{3}v_m t \le x \le 1 - \frac{1}{2}v_m t\right\},$$

bounded by the characteristics from $\xi = 0$ and $\xi = 1$, we solve for ξ the characteristics equation. We get

$$\xi = \frac{6x - 2v_m t}{6 - 5v_m t}, \qquad 0 \le \xi \le 1,$$

from which, u being constant along characteristics,

$$u(x,t) = g(\xi) = \frac{1}{3} + \frac{5}{12} \frac{6x - 2v_m t}{6 - 5v_m t} = \frac{4 + 5x - 5v_m t}{2(6 - 5v_m t)} \qquad \text{in } S.$$

Another way to proceed would be to use the formula

$$u = g\left(x - v_m(1 - 2u)t\right)$$

which gives u in implicit form. Substituting the expression of g in the internal $0 < x < 1$ we find

$$u = \frac{1}{3} + \frac{5}{12}\left(x - (1 - 2u)\, v_m t\right).$$

Solving for u, we obtain the previous formula. In summary:

$$u(x,t) = \begin{cases} \dfrac{1}{3} & x < \min\left\{\dfrac{1}{3}v_m t, \dfrac{1}{2} - \dfrac{1}{12}v_m t\right\} \\[2ex] \dfrac{4 + 5x - 5v_m t}{2(6 - 5v_m t)} & \dfrac{1}{3}v_m t \le x \le 1 - \dfrac{1}{2}v_m t \\[2ex] \dfrac{3}{4} & x > \max\left\{1 - \dfrac{1}{2}v_m t, \dfrac{1}{2} - \dfrac{1}{12}v_m t\right\}. \end{cases}$$

---

* Problem 3.2.6 (Traffic in a tunnel). *A realistic model for the velocity inside a very long tunnel is*

$$v(\rho) = \begin{cases} v_m & 0 \le \rho \le \rho_c \\ \lambda \log\left(\rho_m/\rho\right) & \rho_c \le \rho \le \rho_m \end{cases}$$

where ρ is the vehicles density and $\lambda = \dfrac{v_m}{\log(\rho_m/\rho_c)}$. Note v is continuous also at the point $\rho_c = \rho_m e^{-v_m/\lambda}$, which represents a critical density, below which drivers are free to cruise at the maximum speed allowed. Practical values are $\rho_c = 7$ cars/Km, $v_m = 90$ Km/h, $\rho_m = 110$ cars/Km, $v_m/\lambda = 2.75$.

Suppose the tunnel entrance is placed at $x = 0$, and that prior to the tunnel opening (at time $t = 0$) a queue has formed. The initial datum is

$$\rho(x,0) = g(x) = \begin{cases} \rho_m & x < 0 \\ 0 & x > 0. \end{cases}$$

a) Determine the traffic density and velocity, and sketch the graphs of these functions.

b) Determine and sketch on the xt-plane the path of a car initially at $x = x_0 < 0$, then compute how long it takes it to enter the tunnel.

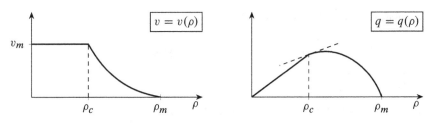

Fig. 3.10 Velocity and flux function for the traffic in a tunnel

Solution. a) By using the usual convective model the problem to solve reads

$$\begin{cases} \rho_t + q'(\rho)\rho_x = 0 & x \in \mathbb{R}, t > 0 \\ \rho(x,0) = g\,(x) = \begin{cases} \rho_m & x < 0 \\ 0 & x > 0, \end{cases} \end{cases}$$

where

$$q(\rho) = \rho v(\rho)$$

and hence $(e^{-v_m/\lambda} = \rho_c/\rho_m)$

$$q'(\rho) = \begin{cases} v_m & 0 \le \rho < \rho_c \\ \lambda\,[\log\,(\rho_m/\rho) - 1] & \rho_c < \rho \le \rho_m. \end{cases}$$

The graphs of v and q in terms of the density ρ are shown in Fig. 3.10. Notice how q' jumps at $\rho = \rho_c$:

$$q'(\rho_c^-) = v_m \quad \text{and} \quad q'(\rho_c^+) = v_m - \lambda.$$

The characteristic from $(\xi, 0)$, i.e. the line $x = \xi + q'(g(\xi))t$, is

$$x = \xi - \lambda t \quad \text{for } \xi < 0, \qquad \text{and } x = \xi + v_m t \quad \text{for } \xi > 0.$$

Therefore we obtain immediately the solution is certain regions:

$$\rho(x,t) = \rho_m \quad \text{for} \quad x < -\lambda t.$$

It remains to find ρ in the sector

$$S = \{(x,t) : -\lambda t \le x \le v_m t\}.$$

For this we recall that q' is discontinuous at $\rho = \rho_c$:

$$q'(\rho_c^-) = v_m \quad \text{and} \quad q'(\rho_c^+) = v_m - \lambda.$$

This suggests writing $S = S_1 \cup S_2$, with

$$S_1 = \{(x,t) : -\lambda t \leq x \leq (v_m - \lambda)t\},$$

where $\rho_c < \rho \leq \rho_m$, and

$$S_2 = \{(x,t) : (v_m - \lambda)t \leq x \leq v_m t\},$$

where $0 < \rho \leq \rho_c$.

In S_1 we proceed as follows. When $\rho_c < \rho \leq \rho_m$ we have

$$q''(\rho) = -\lambda/\rho < 0,$$

so that q is strictly concave. Since the initial datum is decreasing we seek a solution in the form of a rarefaction wave, centred at the origin, that attains continuously the value ρ_m on the line $x = -\lambda t$. The wave is given by $\rho(x,t) = R(x/t)$ where $R = (q')^{-1}$. To find R we solve for ρ the equation

$$q'(\rho) = \lambda \left[\log\left(\frac{\rho_m}{\rho} \right) - 1 \right] = y.$$

This gives

$$R(y) = \rho_m \exp\left(-1 - \frac{y}{\lambda} \right)$$

and hence we find

$$\rho(x,t) = \rho_m \exp\left(-1 - \frac{x}{\lambda t} \right)$$

in the region

$$-\lambda \leq \frac{x}{t} \leq v_m - \lambda.$$

Notice that $\rho = \rho_c$ on the straight line $x = (v_m - \lambda)t$. In S_2, where $\rho \leq \rho_c$, we have $q'(\rho) = v_m$. Thus, q is not strictly convex of concave, and there is no possibility to construct a solution via a rarefaction wave. Changing perspective, we construct the entropic solution by solving the equation in the "quadrant" $\{x > (v_m - \lambda)t, t > 0\}$, prescribing the values $\rho = \rho_c$ on $x = (v_m - \lambda)t$ and 0 on $t = 0$. We have already found $\rho = 0$ when $x > v_m t$ (Fig. 3.11). In the sector S_2 ρ is constant along the characteristics

$$x = v_m t + k,$$

that carry the value $\rho = \rho_c = c^{-v_m/\lambda}$.

To sum up,

$$\rho(x,t) = \begin{cases} \rho_m & x \geq -\lambda t \\ \rho_m e^{-(1+x/(\lambda t))} & -\lambda t \leq x \leq (v_m - \lambda)t \\ \rho_m e^{-v_m/\lambda} & (v_m - \lambda)t \leq x < v_m t \\ 0 & x > v_m t. \end{cases}$$

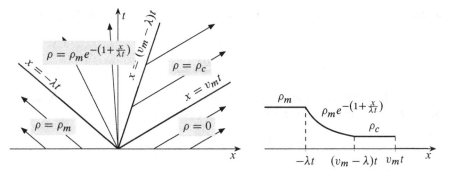

Fig. 3.11 Problem 3.2.6 b): characteristics (*left*); solution at time t (*right*)

In Fig. 3.11 (on the right) we see the density behaviour at a given time: it decreases from its maximum value (at zero speed) to reach the critical density (maximum speed). Note that the solution is discontinuous only along $x = v_m t$. This type of discontinuity is called *contact discontinuity*

b) Consider the vehicle initially placed at $x_0 < 0$. We want to describe its trajectory on the xt-plane. Observe first that the car will not move until time $t_0 = |x_0|/\lambda$ (Fig. 3.12). At that moment it enters the region S where the velocity is

$$v(\rho(x,t)) = \lambda \log \left(e^{1+x/\lambda t} \right) = \lambda + \frac{x}{t}.$$

If $x = x(t)$ denotes the vehicle path, we have

$$\begin{cases} x'(t) = \lambda + \dfrac{x(t)}{t} \\ x(t_0) = x_0. \end{cases}$$

The equation is linear, and integrating gives

$$x(t) = \lambda t \left(\log \frac{\lambda t}{|x_0|} - 1 \right).$$

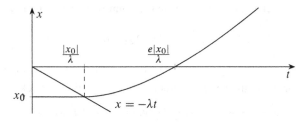

Fig. 3.12 Trajectory of a car in Problem 3.2.6

The car enters the tunnel at the time T such that $x(T) = 0$. The required lapse is then

$$T = \frac{e|x_0|}{\lambda}.$$

Problem 3.2.7 (Shock formation in a traffic model). *Let u, $0 \le u \le 1$ be the normalised density that solves the following traffic problem:*

$$\begin{cases} u_t + v_m(1 - 2u)u_x = 0 & x \in \mathbb{R}, t > 0 \\ u(x, 0) = g(x) & x \in \mathbb{R}. \end{cases}$$

Assume $g \in C^1(\mathbb{R})$, that g' has a unique maximum point x_1 and that

$$g'(x_1) = \max_{\mathbb{R}} g'(x) > 0.$$

a) *Study the qualitative behaviour of the characteristics and deduce that the solution develops a shock.*

b) *Verify that for small times u is defined implicitly by*

$$u = g(x - v_m t(1 - 2u)).$$

Deduce that the first instant t_s at which the shock forms (critical time) is the first time for which
$$1 - 2v_m t g'(x - v_m t(1 - 2u)) = 0.$$

c) *Show that the initial point (x_s, t_s) of the shock belongs to the characteristic Γ_{x_1} emanating from $(x_1, 0)$, and*

$$t_s = \frac{1}{2v_m g'(x_1)}.$$

In case $v_m = 1$, $g(x) = \frac{3}{4}\left[\frac{2}{\pi}\arctan x + 1\right]$, analyse numerically the graph of u at various times and interpret the results.

Solution. a) The characteristic Γ_ξ from the point $(\xi, 0)$ has equation

$$x = \xi + (1 - 2g(\xi))v_m t. \tag{3.10}$$

Under the given hypotheses g is strictly increasing in a neighbourhood of x_1, thus the characteristics starting in the neighbourhood meet, generating a shock.

b) On Γ_ξ we know that $u(x, t) = g(\xi)$, and from (3.10) we find

$$\xi = x - (1 - 2g(\xi))v_m t.$$

Hence

$$u(x,t) = g(x - (1 - 2u(x,t))v_m t).$$

Now we verify when the equation

$$h(x,t,u) = u - g(x - (1 - 2u)v_m t) = 0, \tag{3.11}$$

really defines an implicit function u of x and t. The sufficient conditions provided by the implicit function theorem are the following:

1. h is C^1, true because g is C^1.

2. (3.11) can be solved at some point, in fact

$$h(x, 0, g(x)) = g(x) - g(x) = 0$$

at all points on the x-axis.

3. Finally,

$$h_u(x,t,u) = 1 - 2v_m t g'(x - (1 - 2u)v_m t) \neq 0. \tag{3.12}$$

As g' is either negative, or bounded when positive, equation (3.12) is always true for small times.

As long as (3.12) holds, by the implicit function theorem, equation (3.11) defines *a unique function* $u = u(x,t)$ in $C^1(\mathbb{R})$. This solution cannot develop (shock) discontinuities. On the other hand the same inverse function theorem gives a formula for u_x:

$$u_x(x,t) = -\frac{h_x(x,t,u)}{h_u(x,t,u)} = \frac{g'(x - (1 - 2u)v_m t)}{1 - 2v_m t g'(x - (1 - 2u)v_m t)}. \tag{3.13}$$

So if $t_s > 0$ is the first instant for which h_u is zero (for some $x = x_s$), necessarily

$$u_x(x,t) \to \infty \quad \text{as} \quad (x,t) \to (x_s, t_s)$$

since the numerator of (3.13) does not vanish at (x_s, t_s) (it goes to $(2v_m t_s)^{-1}$). Therefore t_s must be the *critical time*, i.e. when the shock starts.

c) Let us find t_s. Consider the characteristic Γ_ξ. For any $(x,t) \in \Gamma_\xi$

$$x - (1 - 2u(x,t)v_m t) = \xi,$$

so that (3.12) fails when

$$h_u(x,t,u(x,t)) = 1 - 2v_m t g'(\xi) = 0, \quad \text{i.e.} \quad t = \frac{1}{2v_m g'(\xi)}.$$

From part a) we know that t_s is the smallest (positive) t for which the previous equation

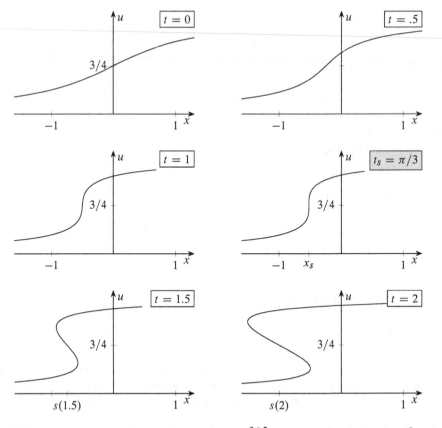

Fig. 3.13 Curve implicitly defined by the equation $u - \frac{3}{4}\left[\frac{2}{\pi}\arctan(x - (1 - 2u)v_m t) + 1\right] = 0$ at various times. The abscissas s (1.5) and s (2) denote the shock positions obtained by the *equal-area rule* [18, Chap. 4, Sect. 4]

holds. By assumption $g'(x_1) \geq g'(\xi)$ for any ξ, therefore (x_s, t_s) belongs to Γ_{x_1}, and moreover

$$t_s = \frac{1}{2v_m g'(x_1)},$$

$$x_s = x_1 + \frac{1}{2g'(x_1)}(1 - 2g(x_1)).$$

In case

$$g(x) = \frac{3}{4}\left[\frac{2}{\pi}\arctan x + 1\right],$$

the curve defined implicitly by (3.11) evolves as in Fig. 3.13.

Problem 3.2.8 (Envelope of characteristics and shock formation). *Consider the Cauchy problem:*

$$\begin{cases} u_t + q(u)_x = 0 & x \in \mathbb{R}, t > 0 \\ u(x,0) = g(x) & x \in \mathbb{R}. \end{cases}$$

Suppose $q \in C^2(\mathbb{R})$, $q'' < 0$ *and* $g \in C^1(\mathbb{R})$, *with*

$$g(x) = \begin{cases} g(x) = 0 & x \leq 0 \\ g'(x) > 0 & 0 < x < 1 \\ g(x) = 1 & x \geq 1. \end{cases}$$

a) *Show that the family of characteristics*

$$x = q'(u)t + \xi = q'(g(\xi))t + \xi, \quad \xi \in (0,1)$$

admit an envelope.

b) *Determine the point* (x_s, t_s) *of the envelope with smallest time coordinate, and show that this is the point where the shock originates from. Recover the result of Problem 3.2.7.*

c) *Show that* (x_s, t_s) *is a singular point for the envelope, meaning that the tangent vector at* (x_s, t_s) *is zero (assume q and g are regular enough.)*

Solution. a) Figure 3.14 shows the envelope of the characteristics

$$x = q'(g(\xi))t + \xi,$$

$\xi \in (0,1)$, in two particular cases.

To check the existence of an envelope, we consider the system

$$\begin{cases} x = q'(g(\xi))t + \xi \\ 0 = q''(g(\xi))g'(\xi)t + 1 = 0 \end{cases}$$

where the second equation is the derivative of the first with respect to ξ. As $q'' < 0$ and $g' > 0$ for $\xi \in (0,1)$, we have $q''(g(\xi))g'(\xi) < 0$ and the envelope is given by the parametric equations

$$x_{inv}(\xi) = \xi - \frac{q'(g(\xi))}{q''(g(\xi))g'(\xi)}, \qquad t_{inv}(\xi) = -\frac{1}{q''(g(\xi))g'(\xi)},$$

obtained by solving for ξ the system in the variables x and t.

b) The shock forms in correspondence to the point (x_s, t_s) of the envelope with smallest time coordinate, because that is the first point where two characteristics meet. As

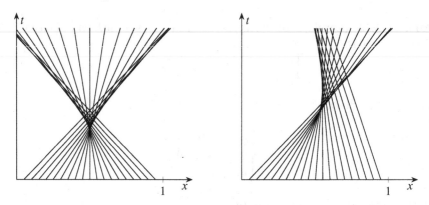

Fig. 3.14 Problem 3.2.8, envelope of characteristics with a cusp, in the case $q(u) = u - u^2$ and: $g(\xi) = (1 - \cos(\pi\xi))/2$ (*left*); $g(\xi) = 5\xi^2 e^{-2\xi}$ (*right*)

$g'(0) = g'(1) = 0$ and $q''(g(\xi))g'(\xi) < 0$ for $0 < \xi < 1$, the function

$$z(\xi) = -q''(g(\xi))g'(\xi)$$

has a positive maximum at some $\xi_M \in (0, 1)$. From the second equation

$$t_s = \min_{\xi \in (0,1)} \frac{1}{z(\xi)} = \frac{1}{z(\xi_M)}.$$

For Problem 3.2.7 (page 168) we have

$$q(u) = v_m(u - u^2)$$
$$q'(u) = v_m(1 - 2u)$$
$$q'' = -2v_m < 0.$$

In a neighbourhood of x_1, the positive maximum of g', we have $g' > 0$, so that the characteristics starting there have an envelope. From

$$z(\xi) = -q''(g(\xi))g'(\xi) = 2v_m g'(\xi)$$

we deduce $\xi_M = x_1$, and the solution has a shock starting at time

$$t_s = \frac{1}{2v_m g'(x_1)},$$

confirming the result in Problem 3.2.7.

c) To check that (x_s, t_s), origin of the shock and "origin" of the envelope, is singular, we need to show that

$$\frac{dx}{d\xi} \quad \text{and} \quad \frac{dt}{d\xi}$$

vanish at $\xi = \xi_M$. Assume q has three derivatives and g two. Then

$$\frac{dx}{d\xi} = \frac{-q'(g(\xi))}{z^2(\xi)} z'(\xi) = -\frac{dt}{d\xi}.$$

Since z has a (positive) maximum at $\xi = \xi_M$, we have

$$z'(\xi_M) = 0$$

and the derivatives vanish. The shock starts at the singular points of the envelope (cusps).

Problem 3.2.9 (Non-homogeneous conservation laws). *Consider the problem*

$$\begin{cases} u_t + q(u)_x = f(u, x, t) & x \in \mathbb{R}, t > 0 \\ u(x, 0) = g(x) & x \in \mathbb{R}. \end{cases}$$

a) Let $x = x(t)$ be a characteristic for the homogeneous equation ($f = 0$) and set

$$z(t) = u(x(t), t).$$

Which Cauchy problems do $x(t)$ and $z(t)$ solve?

b) Supposing f and g bounded, define weak solutions for the problem.

c) Deduce the Rankine-Hugoniot conditions for a shock curve $x = s(t)$.

Solution. a) Set $z = u(x(t), t)$. We have

$$z'(t) = u_t(x(t), t) + u_x(x(t), t) x'(t),$$

and by the conservation law

$$u_t(x(t), t) + u_x(x(t), t) q'(z(t)) = f(z(t), x(t), t).$$

The characteristic from $(\xi, 0)$ solves

$$x'(t) = q'(z(t)), \qquad x(0) = \xi$$

while z satisfies the Cauchy problem

$$z'(t) = f(z(t), x(t), t), \qquad z(0) = g(\xi),$$

which uniquely determines u along the characteristic, under the usual smoothness assumption on f.

b) We mimic the procedure for homogeneous equations. Let us multiply the equation by a test function $\varphi \in C_0^1(D)$, which is C^1 with compact support K contained in

$$D = \{(x,t) \in \mathbb{R}^2 : t \geq 0\},$$

and integrate over D, obtaining

$$\int_D (q(u)_x + u_t)\varphi \, dxdt = \int_D f(u,x,t)\varphi \, dxdt.$$

The integrals are finite because the support of φ is bounded. The notion of weak solution is found essentially by integrating by parts. Interpreting $q(u)_x + u_t$ as the divergence of the vector field $(q(u), u)$ we may apply Green's theorem:

$$\int_K (q(u)_x + u_t)\varphi \, dxdt = -\int_K [q(u)\varphi_x + u\varphi_t] \, dxdt - \int_{\mathbb{R}} u(x,0)\varphi(x,0) \, dx$$

$$+ \int_{\partial K \cap \{t>0\}} [q(u)\varphi n_1 + u\varphi n_2] \, ds$$

where (n_1, n_2) is the outer unit normal to ∂K and ds the infinitesimal length element. The last integral is zero (φ is continuous, hence null on $\partial K \cap \{t > 0\}$). So we define *weak solution* a *locally bounded function* u such that

$$\int_D [q(u)\varphi_x + u\varphi_t] \, dxdt + \int_{\mathbb{R}} u\varphi \, dx + \int_D f(u,x,t)\varphi \, dxdt \qquad \text{for any } \varphi \in C_0^1(D).$$

As for the homogeneous situation, a weak solution which is C^1 in $\mathbb{R} \times \{t \geq 0\}$ is a classical solution as well.

c) Suppose a curve Γ, $x = s(t)$, splits an open set $V \subset \{t > 0\}$ into two disjoint subdomains

$$V^- = \{(x,t) : x < s(t)\} \quad \text{and} \quad V^+ = \{(x,t) \in V : x > s(t)\}.$$

Assume that u is a weak solution, which is C^1 in the closures $\overline{V^-}$ and $\overline{V^+}$ separately, with a jump discontinuity along Γ. In particular, this implies

$$u_t + q(u)_x = f(u,x,t)$$

in V^- and V^+. If $(x,t) \in \Gamma$, write $u^+(x,t)$ for the limit of u when approaching Γ on the right, $u^-(x,t)$ for the limit from the left. Pick a test function φ, with support in V that intersects Γ. From part b)

$$-\int_{V^-} [q(u)\varphi_x + u\varphi_t] \, dxdt - \int_{V^+} [q(u)\varphi_x + u\varphi_t] \, dxdt = \int_V f(u,x,t)\varphi \, dxdt.$$

Since u is regular on V^-, V^+, we can invoke Green's theorem on the integrals on the left. Recalling that $\varphi = 0$ on $\partial V^\pm \setminus \Gamma$:

$$-\int_{V^\pm} [q(u)\varphi_x + u\varphi_t]\, dxdt = \int_{V^\pm} (q(u)_x + u_t)\varphi\, dxdt \mp \int_\Gamma [q(u^\pm)n_1 + u^\pm n_2]\varphi\, ds$$

$$= \int_V f(u,x,t)\varphi\, dxdt \mp \int_\Gamma [q(u^\pm)n_1 + u^\pm n_2]\varphi\, ds$$

where (n_1, n_2) is the outward unit normal to Γ with respect to V^+ (we used the fact that $u_t + q(u)_x = f(u,x,t)$ on $V^\pm$). Substituting into the definition of weak solution, we find

$$\int_\Gamma [(q(u^+) - q(u^-))n_1 + (u^+ - u^-)n_2]\varphi\, ds = 0.$$

Since φ is arbitrary, and the jumps $q(u^+) - q(u^-)$, $u^+ - u^-$ are continuous along Γ, we deduce

$$(q(u^+) - q(u^-))n_1 + (u^+ - u^-)n_2 = 0 \ \text{ along } \Gamma.$$

On the other hand, if $s \in C^1$ we have

$$(n_1, n_2) = \frac{1}{\sqrt{1 + s'(t)^2}}(-1, s'(t)),$$

so

$$s' = \frac{q(u^+) - q(u^-)}{u^+ - u^-}.$$

The Rankine-Hugoniot condition coincides with the one for the non-homogeneous case.

Problem 3.2.10 (Fluid in a porous tube). *Consider a cylindrical tube, infinitely long, placed along the x-axis, containing a fluid moving to the right. Denote by $\rho = \rho(x,t)$ the fluid density, and suppose that the speed at each point depends on the density by $v = \frac{1}{2}\rho$. Assume, further, that the tube wall is made of a porous material that leaks at the rate $H = k\rho^2$ (mass per unit length, per unit time).*

a) Deduce that if ρ is smooth, it satisfies

$$\rho_t + \left(\frac{1}{2}\rho^2\right)_x = -k\rho^2.$$

b) Compute the solution with $\rho(x,0) = 1$ and the corresponding characteristics.

Solution. a) We are dealing with a transport model. The leaking rate H leads to write the conservation law

$$\rho_t + q(\rho)_x = -H = -k\rho^2.$$

Due to the convective nature of motion the flow is described by

$$q(\rho) = v(\rho)\rho = \frac{1}{2}\rho^2,$$

yielding the required equation.

b) From Problem 3.2.9 a), indicating by $x = x(t)$ the characteristic from $(0, \xi)$ and setting $z = \rho(x(t), t)$, we have

$$\begin{cases} x'(t) = z(t) & x(0) = \xi \\ z'(t) = -kz^2(t) & z(0) = 1. \end{cases}$$

From the second equation we get

$$z(t) = 1/(kt + 1);$$

as the latter does not depend on ξ, we may write

$$\rho(x, t) = \frac{1}{1 + kt}.$$

The characteristics are parallel logarithms:

$$x(t) = \frac{1}{k} \ln(1 + kt) + \xi.$$

** **Problem 3.2.11** (A saturation problem). *Suppose a certain substance is poured into a semi-infinite container (aligned along the axis $x \geq 0$) with a solvent; the substance concentration $u = u(x, t)$ is governed by the equation*

$$u_x + (1 + f'(u))u_t = 0 \qquad \text{with } u(x, 0) = 0, \; x > 0, \; t > 0.$$

At the entrance $(x = 0)$ the substance is maintained at the concentration

$$g(t) = \begin{cases} \dfrac{c_0}{\alpha} t & 0 \leq t \leq \alpha \\ c_0 & t \geq \alpha. \end{cases} \qquad (c_0, \alpha > 0)$$

Study the evolution of u if one takes

$$f(u) = \frac{\gamma u}{1 + u} \qquad \text{(Langmuir isothermal } (\gamma > 0))$$

and discuss the case where α tends to zero[a].

---

[a] See [28, Vol. 1, Chap. 6.4], also for the physical-chemical interpretation of the model.

Solution. First of all let us remark that, compared to the conservation laws seen so far, the roles of x and t are exchanged. We have

$$q'(u) = 1 + f'(u) = 1 + \frac{\gamma}{(1 + u)^2},$$

and since q is concave and g increasing, we expect a shock. The characteristics are the lines

$$t = (1 + f'(u))x + k = \left(1 + \frac{\gamma}{(1 + u)^2}\right)x + k \qquad k \in \mathbb{R}.$$

In particular the characteristics from the point $(\xi, 0)$ on the x-axis are the parallel lines

$$t = (1 + \gamma)(x - \xi).$$

Those from $(0, \tau)$, on the t-axis, $0 \le \tau \le \alpha$, are the lines

$$t = \left(1 + \frac{\gamma}{(1 + c_0\tau/\alpha)^2}\right) x + \tau, \quad \text{if } 0 \le \tau \le \alpha,$$

which we rewrite as

$$\left(1 + \frac{c_0}{\alpha}\tau\right)^2 (t - \tau - x) = \gamma x. \tag{3.14}$$

Since

$$1 + \frac{\gamma}{(1 + c_0\tau/\alpha)^2} \le (1 + \gamma)$$

these and the lines from the x-axis will end up meeting along a shock curve. Equations (3.14), moreover, have decreasing slope as τ increases, thus they will have an envelope. The first part of the shock wave is contained in the cusp region bounded by the branches of the envelope of (3.14) (Fig. 3.15), and starts from the point $C = (x_s, t_s)$ of the envelope with minimum time coordinate, i.e. the cusp itself.

The characteristics from $(0, \tau)$, $\tau \ge \alpha$, are the parallel lines:

$$t = (1 + f'(c_0))x + \tau.$$

As

$$1 + f'(c_0) = 1 + \frac{\gamma}{(1 + c_0)^2} < (1 + \gamma),$$

also these interact, along the shock wave, with the characteristics issued from the horizontal axis.

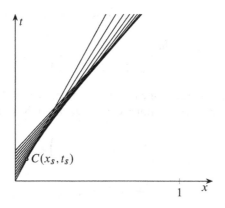

Fig. 3.15 Envelope in Problem 3.2.11 ($\alpha = \gamma = 1$, $c_0 = 7$)

To find the envelope let us differentiate (3.14) in τ; we find

$$-\left(1 + \frac{c_0}{\alpha}\tau\right)^2 + \frac{2c_0}{\alpha}\left(1 + \frac{c_0}{\alpha}\tau\right)(t - \tau - x) = 0.$$

The envelope is found by solving the system

$$\begin{cases} t - \tau - x = \gamma x \left(1 + \frac{c_0}{\alpha}\tau\right)^{-2} \\ \frac{2c_0}{\alpha}(t - \tau - x) = \left(1 + \frac{c_0}{\alpha}\tau\right). \end{cases}$$

The parametric equations $t = t(\tau)$, $x = x(\tau)$ can be determined by dividing the equations by one another (note that the second one implies $t - \tau - x \neq 0$); we find:

$$x = \left(1 + \frac{c_0}{\alpha}\tau\right)^3 \frac{\alpha}{2\gamma c_0},$$

hence

$$t = \left(1 + \gamma\left(1 + \frac{c_0}{\alpha}\tau\right)^{-2}\right)x + \tau$$

$$= \frac{\alpha}{2\gamma c_0}\left(1 + \frac{c_0}{\alpha}\tau\right)\left(\left(1 + \frac{c_0}{\alpha}\tau\right)^2 + \gamma\right) + \tau \equiv h(\tau).$$

The shock wave starts at the time coordinate of the cusp,

$$t_s = \min_{0 \leq \tau \leq \alpha} h(\tau).$$

Since $h(\tau)$ is the product of increasing terms in τ, the minimum of h is reached for $\tau = 0$, whence

$$t_s = \frac{\alpha(1 + \gamma)}{2\gamma c_0}, \qquad x_s = \frac{\alpha}{2\gamma c_0}.$$

The shock curve $t = t(x)$ is found by means of the Rankine-Hugoniot conditions:

$$\frac{dt}{dx} = \frac{[u + f(u)]_-^+}{[u]_-^+} = 1 + \frac{\gamma}{1 + u^-} \qquad (3.15)$$

(recall $u^+ = 0$). It remains to compute $u = u^-$ along the shock. To this purpose, observe that a shock point (x, t) comes, for times close to t_s, from the characteristic issued from the t axis, of equation

$$t = \tau + \left(1 + \frac{\gamma}{(1 + u)^2}\right)x \qquad (3.16)$$

where

$$\tau = g^{-1}(u) = \alpha u/c_0.$$

Then we substitute u, obtained from (3.16), into (3.15). To simplify the computations we think of (3.15), (3.16) as equations in the parameter u along the shock line. Differentiating

(3.16) in u we get

$$\frac{dt}{du} = \frac{d\tau}{du} - \frac{2\gamma}{(1+u)^3}x + \left(1 + \frac{\gamma}{(1+u)^2}\right)\frac{dx}{du} =$$

$$= \frac{\alpha}{c_0} - \frac{2\gamma}{(1+u)^3}x + \left(1 + \frac{\gamma}{(1+u)^2}\right)\frac{dx}{du}.$$

From (3.15), on the other hand,

$$\frac{dt}{du} = \frac{dt}{dx}\frac{dx}{du} = \left(1 + \frac{\gamma}{1+u}\right)\frac{dx}{du},$$

so

$$\left(1 + \frac{\gamma}{1+u}\right)\frac{dx}{du} = \frac{\alpha}{c_0} - \frac{2\gamma}{(1+u)^3}x + \left(1 + \frac{\gamma}{(1+u)^2}\right)\frac{dx}{du}.$$

Simplifying,

$$\frac{u}{(1+u)^2}\frac{dx}{du} + \frac{2}{(1+u)^3}x = \frac{\alpha}{\gamma c_0}, \tag{3.17}$$

which is a linear equation in $x = x(u)$, that we can easily integrate. An alternative way is to observe that

$$\frac{d}{du}\left(\frac{u}{1+u}\right)^2 = \frac{d}{du}\left(1 - \frac{1}{1+u}\right)^2 = \frac{2u}{(1+u)^3}.$$

Multiply (3.17) by u. We find

$$\frac{d}{du}\left[x\left(\frac{u}{1+u}\right)^2\right] = \frac{\alpha u}{\gamma c_0}$$

and integrate using the initial condition $x(0) = x_s = \alpha/(2\gamma c_0)$, to get

$$x(u) = \frac{\alpha}{2\gamma c_0}(1+u)^2$$

and then

$$u = u^- = -1 + \left(\frac{2\gamma c_0 x}{\alpha}\right)^{1/2}, \tag{3.18}$$

along the shock curve. The formula indicates that *the jump in u along the shock (the shock strength) grows like $\sqrt{x}$*. Finally, substituting into (3.16), a little manipulations give

$$t = \frac{\alpha u}{c_0} + \left(1 + \frac{\gamma}{(1+u)^2}\right)x = x + \left(\frac{2\alpha\gamma}{c_0}x\right)^{1/2} - \frac{\alpha}{2c_0} \tag{3.19}$$

showing that the shock curve is a parabola.

From (3.15), the shock starts off with speed

$$\frac{dx}{dt} = (1+\gamma)^{-1}$$

and accelerates until u reaches its maximum c_0. By (3.18), (3.19) this corresponds to the point P of coordinates

$$x_P = \frac{\alpha}{2\gamma c_0}(1 + c_0)^2, \quad t_P = \frac{\alpha}{2\gamma c_0}\left[(1 + c_0)^2 + \gamma(1 + 2c_0)\right].$$

From there on, the shock wave travels with constant speed

$$\frac{dx}{dt} = \left(1 + \frac{\gamma}{1 + c_0}\right)^{-1}$$

along a straight line. Since it passes through P, it is easy to see that the equation of the shock past P is

$$t = \left(1 + \frac{\gamma}{1 + c_0}\right)x + \frac{\alpha}{2}.$$

Note that the first characteristic to carry $u^- = c_0$ on the shock curve corresponds to $\tau = \alpha$ (Fig. 3.16):

$$t = \alpha + \left(1 + \frac{\gamma}{(1 + c_0)^2}\right)x.$$

If $\alpha \to 0$, the shock starts in the origin and travels along

$$t = \left(1 + \frac{\gamma}{1 + c_0}\right)x.$$

The values of u are shown in Fig. 3.16. $u^* = u^*(x, t)$ is the unique positive solution of equation (3.15) with $\tau = \alpha u/c_0$.

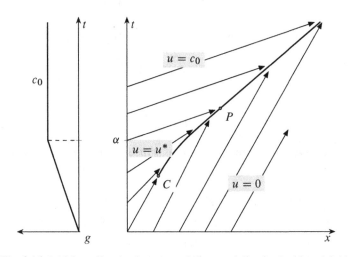

Fig. 3.16 Initial profile, shock curve and characteristics for Problem 3.2.11

3.2.2 Characteristics for linear and quasilinear equations

Problem 3.2.12 (Two-dimensional Cauchy problem). *Solve*

$$\begin{cases} u_x + x u_y = y & (x, y) \in \mathbb{R}^2 \\ u(0, y) = \cos y & y \in \mathbb{R}. \end{cases}$$

Solution. We parametrise the curve carrying (in $\mathbb{R}^3$) the data by

$$f(s) = 0, \ g(s) = s, \ h(s) = \cos s.$$

Set $x = x(t)$, $y = y(t)$ and $z = u(x(t), y(t))$, and write the associated characteristic system

$$\begin{cases} x'(t) = 1 \\ y'(t) = x(t) \\ z'(t) = y(t) \end{cases}$$

with initial condition

$$x(0) = 0, \ y(0) = s, \ z(0) = \cos s.$$

The first equation gives $x(t) = t + c$, i.e. $x(t) = t$ using the initial condition. The second equation becomes

$$y'(t) = t,$$

hence $y(t) = t^2/2 + c$, and then again $y(t) = t^2/2 + s$. The third equation reads

$$z'(t) = \frac{t^2}{2} + s$$

thus $z(t) = t^3/6 + st + \cos s$. In terms of the variable s we have, overall:

$$\begin{cases} x = X(s, t) = t \\ y = Y(s, t) = \dfrac{t^2}{2} + s \\ z = Z(s, t) = \dfrac{t^3}{6} + st + \cos s. \end{cases}$$

At this point we can solve for $s = S(x, y)$, $t = T(x, y)$ in the first two equations, and substitute into the third one to obtain the solution:

$$u(x, y) = z(T(x, y), S(x, y)).$$

The assumption in the implicit function theorem holds for the initial datum: in fact,

$$J(s, t) = \det \begin{pmatrix} X_s & Y_s \\ X_t & Y_t \end{pmatrix} = \det \begin{pmatrix} 0 & 1 \\ 1 & t \end{pmatrix} = -1 \neq 0,$$

for any t, hence also for $t = 0$. Here the computation is direct, since easily

$$t = x, \quad \text{and} \quad s = y - t^2/2 = y - x^2/2,$$

giving

$$u(x, y) = \frac{x^3}{6} + x\left(y - \frac{x^2}{2}\right) + \cos\left(y - \frac{x^2}{2}\right) = xy - \frac{1}{3}x^3 + \cos\left(y - \frac{x^2}{2}\right).$$

The solution is defined in the whole $\mathbb{R}^2$.

Problem 3.2.13 (Two-dimensional Cauchy problem). *Compute, where defined, the solution $u = u(x, y)$ of*

$$\begin{cases} uu_x + yu_y = x \\ u(x, 1) = 2x. \end{cases}$$

Solution. The curve carrying the data is

$$f(s) = s, \, g(s) = 1, \, h(s) = 2s.$$

Write $x = x(t)$, $y = y(t)$, $z = u(x(t), y(t))$. The characteristic system reads

$$\begin{cases} x'(t) = z(t) \\ y'(t) = y(t) \\ z'(t) = x(t) \end{cases}$$

with initial condition

$$x(0) = s, \, y(0) = 1, \, z(0) = 2s.$$

The second equation and $y(0) = 1$ immediately give $y(t) = e^t$. Differentiating the first, and using the third equation we write

$$x''(t) = x(t), \qquad x(0) = s, \, x'(0) = 2s.$$

The general integral is $c_1 e^t + c_2 e^{-t}$, hence $x(t) = \frac{3}{2}se^t - \frac{1}{2}se^{-t}$ and $z(t) = \frac{3}{2}se^t + \frac{1}{2}se^{-t}$. Altogether

$$\begin{cases} x = X(s, t) = \frac{3}{2}se^t - \frac{1}{2}se^{-t} \\ y = Y(s, t) = e^t \\ z = Z(s, t) = \frac{3}{2}se^t + \frac{1}{2}se^{-t}. \end{cases} \qquad (3.20)$$

Let us see if we can make $t = T(x, y)$, $s = S(x, y)$ explicit from the first two. We have

$$J(s, t) = \det\begin{pmatrix} X_s & Y_s \\ X_t & Y_t \end{pmatrix} = \det\begin{pmatrix} \frac{3}{2}e^t - \frac{1}{2}e^{-t} & 0 \\ \frac{3}{2}se^t + \frac{1}{2}se^{-t} & e^t \end{pmatrix} = \frac{3}{2}e^{2t} - \frac{1}{2}.$$

On the initial curve $J(s, 0) = -1$, always non-zero. Yet $J = 0$ for $t = -\log\sqrt{3}$, any s. Hence we expect to encounter some problems near

$$\left(X\left(s, -\log\sqrt{3}\right), Y\left(s, -\log\sqrt{3}\right)\right) = (0, \sqrt{3}/3).$$

At any rate, as $e^t = y$, the first equation in (3.20) implies

$$s\left(\frac{3y}{2} - \frac{1}{2y}\right) = x,$$

i.e.

$$s = \frac{2xy}{3y^2 - 1}, t = \log y.$$

Substituting into the third equation of (3.20) we find

$$u(x, y) = \frac{2xy}{3y^2 - 1}\left(\frac{3y}{2} + \frac{1}{2y}\right) = x\frac{3y^2 + 1}{3y^2 - 1}.$$

As expected, the solution is defined only when

$$y > \frac{\sqrt{3}}{3}$$

(recall that the initial datum is given on $y = 1$).

Problem 3.2.14 (Compatibility conditions on the data). *Consider the Cauchy problem*

$$a(x, y)u_x - u_y = -u$$

on

$$D = \{(x, y) : y > x^2\}$$

subject to

$$u(x, x^2) = g(x).$$

a) Discuss whether the problem has solutions.

b) Study the case $a(x, y) = y/2, g(x) = \exp(-cx^2)$ as $c \in \mathbb{R}$ varies.

Solution. a) We begin by examining the characteristic points on the parabola (s, s^2). Consider the determinant

$$J(s, 0) = \det\begin{pmatrix} X_s(s, 0) & Y_s(s, 0) \\ X_t(s, 0) & Y_t(s, 0) \end{pmatrix} = \det\begin{pmatrix} 1 & 2s \\ a(s, s^2) & -1 \end{pmatrix} = -2a(s, s^2)s - 1,$$

where $X(s, t), Y(s, t), Z(s, t)$ is the solution of the characteristic system

$$\begin{cases} x' = a(x, y) \\ y' = -1 \\ z' = z \end{cases}$$

with $x(0) = s$, $y(0) = s^2$, $z(0) = g(s)$.

If $J(s, 0) \neq 0$, the characteristic directions are never tangent and the problem admits a local C^1 solution in a neighbourhood of the parabola. Suppose now that for some s_0

$$J(s_0, 0) = 2a(s_0, s_0^2)s_0 + 1 = 0.$$

Necessarily $s_0 \neq 0$. If a local C^1 solution around (s_0, s_0^2) exists, then the space curve $(s, s^2, g(s))$ must be characteristic at $s = s_0$. This means that the rank of

$$\begin{pmatrix} X_s(s, 0) & Y_s(s, 0) & Z_s(s, 0) \\ X_t(s, 0) & Y_t(s, 0) & Z_t(s, 0) \end{pmatrix} = \begin{pmatrix} 1 & 2s_0 & g'(s_0) \\ -1/2s_0 & -1 & -g(s_0) \end{pmatrix}$$

is 1, i.e.

$$g'(s_0) = 2s_0 g(s_0). \tag{3.21}$$

If (3.21) does not hold there are *no* C^1 solutions in any neighbourhood of (s_0, s_0^2). There might be less regular solutions of course.

b) In the case $a(x, y) = y/2$ and $g(s) = \exp(-cs^2)$, we have

$$2a(s, s^2)s + 1 = s^3 + 1$$

and there is only one characteristic point: $(-1, 1)$. Since $g'(s) = -2csg(s)$, (3.21) is equivalent to

$$2c = -2$$

and the problem can admit regular solutions around $(-1, 1)$ only if $c = -1$. If so, $g(s) = \exp(s^2)$ and the solution is $u(x, y) = e^y$, as is easy to verify.

Problem 3.2.15 (First integrals). *Use the method of first integrals to find a solution $u = u(x, y)$ for*

$$uu_y = -y$$

in terms of an arbitrary function.

Solution. Set $x = x(t), y = y(t), z = u(x(t), y(t))$ and write (formally) the characteristic system as follows:

$$\frac{dx}{0} = \frac{dy}{z} = -\frac{dz}{y}. \tag{3.22}$$

We have to find two independent first integrals $\varphi(x, y, z)$ and $\psi(x, y, z)$. From (3.22) we infer

$$dx = 0, \qquad y\, dy = -z\, dz.$$

The first one gives

$$x = c_1,$$

so

$$\varphi(x, y, z) = x$$

is a first integral (x remains constant on the trajectories of the characteristic system). From the other one

$$y^2 + z^2 = c_2,$$

showing that

$$\psi(x, y, z) = y^2 + z^2$$

is another first integral (for any y and z, also zero). To check the independence of φ and ψ we must verify that their gradients are linearly independent vector fields. In fact

$$\nabla\varphi(x, y, z) = (1, 0, 0), \qquad \nabla\psi(x, y, z) = (0, 2y, 2z)$$

are independent except at the origin. Overall, the general integral $z = u(x, y)$ is defined implicitly by

$$F(\varphi(x, y, z), \psi(x, y, z)) = F(x, y^2 + z^2) = 0,$$

with F smooth and arbitrary. Supposing $F_x \neq 0$ we may write

$$x = f\left(y^2 + z^2\right)$$

for any smooth f. Hence the solution are surfaces of revolution about the x-axis.

Problem 3.2.16 (Euler equation). *Determine the solutions of the Euler equation*

$$xu_x + yu_y = nu. \qquad (3.23)$$

where $n > 0$.

Solution. Let us define $x = x(t)$, $y = y(t)$, $z = u(x(t), y(t))$ and write the characteristic system in the form

$$\frac{dx}{x} = \frac{dy}{y} = \frac{dz}{nz}.$$

We seek two (independent) first integrals, and observe that each of the equations

$$\frac{dx}{x} = \frac{dy}{y} \quad \text{and} \quad \frac{dx}{x} = \frac{dz}{nz}$$

contain two of the three unknowns only. The first one gives $\log|x| = \log|y| + c$, so $\frac{y}{x} = c_1$ and

$$\varphi(x, y, z) = \frac{y}{x}$$

is then a first integral when $x \neq 0$ (clearly x/y is a first integral when $y \neq 0$). The second equation can be treated similarly (still for $x \neq 0$)

$$\frac{z}{x^n} = c_2,$$

thus another first integral is

$$\psi(x, y, z) = \frac{z}{x^n}.$$

For $x \neq 0$ the first integrals are independent, for their gradients

$$\nabla\varphi = (-x^{-2}y, x^{-1}, 0) \quad \text{and} \quad \nabla\psi = (-nx^{-n-1}z, 0, x^{-n})$$

are linearly independent. The general solution $z = u(x, y)$ is given, implicitly, by the equation

$$F(\varphi, \psi) = F\left(\frac{y}{x}, \frac{z}{x^n}\right) = 0,$$

with F arbitrary and smooth. If $F_\psi \neq 0$ we can write

$$z = u(x, y) = x^n f\left(\frac{y}{x}\right),$$

with f smooth and arbitrary. Thus the solutions are positively homogeneous functions of degree n. Indeed, for any $l > 0$,

$$u(lx, ly) = (lx)^n f\left(\frac{lx}{ly}\right) = l^n u(x, y)$$

As a matter of fact it can be proved in an elementary way that a function is homogeneous of degree n if and only if it satisfies the *Euler equation* (3.23).

Problem 3.2.17 (Cauchy problem). *Solve the problem*

$$u_x + u_y = u \qquad u(x, 0) = \cos x$$

in two different ways: a) using the characteristic system, and b) via first integrals.

Solution. a) The parametric equations of the initial curve are

$$f(s) = s, \quad g(s) = 0, \quad h(s) = \cos s.$$

Write $x = x(t), y = y(t), z = z(x(t), y(t))$, and the characteristic system is

$$\begin{cases} x'(t) = 1 & x(0) = s \\ y'(t) = 1 & y(0) = 0 \\ z'(t) = z(t) & z(0) = \cos s. \end{cases}$$

The solutions

$$\begin{cases} x(t, s) = t + s \\ y(t, s) = t \\ z(t, s) = e^t \cos s, \end{cases}$$

immediately give $t = y, s = x - y$. Hence the solution is

$$z = u(x, y) = e^y \cos(x - y).$$

b) Using first integrals, let us rewrite the characteristic system as

$$dx = dy = \frac{dz}{z}.$$

Regrouping as $dx = dy, dx = dz/z$, we obtain two first integrals

$$\varphi(x, y, z) = x - y, \quad \text{and} \quad \psi(x, y, z) = ze^{-x}.$$

As

$$\nabla \varphi = (1, -1, 0) \quad \text{and} \quad \nabla \psi = (-ze^{-x}, 0, e^{-x})$$

the gradients are linearly independent, making

$$z = u(x, y) = e^x f(x - y)$$

for any smooth f, the general solution. Imposing the initial condition

$$e^x f(x) = \cos x,$$

i.e.

$$f(x) = e^{-x} \cos x,$$

we recover the solution of part a).

Problem 3.2.18 (Characteristic variables). *Consider the first-order linear equation*

$$a(x, y)u_x + b(x, y)u_y + c(x, y)u = f(x, y) \tag{3.24}$$

where a, b, c, f are C^1 functions on the open set $D \subset \mathbb{R}^2$, and $a \neq 0$ on D. Let $\varphi(x, y) = k, k \in \mathbb{R}$, be the family of characteristic curves for the reduced equation

$$au_x + bu_y = 0.$$

Let $\psi = \psi(x, y)$ be a smooth function independent from φ, that is:

$$\det \begin{pmatrix} \varphi_x & \varphi_y \\ \psi_x & \psi_y \end{pmatrix} \neq 0 \qquad in \ D. \tag{3.25}$$

Show that the change of variables

$$\begin{cases} \xi = \varphi(x, y) \\ \eta = \psi(x, y) \end{cases} \tag{3.26}$$

transforms (3.24) in an ordinary equation of the form

$$w_\eta + P(\xi, \eta)w = Q(\xi, \eta).$$

Deduce a formula for the general solution of (3.24).

Solution. By (3.25) the transformation (3.26) is locally invertible, thus we can express x, y as smooth functions of ξ and η:

$$x = \Phi(\xi, \eta), \qquad y = \Psi(\xi, \eta).$$

Set

$$w(\xi, \eta) = u(\Phi(\xi, \eta), \Psi(\xi, \eta))$$

or

$$u(x, y) = w(\varphi(x, y), \psi(x, y)).$$

We have

$$u_x = w_\xi \varphi_x + w_\eta \psi_x, \qquad u_y = w_\xi \varphi_y + w_\eta \psi_y,$$

and substituting into (3.24) we obtain

$$(a\varphi_x + b\varphi_y)w_\xi + (a\psi_x + b\psi_y)w_\eta + C(\xi, \eta)w = F(\xi, \eta), \tag{3.27}$$

where, for simplicity, we have set

$$C(\xi, \eta) = c(\Phi(\xi, \eta), \Psi(\xi, \eta)), \qquad F(\xi, \eta) = f(\Phi(\xi, \eta), \Psi(\xi, \eta)).$$

Now observe the following:

1. $a\varphi_x + b\varphi_y = 0$ on D. In fact, along the characteristics $\varphi(x, y) = k$,

$$\varphi_x \, dx + \varphi_y \, dy = 0.$$

On the other hand, along the characteristics $\varphi = k$ we have the system

$$dx = a \, dt, \quad dy = b \, dt,$$

so that

$$\left[\varphi_x a + \varphi_y b\right] dt = 0,$$

and thus the claim.

2. $a\psi_x + b\psi_y \neq 0$ in D. This is due to the independence of ψ and φ, i.e. equation (3.25), implying

$$-\frac{b}{a} = \frac{\varphi_x}{\varphi_y} \neq \frac{\psi_x}{\psi_y}.$$

Setting

$$D(\xi, \eta) = a(x, y)\psi_x(x, y) + b(x, y)\psi_y(x, y)\big|_{x=\Phi(\xi,\eta),\, y=\Psi(\xi,\eta)},$$

from 1. and 2. we obtain that (3.27) now reads

$$w_\eta + \frac{C(\xi, \eta)}{D(\xi, \eta)} w = \frac{F(\xi, \eta)}{D(\xi, \eta)},$$

which is an (ordinary) linear equation of order one in $w(\xi, \cdot)$, for any given ξ. The general solution is

$$w(\xi, \eta) = \exp\left(-\int \frac{C(\xi, \eta)}{D(\xi, \eta)} \, d\eta\right) \cdot \left[\int \exp\left(\int \frac{C(\xi, \eta)}{D(\xi, \eta)} \, d\eta\right) \frac{F(\xi, \eta)}{D(\xi, \eta)} \, d\eta + G(\xi)\right],$$

with G smooth. Now it suffices to go back to the original variables to obtain the solution

$$u(x, y) = w(\varphi(x, y), \psi(x, y)).$$

Problem 3.2.19 (Using characteristic variables). *Using the method seen in the previous problem determine the solution of*

$$xu_x - yu_y = u - y \qquad \text{in } D = \{(x, y) \in \mathbb{R}^2 : y > 0\}$$

such that $u(x, y) = y$ on the parabola $x = y^2$ $(y > 0)$.

Solution. We adopt the strategy (and notation) of the previous exercise. The characteristics of $xu_x - yu_y = 0$ solve the differential equation

$$\frac{dx}{x} = -\frac{dy}{y},$$

and are then given by $\varphi(x, y) = xy = k$. Let us choose (for example) $\psi(x, y) = y$. This choice is admissible, for

$$\det \begin{pmatrix} \varphi_x & \varphi_y \\ \psi_x & \psi_y \end{pmatrix} = \det \begin{pmatrix} y & x \\ 0 & 1 \end{pmatrix} = y > 0 \qquad \text{on } D,$$

and therefore φ and ψ are independent in D. We now make the change of variables (with $y > 0$)

$$\xi = xy, \ \eta = y$$

with inverse $(\eta > 0)$ $x = \xi/\eta$, $y = \eta$. Then

$$C(\xi, \eta) = -1, \qquad F(\xi, \eta) = -\eta, \qquad D(\xi, \eta) = -\eta$$

and $w(\xi, \eta) = u(\xi/\eta, \eta)$ solves the equation

$$w_\eta + \frac{1}{\eta} w = 1.$$

Hence

$$w(\xi, \eta) = \frac{\eta}{2} + \frac{G(\xi)}{\eta},$$

i.e.

$$u(x, y) = \frac{y}{2} + \frac{G(xy)}{y},$$

for any smooth G. To find G we impose the boundary condition: along $x = y^2$ we must have

$$y = \frac{y}{2} + \frac{G(y^3)}{y},$$

whence $G(y^3) = y^2/2$, i.e. $G(z) = z^{2/3}/2$. The solution is then

$$u(x, y) = \frac{1}{2}(y + x^{2/3} y^{-1/3}).$$

Remark. No neighbourhood of the origin contains a solution to the Cauchy problem. In fact, since $x = 0$ is a characteristic tangent to $x = y^2$ at $(0, 0)$, the origin is a *characteristic point*. Differentiating $u\left(y^2, y\right) = y$ we obtain

$$2y u_x(y^2, y) + u_y(y^2, y) = 1. \tag{3.28}$$

Suppose $y \neq 0$ for the moment. Restricting the PDE on the parabola and simplifying gives

$$y u_x(y^2, y) - u_y(y^2, y) = 0. \tag{3.29}$$

The last two conditions uniquely determine u_x, u_y on the parabola provided the matrices

$$\begin{pmatrix} 2y & 1 \\ y & -1 \end{pmatrix} \quad \text{and} \quad \begin{pmatrix} 2y & 1 & 1 \\ y & -1 & 0 \end{pmatrix}$$

have the same rank. By continuity this condition must hold at $y = 0$ as well. But there the ranks are different, and the solution does not exist.

Problem 3.2.20 (Heat exchanger). *The simplest heat exchanger is modelled by a cylindrical tube at temperature $T_a = T_a(t)$ (in Celsius degrees) immersed in a fluid of temperature $T(x,t)$. The fluid flows in the tube at constant speed v m/s (meters per seconds). Denote by U the heat transfer coefficient through the walls (unit: cal $s^{-1}m^{-2}(^\circ C)^{-1}$), by C_p the heat capacity per unit of volume (in cal $m^{-3}(^\circ C)^{-1}$), by A (m^2) the area and by p (m) the perimeter of the cross section. We assume that U and C_p are constant.*

a) Write down the differential equation governing the evolution of the temperature T.

b) Solve the problem knowing that

$$\begin{cases} T(x,0) = T_0(x) & 0 \le x \le L \\ T(0,t) = g(t) & t > 0. \end{cases}$$

Solution. a) We start by writing the equation of thermal equilibrium between the cross sections at x and $x + \Delta x$, relative to the time interval $(t, t + \Delta t)$. We know that:

• The heat exchanged (better, absorbed) through the walls of the tube is linear (Newton's law of cooling)
$$Up\Delta x(T_a - T)\Delta t \quad \text{(calories)}.$$

• The heat flowing in, due to the temperature difference at x, $x + \Delta x$, is given by (Fourier's law)

$$vAC_p[T(x,t) - T(x + \Delta x,t)]\Delta t \quad \text{(calories)}.$$

• The variation of heat in this portion, between times t and $t + \Delta t$, is

$$A\Delta xC_p[T(x,t) - T(x,t + \Delta t)] \quad \text{(calories)}.$$

All this together gives

$$A\Delta xC_p[T(x,t) - T(x,t + \Delta t)] = \\ vAC_p[T(x,t) - T(x + \Delta x,t)]\Delta t + Up\Delta x(T_a(t) - T)\Delta t.$$

Divide by $\Delta x\Delta t$ and take the limit as $\Delta x, \Delta t \to 0$. We obtain:

$$\frac{\partial T}{\partial t} + v\frac{\partial T}{\partial x} = h(T_a(t) - T), \qquad \text{where } h = \frac{Up}{AC_p}.$$

b) The characteristics (actually, their projections on the xt-plane) are the straight lines

$$x - vt = k.$$

It is natural to divide the region $[0, L] \times [0, +\infty)$ in two parts: the first part between the x-axis and the line $x = vt$, where T is determined by T_0, and corresponding to $vt < x \leq L$; the second one above the line $x = vt$, corresponding to $0 \leq x \leq \min(vt, L)$, where T depends on g.

To determine the solution in $vt < x \leq L$ we consider the characteristic $x = vt + \xi$, $0 \leq \xi \leq L$, and set $z(t) = T(vt + \xi, t)$. Using the differential equation, we see that z solves

$$z'(t) = vT_x(vt + \xi, t) + T_t(vt + \xi, t) = h(T_a(t) - z(t))$$

with

$$z(0) = T(\xi, 0) = T_0(\xi).$$

Therefore

$$z(t) = T_0(\xi)e^{-ht} + h \int_0^t T_a(s)e^{-h(t-s)} ds$$

and since $\xi = x - vt$, we find

$$T(x, t) = T_0(x - vt)e^{-ht} + h \int_0^t T_a(s)e^{-h(t-s)} ds, \qquad vt < x \leq vL.$$

Now let us examine the points (x, t) for which $vt \geq x$. Solve the equation of the characteristic for t

$$t = \frac{x}{v} + \tau$$

and set

$$u(x) = T\left(x, \frac{x}{v} + \tau\right).$$

Then

$$\frac{du}{dx} = T_x\left(x, \frac{x}{v} + \tau\right) + \frac{1}{v}T_t\left(x, \frac{x}{v} + \tau\right) = \frac{h}{v}\left(T_a\left(\frac{x}{v} + \tau\right) - u\right)$$

with

$$u(0) = T(0, \tau) = g(\tau).$$

Therefore

$$u(x) = g(\tau)e^{-hx/v} + \frac{h}{v}\int_0^x T_a\left(\frac{s}{v} + \tau\right)e^{-h(x-s)/v} ds,$$

and, since $\tau = t - \frac{x}{v}$, we have

$$T(x, t) = g\left(t - \frac{x}{v}\right)e^{-hx/v} + \frac{h}{v}\int_0^x T_a\left(t + \frac{s - x}{v}\right)e^{-h(x-s)/v} ds.$$

Finally, setting

$$\eta = t + \frac{s - x}{v}, \qquad d\eta = \frac{1}{v} ds$$

we obtain

$$T(x, t) = g\left(t - \frac{x}{v}\right)e^{-hx/v} + h \int_{t - \frac{x}{v}}^t T_a(\eta)e^{-h(t-\eta)} d\eta, \qquad 0 \leq x \leq \min(vt, L).$$

Remark. In case

$$g(t) = g_0, \qquad T_a(t) = T_a$$

are constant, for $x > vt$ we find

$$T(x,t) = g_0 e^{-hx/v} + T_a \left(1 - e^{-hx/v}\right),$$

the stationary state, which is reached everywhere along the tube as soon as $t \geq L/v$.

Problem 3.2.21 (Three-dimensional Cauchy problem). *Compute, where defined, the solution $u = u(x,y,z)$ to*

$$\begin{cases} u_x + x u_y - u_z = u \\ u(x,y,1) = x + y. \end{cases}$$

Solution. We parametrise the surface carrying the data by

$$f(r,s) = s, \ g(r,s) = r, \ h(r,s) = 1, \ k(r,s) = s + r$$

then write $x = x(t)$, $y = y(t)$, $z = z(t)$, $v = u(x(t), y(t), z(t))$, and the characteristic system reads

$$\begin{cases} x'(t) = 1 & x(0) = s \\ y'(t) = x(t) & y(0) = r \\ z'(t) = -1 & z(0) = 1 \\ v'(t) = v(t) & v(0) = s + r. \end{cases}$$

The first and third equations give $x(t) = t + s$ and $z(t) = -t + 1$. The last one is decoupled, and implies $v(t) = (s + r)e^t$. The second one now reads

$$y'(t) = t + s,$$

and by the initial condition $y(t) = t^2/2 + st + r$. Summarising,

$$\begin{cases} x = X(r,s,t) = t + s \\ y = Y(r,s,t) = t^2/2 + st + r \\ z = Z(r,s,t) = -t + 1 \\ v = V(r,s,t) = (s + r)e^t. \end{cases} \tag{3.30}$$

To have (t, r, s) in terms of (x, y, z), we get $t = 1 - z$ from the third, $s = x - t = x + z - 1$ from the first and

$$r = y - t^2/2 - st = y - (1-z)^2/2 - (1-z)(x + z - 1)$$

from the second. Substituting into the fourth equation we find

$$u(x, y, z) = \left[xz + y + \frac{1}{2}(z^2 - 1) \right] e^{1-z}.$$

The solution is defined and smooth in the whole $\mathbb{R}^3$. In this case the general sufficient condition for solving the first three equations of system (3.30) would have been

$$J(r, s, t) = \det \begin{pmatrix} X_r & Y_r & Z_r \\ X_s & Y_s & Z_s \\ X_t & Y_t & Z_t \end{pmatrix} = \det \begin{pmatrix} 0 & 1 & 0 \\ 1 & t & 0 \\ 1 & t+s & -1 \end{pmatrix} = 1 \neq 0,$$

In fact, in our case $J(r, s, t) = 1$.

Problem 3.2.22 (Clairaut's equation). *Consider the Clairaut equation*

$$xu_x + yu_y + f(u_x, u_y) = u.$$

a) Verify that $u(x, y; a, b) = ax + by + f(a, b)$ is a solution for any $(a, b) \in \mathbb{R}^2$.

b) In the case

$$f(a, b) = \frac{a^2 + b^2}{2}$$

use the envelope of the solutions of part a) to generate a nonlinear solution in x and y and a solution depending on an arbitrary function.

Solution. a) Immediate: $u_x = a, u_y = b$ and hence

$$xu_x + yu_y + f(u_x, u_y) = ax + by + f(a, b) = u.$$

b) With the given f the equation becomes

$$xu_x + yu_y + \frac{u_x^2 + u_y^2}{2} = u,$$

while the family of solutions reads

$$G(x, y, u; a, b) = u - \left[ax + by + \frac{a^2 + b^2}{2} \right] = 0.$$

Let us find its envelope: we have to eliminate a, b from the system

$$G = 0, \qquad G_a = 0, \qquad G_b = 0$$

i.e.

$$\begin{cases} u = ax + by + \frac{a^2 + b^2}{2} \\ x + a = 0 \\ y + b = 0, \end{cases}$$

showing that the envelope is

$$u(x, y) = -\frac{1}{2}(x^2 + y^2).$$

It is easy to see, by direct computation, that u is indeed a solution.

Remark. A two-parameter family of solutions $z = \varphi(x, y; a, b)$ is called *complete integral* if the vector fields

$$\left(\varphi_a, \varphi_{xa}, \varphi_{ya}\right) \quad \text{and} \quad \left(\varphi_b, \varphi_{xb}, \varphi_{yb}\right)$$

are linearly *independent*. Given a complete integral, we can recover a solution depending on a function by setting, for instance, $b = w(a)$ and writing the envelope of $z = \varphi(x, y; a, w(a))$. For instance, choosing $w(a) = a$ gives, for the Clairaut equation with $f(a, b) = \frac{a^2 + b^2}{2}$,

$$\begin{cases} u = a(x + y) + a^2 \\ 0 = x + y + 2a \end{cases} \quad \text{whence } u = -\frac{(x + y)^2}{4}.$$

3.3 Further Exercises

3.3.1. *Solve the global Cauchy problem for the Burgers equation $u_t + u u_x = 0$ with initial datum $g(x)$ given by:*

a) $\begin{cases} 1 & x \le 0 \\ 1 - x & 0 < x < 1 \\ 0 & x \ge 1. \end{cases}$ b) $\begin{cases} 1 & x < 0 \\ 2 & 0 < x < 1 \\ 0 & x < 1. \end{cases}$ c) $\begin{cases} |x| & |x| \le 1 \\ 1 & |x| \ge 1. \end{cases}$

3.3.2. *Consider the traffic equation (Problem 3.2.4 on page 160), with initial density*

$$g(x) = \begin{cases} a\rho_m & x < 0 \\ \rho_m & x > 0. \end{cases}$$

For $0 < a < 1$, determine the characteristics, the possible shock waves, and find the solutions on the half-plane (x, t), $t > 0$.

3.3.3. *Solve the global Cauchy problem for the traffic equation (Problem 3.2.4 on page 160) with initial density*

$$g(x) = \begin{cases} \rho_m/4 & x < -1 \\ \rho_m/3 & -1 < x \le 0 \\ (2 + x)\rho_m/6 & 0 \le x \le 1 \\ \rho_m/2 & x \ge 1. \end{cases}$$

3.3.4. *Solve the following problem, variant of the tunnel Problem 3.2.6 (page 164)*

$$\begin{cases} \rho_t + q(\rho)_x = 0 & x \in \mathbb{R}, \, t > 0 \\ \rho(x, 0) = g(x), \end{cases}$$

where

$$q(\rho) = \rho \log \left(\frac{1}{\rho} \right) \quad and \quad g(x) = \begin{cases} 1 & x < 0 \\ e^{-4} & x > 0. \end{cases}$$

Determine the path of the vehicle initially at $x_0 = -1$.

3.3.5. *Generalise Problem 3.2.7 (page 168) to*

$$\begin{cases} u_t + q(u)_x = 0 & x \in \mathbb{R}, t > 0 \\ u(x,0) = g(x) & x \in \mathbb{R}, \end{cases}$$

with $q \in C^2(\mathbb{R})$ concave.

3.3.6. *Study*

$$\begin{cases} c_x + q\,(c)_t = -c & x > 0, t > 0 \\ c(x,0) = 0 & x > 0 \\ c(0,t) = c_0 & t > 0 \end{cases}$$

where $q\,(c) = \left(c + \frac{c}{c+1}\right)$ and $c_0 > 0$ is constant.

3.3.7. *Compute, where defined, the solutions to the following (two-dimensional) Cauchy problems:*

a) $xu_x + u_y = y, u\,(x,0) = x^2$.
b) $u_x - 2u_y = u, u\,(0,y) = y$.
c) $u_x + 3u^2 u_y = 1, u\,(x,0) = 0$.

3.3.8. *Write the general solution of (depending on an arbitrary function):*

a) $u_x - \sqrt{u}u_y = 0.$ *b)* $\dfrac{1}{y}u_x + u_y = u^2.$

3.3.9. *Find a solution $z = u(x, y)$ of*

$$yu_x - xu_y = 2xyu, \quad u = s^2 \text{ along } \Gamma = \{(s,s) : s \in \mathbb{R}\}.$$

3.3.10. *Compute the general solution of*

$$xu_x - yu_y + u = y \quad on\ D = \{(x, y) \in \mathbb{R}^2 : x, y > 0\}$$

in two ways: a) by using characteristic variables (Problem 3.2.18 on page 187) and b) by using first integrals.

3.3.11. *(Picone) Let $D = \{(x, y) : x^2 + y^2 \le 1\}$ be the closed unit disc in $\mathbb{R}^2$ and $u \in C^1(D)$ be a solution to*

$$a\,(x, y)\,u_x + b\,(x, y)\,u_y = -\gamma u \quad on\ D.$$

Suppose the coefficients a and b are continuous on D and satisfy

$$a\,(x, y)\,x + b\,(x, y)\,y > 0 \quad on\ \partial D.$$

If $\gamma > 0$, show that $u \equiv 0$ on D.

3.3.12. *(Buckley-Leverett equation) Consider the equation*

$$\phi S_t = H'(S)\mathbf{q} \cdot \nabla S,$$

where $S = S(\mathbf{x}, t)$, $\mathbf{x} \in \mathbb{R}^3$ and

$$H(S) = -\frac{k_1(S)}{\mu_1} \left(\frac{k_1(S)}{\mu_1} + \frac{k_2(S)}{\mu_2} \right)^{-1}.$$

Solve the equation when

$$\frac{k_1(S)}{\mu_1} = 1 - S, \qquad \frac{k_2(S)}{\mu_2} = S, \qquad \phi = \frac{1}{2}, \qquad \mathbf{q} = \left(\frac{1}{2}e^{-x_1}, \frac{1}{2}e^{-x_2}, \frac{1}{2}e^{-x_3} \right),$$

with initial condition $S(\mathbf{x}, 0) = g(\mathbf{x})$.

3.3.13. Solve

$$(cy - bz)u_x + (az - cx)u_y + (bx - ay)u_z = 0$$

and interpret geometrically the solution (seek a suitable combination of the characteristic equations that gives integrable equations).

3.3.14. *a)* Let $\mathcal{F}$ be the family of surfaces in $\mathbb{R}^3$ given in implicit form by $w(x, y, z) = c$ depending on the parameter c. Find the family of orthogonal surfaces, those that intersect each member of $\mathcal{F}$ perpendicularly.

b) Determine the family of orthogonal surfaces to the family of paraboloids $z - x^2 - y^2 = c$.

3.3.15. The concentration u of a substance contained in a semi-infinite tube (along the semiaxis $x \geq 0$) solves

$$u_t + vu_x = -ku,$$

where v and k are positive constants. Compute u knowing that

$$u(x, 0) = f(x) \text{ for } x > 0, \qquad u(0, t) = g(t) \text{ for } t > 0.$$

3.3.16. The concentration u of a substance contained in a semi-infinite tube (along the semiaxis $x \geq 0$) solves

$$u_t + vu_x = -ku^\alpha,$$

with $v, k, \alpha \neq 1$ positive constants. Compute u knowing that

$$u(x, 0) = f(x) \text{ for } x > 0, \qquad u(0, t) = g(t) \text{ for } t > 0.$$

3.3.17. (Separation of variables) Consider the PDE

$$a(x)u_x^2 + b(y)u_y^2 = f(x) + g(y),$$

with a, b continuous on some interval I.

a) Determine, at least formally, a complete integral by seeking solutions of the form $u(x, y) = v(x) + w(y)$.

b) Exploit this result for solving $u_x^2 + u_y^2 = 1$.

3.3.18. (Burgers equation, decreasing datum) *Study the problem*

$$\begin{cases} u_t + uu_x = 0 & x \in \mathbb{R}, t > 0 \\ u(x,0) = g(x) & x \in \mathbb{R}, \end{cases} \quad \text{where} \quad g(x) = \begin{cases} 1 & x < 0 \\ 1 - x^2 & 0 \le x \le 1 \\ 0 & x > 1. \end{cases}$$

3.3.19. *Consider the model of Problem 3.2.11 on page 176. Set $\gamma = 1$ and find the solution on the quadrant $x > 0, t > 0$, with $u(x,0) = 0$ and*

$$u(0,t) = g(t) = \begin{cases} c_0 & 0 \le t \le 1 \\ 0 & t > 1. \end{cases}$$

3.3.1 Solutions

Solution 3.3.1. **a)** With the given datum we deduce that the characteristics have equation

$$\begin{cases} x = \xi + t & \text{if } \xi \le 0 \\ x = \xi + (1 - \xi)t & \text{if } 0 < \xi < 1 \\ x = \xi & \text{if } \xi \ge 1. \end{cases}$$

All characteristics based on the segment $0 < \xi < 1$ cross the point $(1, 1)$; therefore, for $0 < t < 1$ they do not collide. This fact means that in this time interval the solution has no discontinuity. A discontinuity originates at the point $(1, 1)$ because of the collision between the vertical characteristics (carrying the datum $u^- = 0$) and the characteristics $x = \xi + t$ (carrying $u^+ = 1$). Observing that $q^- = 0$ and $q^+ = 1/2$ are the fluxes from the left and the right of the discontinuity, respectively, according to the Rankine-Hugoniot condition, the shock curve solves the problem

$$\begin{cases} \dot{s} = 1/2 \\ s(1) = 1, \end{cases}$$

and therefore $s(t) = (t + 1)/2$.

In the region $S = \{0 \le x < 1, 0 \le t < x\}$ the solution is implicitly defined by the equation $u = g(x - q'(u)t)$. In this case, given that $g(x) = 1 - x$, we deduce

$$u(x,t) = \frac{1 - x}{1 - t}.$$

The solution of the problem is

$$u(x,t) = \begin{cases} 1 & \text{if } x \le t < 1 \text{ or } x < (t + 1)/2 \text{ with } t \ge 1 \\ (1 - x)/(1 - t) & \text{if } 0 \le t \le x < 1 \\ 0 & \text{if } t < 2x - 1 \text{ with } x > 1. \end{cases}$$

b) For the assigned initial condition, the characteristics have equation

$$\begin{cases} x = \xi + t & \text{if } \xi < 0 \\ x = \xi + 2t & \text{if } 0 < \xi < 1 \\ x = \xi & \text{if } \xi > 1. \end{cases}$$

The domain $S = \{t < x < 2t\}$ is reached by no characteristics; we can define the solution u as a rarefaction wave that connects the values $u = 1$ and $u = 2$, transported by the characteristics

$x = t$ and $x = 2t$. In this case, the rarefaction wave based at the point $(0, 0)$ is $u = x/t$. Along the straight lines $x = ht$, with $1 < h < 2$, u is constant. The collision of the characteristics creates a discontinuity $x = s(t)$ that propagates following the Rankine-Hugoniot condition. At least for small times t, the shock curve is due to a discontinuity created by the impact between the value $u^+ = 2$ (transported along the characteristics $x = \xi + 2t$) and the value $u^- = 0$ (along the vertical characteristics). Taking into account that $q(u) = u^2/2$, those data are respectively associated to the fluxes $q^+ = q(u^+) = 2$ and $q^- = q(u^-) = 0$. Therefore, the shock curve is given by the solution of the problem

$$\begin{cases} \dot{s} = 1 \\ s(0) = 1 \end{cases} \qquad \text{i.e.} \qquad s(t) = t + 1.$$

For times $t > 1$, namely, from the point $(2, 1)$ the vertical characteristics $x = \xi$, with $\xi > 2$, interact with the rarefaction wave $u = x/t$. In this situation, the Rankine-Hugoniot condition has to be solved using $u^+ = s/t$ and $q^+ = s^2/2t^2$; hence, the shock curve is determined by the Cauchy problem

$$\begin{cases} \dot{s} = s/2t \\ s(1) = 2. \end{cases}$$

The solution of the Cauchy problem is $s(t) = 2\sqrt{t}$. For times $t > 4$, i.e. from the point $(4, 4)$, the shock curve interacts with the characteristics $x = \xi + t$, $\xi < 0$, that carry the data $u = 1$. Therefore it is deviated again; in this case we have $u^+ = 1$, $q^+ = q(u^+) = 1/2$, and

$$\begin{cases} \dot{s} = 1/2 \\ s(4) = 4 \end{cases} \qquad \text{i.e.} \qquad s(t) = \frac{t}{2} + 2.$$

Then the shock curve is made of three arcs connected at the points $(2, 1)$ and $(4, 4)$. The solution is

$$u(x, t) = \begin{cases} 1 & \text{if } x < t < 4 \text{ or } x < t/2 + 2 \text{ with } t > 4 \\ x/t & \text{if } t \le x \le 2t, \text{ with } x \le 2\sqrt{t} \\ 2 & \text{if } 2t \le x < t + 1 \\ 0 & \text{if } t + 1 < x < 2 \text{ or } t < x^2/4 \text{ with } 2 \le x < 4 \\ & \text{or } t < 2x - 4 \text{ with } x \ge 4. \end{cases}$$

c) Reasoning as above we find

$$u(x, t) = \begin{cases} 1 & x \le t - 1 \text{ for } t < 1 \\ 1 & x < 1 + t - \sqrt{2 + 2t} \text{ for } t \ge 1 \\ \frac{x}{t-1} & t - 1 \le x \le 0 \text{ for } t < 1 \\ \frac{x}{t+1} & \max(0, 1 + t - \sqrt{2 + 2t}) < x \le t + 1 \\ 1 & x \ge t + 1. \end{cases}$$

Solution 3.3.2. The characteristic from $(\xi, 0)$ has equation

$$x = \xi + q'(g(\xi))t = \xi + v_m \left(1 - \frac{2\rho}{\rho_m} \right) t.$$

Substituting the initial datum we find

$$x = \xi + v_m(1 - 2a)t, \qquad \text{for } \xi < 0$$

and

$$x = \xi - v_m t, \qquad \text{for } \xi > 0.$$

As $0 < a < 1$, $v_m(1-2a) > -v_m$, and thus a shock wave originates from (0,0). From the Rankine-Hugoniot condition we find

$$s' = \frac{q(\rho^+) - q(\rho^-)}{\rho^+ - \rho^-} = \frac{v_m}{\rho_m} \frac{\rho^+(\rho_m - \rho^+) - \rho^-(\rho_m - \rho^-)}{\rho^+ - \rho^-} = \frac{v_m}{\rho_m}\left(\rho_m - \rho^+ - \rho^-\right).$$

Substituting

$$\rho^+ = \rho_m, \rho^- = a\rho_m$$

we obtain

$$\begin{cases} s' = -av_m \\ s(0) = 0, \end{cases}$$

and then

$$s(t) = -av_m t.$$

Therefore, for any $0 < a < 1$, we have

$$u(x,t) = \begin{cases} a\rho_m & x < -av_m t \\ \rho_m & x > -av_m t. \end{cases}$$

Solution 3.3.3. Since $q(\rho) = v_m \rho(1 - \rho/\rho_m)$ is concave and g increasing we expect multiple shocks. The characteristics have equation

$$x = q'(g(\xi))t + \xi = v_m\left(1 - \frac{2g(\xi)}{\rho_m}\right)t + \xi = \begin{cases} v_m t/2 + \xi & \xi < -1 \\ v_m t/3 + \xi & -1 < \xi \le 0 \\ v_m t(1-\xi)/3 + \xi & 0 \le \xi \le 1 \\ \xi & \xi \ge 1. \end{cases}$$

The Rankine-Hugoniot condition for the shock wave $x = s(t)$ may be written

$$s'(t) = \frac{q(\rho^+(s(t),t)) - q(\rho^-(s(t),t))}{\rho^+(s(t),t) - \rho^-(s(t),t)} = v_m\left[1 - \frac{\rho^+(s(t),t) + \rho^-(s(t),t)}{\rho_m}\right].$$

The first shock is determined by the collision of characteristics coming from points $\xi < -1$ on the left and points $-1 < \xi < 0$ on the right. The starting point is $S_1 = (-1, 0)$. Substituting $\rho^- \equiv \rho_m/4$ and $\rho^+ \equiv \rho_m/3$, we obtain

$$\begin{cases} s_1' = 5v_m/12 \\ s_1(0) = -1, \end{cases} \qquad \text{and hence } x = s_1(t) = \frac{5}{12}v_m t - 1.$$

Solving for ξ the equation of the characteristics corresponding to $0 \le \xi \le 1$, we find

$$\xi = \frac{v_m t - 3x}{v_m t - 3}, \qquad \text{and hence } \rho(x,t) = g(\xi) = \frac{v_m t - x - 2}{2(v_m t - 3)}\rho_m.$$

This expression holds when $t < 3/v_m$, and all these characteristics meet at

$$S_2 = \left(1, \frac{3}{v_m}\right).$$

From S_2 another shock starts, for which $\rho^+ \equiv \rho_m/2$ and $\rho^- \equiv \rho_m/3$. The latter follows from $s_1(3/v_m) = 1/4 < 1$, whence the wave with density $\rho_m/4$ reaches $x = 1$ in a subsequent time.

We find:

$$\begin{cases} s_2' = v_m/6 \\ s_2(3/v_m) = 1, \end{cases} \qquad \text{hence } x = s_2(t) = \frac{1}{6}v_m t + \frac{1}{2}.$$

The two shock curves meet at a point (x, t) solution of the system

$$\begin{cases} x = \frac{5}{12}v_m t - 1 \\ x = \frac{1}{6}v_m t + \frac{1}{2}, \end{cases} \qquad \text{i.e. at } S_3 = \left(\frac{3}{2}, \frac{6}{v_m}\right).$$

The third and final shock is born at S_3, where $\rho^+ \equiv \rho_m/2$ and $\rho^- \equiv \rho_m/4$:

$$\begin{cases} s_3' = v_m/4 \\ s_3(6/v_m) = 3/2, \end{cases} \qquad \text{yielding } x = s_3(t) = \frac{1}{4}v_m t.$$

Summing up, the solution is:

- For $0 \le t < \dfrac{3}{v_m}$, $\rho(x,t) = \begin{cases} \rho_m/4 & x < s_1(t) \\ \rho_m/3 & s_1(t) < x \le v_m t/3 \\ \rho_m(v_m t - x - 2)/(2v_m t - 6) & v_m t/3 \le x \le 1 \\ \rho_m/2 & x \ge 1. \end{cases}$

- For $\dfrac{3}{v_m} \le t < \dfrac{6}{v_m}$, $\rho(x,t) = \begin{cases} \rho_m/4 & x < s_1(t) \\ \rho_m/3 & s_1(t) < x < s_2(t) \\ \rho_m/2 & x > s_2(t). \end{cases}$

- For $t \ge \dfrac{6}{v_m}$, $\rho(x,t) = \begin{cases} \rho_m/4 & x < s_3(t) \\ \rho_m/2 & x > s_3(t). \end{cases}$

Solution 3.3.4. Since

$$q'(\rho) = -1 - \log \rho,$$

the problem to solve is

$$\begin{cases} \rho_t - (1 + \log \rho)\rho_x = 0 & x \in \mathbb{R}, \, t > 0 \\ \rho(x, 0) = \begin{cases} 1 & x < 0 \\ e^{-4} & x > 0. \end{cases} \end{cases}$$

The characteristic from $(\xi, 0)$, the line $x = \xi + q'(g(\xi))t$, coincides with

$$x = \xi - t \qquad \text{for } \xi < 0$$

and with

$$x = \xi + 3t \qquad \text{for } \xi > 0.$$

Therefore, using the initial data, we get

$$\rho(x, t) = \begin{cases} 1 & x < -t \\ e^{-4} & x > 3t. \end{cases}$$

There remains to find ρ in the region

$$\{(x, t) : -t \le x \le 3t\}.$$

We have

$$q''(\rho) = -1/\rho < 0,$$

and since the initial datum is decreasing, we look for an entropic solution in the form of a rarefaction wave centred at the origin. If we set $R(y) = (q')^{-1}(y)$, the wave is defined by

$$\rho(x,t) = R\left(\frac{x}{t}\right).$$

Solving for ρ the equation

$$q'(\rho) = -1 - \log \rho = y$$

we find

$$R(y) = e^{-(1+y)}$$

and hence

$$\rho(x,t) = e^{-(1+x/t)} \qquad \text{for } -1 \leq \frac{x}{t} \leq 3.$$

Summing up,

$$\rho(x,t) = \begin{cases} 1 & x \geq -t \\ e^{-(1+x/t)} & -t \leq x \leq 3t \\ e^{-4} & x > 3t. \end{cases}$$

To study the path of a vehicle, recall that the speed at a given density ρ is given by

$$v(\rho) = q(\rho)/\rho = -\log \rho.$$

Suppose now that a car is initially at $x_0 = -1$. It will not move (maximum density, $v(1) = 0$) until $x_0 = -1 < -t$, i.e. until time $t_0 = 1$. At that instant the car path enters the region of the rarefaction wave, where the speed is

$$v(\rho(x,t)) = -\log(e^{-(1+x/t)}) = 1 + \frac{x}{t}.$$

If $x = x(t)$ denotes the car path, we have

$$\begin{cases} x' = 1 + \dfrac{x}{t} \\ x(1) = -1. \end{cases}$$

The equation is linear, and integrating we have

$$x(t) = t(\log t - 1).$$

This holds until

$$x(t) = t(\log t - 1) < 3t$$

that is, up to $t = t_1 = e^4$. At this moment the car enters (and remains) in the region where it travels at constant speed $v(e^{-4}) = 4$, therefore the law of motion is

$$x(t) = 4t - e^4.$$

Therefore

$$x(t) = \begin{cases} -1 & 0 \leq t \leq 1 \\ t(\log t - 1) & 1 \leq t \leq e^4 \\ 4t - e^4 & t \geq e^4. \end{cases}$$

Note that the trajectory is not only continuous, but also C^1.

Solution 3.3.5. The characteristic Γ_ξ, from the point $(\xi, 0)$, is

$$x = \xi + q'(g(\xi))t,$$

and on Γ_ξ we have $u(x,t) = g(\xi)$. Hence

$$\xi = x - q'(g(\xi))t = x - q'(u(x,t))t$$

can be substituted to give

$$u(x,t) = g(x - q'(u(x,t))t).$$

But g is C^1, and $u(x,0) = g(x)$, hence we can use the implicit function theorem to solve for u. Set

$$h(x,t,u) = u - g(x - q'(u)t) = 0.$$

Then the theorem applies provided

$$h_u(x,t,u) = 1 + g'(x - q'(u)t)q''(u)t \neq 0.$$

Consider the characteristic Γ_ξ. For any $(x,t) \in \Gamma_\xi$

$$x - q'(u)t = \xi,$$

hence

$$h_u(x,t,u(x,t)) = 1 + g'(\xi)q''(u)t = 1 + g'(\xi)q''(g(\xi))t.$$

We deduce that $h_u = 0$ if and only if

$$t = -\frac{1}{g'(\xi)q''(g(\xi))}.$$

The starting time t_s for the shock is the smallest t for which the above equation holds. We find t_s on the characteristic corresponding to ξ_m, where ξ_m realises

$$\max_\xi g'(\xi)q''(g(\xi)).$$

Therefore the sufficient conditions for the shock to begin are

$$q'' < 0, \quad g' > 0$$

(q is concave by assumption).

Solution 3.3.6.   Let us begin finding the characteristics. Since

$$q(c) = c + \frac{c}{c+1},$$

we have

$$q'(c) = 1 + \frac{1}{(c+1)^2}.$$

The characteristic system, with datum on $x > 0$, is

$$\begin{cases} x'(s) = 1 & x(0) = \xi \\ t'(s) = 1 + \dfrac{1}{(c+1)^2} & t(0) = 0 \\ c'(s) = -c & c(0) = 0. \end{cases}$$

Then

$$x = s + \xi, t = 2s \text{ and } c = 0.$$

The characteristics from the x-axis are the lines

$$t = 2(x - \xi), \quad \text{with } \xi > 0,$$

and carry

$$c^+ = 0.$$

Consider now the datum on $t > 0$. The characteristic system reads

$$\begin{cases} x'(s) = 1 & x(0) = 0 \\ t'(s) = 1 + \dfrac{1}{(c+1)^2} & t(0) = \tau \\ c'(s) = -c & c(0) = c_0. \end{cases}$$

Immediately $x = s$ and $c = c_0 e^{-s}$, whence

$$c^-(x,t) = c_0 e^{-x}.$$

The corresponding characteristics (aside, the precise expression of the datum is unnecessary to determine the solution) can be found dividing the second and third equations and integrating. We find

$$-dt = \left(\frac{1}{c} + \frac{1}{c(c+1)^2} \right) dc$$

and then

$$\int \frac{dv}{v(v+1)^2} = \int \left(\frac{1}{v} - \frac{1}{v+1} - \frac{1}{(v+1)^2} \right) dv = \ln \frac{v}{v+1} + \frac{1}{v+1} + \text{constant},$$

so

$$t = \tau - \ln \frac{c^2}{c+1} - \frac{1}{c+1} + k = \tau + 2x - \ln \frac{1}{c_0 e^{-x} + 1} - \frac{1}{c_0 e^{-x} + 1} + k'$$

(where k' is chosen so that $t(0) = \tau$). In any case, as already noticed, the computation was superfluous. Indeed, from the characteristic system we see directly that along the characteristics from $t > 0$

$$\frac{dt}{dx}\bigg|_{x=0} = 1 + \frac{1}{(c_0+1)^2} < 2$$

for any $c_0 > 0$. As a consequence the characteristics from $t > 0$ meet those coming from $x > 0$, at least for τ and ξ small. Since the two characteristics families carry pointwise-distinct data, there is a shock curve $t = s(x)$. From the Rankine-Hugoniot conditions we get (see Problem 3.2.9)

$$s'(x) = \frac{q(c^+) - q(c^-)}{c^+ - c^-} = 1 + \frac{1}{c_0 e^{-x} + 1}.$$

Thus we have

$$\int \frac{dx}{c_0 e^{-x} + 1} = \int \frac{e^x dx}{c_0 + e^x} = \log (c_0 + e^x) + k,$$

and since $s(0) = 0$, the shock wave has equation

$$t = s(x) = x + \log \frac{c_0 + e^x}{c_0 + 1}.$$

In conclusion, the solution is

$$c(x,t) = \begin{cases} 0 & 0 \le t < x + \log \dfrac{c_0 + e^x}{c_0 + 1} \\[2ex] c_0 e^{-x} & t > x + \log \dfrac{c_0 + e^x}{c_0 + 1}. \end{cases}$$

Solution 3.3.7. **a)** Set $x = x(t)$, $y = y(t)$, $z = u(x(t), y(t))$. The characteristic system is

$$\begin{cases} x'(t) = x(t) & x(0) = s \\ y'(t) = 1 & y(0) = 0 \\ z'(t) = y(t) & z(0) = s^2 \end{cases} \qquad \text{from which} \qquad \begin{cases} x = X(s,t) = se^t \\ y = Y(s,t) = t \\ z = Z(s,t) = \frac{t^2}{2} + s^2. \end{cases}$$

We find $t = y$, $s = xe^{-y}$, and then $u(x,t) = \frac{y^2}{2} + x^2 e^{-2y}$.

b) The characteristic system is

$$\begin{cases} x'(t) = 1 & x(0) = 0 \\ y'(t) = -2 & y(0) = s \\ z'(t) = z(t) & z(0) = s \end{cases} \qquad \text{from which} \qquad \begin{cases} x = X(s,t) = t \\ y = Y(s,t) = -2t + s \\ z = Z(s,t) = se^t. \end{cases}$$

This gives $t = x$, $s = 2x + y$ and $u(x,t) = (2x + y)e^x$.

c) The characteristic system is

$$\begin{cases} x'(t) = 1 & x(0) = s \\ y'(t) = 3z^2(t) & y(0) = 0 \\ z'(t) = 1 & z(0) = 0 \end{cases} \qquad \text{from which} \qquad \begin{cases} x = X(s,t) = t + s \\ y = Y(s,t) = t^3 \\ z = Z(s,t) = t. \end{cases}$$

We obtain, $t = y^{1/3}$, and then $u(x,t) = y^{1/3}$.

Solution 3.3.8. **a)** Set $x = x(t)$, $y = y(t)$, $z = u(x(t), y(t))$, and write the characteristic system in the form

$$\frac{dx}{1} = \frac{dy}{\sqrt{z}} = \frac{dz}{1}, \qquad \text{so that} \qquad dy = \sqrt{z}\,dz, \qquad dx = dz.$$

We find the first integrals

$$\varphi(x, y, z) = \frac{2}{3}z^{3/2} - y, \qquad \psi(x, y, z) = x - z.$$

The general solution $z = u(x, y)$ is given implicitly by the equation

$$F\left(\frac{2}{3}z^{3/2} - y, x - z\right) = 0,$$

for some smooth F (and $z > 0$).

b) The characteristic system

$$\frac{dx}{1/y} = \frac{dy}{1} = \frac{dz}{z^2}, \qquad \text{gives} \qquad dy = \frac{dz}{z^2}, \qquad dx = \frac{dy}{y}.$$

We find the first integrals

$$\varphi(x, y, z) = y + \frac{1}{z}, \qquad \psi(x, y, z) = x - \log|y|,$$

and the general solution $z = u(x, y)$ is given implicitly by the equation

$$F\left(y + \frac{1}{z}, x - \log|y|\right) = 0,$$

for arbitrary F, smooth (and $yz \neq 0$).

Solution 3.3.9. The characteristic system

$$\begin{cases} x'(t) = y(t) & x(0) = s \\ y'(t) = -x(t) & y(0) = s \\ z'(t) = 2x(t)y(t)z(t) & z(0) = s^2, \end{cases} \quad \text{is solved by} \quad \begin{cases} x = X(s,t) = s(\cos t + \sin t) \\ y = Y(s,t) = s(\cos t - \sin t) \\ z = Z(s,t) = s^2 e^{s^2 \sin 2t}. \end{cases}$$

One sees that $x^2 + y^2 = 2s^2$, while $x^2 - y^2 = 2s^2 \sin 2t$. Therefore

$$u(x,t) = \frac{x^2 + y^2}{2} e^{(x^2-y^2)/2}.$$

Solution 3.3.10. **a)** The characteristics for

$$xu_x - yu_y = 0$$

are (see Problem 3.2.19 on page 188)

$$\varphi(x,y) = xy = k.$$

Let us choose, again, $\psi(x,y) = y$. We know that φ and ψ are independent. Choose new variables (for $x, y > 0$)

$$\begin{cases} \xi = xy \\ \eta = y \end{cases}$$

with inverse $(\xi, \eta > 0)$ $x = \xi/\eta$, $y = \eta$. According to the notation of Problem 3.2.18 (page 187) we have

$$C(\xi, \eta) = 1, \qquad F(\xi, \eta) = \eta, \qquad D(\xi, \eta) = -\eta,$$

and then

$$w(\xi, \eta) = u(\xi/\eta, \eta)$$

solves

$$w_\eta - \frac{1}{\eta} w = -1.$$

In this way

$$w(\xi, \eta) = \eta(G(\xi) - \log \eta),$$

i.e.

$$u(x,y) = y(G(xy) - \log y)$$

for an arbitrary function G (smooth).

b) We write the characteristic system in the form

$$\frac{dx}{x} = -\frac{dy}{y} = \frac{du}{y - u}.$$

The first equality immediately gives the first integral

$$\varphi(x, y, z) = xy,$$

while the second, written as

$$\frac{du}{dy} = -1 + \frac{u}{y},$$

gives the first integral, independent from the previous one,

$$\psi(x, y, z) = \frac{u}{y} + \log y.$$

As φ does not depend on u we can write the solution in the form $\psi = G(\varphi)$ directly,

$$\frac{u}{y} + \log y = G(xy).$$

This completely agrees with part a).

Solution 3.3.11. Let (x_0, y_0) be a global maximum point for u on D. If (x_0, y_0) is inside D then $\nabla u(x_0, y_0) = 0$, and from the differential equation $u(x_0, y_0) = 0$. If, instead, $(x_0, y_0) \in \partial D$, then $\nabla u(x_0, y_0)$ may not vanish, but it certainly points outwards. On the other hand, the unit outward normal to ∂D is the vector (x_0, y_0) itself, and $\mathbf{v} \cdot (a, b) = x_0 a(x_0, y_0) + y_0 b(x_0, y_0) > 0$ says that (a, b), too, points outwards. Therefore

$$0 \leq \nabla u(x_0, y_0) \cdot (a(x_0, y_0), b(x_0, y_0)) = -\gamma u(x_0, y_0),$$

and since $\gamma > 0$, in either case

$$\max_{D} u \leq 0.$$

In the same way (or writing $-u$ instead of u and repeating the argument) we see

$$\min_{D} u \geq 0,$$

making u identically zero.

Solution 3.3.12. We obtain $H(S) = S - 1$, thus the problem to solve reads

$$\begin{cases} \sum_{i=1}^{3} e^{-x_i} S_{x_i} - S_t = 0 & \mathbf{x} \in \mathbb{R}^3, t > 0 \\ S(\mathbf{x}, 0) = f(\mathbf{x}). \end{cases}$$

The associated characteristic system (for $z(v) = S(\mathbf{x}(v), t(v))$) is

$$\begin{cases} x_i'(v) = e^{-x_i(v)} & x_i(0) = s_i, \quad i = 1, 2, 3 \\ t'(v) = -1 & t(0) = 0 \\ z'(v) = 0 & z(0) = g(s_1, s_2, s_3), \end{cases}$$

from which

$$\begin{cases} x_i(v) = \log(v + e^{s_i}), \quad i = 1, 2, 3 \\ t(v) = -v \\ z(v) = g(s_1, s_2, s_3). \end{cases}$$

For any i

$$s_i = \log(t + e^{x_i}),$$

and finally

$$S(\mathbf{x}, t) = g(\log(t + e^{x_1}), \log(t + e^{x_2}), \log(t + e^{x_3})).$$

Solution 3.3.13. As the equation does not contain zero-order terms, u must be constant along the solutions of the reduced characteristic system, which can be written as

$$\begin{cases} dx = (cy - bz) \, dt \\ dy = (az - cx) \, dt \\ dz = (bx - ay) \, dt. \end{cases}$$

In order to decouple the equations and find two remaining first integrals, let us multiply the first one by a, the second by b and the third one by c, then add. We find

$$a\,dx + b\,dy + c\,dz = 0,$$

which gives the first integral

$$\varphi(x, y, z) = ax + by + cz.$$

Multiplying the equations by x, y, z, respectively, and adding up, yields

$$x\,dx + y\,dy + z\,dz = 0,$$

and thus a first integral, independent from the previous one (for $(x, y, z) \neq \mathbf{0}$), is given by

$$\psi(x, y, z) = x^2 + y^2 + z^2.$$

The general solution is then

$$u(x, y, z) = F\left(\varphi(x, y, z), \psi(x, y, z)\right) = F(ax + by + cz, x^2 + y^2 + z^2),$$

for arbitrary smooth function F.

Geometrical interpretation. The level sets of the first integral φ

$$ax + by + cz = c_1$$

are parallel planes, orthogonal to the unit vector

$$(a', b', c') = \frac{(a, b, c)}{\sqrt{a^2 + b^2 + c^2}}.$$

The level sets of ψ

$$x^2 + y^2 + z^2 = c_2$$

are spheres centred at the origin. The solution u is constant on the intersection curves (characteristics) of the two families, which are circles lying on the given planes and centred along the line r

$$\frac{x}{a'} = \frac{y}{b'} = \frac{z}{c'},$$

orthogonal to every plane. Therefore the level surfaces of u are, in implicit form,

$$a'x + b'y + c'z = G(x^2 + y^2 + z^2).$$

This equation defines, for any given G, a surface of revolution S about the line r. Given $P = (x, y, z)$ on S, in fact, the modulus of the left-hand side $|a'x + b'y + c'z|$ equals the distance of P from r, and the equation shows that this distance depends on the norm of P.

Solution 3.3.14. **a)** The simplest way to express the orthogonality of two surfaces is to impose that the normal vector fields be orthogonal. For level surfaces defined by a function w a normal is simply given by the gradient of w. We indicate with $\varphi(x, y, z) = c$ the equation of the orthogonal family. We must have

$$0 = \nabla w \cdot \nabla \varphi = w_x \varphi_x + w_y \varphi_y + w_z \varphi_z.$$

The reduced characteristic system is

$$\frac{dx}{w_x} = \frac{dy}{w_y} = \frac{dz}{w_z}.$$

If φ_1, φ_2 are first integrals for this system, then the family we are looking for is

$$F\left(\varphi_1\left(x, y, z\right), \varphi_2\left(x, y, z\right)\right) = 0,$$

with F arbitrary and smooth.

b) This is a family of paraboloids of revolution about the z-axis. The equation of the orthogonal surfaces

$$\varphi\left(x, y, z\right) = c$$

is

$$-2x\varphi_x - 2y\varphi_y + \varphi_z = 0$$

whose reduced characteristic system is

$$-\frac{dx}{2x} = -\frac{dy}{2y} = dz.$$

The first integrals are

$$\varphi_1(x, y, z) = \frac{y}{x}, \qquad \varphi_2(x, y, z) = 2z + \log y,$$

and since φ_1 does not depend on z, the solution can be written in the form

$$z = -\frac{1}{2}\log y + G\left(\frac{y}{x}\right)$$

for arbitrary smooth G.

Solution 3.3.15. The characteristics of the reduced equation are the straight lines

$$x - vt = \xi, \qquad \xi \in \mathbb{R}.$$

If we set

$$z(t) = u(vt + \xi, t)$$

then z solves the ODE

$$z'(t) = vu_x(vt + \xi, t) + u_t(vt + \xi, t) = -ku(vt + \xi, t) = -kz(t),$$

from which

$$z(t) = Ce^{-kt}.$$

We have to recover C, as ξ varies, in terms of the initial datum. If $\xi > 0$ (i.e. in the region $x > vt$) the corresponding characteristic is issued from $(\xi, 0)$, so

$$f(\xi) = z(0) = C.$$

In this case

$$u(x, t) = f(\xi)e^{-kt} = f(x - vt)e^{-kt}.$$

Conversely, if $\xi < 0$, it is more convenient to write

$$\tau = -\frac{\xi}{v}.$$

The characteristic issued from $(0, \tau)$ is given by

$$t - \frac{x}{v} = \tau, \qquad \text{with } \tau > 0.$$

Since
$$g(\tau) = z(\tau) = Ce^{-k\tau},$$

we see that
$$u(x,t) = g(\tau) e^{k\tau} e^{-kt} = g\left(t - \frac{x}{v}\right) e^{-kx/v}.$$

Thus, the solution is
$$u(x,t) = \begin{cases} f(x - vt)\exp(-kt) & x > vt \\ g\left(t - \frac{x}{v}\right)\exp\left(-\frac{kx}{v}\right) & x < vt. \end{cases}$$

Solution 3.3.16. Exactly as in the previous exercise (which we follow), the characteristics of the reduced equation are the lines
$$x - vt = \xi, \qquad \xi \in \mathbb{R}.$$

Set
$$z(t) = u(vt + \xi, t)$$

so that
$$z'(t) = vu_x(vt + \xi, t) + u_t(vt + \xi, t) = -ku^\alpha(vt + \xi, t) = -kz^\alpha(t).$$

Integrating (the equation is an ODE with separable variables) gives ($\alpha \neq 1$)
$$\frac{z^{1-\alpha}(t)}{1 - \alpha} = -kt + C,$$

whence (for some other C)
$$z(t) = u(vt + \xi, t) = ((\alpha - 1)kt + C)^{1/(1-\alpha)}.$$

If $\xi > 0$ the characteristic is issued from $(\xi, 0)$, so
$$f(\xi) = z(0) = C^{1/(1-\alpha)}.$$

If $\xi < 0$, the characteristic from $(0, \tau)$ is given by
$$t - \frac{x}{v} = \tau, \qquad \text{with } \tau > 0.$$

We find
$$g(\tau) = z(\tau) = (k(\alpha - 1)\tau + C)^{1/(1-\alpha)}.$$

Solving for C in the two cases and substituting, we finally obtain
$$u(x,t) = \begin{cases} \left\{(\alpha - 1)kt + [f(x - vt)]^{1-\alpha}\right\}^{1/(1-\alpha)} & 0 \le vt < x \\ \left\{(\alpha - 1)\frac{kx}{v} + [g\left(t - \frac{x}{v}\right)]^{1-\alpha}\right\}^{1/(1-\alpha)} & 0 \le x < vt. \end{cases}$$

Solution 3.3.17. a) Substitute
$$u(x, y) = v(x) + w(y)$$

into the equation to find
$$a(x)(v')^2 + b(y)(w')^2 = f(x) + g(y)$$

i.e.
$$a(x)(v')^2 - f(x) = -b(y)(w')^2 + g(y).$$

The two sides of the equation depend on different variables, and hence they are constant:
$$a(x)(v')^2 - f(x) = -b(y)(w')^2 + g(y) = \alpha.$$

We then solve separately

$$a(x)(v')^2 - f(x) = \alpha$$

and

$$-b(y)(w')^2 + g(y) = \alpha,$$

to obtain four solutions

$$u(x,y) = \pm \left[\int_{x_0}^x \sqrt{\frac{f(s)+\alpha}{a(s)}}\,ds \pm \int_{x_0}^x \sqrt{\frac{g(s)-\alpha}{b(s)}}\,ds \right] + \beta \tag{3.31}$$

with α, β constant.

b) In case

$$u_x^2 + u_y^2 = 1,$$

then $a = b = 1$ and we can take $f = 0$, $g = 1$. Then

$$(v'(x))^2 = 1 - (w'(y))^2 = \alpha$$

so necessarily $\alpha \geq 0$. Setting $\alpha = \sin^2 \gamma$ from (3.31) we find the complete integral

$$u(x,y) = x \sin \gamma + y \cos \gamma + \beta,$$

with γ, β arbitrary constants.

Solution 3.3.18. The equation is of the type

$$u_t + q(u)_x = 0$$

with

$$q(u) = u^2/2 \quad \text{and} \quad q'(u) = u.$$

The function q is convex, and g is decreasing, thus we expect that the solution will present a shock wave originating from the point with smallest time coordinate t_0, on the envelope of characteristics. The characteristic line from the point $(\xi, 0)$ is

$$x = \xi + q'(g(\xi))t = \xi + g(\xi)t.$$

For small times, then, the characteristics carry the datum $u = 1$ for $\xi < 0$, and $u = 0$ for $\xi > 1$. When $0 \leq \xi \leq 1$, instead, we have

$$x = \xi + (1 - \xi^2)t. \tag{3.32}$$

To find the envelope we differentiate the previous equation in ξ, obtaining

$$1 - 2\xi t = 0, \quad \text{i.e. } \xi = \frac{1}{2t}.$$

As $0 \leq \xi \leq 1$, we have $t \geq 1/2$. Substituting into (3.32) we obtain

$$x = t + \frac{1}{4t}.$$

The point with minimum time coordinate ($t_0 = 1/2$) is $x_0 = 1$, the starting point of the shock curve. If $x = s(t)$ denotes the equation of the shock curve, the Rankine-Hugoniot conditions give

$$s'(t) = \frac{q(u^+(s(t),t)) - q(u^-(s(t),t))}{u^+(s(t),t) - u^-(s(t),t)} = \frac{1}{2}[u^+(s(t),t) + u^-(s(t),t)].$$

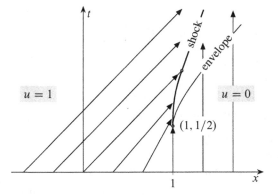

Fig. 3.17 Characteristics for Exercise 3.3.18.

Near the point $(1, 1/2)$ we have $u^+ = 0$. To find u^- note that (x, t), lying on the shock curve, belongs to the characteristic

$$x = \xi + (1 - \xi^2)t, \qquad \text{with} \quad \xi = \frac{1 - \sqrt{1 + 4t^2 - 4xt}}{2t}.$$

(And not $\xi = (1 + \sqrt{1 + 4t^2 - 4tx})/(2t)$. Why?) On this characteristic,

$$u^-(x, t) = 1 - \xi^2 = \frac{2tx - 1 + \sqrt{1 + 4t^2 - 4xt}}{2t^2}.$$

The shock wave, until it meets the characteristic $x = t$, solves the Cauchy problem

$$s'(t) = \frac{2ts - 1 + \sqrt{1 + 4t^2 - 4st}}{4t^2}, \qquad s(1/2) = 1$$

which needs to be solved numerically[1].

When the shock wave from $(1, 1/2)$ intersects the characteristic $x = t$, u^- equals 1 and the shock wave travels with velocity

$$s'(t) = 1/2$$

and is a half-line (Fig. 3.17).

Solution 3.3.19. Comparing with Problem 3.2.11 on page 176 we have

$$q(u) = u + f(u) = u + \frac{u}{1 + u}$$

and the problem to solve is

$$\begin{cases} u_x + \left(1 + \dfrac{1}{(1 + u)^2}\right) u_t = 0 & x > 0, \, t > 0 \\ u(x, 0) = 0 & x > 0 \\ u(0, t) = g(t) & t > 0, \end{cases}$$

---

[1] Despite the function is not Lipschitz, it can be proved that the Cauchy problem with initial condition $(1, 1/2)$ has exactly one solution.

where

$$g(t) = \begin{cases} c_0 & 0 \le t < 1 \\ 0 & t > 1. \end{cases}$$

The characteristics are the lines

$$t = \left(1 + \frac{1}{(1+u)^2}\right) x + \tau.$$

In particular, the equation of the characteristics are:

- $t = 2(x - \xi)$, if based on $(\xi, 0)$.
- $t = 2x + \tau$, if based on $(0, \tau), \tau > 1$.
- $t = \left(1 + \frac{1}{(1+c_0)^2}\right) x + \tau$, if based on $(0, \tau), 0 < \tau < 1$.

Notice that the slope of the characteristics of the last family is smaller than 2. This reveals the presence of a rarefaction wave centred at $(0, 1)$ and of a shock curve from the origin. The rarefaction wave is defined by

$$u(x,t) = R\left(\frac{x}{t-1}\right), \qquad \text{with } R(s) = (q')^{-1}(s)$$

in the region on the left of the shock curve:

$$2 < \frac{t-1}{x} < 1 + \frac{1}{(1+c_0)^2}.$$

As

$$(q')^{-1}(s) = (s-1)^{-1/2} - 1,$$

we have

$$u(x,t) = R\left(\frac{x}{t-1}\right) = \left(\frac{x}{t-1} - 1\right)^{-1/2} - 1 = \sqrt{\frac{x}{t-x-1}} - 1.$$

Concerning the shock, near the origin

$$u^+ = 0 \quad \text{and} \quad u^- = c_0.$$

The Rankine-Hugoniot condition then gives

$$\frac{dt}{dx} = \frac{q(u^+) - q(u^-)}{u^+ - u^-} = 1 + \frac{1}{1 + c_0}.$$

Therefore the equation of the shock curve is

$$t = \left(1 + \frac{1}{1 + c_0}\right) x.$$

The shock travels along a straight line until it hits the characteristic from $(0, 1)$, given by

$$t = \left(1 + \frac{1}{(1 + c_0)^2}\right) x + 1.$$

The intersection point is

$$(x_0, t_0) = \left(\frac{(1 + c_0)^2}{c_0}, \frac{2 + 3c_0 + c_0^2}{c_0}\right).$$

After that point, in particular when $t > x + 1$, we always have $u^+ = 0$; this time, though, the shock interacts on the left with the rarefaction wave

$$u^- = \sqrt{\frac{x}{t-x-1}} - 1.$$

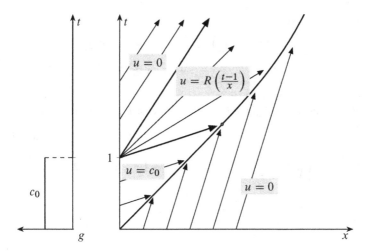

Fig. 3.18 Initial profile, shock and characteristics for Exercise 3.3.19

The ODE of the shock is then

$$\frac{dt}{dx} = 1 + \frac{1}{1 + u^-} = 1 + \sqrt{\frac{t - x - 1}{x}}.$$

The right-hand side suggests the substitution

$$z(x) = t(x) - x - 1$$
$$z'(x) = t'(x) - 1.$$

Remembering

$$t(x_0) = t_0,$$

we obtain that z solves

$$\begin{cases} \dfrac{dz}{dx} = \sqrt{\dfrac{z}{x}} \\ z\left(\dfrac{(1 + c_0)^2}{c_0}\right) = \dfrac{1}{c_0}. \end{cases}$$

Separating variables and integrating between x_0 and x, we find:

$$\sqrt{z} - \sqrt{\frac{1}{c_0}} = \sqrt{x} - \sqrt{\frac{(1 + c_0)^2}{c_0}},$$

that is to say

$$\sqrt{t - x - 1} = \sqrt{x} - \sqrt{c_0}$$

and then

$$t = 2x - 2\sqrt{c_0 x} + 1 + c_0.$$

To sum up, the shock curve Γ is

$$
t = \begin{cases} \left(1 + \frac{1}{1+c_0}\right)x & 0 \le x \le \frac{(1+c_0)^2}{c_0} \\ 2x - 2\sqrt{c_0 x} + 1 + c_0 & x \ge \frac{(1+c_0)^2}{c_0}. \end{cases}
$$

It is not hard to see Γ belongs to $C^1\left(\mathbb{R}^+\right)$.

The solution is *zero on the right* of Γ. On the *left* we have (see Fig. 3.18):

$$
u(x,t) = \begin{cases} c_0 & \left(1 + \frac{1}{1+c_0}\right)x < t \le \left(1 + \frac{1}{(1+c_0)^2}\right)x + 1, \\ \sqrt{\dfrac{x}{t - x - 1}} - 1 & \left(1 + \frac{1}{(1+c_0)^2}\right)x + 1 \le t \le 2x + 1, \\ 0 & t \ge 2x + 1. \end{cases}
$$

4

Waves

4.1 Backgrounds

We shall recall some basic and frequent results in the general theory of second-order wave equations in two variables.

- A *global Cauchy problem* $(n = 1)$ asks to determine $u = u(x,t)$ such that

$$\begin{cases} u_{tt} - c^2 u_{xx} = 0 & x \in \mathbb{R}, t > 0 \\ u(x,0) = g(x), \ u_t(x,0) = h(x) & x \in \mathbb{R}. \end{cases}$$

The solution is given by *d'Alembert's formula*

$$u(x,t) = \frac{1}{2}[g(x+ct) + g(x-ct)] + \frac{1}{2c}\int_{x-ct}^{x+ct} h(y)\, dy. \tag{4.1}$$

If $g \in C^2(\mathbb{R})$ and $h \in C^1(\mathbb{R})$, u is C^2 on the half-space $\mathbb{R} \times [0, \infty)$: there is no regularising effect. The solution has the form

$$F(x - ct) + G(x + ct),$$

that is, it is the superposition of two waves moving with speed c in opposite directions. The information provided by the initial data propagates along the *characteristics*

$$x \pm ct = \text{constant}.$$

In particular, the solution at the point (x,t) depends only on the value of h on the entire interval $[x - ct, x + ct]$ and those of g at the endpoints.

- *Classification of linear equations of order two.* Given an equation

$$au_{xx} + 2bu_{xy} + cu_{yy} + du_x + eu_y + hu = f$$

© Springer International Publishing Switzerland 2015
S. Salsa, G. Verzini, *Partial Differential Equations in Action. Complements and Exercises*,
UNITEXT – La Matematica per il 3+2 87, DOI 10.1007/978-3-319-15416-9_4

one calls *principal part* of the differential operator on the left the bulk of second-order terms

$$a(x, y)\, \partial_{xx} + 2b(x, y)\, \partial_{xy} + c(x, y)\, \partial_{yy}.$$

Given a domain Ω in the xt-plane, the equation is said to be:

a) *Hyperbolic* if $b^2 - ac > 0$.

b) *Parabolic* if $b^2 - ac = 0$.

c) *Elliptic* if $b^2 - ac < 0$.

A hyperbolic equation has two families of real characteristics $\phi(x, y) = \text{constant}$, where ϕ solve

$$a\phi_x^2 + 2b\phi_x\phi_y + c\phi_y^2 = 0.$$

A parabolic equation admits one family of real characteristics, solutions to the same equation. If the equation is elliptic there are no real characteristics.

• *Global Cauchy problem* $(n \geq 2)$. In dimension $n \geq 2$ the global Cauchy problem reads

$$\begin{cases} u_{tt} - c^2 \Delta u = 0 & \mathbf{x} \in \mathbb{R}^n, t > 0 \\ u(\mathbf{x}, 0) = g(\mathbf{x}), \quad u_t(\mathbf{x}, 0) = h(\mathbf{x}) & \mathbf{x} \in \mathbb{R}^n. \end{cases}$$

If $n = 3$, $g \in C^3(\mathbb{R}^3)$ and $h \in C^2(\mathbb{R}^3)$, then the only C^2 solution on $\mathbb{R}^3 \times [0, +\infty)$ is provided by *Kirchhoff's formula*

$$u(\mathbf{x}, t) = \frac{\partial}{\partial t}\left[\frac{1}{4\pi c^2 t}\int_{\{|\mathbf{x}-\sigma|=ct\}} g(\sigma)\, d\sigma\right] + \frac{1}{4\pi c^2 t}\int_{\{|\mathbf{x}-\sigma|=ct\}} h(\sigma)\, d\sigma.$$

In case $n = 2$, $g \in C^3(\mathbb{R}^2)$ and $h \in C^2(\mathbb{R}^2)$, the only C^2 solution on $\mathbb{R}^2 \times [0, +\infty)$ is determined by *Poisson's formula*

$$u(\mathbf{x}, t) = \frac{1}{2\pi c}\left\{\frac{\partial}{\partial t}\int_{\{|\mathbf{x}-\mathbf{y}|\leq ct\}}\frac{g(\mathbf{y})\, d\mathbf{y}}{\sqrt{c^2 t^2 - |\mathbf{x}-\mathbf{y}|^2}} + \int_{\{|\mathbf{x}-\mathbf{y}|\leq ct\}}\frac{h(\mathbf{y})\, d\mathbf{y}}{\sqrt{c^2 t^2 - |\mathbf{x}-\mathbf{y}|^2}}\right\}.$$

• *Domain dependence.* For $n = 3$, Kirchhoff's formula shows that $u(\mathbf{x}, t)$ depends only on the values of the data assumed on the *sphere*

$$\{\sigma \in \mathbb{R}^3 : |\mathbf{x} - \sigma| = ct\}.$$

In dimension $n = 2$, the solution at $(\mathbf{x}, t)$ depends on the values of the data assumed on the disc

$$\{\mathbf{y} \in \mathbb{R}^2 : |\mathbf{x} - \mathbf{y}| \leq ct\}.$$

4.2 Solved Problems

- **4.2.1 – 4.2.10** : One-dimensional waves and vibrations.
- **4.2.11 – 4.2.16** : Canonical forms. Cauchy and Goursat problems.
- **4.2.17 – 4.2.22** : Higher-dimensional problems.

4.2.1 One-dimensional waves and vibrations

Problem 4.2.1 (Pinched string). *A guitar string (initially at rest) is pinched at the mid-point and released. Denoting the string density by ρ and the tension by τ, formulate the mathematical model and write the solution as superposition of standing waves.*

Solution. Let L be the string length and suppose that the string at rest lies along the x-axis between 0 and L. Denote by $u(x,t)$ the displacement from the rest position of the point x at time t, and let a be the initial displacement of $x = L/2$. The initial configuration of the string, once it is pinched in the middle, is described by the function

$$g(x) = a - \frac{2a}{L}\left|x - \frac{L}{2}\right| = \begin{cases} 2ax/L & 0 \le x \le L/2 \\ 2a(L-x)/L & L/2 \le x \le L. \end{cases}$$

If a is small with respect to the length and we ignore the string weight, u solves

$$u_{tt} - c^2 u_{xx} = 0$$

where $c = \sqrt{\tau/\rho}$ is the travelling speed of waves along the string. The fixed endpoints impose homogeneous Dirichlet conditions at the boundary of the interval, while the initial rest status means that the initial velocity is zero. All this gives the following model:

$$\begin{cases} u_{tt} - c^2 u_{xx} = 0 & 0 < x < L, \, t > 0 \\ u(0,t) = u(L,t) = 0 & t \ge 0 \\ u(x,0) = g(x), \, u_t(x,0) = 0 & 0 \le x \le L. \end{cases}$$

In order to write the solution as linear combination of standing waves we separate variables, and seek (non-zero) solutions $u(x,t) = v(x)w(t)$. Substituting into the equation and separating the variables gives

$$\frac{1}{c^2}\frac{w''(t)}{w(t)} = \frac{v''(x)}{v(x)}.$$

The two sides must be equal to a constant $\lambda \in \mathbb{R}$. In particular, keeping in mind the Dirichlet conditions, v will solve the eigenvalue problem

$$\begin{cases} v''(x) - \lambda v(x) = 0 & 0 < x < L \\ v(0) = v(L) = 0. \end{cases}$$

It is easy to check that this has only the zero solutions if $\lambda \geq 0$. For $\lambda < 0$ the general integral is

$$v(x) = C_1 \cos\left(x\sqrt{-\lambda}\right) + C_2 \sin\left(x\sqrt{-\lambda}\right).$$

Imposing the boundary conditions gives $C_1 = 0$ and

$$\sin\left(L\sqrt{-\lambda}\right) = 0.$$

Thus there are non-trivial solutions if and only if $L\sqrt{-\lambda} = k\pi$, with $k = 1, 2 \ldots$. To summarise, we have infinitely many solutions

$$v_k(x) = \sin\left(\frac{k\pi}{L}x\right), \qquad \text{with } \lambda_k = -\frac{k^2\pi^2}{L^2}, \quad k = 1, 2, \ldots .$$

The corresponding w_k then solve

$$w_k''(t) + \frac{c^2 k^2 \pi^2}{L^2} w_k(t) = 0,$$

whence

$$w_k(t) = a_k \cos\left(\frac{ck\pi}{L}t\right) + b_k \sin\left(\frac{ck\pi}{L}t\right).$$

The standing waves

$$u_k(x, t) = w_k(t)v_k(x)$$

are *the normal modes of vibration of the string*, each with frequency $\nu_k = ck/2L$. We can then represent the solution as superposition of normal modes:

$$u(x, t) = \sum_{k=1}^{+\infty}\left[a_k \cos\left(\frac{ck\pi}{L}t\right) + b_k \sin\left(\frac{ck\pi}{L}t\right)\right]\sin\left(\frac{k\pi}{L}x\right).$$

Supposing we can differentiate the series, we find

$$u_t(x, t) = \sum_{k=1}^{+\infty}\frac{ck\pi}{L}\left[-a_k \sin\left(\frac{ck\pi}{L}t\right) + b_k \cos\left(\frac{ck\pi}{L}t\right)\right]\sin\left(\frac{k\pi}{L}x\right).$$

The initial conditions force, for $0 \leq x \leq L$,

$$\sum_{k=1}^{+\infty} a_k \sin\left(\frac{k\pi}{L}x\right) = g(x), \qquad \sum_{k=1}^{+\infty}\frac{ck\pi}{L}b_k \sin\left(\frac{k\pi}{L}x\right) = 0.$$

Consequently $b_k = 0$ for any k, while the a_k are found by expanding in Fourier series the function g on $[0, L]$. A formal solution is

$$u(x, t) = \sum_{k=1}^{+\infty} a_k \cos\left(\frac{ck\pi}{L}t\right)\sin\left(\frac{k\pi}{L}x\right) \tag{4.2}$$

with

$$a_k = \frac{2}{L} \int_0^L g(x) \sin\left(\frac{k\pi}{L}x\right) dx.$$

As the initial datum is symmetric with respect to $x = L/2$, and using the elementary identity

$$\sin(k\pi - \alpha) = (-1)^{k+1} \sin\alpha,$$

we find

$$a_k = \begin{cases} \dfrac{4}{L} \displaystyle\int_0^{L/2} \dfrac{2a}{L} x \sin\left(\dfrac{k\pi}{L}x\right) dx & k \text{ odd} \\ 0 & k \text{ even}, \end{cases}$$

thus, with $k = 2h + 1$, we have

$$a_{2h+1} =$$

$$= \frac{8a}{L^2} \left[\frac{-Lx}{(2h+1)\pi} \cos\left(\frac{(2h+1)\pi}{L}x\right)\bigg|_0^{L/2} + \frac{L}{(2h+1)\pi} \int_0^{L/2} \cos\left(\frac{(2h+1)\pi}{L}x\right) dx \right]$$

$$= \frac{8a}{L^2} \left[-\frac{L^2}{2(2h+1)\pi} \cos\left((2h+1)\frac{\pi}{2}\right) + \frac{L^2}{(2h+1)^2\pi^2} \sin\left((2h+1)\frac{\pi}{2}\right) \right]$$

$$= \frac{8a}{(2h+1)^2\pi^2}(-1)^h.$$

Altogether

$$u(x,t) = \frac{8a}{\pi^2} \sum_{h=0}^{+\infty} \frac{(-1)^h}{(2h+1)^2} \cos\left(\frac{c\pi(2h+1)}{L}t\right) \sin\left(\frac{(2h+1)\pi}{L}x\right).$$

This expression says that only modes that are symmetric with respect to $x = L/2$ can be activated by the initial profile. Note how the series converges uniformly in $[0, L]$ (Weierstrass criterion), but the second derivatives in x and t cannot be computed by swapping derivation and series, because g is not regular, having a non-smooth point at $x = L/2$. The solution thus found is classical only formally; the correct way to interpret it is in a proper (distributional) weak sense (see Chap. 6).

Problem 4.2.2 (Reflection of waves). *Consider the problem*

$$\begin{cases} u_{tt} - c^2 u_{xx} = 0 & 0 < x < L, t > 0 \\ u(x,0) = g(x), \quad u_t(x,0) = 0 & 0 \leq x \leq L \\ u(0,t) = u(L,t) = 0 & t \geq 0. \end{cases}$$

a) Define suitably the datum g outside the interval $[0, L]$, and use d'Alembert's formula to represent the solution as superposition of traveling waves.

b) Examine the physical meaning of the result and the relationship with the method of separation of variables.

Solution. a) The idea is to extend the Cauchy data to the whole $\mathbb{R}$, so that the corresponding global Cauchy problem has a solution vanishing on the lines $x = 0$, $x = L$. If we restrict this solution to $[0, L] \times \{t > 0\}$ we will find what we want. We indicate by $\tilde{g}$ and $\tilde{h}$ the extended data and set out to solve

$$\begin{cases} u_{tt} - c^2 u_{xx} = 0 & x \in \mathbb{R},\, t > 0 \\ u(x, 0) = \tilde{g}(x) & x \in \mathbb{R} \\ u_t(x, 0) = \tilde{h}(x) & x \in \mathbb{R} \end{cases}$$

via d'Alembert's formula

$$u(x, t) = \frac{1}{2}[\tilde{g}(x - ct) + \tilde{g}(x + ct)] + \frac{1}{2c}\int_{x-ct}^{x+ct} \tilde{h}(s)\,ds.$$

As u automatically satisfies the vibrating string equation (at least formally), we should choose $\tilde{g}$ and $\tilde{h}$ so to satisfy the initial/boundary conditions. In our case the simplest extension of h is $\tilde{h} = 0$. As for g, we must have

$$\begin{cases} u(x, 0) = \tilde{g}(x) = g(x) & 0 \leq x \leq L \\ u(0, t) = \dfrac{1}{2}[\tilde{g}(-ct) + \tilde{g}(ct)] = 0 & t > 0 \\ u(L, t) = \dfrac{1}{2}[\tilde{g}(L - ct) + \tilde{g}(L + ct)] = 0 & t > 0. \end{cases}$$

Therefore, for any s,

$$\tilde{g}(s) = -\tilde{g}(-s), \qquad \tilde{g}(L + s) = -\tilde{g}(L - s). \tag{4.3}$$

The first condition implies that $\tilde{g}$ *must* be an *odd* function. Moreover

$$\tilde{g}(s + 2L) = \tilde{g}(L + (L + s)) = -\tilde{g}(L - (L + s)) = -\tilde{g}(-s) = \tilde{g}(s),$$

so $\tilde{g}$ is $2L$-*periodic*. Then we may define $\tilde{g}$ to be the $2L$-periodic function whose restriction to $[-L, L]$ is given by

$$\tilde{g}(s) = \begin{cases} g(s) & 0 < s < L \\ -g(-s) & -L < s < 0. \end{cases}$$

The solution of the initial problem is then

$$u(x, t) = \frac{1}{2}[\tilde{g}(x - ct) + \tilde{g}(x + ct)] \qquad \text{for } 0 \leq x \leq L,\, t \geq 0. \tag{4.4}$$

• *Physical meaning of* (4.4). Let us divide the strip $[0, L] \times (0, +\infty)$ in regions separated by characteristic segments as in Fig. 4.1.

We analyse (4.4) starting from points $P = (x_0, t_0)$ in region 1. The *direct* and *inverse* characteristic emanating from P meet the x-axis at $a = x_0 - ct_0$ and $b = x_0 + ct_0$

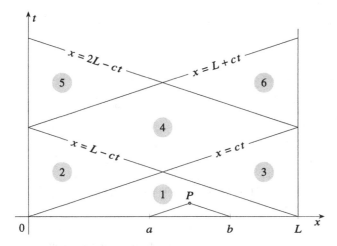

Fig. 4.1 Regions in the xt-plane for Problem 4.2.2

respectively. Since a, b lie in $[0, L]$,

$$u(x_0, t_0) = \frac{1}{2} [\tilde{g}(x_0 + ct_0) + \tilde{g}(x_0 - ct_0)] = \frac{1}{2} [g(x_0 + ct_0) + g(x_0 - ct_0)]$$

and the perturbation at P is the average of a *direct* and an *inverse* wave determined by the datum g (at a and b).

Take now $P = (x_0, t_0)$ in region 2 (Fig. 4.2). The point $b = x_0 + ct_0$, foot of the *inverse* characteristic emanating from P, belongs to $[0, L]$, so $\tilde{g}(x_0 + ct_0) = g(x_0 + ct_0)$. The *direct* characteristic through P meets the x-axis at $a = x_0 - ct_0 < 0$. But $\tilde{g}$ is odd, so $\tilde{g}(a) = -\tilde{g}(-a) = -g(-a)$ i.e.

$$\tilde{g}(x_0 - ct_0) = -g(-x_0 + ct_0).$$

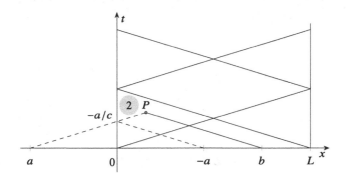

Fig. 4.2 Reflection of an inverse wave

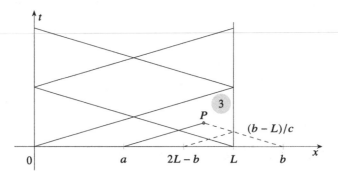

Fig. 4.3 Reflection of a direct wave

This means that the value $\tilde{g}(x_0 - ct_0)$ comes from an *inverse* wave determined by the datum at $-a \in [0, L]$ that reaches the left end at $t = -a/c$, is reflected, changes sign, and then maintains its value until t_0. Thus $u(x_0, t_0)$ is *the superposition of an inverse wave from b and an inverse wave from $-a$, reflected at $(0, -a/c)$ and with opposite sign.*

The argument is similar for any $P = (x_0, t_0)$ in region 3. The point $a = x_0 - ct_0$, foot of the *direct* characteristic through P, belongs to $[0, L]$, so $\tilde{g}(x_0 - ct_0) = g(x_0 - ct_0)$. The *inverse* characteristic from P intersects the x-axis at $b = x_0 + ct_0 > L$. The second relation in (4.3) implies

$$\tilde{g}(x_0 + ct_0) = -\tilde{g}(2L - x_0 - ct_0) = -g(2L - x_0 - ct_0).$$

This time $\tilde{g}(x_0 - ct_0)$ arises from a *direct* wave determined by the datum at $2L - b$, that reaches the right endpoint at time $t = (b - L)/c$, gets reflected, changes sign and stays constant until t_0. Hence $u(x_0, t_0)$ is *the superposition of a direct wave from a and a direct wave from $2L - b$, reflected at $(0, (b - L)/c)$ and with opposite sign* (Fig. 4.3).

Figure 4.4 should clarify the meaning of u in other regions. For instance, at a point in region 5 the wave is the superposition of a direct wave that is reflected and changes sign at the right end, and a direct wave that is reflected twice, changing sign first at the right end then at the left end.

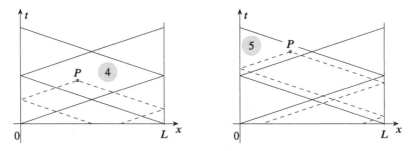

Fig. 4.4 More reflections

• *Relation with the method of separation of variables.* Formula (4.4) can be obtained also by separating variables. In fact, if we proceed as in Problem 4.2.1, and assuming that g is expanded in series of sines on $[0, L]$, the solution is

$$u(x,t) = \sum_{k=1}^{+\infty} a_k \cos\left(\frac{ck\pi}{L}t\right) \sin\left(\frac{k\pi}{L}x\right), \quad \text{with } a_k = \frac{2}{L}\int_0^L g(x)\sin\left(\frac{k\pi}{L}x\right)dx.$$

(4.5)

The a_k are the Fourier coefficients of the $2L$-periodic odd function $\tilde{g}$, which coincides with g on $[0, L]$. With the notation introduced, this means

$$\sum_{k=1}^{+\infty} a_k \sin\left(\frac{k\pi}{L}s\right) = \tilde{g}(s), \quad \text{for any } s \in \mathbb{R}. \quad (4.6)$$

On the other hand, known trigonometric addition formulas transform (4.5) into

$$u(x,t) = \sum_{k=1}^{+\infty} a_k \frac{1}{2}\left[\sin\left(\frac{k\pi}{L}x + \frac{ck\pi}{L}t\right) + \sin\left(\frac{k\pi}{L}x - \frac{ck\pi}{L}t\right)\right] =$$

$$= \frac{1}{2}\left[\sum_{k=1}^{+\infty} a_k \sin\left(\frac{k\pi}{L}(x+ct)\right) + \sum_{k=1}^{+\infty} a_k \sin\left(\frac{k\pi}{L}(x-ct)\right)\right],$$

and from (4.6) we deduce (4.4).

Problem 4.2.3 (Equipartition of energy). *Let u denote the solution to the following global Cauchy problem for the vibrating string:*

$$\begin{cases} \rho u_{tt} - \tau u_{xx} = 0 & x \in \mathbb{R}, \, t > 0 \\ u(x,0) = g(x) & x \in \mathbb{R} \\ u_t(x,0) = h(x) & x \in \mathbb{R}. \end{cases}$$

Assume g and h are regular functions that vanish outside a compact interval $[a,b]$. Prove that after a sufficiently long time T

$$E_{cin}(t) = E_{pot}(t) \quad \text{for any } t \geq T.$$

Solution. Let us recall the expressions for the kinetic and potential energy for small transverse vibrations of an infinite elastic string:

$$E_{cin} = \frac{1}{2}\int_{\mathbb{R}} \rho u_t^2 \, dx, \quad E_{pot} = \frac{1}{2}\int_{\mathbb{R}} \tau u_x^2 \, dx,$$

where ρ is the linear density of mass, τ the tension (constant along the string), and

$c = \sqrt{\tau/\rho}$ is the wave speed along the string. D'Alembert's formula reads

$$u(x,t) = F(x + ct) + G(x - ct),$$

where

$$F(s) = \frac{1}{2}\left[g(s) + \frac{1}{c}\int_0^s h(v)\,dv\right], \qquad G(s) = \frac{1}{2}\left[g(s) - \frac{1}{c}\int_0^s h(v)\,dv\right].$$

So we have

$$u_x(x,t) = F'(x + ct) + G'(x - ct), \qquad u_t(x,t) = c\left[F'(x + ct) - G'(x - ct)\right],$$

and then

$$E_{pot} = \frac{1}{2}\int_{\mathbb{R}} \tau\left[F'(x + ct) + G'(x - ct)\right]^2 dx =$$

$$= \frac{1}{2}\int_{\mathbb{R}} \tau\left[(F'(x + ct))^2 + (G'(x - ct))^2 + 2F'(x + ct)G'(x - ct)\right] dx$$

$$E_{cin} = \frac{1}{2}\int_{\mathbb{R}} c^2\rho_0 \left[F'(x + ct) - G'(x - ct)\right]^2 dx =$$

$$= \frac{1}{2}\int_{\mathbb{R}} \tau\left[(F'(x + ct))^2 + (G'(x - ct))^2 - 2F'(x + ct)G'(x - ct)\right] dx.$$

To prove the claim it suffices, for t large enough, that the product

$$F'(x + ct)G'(x - ct) = \frac{1}{4}\left[g'(x + ct) + \frac{1}{c}h(x + ct)\right] \cdot \left[g'(x - ct) - \frac{1}{c}h(x - ct)\right]$$

vanishes identically. We exploit the fact that the data are zero outside $[a, b]$: if $F'(x + ct) \neq 0$ then

$$a < x + ct < b; \tag{4.7}$$

in the same way $G'(x - ct) \neq 0$ forces

$$a < x - ct < b. \tag{4.8}$$

Therefore $F'(x + ct)G'(x - ct) \neq 0$ implies, by subtracting (4.7) and (4.8), $a - b < 2ct < b - a$. Consequently, if

$$t > T = \frac{b - a}{2c},$$

the product is zero and the kinetic energy equals the potential energy.

Problem 4.2.4 (Global Cauchy problem – impulses). *Find the formal solution to the problem*

$$\begin{cases} u_{tt} - c^2 u_{xx} = 0 & x \in \mathbb{R}, t > 0 \\ u(x,0) = g(x) & x \in \mathbb{R} \\ u_t(x,0) = h(x) & x \in \mathbb{R} \end{cases}$$

with the following initial data:

a) $g(x) = 1$ *if* $|x| < a$, $g(x) = 0$ *if* $|x| > a$; $h(x) = 0$.

b) $g(x) = 0$; $h(x) = 1$ *if* $|x| < a$, $h(x) = 0$ *if* $|x| > a$.

Solution. a) As h is identically zero, d'Alembert's formula reads

$$u(x,t) = \frac{1}{2}\left[g(x+ct) + g(x-ct)\right].$$

We then need to distinguish the regions in the plane where $|x \pm ct| \gtrless a$. The possible cases are described below (see the corresponding regions in Fig. 4.5 starting from the right):

- $x > a + ct$. *A fortiori*, then, $x > a - ct$, and $u(x,t) = 0$.
- $\max\{a - ct, -a + ct\} < x < a + ct$. Here $g(x - ct) = 1$ and $g(x + ct) = 0$. Therefore $u(x,t) = 1/2$.
- $\min\{a - ct, -a + ct\} < x < \max\{a - ct, -a + ct\}$. Both contributions are positive and $u(x,t) = 1$.
- $-a + ct < x < a - ct$ (so $t < a/c$). Both contributions are positive and $u(x,t) = 1$.
- $a - ct < x < -a + ct$ (so $t > a/c$). Both contributions vanish and $u(x,t) = 0$.
- $-a - ct < x < \min\{a - ct, -a + ct\}$. Now $g(x - ct) = 0$ and $g(x + ct) = 1$, so $u(x,t) = 1/2$.
- $x < -a - ct$. This implies $x < -a + ct$ and $u(x,t) = 0$.

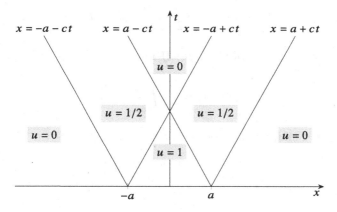

Fig. 4.5 Solution of Problem 4.2.4 a)

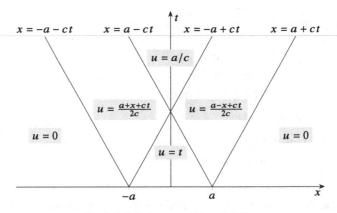

Fig. 4.6 Solution of Problem 4.2.4 b)

b) This time

$$u(x,t) = \frac{1}{2c} \int_{x-ct}^{x+ct} h(s)\,ds,$$

and arguing case by case as before, we obtain (Fig. 4.6):

- $x > a + ct$. Then $u(x,t) = 0$.
- $\max\{a - ct, -a + ct\} < x < a + ct$. We have

$$u(x,t) = \frac{1}{2c} \int_{x-ct}^{a} ds = \frac{a - x + ct}{2c}.$$

- $-a + ct < x < a - ct$ (so $t < a/c$). We have

$$u(x,t) = \frac{1}{2c} \int_{x-ct}^{x+ct} ds = t.$$

- $a - ct < x < -a + ct$ (hence $t > a/c$). Here

$$u(x,t) = \frac{1}{2c} \int_{-a}^{a} ds = \frac{a}{c}.$$

- $-a - ct < x < \min\{a - ct, -a + ct\}$. Then

$$u(x,t) = \frac{1}{2c} \int_{-a}^{x+ct} ds = \frac{a + x + ct}{2c}.$$

- $x < -a - ct$. It follows $u(x,t) = 0$.

Problem 4.2.5 (Forced vibrations). *Consider the problem*

$$\begin{cases} u_{tt} - u_{xx} = f(x,t) & 0 < x < L, \, t > 0 \\ u(x,0) = u_t(x,0) = 0 & 0 \le x \le L \\ u(0,t) = u(L,t) = 0 & t \ge 0, \end{cases} \tag{4.9}$$

with $f \in C^3([0,L] \times [0,+\infty))$ *and* $f(0,t) = f(L,t) = f_{xx}(0,t) = f_{xx}(L,t)$ *for any* $t \ge 0$.

a) Solve the problem by the separation of variables. Show that the expression found is the only classical solution.

b) Study in detail the case

$$f(x,t) = g(t) \sin\left(\frac{\pi x}{L}\right).$$

Solution. a) The Dirichlet conditions are homogeneous so, recalling the computations of Problem 4.2.1 on page 217, we look for solutions of the form

$$u(x,t) = \sum_{k=1}^{+\infty} w_k(t) \sin(klx), \tag{4.10}$$

where $l = \pi/L$ for short. Note that u satisfies automatically the boundary conditions. Assuming we can differentiate term by term, twice, we find

$$u_{tt}(x,t) = \sum_{k=1}^{+\infty} w_k''(t) \sin(klx),$$

$$u_{xx}(x,t) = -\sum_{k=1}^{+\infty} k^2 l^2 w_k(t) \sin(klx).$$

By a formal substitution, then,

$$\sum_{k=1}^{+\infty} \left[w_k''(t) + k^2 l^2 w_k(t) \right] \sin(klx) = f(x,t). \tag{4.11}$$

So now we express f in series of sines in x. For that, we extend f to the strip $[-L,L] \times \{t > 0\}$ as an odd function of x. We find

$$f(x,t) = \sum_{k=1}^{+\infty} f_k(t) \sin(klx) \quad \text{where} \quad f_k(t) = \frac{2}{L} \int_0^L f(\xi,t) \sin(kl\xi) \, d\xi. \tag{4.12}$$

By the Cauchy conditions (applied to each summand of the Fourier series), equation (4.11) is equivalent to the system of infinitely many ODEs

$$\begin{cases} w_k''(t) + k^2 l^2 w_k(t) = f_k(t) \\ w_k(0) = 0, \quad w_k'(0) = 0 \end{cases} \quad (k \ge 1). \tag{4.13}$$

The general integral of the homogeneous equation is

$$w_k(t) = C_1 \cos(klt) + C_2 \sin(klt).$$

To find a particular solution of the complete equation we use the method of variation of constants, by seeking solution of the type

$$w_k(t) = C_1(t) \cos(klt) + C_2(t) \sin(klt)$$

together with

$$C_1'(t) \cos(klt) + C_2'(t) \sin(klt) = 0.$$

Substituting into equation (4.13) we find that C_1, C_2 solve

$$\begin{cases} C_1'(t) \cos(klt) + C_2'(t) \sin(klt) = 0 \\ -kl C_1'(t) \sin(klt) + kl C_2'(t) \cos(klt) = f_k(t). \end{cases}$$

With a little patience we obtain the kth solution of system (4.13), given by

$$w_k(t) = \frac{1}{kl} \int_0^t \sin(kl\tau) f_k(t-\tau)\, d\tau.$$

Substituing back in (4.10) provides

$$u(x,t) = \frac{2}{\pi} \sum_{k=1}^{+\infty} \frac{1}{k} \sin(klx) \int_0^t \int_0^L f(\xi, t-\tau) \sin(kl\xi) \sin(kl\tau)\, d\xi d\tau. \qquad (4.14)$$

• *Analysis of the solution.* With the assumptions made on f, after three integrations by parts of (4.12) we obtain

$$f_k(t) = \frac{2L^2}{\pi^3 k^3} \int_0^L f_{xxx}(\xi, t) \cos(kl\xi)\, d\xi$$

so that for any $t \in [0, T]$,

$$|f_k(t)| \le \frac{2L^3}{\pi^3 k^3} \max_{[0,L]\times[0,T]} |f_{xxx}|, \qquad |w_k(t)| \le \frac{2L^4 T}{\pi^4 k^4} \max_{[0,L]\times[0,T]} |f_{xxx}|.$$

Therefore the Fourier series (4.12) converges uniformly, and (4.14) can be differentiated twice term by term. The expression (4.14) is then the unique solution of problem (4.9).

b) The forcing term corresponds to the first fundamental oscillation mode, and the dependence of the amplitude upon time is given by g. The regularity assumptions hold. We have

$$f_k(t) = \frac{2}{L} g(t) \int_0^L \sin(lx) \sin(klx)\, dx = 0 \quad \text{if } k \ge 2$$

while

$$f_1(t) = \frac{2}{L} g(t) \int_0^L \sin^2(lx)\, dx = g(t).$$

Therefore $w_k(t) = 0$, $k \geq 2$, while

$$w_1(t) = \frac{L}{\pi} \int_0^t \sin(l\tau)\, g(t - \tau)\, d\tau.$$

From (4.10) we find

$$u(x,t) = \frac{L}{\pi} \sin\left(\frac{\pi}{L}x\right) \left(\int_0^t \sin\left(\frac{\pi}{L}\tau\right) g(t - \tau)\, d\tau \right).$$

The string reacts to the forcing term by vibrating with the first fundamental mode, whose amplitude depends upon the convolution integral.

Problem 4.2.6 (Semi-infinite string with fixed end). *Consider the problem*

$$\begin{cases} u_{tt} - c^2 u_{xx} = 0 & x > 0,\ t > 0 \\ u(x,0) = g(x),\ u_t(x,0) = h(x) & x \geq 0 \\ u(0,t) = 0 & t \geq 0, \end{cases}$$

with g, h regular, $g(0) = 0$.

a) Extend suitably the initial data to $\mathbb{R}$ and use d'Alembert's formula to write a representation formula for the solution.

b) Interpret the solution in the case $h(x) = 0$ and

$$g(x) = \begin{cases} \cos(x - 4) & |x - 4| \leq \dfrac{\pi}{2} \\ 0 & \text{otherwise.} \end{cases}$$

Solution. a) Let us look for $\tilde{g}$ and $\tilde{h}$, defined on $\mathbb{R}$ and extending g and h to $x < 0$. The d'Alembert solution

$$u(x,t) = \frac{1}{2}[\tilde{g}(x + ct) + \tilde{g}(x - ct)] + \frac{1}{2c} \int_{x-ct}^{x+ct} \tilde{h}(s)\, ds \qquad (4.15)$$

must satisfy

$$u(0,t) = 0$$

for any $t > 0$, and therefore we have the necessary condition

$$\frac{1}{2}[\tilde{g}(ct) + \tilde{g}(-ct)] + \frac{1}{2c} \int_{-ct}^{ct} \tilde{h}(s)\, ds = 0.$$

The easiest way to satisfy this condition is to require the $\tilde{g}$- and $\tilde{h}$-summands to vanish separately, which happens if we extend g and h in an odd way.

Therefore the solution is given by (4.15) with

$$\tilde{g}(s) = \begin{cases} g(s) & s \geq 0 \\ -g(-s) & s < 0 \end{cases}$$

and

$$\tilde{h}(s) = \begin{cases} h(s) & s \geq 0 \\ -h(-s) & s < 0. \end{cases}$$

b) As $h = 0$, the solution reduces to

$$u(x,t) = \frac{1}{2}[\tilde{g}(x+ct) + \tilde{g}(x-ct)].$$

The initial datum can be understood as the superposition of two sinusoidal waves (with compact support and amplitude $1/2$) that at $t = 0$ start to travel in opposite directions with speed c. Since $x + ct$ is always positive, we have

$$\tilde{g}(x+ct) = g(x+ct)$$

for any (x,t), while

$$x - ct \geq 0 \text{ for any } x \in \left[4 - \frac{\pi}{2}, 4 + \frac{\pi}{2}\right] \text{ if } t \leq \frac{8 - \pi}{2c},$$

$$x - ct \leq 0 \text{ for any } x \in \left[4 - \frac{\pi}{2}, 4 + \frac{\pi}{2}\right] \text{ if } t \geq \frac{8 + \pi}{2c}.$$

Therefore we distinguish several intervals of time:

- $0 < t < \frac{\pi}{2c}$. The impulses start as opposite, but continue to interact in a neighbourhood of the point $x = 4$.

- $\frac{\pi}{2c} < t < \frac{8-\pi}{2c}$. Same as above, but the impulses do not interfere with each other.

- $\frac{8-\pi}{2c} < t < \frac{8+\pi}{2c}$. The impulse heading left reaches the fixed end and is reflected, turning upside down[1] (and interfering with itself).

- $t > \frac{8+\pi}{2c}$. The impulse moving leftwards has turned completely upside down. The string profile is given by two impulses of same shape, one positive and one negative, at a distance of 8, travelling towards the right at speed c.

These phases are shown in Fig. 4.7.

Problem 4.2.7 (Forced vibrations of a semi-infinite string). *A semi-infinite string is initially at rest along the axis $x \geq 0$, and fixed at $x = 0$. An external force $f = f(t)$ sets it in motion.*

a) *Write the mathematical model governing the vibrations.*

b) *Solve the problem using the Laplace transform in t, assuming that the transform of u is bounded as s tends to $+\infty$.*

---

[1] Special case of Problem 4.2.2 (page 219).

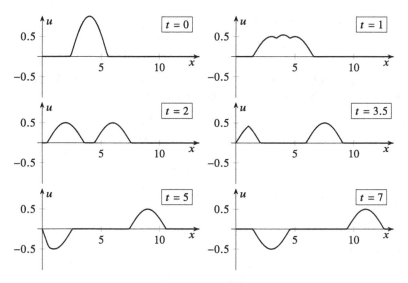

Fig. 4.7 The solution to Problem 4.2.6 ($c = 1$) at different values of t

Solution. a) Indicating with $u(x,t)$ the string profile at time t, the model reads

$$\begin{cases} u_{tt} - c^2 u_{xx} = f(t) & x > 0, t > 0 \\ u(x,0) = u_t(x,0) = 0 & x \geq 0 \\ u(0,t) = 0 & t \geq 0. \end{cases}$$

b) Set

$$U(x,s) = \mathcal{L}(u(x,\cdot))(s) = \int_0^{+\infty} u(x,t)e^{-st}\,dt, \qquad s \geq 0,$$

the t-Laplace transform of u. Transforming the equation and using the initial conditions we find[2]

$$-c^2 U_{xx}(x,s) + s^2 U(x,s) = F(s), \tag{4.16}$$

(where $F = \mathcal{L}(f)$), a second order ODE for the function $x \mapsto U(x,s)$, with constant coefficients. The general integral of the homogeneous equation is

$$A(s)e^{-sx/c} + B(s)e^{sx/c} \qquad (s \geq 0).$$

As $F(s)$ doe not depend on x it is easy to check that

$$F(s)/s^2$$

_____

[2] Recall that $\mathcal{L}(u_t) = sU(x,s) - u(x,0)$, $\mathcal{L}(u_{tt}) = s^2 U(x,s) - su(x,0) - u_t(x,0)$.

is a particular solution of (4.16). Hence

$$U(x, s) = \frac{F(s)}{s^2} + A(s)e^{-sx/c} + B(s)e^{sx/c}.$$

But since we need $U(x, s)$ bounded as $s \to +\infty$, necessarily $B = 0$. At the same time the homogeneous condition at the boundary point $x = 0$ implies

$$0 = U(0, s) = A(s) + \frac{F(s)}{s^2}$$

whence

$$A(s) = -\frac{F(s)}{s^2}.$$

Therefore

$$U(x, s) = F(s) \cdot \frac{1 - e^{-sx/c}}{s^2}.$$

To transform back we recall that

$$\mathcal{L}[t](s) = 1/s^2,$$

so by the 'time-shift' formula (Appendix B) we obtain

$$\mathcal{L}[(t - a)\mathcal{H}(t - a)](s) = \frac{e^{-as}}{s^2}$$

($\mathcal{H}$ is Heaviside's step function). Setting $a = x/c$, we finally obtain

$$u(x, t) = \int_0^t f(t - \tau) \left[\tau - \left(\tau - \frac{x}{c} \right) \mathcal{H} \left(\tau - \frac{x}{c} \right) \right] d\tau. \qquad (4.17)$$

Problem 4.2.8 (Vibrations of a hanging chain). *In this problem we shall find the equation governing the small (plane) vibrations of a hanging chain of length L. Call $u = u(x, t)$ the displacement from the horizontal position and ρ the linear density of mass (a constant). Let us assume that the chain is completely flexible (that is, no resistance to deformations) and that the oscillations are only transverse (the chain moves on a vertical plane).*

a) Denote by $\tau(x + \Delta x)$ and $\tau(x)$ the tensions at points $x + \Delta x$ and x relatively to some small interval $(x, x + \Delta x)$ on the chain; these tensions are the forces acting on that portion of chain from below and above respectively. Argue as for the vibrating string and show that, up to first order approximation,

$$|\tau(x)| = \tau(x) = \rho g x$$

(where g is the acceleration of gravity).

b) Show that small vibrations are governed by the equation

$$u_{tt} = g(x u_{xx} + u_x).$$

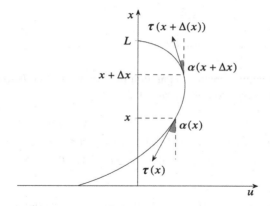

Fig. 4.8 Tension vectors for the hanging chain

Solution. a) Indicate by $\alpha(x)$, $\alpha(x + \Delta x)$ the angles between the vertical direction and the tension vector at $x, x + \Delta x$, respectively (Fig. 4.8). There is no vertical motion, so the resulting force cannot have a vertical component:

$$\tau(x + \Delta x) \cos (\alpha(x + \Delta x)) - \rho g(x + \Delta x) = \tau(x) \cos (\alpha(x)) - \rho g x$$

i.e.,

$$\tau(x + \Delta x) \cos (\alpha(x + \Delta x)) - \tau(x) \cos (\alpha(x)) = \rho g \Delta x.$$

Divide by Δx and let it tend to 0, so that

$$\frac{d}{dx} [\tau(x) \cos (\alpha(x))] = \rho g$$

and then

$$\tau(x) \cos (\alpha(x)) = \rho g x.$$

Having assumed the vibrations are small, we may write $\alpha(x) \approx 0$, and at first-order approximation we may write

$$\tau(x) = \rho g x.$$

b) Use Newton's law for the horizontal component of the force; noting that $\sin (\alpha(x)) \approx \tan (\alpha(x)) \approx u_x$, we infer

$$\rho \Delta x u_{tt} = \tau(x + \Delta x) \sin (\alpha(x + \Delta x)) - \tau(x) \sin (\alpha(x)) =$$
$$= \tau(x + \Delta x) u_x(x + \Delta x) - \tau(x) u_x(x).$$

Dividing by Δx and taking the limit $\Delta x \longrightarrow 0$, we obtain

$$\rho u_{tt} = \frac{d}{dx} [\tau(x) u_x] = \tau'(x) u_x + \tau(x) u_{xx};$$

as $\tau(x) = \rho g x$ (by part a)), u satisfies

$$u_{tt} = g u_x + g x u_{xx} = g\,(u_x + x u_{xx}).$$

Problem 4.2.9 (Hanging chain – separation of variables). *In relation to the previous problem solve (by separation of variables)*

$$\begin{cases} u_{tt} = g(xu_{xx} + u_x) & 0 < x < L,\, t > 0 \\ u(x,0) = f(x),\, u_t(x,0) = h(x) & 0 \le x \le L \\ u(L,t) = 0,\, |u(0,t)|\ bounded & t \ge 0. \end{cases}$$

Solution. Set $u(x,t) = v(x)w(t)$ and substitute into the differential equation to find

$$v(x)w''(t) = g(xv''(x) + v'(x))w(t),$$

whence

$$\frac{1}{g}\frac{w''(t)}{w(t)} = \frac{xv''(x) + v'(x)}{v(x)} = \lambda.$$

If $\lambda \ge 0$ we get only the zero solution (check this), so we might as well set $\lambda = -\mu^2$. This gives the two problems

$$w''(t) + \mu^2 g w(t) = 0, \qquad w(t)\text{ bounded} \tag{4.18}$$

and

$$xv''(x) + v'(x) + \mu^2 v(x) = 0, \qquad v(L) = 0,\, v(0)\text{ bounded.} \tag{4.19}$$

The general integral of (4.18) is

$$w(t) = A\cos\left(\sqrt{g}\mu t\right) + B\sin\left(\sqrt{g}\mu t\right).$$

The equation for v is

$$\left(xv'(x)\right)' + \mu^2 v(x) = 0, \qquad 0 < x < L.$$

As $p(x) = x$ vanishes at the origin, we are looking at a *singular Sturm-Liouville* problem[3]. A change of variables reduces this equation to a Bessel equation. Define $s = 2\sqrt{x}$ and $V(s) = v(s^2/4)$. Then

$$v'(x) = \frac{1}{\sqrt{x}}V'(s), \qquad v''(x) = \frac{1}{x}V''(s) - \frac{1}{2x\sqrt{x}}V'(s).$$

---

[3] Appendix A.

By substituting into (4.19) we see that V solves

$$s^2 V'' + s V' + \mu^2 s^2 V = 0, \qquad V(2\sqrt{L}) = 0, \ V(0) \text{ bounded,}$$

which in turn (once simplified) is a parametric Bessel equation[4] of order 0. The eigenfunctions are

$$V_j(s) = J_0\left(\frac{\alpha_j}{2\sqrt{L}}s\right) \qquad \text{with corresponding eigenvalues} \quad \mu_j = \alpha_j/2\sqrt{L},$$

where the α_j are the infinitely many zeroes of J_0, ordered increasingly. The eigenfunctions form a Hilbert basis for the space $L_w^2(0, 2\sqrt{L})$, with weight $w(x) = x$, and moreover

$$\int_0^{2\sqrt{L}} J_0\left(\frac{\alpha_j}{2\sqrt{L}}s\right) J_0\left(\frac{\alpha_k}{2\sqrt{L}}s\right) s \, ds = 0, \qquad \text{for } j \neq k,$$

and

$$\int_0^{2\sqrt{L}} J_0^2\left(\frac{\alpha_j}{2\sqrt{L}}s\right) s \, ds = 2L J_1^2(\alpha_j).$$

Substituting back we obtain

$$v_j(x) = J_0\left(\alpha_j \sqrt{\frac{x}{L}}\right), \qquad j = 1, 2, \ldots$$

and

$$\frac{1}{L J_1^2(\alpha_j)} \int_0^L J_0\left(\alpha_j \sqrt{\frac{x}{L}}\right) J_0\left(\alpha_k \sqrt{\frac{x}{L}}\right) dx = \delta_{jk},$$

so the v_j form a Hilbert basis in $L^2(0, L)$. The functions

$$u_j(x,t) = J_0\left(\alpha_j \sqrt{\frac{x}{L}}\right)\left[A_j \cos\left(\sqrt{\frac{g}{L}}\frac{\alpha_j}{2}t\right) + B_j \sin\left(\sqrt{\frac{g}{L}}\frac{\alpha_j}{2}t\right)\right]$$

are called the *normal modes* of the vibrating chain. The value $j = 1$ gives the *fundamental mode*, with *fundamental frequency*

$$v_1 = \frac{\alpha_1}{4\pi}\sqrt{\frac{g}{L}}.$$

The general solution is then

$$u(x,t) = \sum_{j=1}^{+\infty} u_j(x,t).$$

---

[4] Appendix A.

The coefficients A_j and B_j appear in the power series expansion of f, h in the basis $\{v_j\}$. From $u(x, 0) = f(x)$ we have

$$A_j = \frac{1}{L J_1^2(\alpha_j)} \int_0^L f(x) J_0 \left(\alpha_j \sqrt{\frac{x}{L}} \right) dx,$$

while $u_t(x, 0) = h(x)$ gives

$$B_j = \frac{2}{\alpha_j J_1^2(\alpha_j)} \frac{1}{\sqrt{gL}} \int_0^L h(x) J_0 \left(\alpha_j \sqrt{\frac{x}{L}} \right) dx.$$

Problem 4.2.10 (Sound waves in a pipe). *Let P_1, P_2 be two identical, cylindrical organ pipes of length L. Assume that its axis is the segment $[0, L]$ along the z-direction. Pipe P_1 is stopped (closed) at $z = 0$ and open at $z = L$, whereas P_2 is open at both ends.*
   Pressing a key makes pressurised air move through the pipes. Which pipe produces the note of higher pitch (i.e., higher frequency)?

Solution. We choose to describe the movement of air inside the pipe, assumed one-dimensional, by means of the *condensation*[5] s. Recall that if ρ_0 denotes the air density at rest, we define the condensation to be the ratio

$$s(z, t) = \frac{\rho(z, t) - \rho_0}{\rho_0}$$

which is a solution to

$$s_{tt} - c^2 s_{zz} = 0,$$

where $c = \sqrt{dp/d\rho}$ and p is the pressure. If $u = u(z, t)$ is the air velocity, we also have the relation[6]

$$u_t = -c^2 s_z. \tag{4.20}$$

Let us consider P_1: at the stopped end $s(0, t) = 0$, while at $z = L$ the velocity u is zero, so (4.20) implies $s_z(L, t) = 0$. To sum up, the condensation for P_1 solves:

$$\begin{cases} s_{tt} - c^2 s_{zz} = 0 & 0 < z < L \\ s(0, t) = 0, \ s_z(0, t) = 0 & t > 0 \end{cases}$$

under suitable initial conditions. Let us separate variables and set $s(z, t) = v(z) w(t)$. Then

$$\frac{v''(z)}{v(z)} = \frac{w''(t)}{c^2 w(t)} = -\lambda^2$$

---

[5] Alternatively one could use the velocity potential.
[6] [18, Chap. 5, Sect. 7].

with λ a positive constant. Keeping into account the boundary constraints, v solves the eigenvalue problem

$$v'' + \lambda^2 v = 0, \quad v'(0) = v(L) = 0.$$

There are infinitely many independent eigenfunctions

$$v_k(z) = \cos\left[\left(k + \frac{1}{2}\right)\frac{\pi z}{L}\right], \quad k = 0, 1, 2, \ldots$$

For each k we have a $w_k(t)$, solution of $w_k''(t) + \lambda_k^2 c^2 w_k(t) = 0$, which we write as:

$$w_k(t) = \cos\left[\left(k + \frac{1}{2}\right)\frac{\pi c t}{L} + \beta_k\right], \quad k = 0, 1, 2, \ldots$$

The general solution is found by superposing the $s_k(z, t) = v_k(z) w_k(t)$:

$$s(z, t) = \sum_{k=0}^{\infty} A_k \cos\left[\left(k + \frac{1}{2}\right)\frac{\pi z}{L}\right] \cos\left[\left(k + \frac{1}{2}\right)\frac{\pi c t}{L} + \beta_k\right]$$

in which the coefficients are determined by the initial excitation mode of the air. At any rate the amplitudes A_k, i.e. the Fourier coefficients of the initial profile, tend to zero as $k \to \infty$, so the pitch of P_1 is determined by the fundamental harmonic, corresponding to

$$s_0(z, t) = v_0(z) w_0(t) = A_0 \cos\left(\frac{\pi z}{2L}\right) \cos\left(\frac{\pi c t}{2L} + \beta_0\right)$$

with frequency

$$f_0 = \frac{c}{4L}.$$

Concerning P_2, s solves

$$\begin{cases} s_{tt} - c^2 s_{zz} = 0 & 0 < z < L \\ s(0, t) = 0, \, s(0, t) = 0 & t > 0 \end{cases}$$

again with suitable initial conditions. We proceed in the same way and find the solution

$$s(z, t) = \sum_{k=1}^{\infty} B_k \cos\left(\frac{k \pi z}{L}\right) \cos\left(\frac{k \pi c t}{L} + \gamma_k\right),$$

whose fundamental harmonic is

$$s_1(z, t) = B_1 \cos\left(\frac{\pi z}{L}\right) \cos\left(\frac{\pi c t}{L} + \gamma_1\right)$$

with fundamental frequency

$$g_0 = \frac{c}{2L}.$$

In conclusion, the *open pipe produces a sound of double frequency*, essentially because the closed end allows for twice as many wavelengths inside the pipe, and therefore halves the frequency.

4.2.2 Canonical forms. Cauchy and Goursat problems

Problem 4.2.11 (Characteristics and general solution). *Determine the type of the following linear equation of order two (in two variables)*

$$2u_{xx} + 6u_{xy} + 4u_{yy} + u_x + u_y = 0$$

and compute its characteristics. Reduce it to canonical form and find the general solution.

Solution. As

$$3^2 - 2 \cdot 4 = 1 > 0$$

the equation is hyperbolic. The principal part factorises as

$$2u_{xx} + 6u_{xy} + 4u_{yy} = 2(\partial_x - 2\partial_y)(\partial_x - \partial_y)u.$$

Using the techniques of Chap. 3 we solve

$$\phi_x - 2\phi_y = 0$$

to get the family of (real) characteristics $\phi(x, y) = 2x - y = $ constant. Moreover,

$$\psi_x - \psi_y = 0$$

gives the characteristic family $\psi(x, y) = x - y = $ constant.
   Another way to proceed would be to look for $y = y(x)$ and to solve the characteristics ODE

$$2\left(\frac{dy}{dx}\right)^2 - 6\frac{dy}{dx} + 4 = 0,$$

giving

$$\frac{dy}{dx} = 1 \quad \text{or} \quad \frac{dy}{dx} = 2$$

so $y - x = c_1$ or $y - 2x = c_2$.
   To write the equation in normal form we change coordinates by setting

$$\begin{cases} \xi = 2x - y \\ \eta = x - y \end{cases} \quad \text{i.e.} \quad \begin{cases} x = \xi - \eta \\ y = \xi - 2\eta. \end{cases}$$

Set now $U(\xi, \eta) = u(\xi - \eta, \xi - 2\eta)$, so that $u(x, y) = U(2x - y, x - y)$, and then

$$u_x = 2U_\xi + U_\eta, \quad u_y = -U_\xi - U_\eta,$$

$$u_{xx} = 4U_{\xi\xi} + 4U_{\xi\eta} + U_{\eta\eta}, \quad u_{xy} = -2U_{\xi\xi} - 2U_{\xi\eta} - U_{\eta\eta}, \quad u_{yy} = U_{\xi\xi} + 2U_{\xi\eta} + U_{\eta\eta}.$$

The equation for U is $4U_{\xi\eta} + U_\xi = 0$, or equivalently

$$(U_\xi)_\eta = \frac{1}{4}U_\xi.$$

Integrating in η first, we obtain $U_\xi(\xi, \eta) = e^{\eta/4} f(\xi)$ with f arbitrary (and regular), and then, integrating in ξ, we find

$$U(\xi, \eta) = e^{\eta/4} F(\xi) + G(\eta),$$

for some regular functions F, G. Returning to the original variables, the general solution reads

$$u(x, y) = e^{(x-y)/4} F(2x - y) + G(x - y).$$

Problem 4.2.12 (Euler-Tricomi equation). *Determine the characteristics of the Tricomi equation*

$$u_{tt} - t u_{xx} = 0.$$

Solution. First of all we note that the equation is hyperbolic, so the characteristics are real, only if $t > 0$ (it is parabolic for $t = 0$, a set with empty interior and not characteristic at any point). In the hyperbolic situation, the Tricomi operator factorises as

$$u_{tt} - t u_{xx} = (\partial_t - \sqrt{t}\partial_x)(\partial_t + \sqrt{t}\partial_x)u,$$

and the characteristics are $\phi(x, t) = $ constant, $\psi(x, t) = $ constant, with

$$\phi_t - \sqrt{t}\phi_x = 0 \qquad \psi_t + \sqrt{t}\psi_x = 0.$$

The methods of the previous chapter lead to the general solution of these first-order equations:

$$\phi(x, t) = F\left(3x + 2t^{3/2}\right) \qquad \psi(x, t) = G\left(3x - 2t^{3/2}\right)$$

with F, G arbitrary. Therefore the characteristic curves have equation

$$3x \pm 2t^{3/2} = \text{constant} \qquad \text{for } t \geq 0.$$

Problem 4.2.13 (Cauchy problem). *Solve, if possibile, the Cauchy problem*

$$\begin{cases} u_{yy} - 2u_{xy} + 4e^x = 0 & (x, y) \in \mathbb{R}^2 \\ u(x, 0) = \varphi(x) & x \in \mathbb{R} \\ u_y(x, 0) = \psi(x) & x \in \mathbb{R}. \end{cases}$$

Solution. First let us try to put the equation in normal form in order to write the general integral, and then we shall discuss the initial condition. It is easy to see that the equation is hyperbolic and its principal part decomposes

$$u_{yy} - 2u_{xy} = \partial_y(\partial_y - 2\partial_x)u.$$

Thus we can compute immediately the two characteristic families

$$x = \text{constant} \quad \text{and} \quad x + 2y = \text{constant}.$$

To reduce to normal form we change the coordinates by putting

$$\begin{cases} \xi = x \\ \eta = x + 2y \end{cases} \quad \text{i.e.} \quad \begin{cases} x = \xi \\ y = \dfrac{-\xi + \eta}{2}. \end{cases}$$

Set $u(x, y) = U(x, x + 2y)$, so

$$u_y = 2U_\eta, \quad u_{xy} = 2U_{\xi\eta} + 2U_{\eta\eta}, \quad u_{yy} = 4U_{\eta\eta}.$$

The equation for U is

$$U_{\xi\eta} = e^\xi,$$

hence

$$U_\eta = e^\xi + f(\eta) \quad \text{and} \quad U(\xi, \eta) = \eta e^\xi + F(\eta) + G(\xi)$$

with F and G arbitrary. Back in the original variables, the general integral is

$$u(x, y) = 2ye^x + F(x + 2y) + G(x)$$

where the summand xe^x has been absorbed into the function G.

Now we come to the Cauchy conditions. To begin with, these are given on the straight line $y = 0$, which is not characteristic at any of its points, so the problem is well posed (at least locally). Substituting, we have

$$\begin{cases} \varphi(x) = u(x, 0) = F(x) + G(x) \\ \psi(x) = u_y(x, 0) = 2e^x + 2F'(x). \end{cases}$$

The second relation gives

$$F(x) = -e^x + \frac{1}{2}\int_c^x \psi(s)\, ds \quad (c \text{ arbitrary}),$$

whence

$$G(x) = \varphi(x) + e^x - \frac{1}{2}\int_c^x \psi(s)\, ds$$

and finally

$$u(x, y) = 2ye^x - e^{x+2y} + \frac{1}{2}\int_c^{x+2y} \psi(s)\,ds + \varphi(x) + e^x - \frac{1}{2}\int_c^x \psi(s)\,ds =$$

$$= \left(1 + 2y - e^{2y}\right)e^x + \varphi(x) + \frac{1}{2}\int_x^{x+2y} \psi(s)\,ds.$$

Problem 4.2.14 (Characteristics and general integral). *Determine the characteristics of*

$$t^2 u_{tt} + 2t u_{xt} + u_{xx} - u_x = 0.$$

Reduce the equation to canonical form and find the general solution.

Solution. The equation is parabolic, since the principal part decomposes as

$$t^2 u_{tt} + 2t u_{xt} + u_{xx} = (t\partial_t + \partial_x)^2 u.$$

Thus the unique characteristic family is $\phi(x, t) = $ constant, where ϕ solves the first-order equation

$$t\phi_t + \phi_x = 0. \tag{4.21}$$

Using the methods of the previous chapter, we compute

$$\phi(x, t) = g\left(te^{-x}\right),$$

g arbitrary, so $\phi(x, t) = $ constant reads

$$te^{-x} = \text{constant}.$$

Let $\psi = \psi(x)$ be a smooth function, to be chosen later on, such that $\psi' > 0$. Thus, its inverse ψ^{-1} is well defined, and we can set

$$\begin{cases} \xi = te^{-x} \\ \eta = \psi(x) \end{cases} \quad \text{i.e.} \quad \begin{cases} x = \psi^{-1}(\xi) \\ t = \xi \exp[\psi^{-1}(\xi)]. \end{cases}$$

Define $U(\xi, \eta) = u(\psi^{-1}(\xi), \xi \exp[\psi^{-1}(\xi)])$, or in other words

$$u(x, y) = U(te^{-x}, \psi(x)).$$

Then

$$u_x = -te^{-x}U_\xi + \psi'U_\eta, \qquad u_t = e^{-x}U_\xi,$$

$$u_{xx} = te^{-x}U_\xi + t^2e^{-2x}U_{\xi\xi} - 2\psi'te^{-x}U_{\xi\eta} + (\psi')^2U_{\eta\eta} + \psi''U_\eta,$$

$$u_{xt} = -e^{-x}U_\xi - te^{-2x}U_{\xi\xi} + \psi'e^{-x}U_{\xi\eta}, \qquad u_{tt} = e^{-2x}U_{\xi\xi}.$$

Substituting into the original equation, we get

$$(\psi')^2 U_{\eta\eta} + (\psi'' - \psi')U_\eta = 0.$$

Now, if we pick $\eta = \psi(x) = e^x$, the second summand vanishes, and $\psi' > 0$. Therefore U solves $U_{\eta\eta} = 0$, and then

$$U(\xi, \eta) = F(\xi) + G(\xi)\eta,$$

with F and G arbitrary. Returning to the original variables, we finally find

$$u(x,t) = F(te^{-x}) + G(te^{-x})e^x.$$

Problem 4.2.15 (A maximum principle). *Let $u = u(x,t)$ be a function such that*

$$Lu = u_{tt} - u_{xx} \le 0 \qquad\qquad (4.22)$$

on the characteristic (triangular) domain

$$T = \{(x,t) : x > t, x + t < 1, t > 0\}.$$

Assume that $u \in C^2\left(\overline{T}\right)$ and prove that:

a) If $u_t(x,0) \le 0$, for $0 \le x \le 1$, then

$$\max_{\overline{T}} u = \max_{0 \le x \le 1} u(x,0).$$

b) If $u_t(x,0) < 0$, for $0 \le x \le 1$, or $Lu < 0$ on T, then

$$u(x,t) < \max_{0 \le x \le 1} u(x,0), \quad \text{for any } (x,t) \in T.$$

Solution. a) Fix a point C in $\overline{T}$ and consider the characteristic triangle T_C of vertices A, B, C, as in Fig. 4.9. We integrate on T_C the inequality $Lu \le 0$:

$$\int_{T_C} (u_{tt} - u_{xx}) \, dx dt \le 0.$$

From Green's formula

$$\int_{T_C} (u_{tt} - u_{xx}) \, dx dt = \int_{\partial + T_C} (-u_t \, dx - u_x \, dt)$$

where $\partial^+ T_C$ is the boundary of T_C, oriented counter-clockwise. Notice that $dt = 0$ along AB, $dx = -dt$ on BC and $dx = dt$ on CA. Therefore:

$$\int_{\partial^+ T_C} (-u_t \, dx - u_x \, dt) =$$

$$= -\int_{AB} u_t \, dx + \int_{BC} (u_t \, dt + u_x \, dx) - \int_{CA} (u_t \, dt + u_x \, dx)$$

$$= -\int_{AB} u_t \, dx + \int_{BC} du - \int_{CA} du$$

$$= -\int_{x_A}^{x_B} u_t(x,0) \, dx + (u(C) - u(B)) - (u(A) - u(C))$$

$$= -\int_{x_A}^{x_B} u_t(x,0) \, dx + 2u(C) - u(B) - u(A).$$

So we may write[7]

$$u(C) = \frac{u(B) + u(A)}{2} + \int_{x_A}^{x_B} u_t(x,0) \, dx + \int_{T_C} (u_{tt} - u_{xx}) \, dx dt. \qquad (4.23)$$

As $u_t(x,0) \le 0$ and $u_{tt} - u_{xx} \le 0$, we deduce

$$u(C) \le \frac{u(B) + u(A)}{2} \le \max_{0 \le x \le 1} u.$$

But C was chosen arbitrarily in $\overline{T}$, so the claim follows.

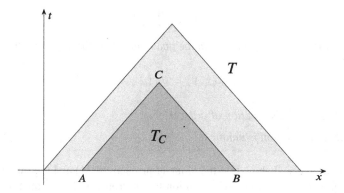

Fig. 4.9 Characteristic triangles

---

[7] By this method one also recovers d'Alembert's formula.

b) If $u_t(x, 0) < 0$ for $0 \leq x \leq 1$ or $Lu < 0$ on T, then (4.23) implies

$$u(C) < \frac{u(B) + u(A)}{2} \leq \max_{0 \leq x \leq 1} u$$

for $C \in T$.

Problem 4.2.16 (Goursat problem). *Consider the problem*

$$\begin{cases} u_{xy} + \lambda u = 0 & x > 0, \ y > 0 \\ u(x, 0) = u(0, y) = 1 & x \geq 0, \ y \geq 0 \end{cases} \tag{4.24}$$

where $\lambda > 0$, on the quadrant $\mathbb{R}_+^2 = \{(x, y) : x \geq 0, \ y \geq 0\}$. Observe that the data are defined on the characteristics $x = 0$ and $y = 0$, and for this reason the problem is called characteristic or Goursat problem.

a) *Assume that $u \in C^2(\mathbb{R}_+^2)$ and reduce the problem to the integral equation*

$$u(x, y) = 1 - \lambda \int_0^x \int_0^y u(\xi, \eta) \, d\xi d\eta. \tag{4.25}$$

b) *Show that the recursive sequence*

$$\begin{cases} u_0(x, y) = 1 \\ u_{n+1}(x, y) = 1 - \lambda \int_0^x \int_0^y u_n(\xi, \eta) \, d\xi d\eta \end{cases} \tag{4.26}$$

is uniformly convergent on compact sets in $\mathbb{R}_+^2$.

c) *Set*

$$u_\infty = \lim_{n \to +\infty} u_n.$$

Show that u_∞ is a solution (C^2 on the quadrant) of (4.25), and also that

$$u_\infty(x, y) = J_0\left(2\sqrt{\lambda xy}\right),$$

where J_0 is Bessel's function of order 0.

d) *Prove that the solution is unique.*

Solution. a) We will show that the Goursat problem (4.24) and the integral equation (4.25) are equivalent. Let u be a C^2 solution to the problem (4.24). Integrating in x between 0 and x gives

$$u_y(x, y) - u_y(0, y) = -\lambda \int_0^x u(\xi, y) \, d\xi.$$

Integrating once more, this time from 0 to y in the other variable, gives

$$u(x, y) - u(0, y) - u(x, 0) + u(0, 0) = -\lambda \int_0^x \int_0^y u(\xi, \eta) \, d\xi d\eta.$$

As $u(x, 0) = u(0, y) = 1$, we obtain (4.25). Vice versa, if $u \in C\left(\mathbb{R}_+^2\right)$ satisfies (4.25), a direct computation shows that $u(x, 0) = u(0, y) = 1$, and by the fundamental theorem of calculus $u \in C^2\left(\mathbb{R}_+^2\right)$. By differentiating (4.25) twice, we prove the claim.

b) Let us compute the first few terms of the sequence explicitly:

$$u_1(x, y) = 1 - \lambda \int_0^x \int_0^y d\xi d\eta = 1 - \lambda xy$$

$$u_2(x, y) = 1 - \lambda \int_0^x \int_0^y (1 - \lambda \xi \eta) \, d\xi d\eta = 1 - \lambda xy + \lambda^2 \frac{x^2 y^2}{4} = u_1(x, y) + \lambda^2 \frac{x^2 y^2}{(2!)^2}$$

$$u_3(x, y) = 1 - \lambda \int_0^x \int_0^y \left(1 - \lambda xy + \lambda^2 \frac{x^2 y^2}{4}\right) d\xi d\eta = u_2(x, y) - \lambda^3 \frac{x^3 y^3}{(3!)^2}.$$

Using an induction argument, we have

$$u_{n+1}(x, y) = u_n(x, y) + (-1)^{n+1} \lambda^{n+1} \frac{x^{n+1} y^{n+1}}{((n+1)!)^2}$$

so

$$u_\infty(x, y) = \sum_{n=0}^{+\infty} \frac{(-\lambda)^n}{(n!)^2} x^n y^n.$$

This expression is a power series $\sum a_n x^n y^n$ with convergence radius

$$\lim_{n \to +\infty} \left|\frac{a_n}{a_{n+1}}\right| = \lim_{n \to +\infty} \frac{1}{\lambda}(n+1)^2 = +\infty \qquad (\lambda > 0).$$

The uniform convergence theorem for power series guarantees uniform convergence on any set of the type $\{(x, y) : 0 \le xy \le M)\}$, and therefore on any compact subset of $\mathbb{R}_+^2$. In particular, $u_\infty \in C\left(\mathbb{R}_+^2\right)$.

c) Exploiting the uniform convergence on compact subsets in $\mathbb{R}_+^2$ once more, we may, for any given (x, y), pass to the limit in the recursive relation (4.26), obtaining

$$u_\infty(x, y) = 1 - \lambda \int_0^x \int_0^y u_\infty(\xi, \eta) \, d\xi d\eta.$$

By a), $u_\infty \in C^2\left(\mathbb{R}_+^2\right)$ and it solves the Goursat problem. To finish, since by definition (see Appendix A)

$$J_0(z) = \sum_{n=0}^{+\infty} \frac{(-1)^n}{((n)!)^2} z^{2n}$$

we may write

$$u_\infty(x, y) = \sum_{n=0}^{+\infty} \frac{(-1)^n}{((n)!)^2} (\sqrt{\lambda xy})^{2n} = J_0(\sqrt{\lambda xy}).$$

d) Suppose u_1 and u_2 are Goursat solutions (for the same λ). Then $w = u_1 - u_2$ solves the integral equation

$$w(x, y) = -\lambda \int_0^x \int_0^y w(\xi, \eta) \, d\xi d\eta.$$

Therefore we can write

$$w(x, y) = -\lambda \int_0^x \int_0^y \left[-\lambda \int_0^s \int_0^t w(\xi, \eta) \, d\xi d\eta \right] ds dt. \tag{4.27}$$

Integrating by parts we have

$$\int_0^y \left[\int_0^t w(\xi, \eta) \, d\eta \right] dt = \left[t \int_0^t w(\xi, \eta) \, d\eta \right]_0^y - \int_0^y t w(\xi, t) \, dt$$

$$= \int_0^y (y - t) w(\xi, t) \, dt$$

and, similarly,

$$\int_0^x \left[\int_0^s w(\xi, t) \, d\xi \right] dt = \int_0^x (x - s) w(s, t) \, ds.$$

Substituting into (4.27), we get

$$w(x, y) = \lambda^2 \int_0^x \int_0^y (x - s)(y - t) w(s, t) \, ds dt.$$

By iterating the argument we obtain

$$w(x, y) = -\lambda^3 \int_0^x \int_0^y \frac{(x - s)^2 (y - t)^2}{2 \cdot 2} w(s, t) \, ds dt$$

and, by induction,

$$w(x, y) = (-\lambda)^n \int_0^x \int_0^y \frac{(x - s)^n (y - t)^n}{(n!)^2} w(s, t) \, ds dt$$

for any $n \geq 1$. Now let us fix a compact set

$$K = \{(x, y) \in \mathbb{R}_+^2 : 0 \leq x \leq k, 0 \leq y \leq k\}$$

and set $M = \max_K |w|$. As $(x - s)^n (y - t)^n \leq k^{2n}$, we have

$$|w(x, y)| \leq \frac{|\lambda|^n k^{2n+2} M}{(n!)^2} \to 0 \qquad \text{for } n \to +\infty.$$

Therefore $w \equiv 0$ on any compact set in $\mathbb{R}_+^2$, and therefore everywhere in $\mathbb{R}_+^2$.

4.2.3 Higher-dimensional problems

> **Problem 4.2.17** (Circular membrane). *Consider a membrane*
>
> $$B_1 = \{(x, y) \in \mathbb{R}^2 : x^2 + y^2 \leq 1\}$$
>
> *at rest with fixed boundary. If the weight is negligible and there are no external forces acting on it, the membrane vibrates according to the following problem (written in polar coordinates, given the round symmetry):*
>
> $$\begin{cases} u_{tt} - c^2 \left(u_{rr} + \dfrac{1}{r} u_r + \dfrac{1}{r^2} u_{\theta\theta} \right) = 0 & 0 < r < 1, \, 0 \leq \theta \leq 2\pi, \, t > 0 \\ u(r, \theta, 0) = g(r, \theta), \quad u_t(r, \theta, 0) = h(r, \theta) & 0 < r < 1, \, 0 \leq \theta \leq 2\pi \\ u(1, \theta, t) = 0 & 0 \leq \theta \leq 2\pi, \, t > 0. \end{cases}$$
>
> *Using separation of variables, find a representation formula for the solution u in the case $h = 0$, $g = g(r)$.*

Solution. To start with, we remark that the solution (unique under reasonable assumptions on the data) is radially symmetric: indeed, if not, we might construct different solutions to the same problem simply by rotating a given solution (all data have a spherical symmetry!), contradicting uniqueness. Therefore $u = u(r, t)$ solves

$$\begin{cases} u_{tt} - c^2 \left(u_{rr} + \dfrac{1}{r} u_r \right) = 0 & 0 < r < 1, \, t > 0 \\ u(r, 0) = g(r), \quad u_t(r, 0) = 0 & 0 < r < 1 \\ u(1, t) = 0 & t > 0. \end{cases}$$

We seek solution of the type $u(r, t) = v(r)w(t)$ and, for the time being, such that $w'(0) = 0$ and

$$v(0) \text{ finite}, \quad v(1) = 0.$$

Substituting in the equation, we get

$$v(r)w''(t) - c^2 \left(v''(r) + \frac{1}{r} v'(r) \right) w(t) = 0.$$

Now we separate the variables and deduce

$$\frac{1}{c^2} \frac{w''(t)}{w(t)} = \frac{rv''(r) + v'(r)}{rv(r)} = \mu,$$

with μ constant. In particular v solves

$$\begin{cases} (rv')' - \mu r v = 0 \\ v(0) \text{ bounded} \quad v(1) = 0. \end{cases}$$

If $\mu \geq 0$ the only solution is $v \equiv 0$. In fact, by multiplying the ODE by v and integrating by parts on the interval $(0, 1)$, then taking into account the boundary conditions, we eventually find

$$0 = \int_0^1 (rv')'v\, dr - \mu \int_0^1 rv^2\, dr = -\int_0^1 [(v')^2 + \mu v^2]r\, dr.$$

If $\mu \geq 0$, the last integrand is non-negative and therefore $v \equiv 0$ on $(0, 1)$.
    If, conversely, $\mu = -\lambda^2$, the ODE is

$$v'' + \frac{v'}{r} + \lambda^2 v = 0,$$

that is a parametric Bessel equation of order 0 (Appendix A). The eigenfunctions, forming a basis of the Hilbert space $L_w^2(0, 1)$ with weight $w(r) = r$, are

$$u_n(r) = J_0(\lambda_n r),$$

where J_0 is the zero-order Bessel function and $\lambda_1, \lambda_2, \ldots$ are its zeroes. On the other hand the equation for w is

$$w''(t) + c^2 \lambda_n^2 w(t) = 0;$$

this equation, together with the initial condition $w'(0) = 0$ is solved by the one-parameter family

$$w_n(t) = a_n \cos(c\lambda_n t).$$

Thus we have found the fundamental modes of vibration of the membrane:

$$u_n(r, t) = a_n \cos(c\lambda_n t) J_0(\lambda_n r).$$

The solution of the original problem is then obtained by superposing those modes, that is

$$u(r, t) = \sum_{n=1}^{+\infty} a_n \cos(c\lambda_n t) J_0(\lambda_n r).$$

To satisfy the initial condition $u(r, 0) = g(r)$, as well, the coefficients a_n must satisfy the equation

$$\sum_{n=1}^{+\infty} a_n J_0(\lambda_n r) = g(r)$$

whence

$$a_n = \frac{2}{J_1^2(\lambda_n)} \int_0^1 sg(s) J_0(\lambda_n s)\, ds$$

in terms of the Bessel function of order 1, J_1.

Problem 4.2.18 (Dissipative term). *Consider the problem*

$$\begin{cases} u_{tt}(\mathbf{x},t) + k u_t(\mathbf{x},t) = c^2 \Delta u(\mathbf{x},t) & \mathbf{x} \in \mathbb{R}^2,\, t > 0 \\ u(\mathbf{x},0) = 0, \quad u_t(\mathbf{x},0) = g(\mathbf{x}) & \mathbf{x} \in \mathbb{R}^2. \end{cases}$$

a) Determine $\alpha \in \mathbb{R}$ so that

$$v(\mathbf{x},t) = e^{\alpha t} u(\mathbf{x},t)$$

solves an equation without first-order term (but with zero-order one) on $\mathbb{R}^2 \times \{t > 0\}$.

b) Find $\beta \in \mathbb{R}$ so that

$$w(x_1, x_2, x_3, t) = w(\mathbf{x}, x_3, t) = e^{\beta x_3} v(\mathbf{x}, t)$$

solves an equation with second-order terms only, on $\mathbb{R}^3 \times \{t > 0\}$.

c) Determine the solution u of the original problem.

Solution. a) Let us write

$$u(\mathbf{x},t) = e^{-\alpha t} v(\mathbf{x},t),$$

so that

$$u_t = -\alpha e^{-\alpha t} v + e^{-\alpha t} v_t, \quad u_{tt} = \alpha^2 e^{-\alpha t} v - 2\alpha e^{-\alpha t} v_t + e^{-\alpha t} v_{tt}, \quad \Delta u = e^{-\alpha t} \Delta v.$$

Substituting into the equation, we find

$$e^{-\alpha t} \left(v_{tt} + (k - 2\alpha) v_t + (\alpha^2 - k\alpha) v \right) = c^2 e^{-\alpha t} \Delta v.$$

Consequently, if we put $\alpha = k/2$, then

$$v(\mathbf{x},t) = e^{kt/2} u(\mathbf{x},t) \qquad \text{solves} \qquad v_{tt} - \frac{k^2}{4} v = c^2 \Delta v.$$

b) We write

$$w(\mathbf{x}, t_3, t) = e^{-\beta x_3} w(\mathbf{x},t)$$

and then

$$w_{tt} = e^{\beta x_3} v_{tt}, \qquad w_{x_1 x_1} + w_{x_2 x_2}) = e^{\beta x_3} \left(v_{x_1 x_1} + v_{x_2 x_2} \right).$$

Substituting gives

$$e^{-\beta x_3} w_{tt} = c^2 e^{-\beta x_3} \left(w_{x_1 x_1} + w_{x_2 x_2} + \frac{k^2}{4c^2} w \right).$$

Now notice that $w_{x_3 x_3} = \beta^2 w$. If we choose $\beta = k/(2c)$, we obtain that

$$w(\mathbf{x}, x_3, t) = e^{kx_3/(2c)} v(\mathbf{x},t) \qquad \text{solves} \qquad w_{tt} = c^2 \Delta_3 w,$$

where the operator Δ_3 is the Laplacian in three dimensions.

c) Since

$$w(\mathbf{x}, x_3, t) = e^{k(x_3+tc)/2c} u(\mathbf{x}, t),$$

we can find the initial conditions for w. Since

$$w(\mathbf{x}, x_3, 0) = e^{kx_3/2c} u(\mathbf{x}, 0) = 0$$

$$w_t(\mathbf{x}, x_3, 0) = \frac{k}{2} e^{kx_3/2c} u(\mathbf{x}, 0) + e^{kx_3/2c} u(\mathbf{x}, 0) = e^{kx_3/2c} g(\mathbf{x}),$$

from Kirchhoff's formula we obtain

$$v(\mathbf{x}, t) = w(\mathbf{x}, 0, t) = \frac{1}{4\pi c^2 t} \int_{\partial B_{ct}(\mathbf{x},0)} e^{k\sigma_3/2c} g(\sigma_1, \sigma_2) \, d\sigma,$$

where $d\sigma$ is the surface element of $\partial B_{ct}(\mathbf{x}, 0)$. We can attain a more significant formula by reducing the integral on the sphere $\partial B_{ct}(\mathbf{x}, 0)$ to a double integral on the disc

$$K_{ct}(\mathbf{x}) = \left\{ \mathbf{y} = (y_1, y_2) : |\mathbf{y} - \mathbf{x}|^2 < c^2 t^2 \right\}.$$

The equations on the upper and lower hemispheres are

$$y_3 = \pm\sqrt{c^2 t^2 - r^2} \qquad \left(r^2 = |\mathbf{y} - \mathbf{x}|^2 \right)$$

where

$$d\sigma = \sqrt{1 + \frac{r^2}{c^2 t^2 - r^2}} dy_1 dy_2 = \frac{ct}{\sqrt{c^2 t^2 - r^2}} dy_1 dy_2.$$

Hence

$$v(\mathbf{x}, t) = \frac{1}{4\pi c} \int_{K_{ct}(\mathbf{x})} \left[e^{k\sqrt{c^2 t^2 - r^2}/2c} + e^{-k\sqrt{c^2 t^2 - r^2}/2c} \right] \frac{g(y_1, y_2)}{\sqrt{c^2 t^2 - r^2}} dy_1 dy_2$$

$$= \frac{1}{2\pi c} \int_{K_{ct}(\mathbf{x})} \cosh\left(\frac{k\sqrt{c^2 t^2 - r^2}}{2c} \right) \frac{g(y_1, y_2)}{\sqrt{c^2 t^2 - r^2}} dy_1 dy_2.$$

and finally

$$u(\mathbf{x}, t) = e^{-kt/2} v(\mathbf{x}, t).$$

Problem 4.2.19 (Fundamental solution in dimension 2). *Write the fundamental solution to the two-dimensional wave equation, with and without dissipation.*

Solution. In order to determine the fundamental solution we remind that the Cauchy problem

$$\begin{cases} u_{tt}(\mathbf{x}, t) - c^2 \Delta u(\mathbf{x}, t) = 0 & \mathbf{x} \in \mathbb{R}^2, \ t > 0 \\ u(\mathbf{x}, 0) = 0, \quad u_t(\mathbf{x}, 0) = h(\mathbf{x}) & \mathbf{x} \in \mathbb{R}^2 \end{cases}$$

is solved by the Poisson integral

$$u(\mathbf{x}, t) = \frac{1}{2\pi c} \int_{B_{ct}(\mathbf{x})} \frac{h(\mathbf{y})}{\sqrt{c^2 t^2 - |\mathbf{x} - \mathbf{y}|^2}} \, d\mathbf{y} =$$

$$= \frac{1}{2\pi c} \int_{\mathbb{R}^2} \frac{\mathcal{H}\left(c^2 t^2 - |\mathbf{x} - \mathbf{y}|^2\right)}{\sqrt{c^2 t^2 - |\mathbf{x} - \mathbf{y}|^2}} h(\mathbf{y}) d\mathbf{y}$$

where $\mathcal{H}$ is Heaviside's function. The expression for the solution $K(\mathbf{x}, \mathbf{z}, t)$ is found by choosing h to be the Dirac distribution in the variable $\mathbf{z}$. We find

$$K(\mathbf{x}, \mathbf{z}, t) = \frac{1}{2\pi c} \frac{\mathcal{H}\left(c^2 t^2 - |\mathbf{x} - \mathbf{z}|^2\right)}{\sqrt{c^2 t^2 - |\mathbf{x} - \mathbf{z}|^2}}$$

$$= \begin{cases} \dfrac{1}{2\pi c} \dfrac{1}{\sqrt{c^2 t^2 - |\mathbf{x} - \mathbf{z}|^2}} & \text{for } |\mathbf{x} - \mathbf{z}|^2 < c^2 t^2 \\ 0 & \text{for } |\mathbf{x} - \mathbf{z}|^2 > c^2 t^2. \end{cases}$$

This solution describes the perturbation created by an initial unit impulse, corresponding to the data

$$u(\mathbf{x}, 0) = 0 \quad \text{and} \quad u_t(\mathbf{x}, 0) = \delta(\mathbf{x} - \mathbf{z}).$$

Note that at a given time t every point inside the disc

$$\{(\mathbf{x}, t) : |\mathbf{x} - \mathbf{z}|^2 < c^2 t^2\}$$

is affected by the oscillation. In the case the equation has a dissipative term $k u_t$ as well, from the previous problem we recover

$$K_{diss}(\mathbf{x}, \mathbf{z}, t) = \frac{e^{-kt/2}}{2\pi c} \cosh\left(\frac{k\sqrt{c^2 t^2 - |\mathbf{x} - \mathbf{z}|^2}}{2c}\right) \frac{\mathcal{H}\left(c^2 t^2 - |\mathbf{x} - \mathbf{z}|^2\right)}{\sqrt{c^2 t^2 - |\mathbf{x} - \mathbf{z}|^2}}$$

as the fundamental solution.

Problem 4.2.20 (An application of Kirchhoff's formula). *Let us consider*

$$\begin{cases} u_{tt} - c^2 \Delta u = 0 & \mathbf{x} \in \mathbb{R}^3, t > 0 \\ u(\mathbf{x}, 0) = g(\mathbf{x}), \quad u_t(\mathbf{x}, 0) = h(\mathbf{x}) & \mathbf{x} \in \mathbb{R}^3, \end{cases}$$

where g and h are respectively C^3 and C^2 in $\mathbb{R}^3 \times [0, +\infty)$. Suppose that g, h have support contained in the ball $B_\rho(0)$. Describe the support of the solution u at each time instant t.

Solution. By Kirchhoff's formula the problem is solved by

$$u(\mathbf{x}, t) = \frac{\partial}{\partial t} \left[\frac{1}{4\pi c^2 t} \int_{\partial B_{ct}(\mathbf{x})} g(\sigma) \, d\sigma \right] + \frac{1}{4\pi c^2 t} \int_{\partial B_{ct}(\mathbf{x})} h(\sigma) \, d\sigma$$

The support condition on the initial data may be written as

$$|\mathbf{y}| \geq \rho \qquad \Longrightarrow \qquad g(\mathbf{y}) = 0, \ h(\mathbf{y}) = 0.$$

From the solution expression we infer that u vanishes at the point $(\mathbf{x}, t)$ if

$$\partial B_{ct}(\mathbf{x}) \cap B_\rho(0) = \emptyset.$$

This happens when either

$$|\mathbf{x}| + \rho < ct \qquad \text{or} \qquad |\mathbf{x}| - \rho > ct.$$

The solution support at time t is therefore contained in the ball $|\mathbf{x}| \leq ct + \rho$ for $t \leq \rho/c$, and in the spherical shell

$$\{\mathbf{x} \in \mathbb{R}^3 : ct - \rho \leq |\mathbf{x}| \leq ct + \rho\} \quad \text{for} \quad t > \rho/c,$$

which grows in any radial direction at speed c.

Problem 4.2.21 (Focalised discontinuity). *Determine the solution to*

$$\begin{cases} u_{tt} - c^2 \Delta u = 0 & \mathbf{x} \in \mathbb{R}^3, t > 0 \\ u(\mathbf{x}, 0) = 0, \ u_t(\mathbf{x}, 0) = h(|\mathbf{x}|) & \mathbf{x} \in \mathbb{R}^3, \end{cases}$$

where $r = |\mathbf{x}|$ and

$$h(r) = \begin{cases} 1 & 0 \leq r \leq 1 \\ 0 & r > 1 \end{cases}.$$

Solution. Note that initially the solution is continuous at $\mathbf{x} = 0$. If we imagine h extended to the entire real axis in an *even* way, w, being radial, has the form

$$w(r, t) = \frac{F(r + ct)}{r} + \frac{G(r - ct)}{r}.$$

Now, $w(r, 0) = 0$ implies $F(r) = -G(r)$. The second initial condition then gives $G'(r) = -rh(r)/2c$, whence

$$G(r) = -\frac{1}{2c} \int_0^r sh(s) \ ds + G(0)$$

and then

$$w(r, t) = \frac{F(r + ct)}{r} + \frac{G(r - ct)}{r} = \frac{1}{2cr} \int_{r-ct}^{r+ct} sh(s) \ ds.$$

To compute w we must consider various regions in the rt-plane (Fig. 4.10). On the triangle

$$\{(r, t) : -1 < r - ct < r + ct < 1\}$$

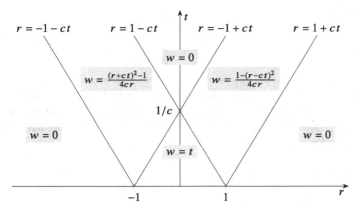

Fig. 4.10 Solution to Problem 4.2.21

(where, in particular, $t < 1/c$) we have $h = 1$, so

$$\int_{r-ct}^{r+ct} sh\,(s)\,ds = \left[\frac{s^2}{2}\right]_{r-ct}^{r+ct} = 2crt.$$

If $r \neq 0$ then $w\,(r,t) = t$, which holds at $r = 0$, too, by continuity. In particular

$$w\,(0,t) \to \frac{1}{c} \quad \text{if } t \to \left(\frac{1}{c}\right)^{-}.$$

If $t > 1/c$, on the region

$$\{(r,t) : r - ct < -1, r + ct > 1\}$$

we have

$$\int_{r-ct}^{r+ct} sh\,(s)\,ds = \int_{-1}^{1} s\,ds = 0.$$

Therefore $w\,(r,t) = 0$ and w is discontinuous at $(0, 1/c)$: *this fact is a consequence of initial discontinuity of w_t on the sphere $r = 1$ focalized at $r = 0$ at time $t = 1/c$.*

On the sector $\{(r,t) : r - ct > 1\}$ and $\{(r,t) : r + ct < -1\}$ h is null, so $w\,(r,t) = 0$. Finally, on the strip

$$\{(r,t) : -1 < r - ct < 1, r + ct > 1\}$$

we have

$$w\,(r,t) = \frac{1}{2cr} \int_{r-ct}^{1} sh\,(s)\,ds = \frac{1}{2cr} \int_{r-ct}^{1} s\,ds = \frac{1 - (r - ct)^2}{4cr},$$

while on the strip

$$\{(r,t) : r - ct < -1, -1 < r + ct < 1\}$$

we have (Fig. 4.10)

$$w(r,t) = \frac{1}{2cr} \int_{-1}^{r+ct} sh(s)\, ds = \frac{1}{2cr} \int_{-1}^{r+ct} s\, ds = \frac{(r+ct)^2 - 1}{4cr}.$$

Problem 4.2.22 (Reflection of acoustic waves). *A plane harmonic of amplitude A and frequency ω_I, moving at speed c on the plane xz, hits the plane z = 0 and is reflected, see Fig. 4.11. Determine the (velocity) potentials of the incoming and the reflected waves (use complex variables).*

Solution. In general, a plane harmonic has a potential of the form

$$\phi(x,z,t) = A \exp\{i(k_1 x + k_2 y + k_3 z - \omega t)\} \qquad (\omega > 0).$$

The wave moves in the direction of the vector $\mathbf{k} = (k_1, k_2, k_3)$, called *wavenumber* vector, with velocity $\omega/|\mathbf{k}|$. In the present case $\mathbf{k}_I = (k_1, 0, k_3)$, and the potential of the wave is

$$\phi_I(x,z,t) = A \exp\{i(k_1 x + k_3 z - \omega_I t)\}.$$

Let c denote the wave's speed. Then

$$|\mathbf{k}_I| = \frac{\omega_I}{c},$$

so, looking at Fig. 4.11, we obtain

$$k_1 = -\frac{\omega_I}{c} \sin\theta_I, \quad k_2 = \frac{\omega_I}{c} \cos\theta_I$$

and then

$$\phi_I(x,z,t) = A \exp\left\{i\omega_I\left(-\frac{z}{c}\cos\theta_I + \frac{x}{c}\sin\theta_I - t\right)\right\}.$$

To determine the reflected potential $\phi_R(x,z,t)$ we look for a solution to

$$\phi_{tt} - c^2(\phi_{xx} + \phi_{zz}) = 0$$

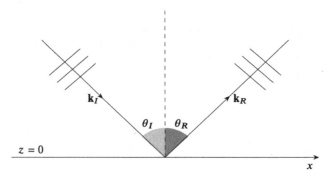

Fig. 4.11 Reflection of acoustic waves

of the form

$$\phi(x,z,t) = \phi_I(x,z,t) + \phi_R(x,z,t)$$

together with the condition

$$\frac{\partial \phi}{\partial z} = 0, \quad \text{i.e.} \quad \frac{\partial \phi_R}{\partial z} = -\frac{\partial \phi_I}{\partial z}, \quad \text{on the plane } z = 0,$$

due to the reflection. Then

$$\phi_R(x,z,t) = B \exp\left\{i\omega_R\left(\frac{z}{c}\cos\theta_R + \frac{x}{c}\sin\theta_R - t\right)\right\}$$

works. The reflection condition imposes

$$-A\cos\theta_I \exp\left\{i\omega_I\left(\frac{z}{c}\cos\theta_I - t\right)\right\} + B\cos\theta_R \exp\left\{i\omega_R\left(\frac{z}{c}\cos\theta_R - t\right)\right\} = 0.$$

This equation must hold for $z \in \mathbb{R}$ and $t \geq 0$, so

$$A = B, \theta_I = \theta_R, \omega_I = \omega_R.$$

The reflected wave has the same amplitude or frequency, and the angle of incidence equals the angle of reflection.

4.3 Further Exercises

4.3.1. *Consider the problem*

$$\begin{cases} u_{tt} - c^2 u_{xx} = 0 & 0 < x < L, t > 0 \\ u(x,0) = g(x), u_t(x,0) = h(x) & 0 \leq x \leq L \\ u(0,t) = 0, u(L,t) = B & t > 0. \end{cases}$$

a) Determine the stationary solution u_0.
b) Find a formal expression for u.
c) Establish whether u tends to u_0 as $t \to \infty$.

4.3.2. *Consider the problem*

$$\begin{cases} u_{tt} - c^2 u_{xx} = 0 & 0 < x < L, t > 0 \\ u(x,0) = g(x), u_t(x,0) = 0 & 0 \leq x \leq L \\ u_x(0,t) = u_x(L,t) = 0 & t \geq 0. \end{cases}$$

a) Define suitably g outside the interval $[0, L]$, and use d'Alembert's formula to represent the solution as superposition of travelling waves.
b) Explain the physical meaning of the result.

4.3.3. *Solve the (Cauchy-Dirichlet) problem*

$$\begin{cases} u_{tt} - c^2 u_{xx} = f(x,t) & 0 < x < L, t > 0 \\ u(x,0) = u_t(x,0) = 0 & 0 \le x \le L \\ u(0,t) = u(L,t) = 0 & t \ge 0, \end{cases}$$

formally, in the following cases:

a) $f(x,t) = e^{-t} \sin\left(\dfrac{\pi x}{L}\right).$

b) $f(x,t) = xe^{-t}.$

4.3.4. *Find the formal solution to the (Cauchy-Neumann) problem*

$$\begin{cases} u_{tt} - c^2 u_{xx} = f(x,t) & 0 < x < L, t > 0 \\ u(x,0) = u_t(x,0) = 0 & 0 \le x \le L \\ u_x(0,t) = u_x(L,t) = 0 & t \ge 0 \end{cases}$$

where

$$f(x,t) = e^{-t} \cos\left(\frac{\pi x}{2L}\right).$$

4.3.5. *Let $u = u(x,t)$ solve*

$$Lu = u_{tt} - u_{xx} - h^2 u \le 0, \quad h > 0 \tag{4.28}$$

on the characteristic triangle

$$T = \{(x,t) : x > t, x + t < 1, t > 0\}.$$

Prove that if $u \in C^2(\overline{T})$, and

$$u(x,0) \le M < 0, \quad u_t(x,0) \le 0$$

for $0 \le x \le 1$, then

$$u < 0 \quad \text{on } T.$$

4.3.6. *Consider the free transverse vibrations of a semi-infinite string. Assume that the initial position and velocity are known, and that the end slides along a transverse straight guide perpendicular to the strip. Find a mathematical model describing the motion and write the expression of the solution (see also Problem 4.2.6 on page 229).*

4.3.7. *Solve Problem 4.2.7 (page 230) with*

$$f(t) = -g$$

(that is, when the oscillations are forced by gravity).

4.3.8. *Solve the previous problem in case the string is not initially at rest, but rather*

$$u_t(x,0) = 1.$$

4.3.9. *Analyse numerically Problems 4.2.8, 4.2.9 in case* $L = 1$ m, $g = 9.8$ m/s^2, $h = 0$ *and*

$$f(x) = \begin{cases} \dfrac{x}{100} & 0 \le x \le \dfrac{1}{2} \\ \dfrac{(1-x)}{100} & \dfrac{1}{2} \le x \le 1. \end{cases}$$

In particular:

a) *Determine the first three frequencies of the vibration.*
b) *Compute the amplitudes of the first three modes.*
c) *Plot, for some values of* t, *the approximate sum of the first three modes.*

4.3.10. *Determine the type of the second-order equation*

$$u_{yy} - 2u_{xy} + 2u_x - u_y = 4e^x$$

and compute the characteristics. Determine, if possible, the general solution.

4.3.11. *Determine the type of the second-order equation*

$$u_{xy} + yu_y - u = 0,$$

and compute the characteristics. Determine, if possible, the general integral.

4.3.12. *Let* Γ *denote the curve* $x = \varphi(y)$ *with* $\varphi \in C^2(0, \infty)$, *continuous at* $y = 0$, $\varphi(0) = 0$, *and such that* $\varphi'(x) \ge 0$ *for* $x > 0$. *Consider the problem:*

$$\begin{cases} u_{xy} = F(x, y) & (x, y) \in Q = \{x > \varphi(y), y > 0\} \\ u(x, 0) = g(x) & x > 0 \\ u(\varphi(y), y) = h(y) & y > 0, \end{cases}$$

where $F \in C(\overline{Q})$, $h, g \in C^2(0, \infty)$ *continuous at* $y = 0$.

a) *Integrate on the rectangle* R *of Fig. 4.12 and find a representation for* u *in terms of* F, g, h.
b) *Determine conditions on* g, h *that make* u *a solution belonging to* $C^2(Q) \cap C(\overline{Q})$.
c) *Discuss uniqueness.*

4.3.13. *Consider the domain*

$$S = \{(x, t) : -t < x < t, t > 0\}.$$

Solve

$$\begin{cases} u_{tt} - u_{xx} = 0 & \text{on } S \\ u(x, x) = g(x), \ u(x, -x) = h(x) & x \ge 0 \\ g(0) = h(0). \end{cases}$$

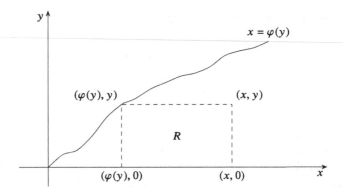

Fig. 4.12 Fig. for Exercise 4.3.12

4.3.14. *Establish whether the following problems are well or ill posed.*

a) *Characteristic Cauchy problem*[8]:

$$
\begin{cases}
u_{tt} - u_{xx} = 0 & \text{on the half-space } x > t \\
u(x,x) = f(x) & x \in \mathbb{R} \\
\partial_\nu u(x,x) = g(x) & x \in \mathbb{R},
\end{cases}
$$

where

$$
\nu = \frac{1}{\sqrt{2}}(1,-1)
$$

(unit normal to the characteristic $x = t$).

b) *Boundary values on the quadrant:*

$$
\begin{cases}
u_{tt} - u_{xx} = 0 & x > 0, t > 0 \\
u(x,0) = f(x) & x \geq 0 \\
u(0,t) = g(t) & t \geq 0.
\end{cases}
$$

4.3.15. *Consider the following model for a string with internal friction:*

$$
\begin{cases}
\rho u_{tt} - \tau u_{xx} = \gamma u_{xxt} & 0 < x < L, t > 0 \\
u(x,0) = f(x), u_t(x,0) = g(x) & 0 < x < L \\
u(0,t) = u(L,t) = 0 & t \geq 0,
\end{cases}
\tag{4.29}
$$

where ρ, τ, γ are positive constants.

a) *Define*

$$
E(t) = \frac{1}{2} \int_0^L \left[\rho u_t^2 + \tau u_x^2 \right] dx.
$$

---

[8] The Cauchy data are assigned on a characteristic curve.

Assuming u sufficiently regular, prove that

$$E'(t) \leq 0$$

and interpret the result.

b) *Use part a) to prove a uniqueness theorem for problem (4.29).*

c) *Using separation of variables establish whether $u(x,t) \to 0$ as $t \to +\infty$.*

4.3.16. *Given $\lambda > 0$, consider the problem*

$$\begin{cases} u_{tt}(x,y,t) = u_{xx}(x,y,t) + u_{yy}(x,y,t) + \lambda u(x,y,t) & (x,y) \in \mathbb{R}^2, \, t > 0 \\ u(x,y,0) = f(x,y), \quad u_t(x,y,0) = g(x,y) & (x,y) \in \mathbb{R}^2. \end{cases}$$

a) *Determine k so that the function*

$$v(x,y,z,t) = e^{kz} u(x,y,t)$$

solves

$$v_{tt} = v_{xx} + v_{yy} + v_{zz}.$$

b) *Use part a) to represent the solution.*

4.3.17. (Plane acoustic waveguide) *Consider the region in space bounded by the planes $x = 0$, $x = d$ (acoustic waveguide). Which are the plane harmonics, polarised on the plane xz, that can propagate in the region? Analyse their properties.*

4.3.18. (Sound waves in a pipe) *Study acoustic waves of speed c and angular frequency ω that travel in a semi-infinite cylindrical pipe of radius a. Decide, in particular, which modes can propagate without damping effects as a varies.*

4.3.1 Solutions

Solution 4.3.1. **a)** The stationary solution is the time-independent function that solves the problem with the Dirichlet conditions only. Thus $u_0 = u_0(x)$ solves

$$\begin{cases} u_0'' = 0 & 0 < x < L \\ u_0(0) = 0, \, u_0(L) = B, \end{cases}$$

i.e. $u_0(x) = \frac{B}{L}x$.

b) To apply variable separation we need homogeneous Dirichlet conditions. To this purpose set $U(x,t) = u(x,t) - u_0(x)$. Then U solves

$$\begin{cases} U_{tt} - c^2 U_{xx} = 0 & 0 < x < L, \, t > 0 \\ U(x,0) = g(x) - \dfrac{B}{L}x, \, U_t(x,0) = h(x) & 0 \leq x \leq L \\ U(0,t) = U(L,t) = 0 & t > 0. \end{cases}$$

We want to find solutions $U(x, t) = v(x)w(t)$. Following Problem 4.2.1 (page 217) we can write

$$U(x,t) = \sum_{k=1}^{+\infty} \left[a_k \cos\left(\frac{ck\pi}{L}t\right) + b_k \sin\left(\frac{ck\pi}{L}t\right) \right] \sin\left(\frac{k\pi}{L}x\right).$$

For $t = 0$ and $0 \le x \le L$ we have

$$\sum_{k=1}^{+\infty} a_k \sin\left(\frac{k\pi}{L}x\right) = g(x) - \frac{B}{L}x, \qquad \sum_{k=1}^{+\infty} \frac{ck\pi}{L} b_k \sin\left(\frac{k\pi}{L}x\right) = h(x),$$

so the Fourier coefficients are determined by

$$a_k = \frac{2}{L} \int_0^L \left(g(x) - \frac{B}{L}x\right) \sin\left(\frac{k\pi}{L}x\right) dx, \qquad b_k = \frac{2}{ck\pi} \int_0^L h(x) \sin\left(\frac{k\pi}{L}x\right) dx.$$

c) The function u found in part b) is $(2L/c)$-periodic in time, so it admits limit for $t \to \infty$ if and only if it is constant in time, i.e. if and only if $g(x) = u_0(x)$ and $h(x) = 0$. In all other cases the solution oscillates indefinitely and does not converge to the stationary solution.

Solution 4.3.2. Let us proceed as in Problem 4.2.2 on page 219, by extending the Cauchy data to $\mathbb{R}$, so that the corresponding global Cauchy problem has a solution with vanishing spatial derivative along $x = 0$, $x = L$. By restricting this solution to the strip $[0, L] \times \{t > 0\}$ we get the required solution. Indicate by $\tilde{g}$ and $\tilde{h}$ the extended data. The global solution is given by d'Alembert's formula

$$u(x,t) = \frac{1}{2}[\tilde{g}(x - ct) + \tilde{g}(x + ct)] + \frac{1}{2c} \int_{x-ct}^{x+ct} \tilde{h}(s)\, ds.$$

We must choose $\tilde{g}, \tilde{h}$ to satisfy both the initial and the boundary Neumann conditions. The simplest possibility for h is to set $\tilde{h} = 0$. As for g we must have

$$\begin{cases} u(x,0) = \tilde{g}(x) = g(x) & 0 \le x \le L \\ u_x(0,t) = \dfrac{1}{2}[\tilde{g}'(-ct) + \tilde{g}'(ct)] = 0 & t > 0 \\ u(L,t) = \dfrac{1}{2}[\tilde{g}'(L - ct) + \tilde{g}'(L + ct)] = 0 & t > 0. \end{cases}$$

Therefore, for any s,

$$\tilde{g}'(s) = -\tilde{g}'(-s), \qquad \tilde{g}'(L + s) = -\tilde{g}'(L - s). \tag{4.30}$$

The first condition says that $\tilde{g}'$ must be odd, hence $\widetilde{g}$ is even. By the second condition

$$\tilde{g}'(s + 2L) = \tilde{g}'(L + (L + s)) = -\tilde{g}'(L - (L + s)) = -\tilde{g}'(-s) = \tilde{g}'(s),$$

so $\tilde{g}'$ is $2L$-periodic. But $\tilde{g}'$ is odd, so its mean value on the period interval is zero: consequently also $\widetilde{g}$ is $2L$-periodic, and

$$\tilde{g}(s) = \begin{cases} g(s) & 0 < s < L, \\ -g(-s) & -L < s < 0. \end{cases}$$

The solution to the original problem is then

$$u(x,t) = \frac{1}{2} [\tilde{g}(x - ct) + \tilde{g}(x + ct)] \qquad \text{for } 0 \le x \le L,\ t \ge 0. \tag{4.31}$$

The interpretation of (4.31) follows from arguments similar to those of Problem 4.2.2 (page 219). The only difference is that, here, the reflection at the ends does *not* change the overall sign, because the prolongation of g is even.

Solution 4.3.3. Formula (4.14) gives

a) $u(x,t) = \dfrac{L^2}{L^2 + c^2\pi^2} \left[e^{-t} - \cos\left(\dfrac{c\pi}{L}t\right) + \dfrac{L}{c\pi} \sin\left(\dfrac{c\pi}{L}t\right) \right] \sin\left(\dfrac{\pi}{L}x\right).$

b) $u(x,t) = \dfrac{2L}{\pi} \displaystyle\sum_{k=1}^{+\infty} \dfrac{(-1)^{k+1}L^2}{k(L^2 + c^2\pi^2 k^2)} \left[e^{-t} - \cos\left(\dfrac{ck\pi}{L}t\right) + \dfrac{L}{ck\pi} \sin\left(\dfrac{ck\pi}{L}t\right) \right] \sin\left(\dfrac{k\pi}{L}x\right).$

Solution 4.3.4. Following the solution of Problem 4.2.5 on page 227:

$$u(x,t) = \frac{4L^2}{4L^2 + c^2\pi^2} \left[e^{-t} - \cos\left(\frac{c\pi}{2L}t\right) + \frac{2L}{c\pi} \sin\left(\frac{c\pi}{2L}t\right) \right] \cos\left(\frac{\pi}{2L}x\right).$$

Solution 4.3.5. Let C be a point in T and consider the characteristic triangle T_C determined by A, B, C, see Fig. 4.9 (page 243). We proceed as in Problem 4.2.15 (page 242) and integrate on T_C the inequality $Lu \le 0$. We find

$$u(C) \le \frac{u(B) + u(A)}{2} + \int_0^1 u_t(x,0)\,dx + \frac{1}{2}\int_{T_C} h^2 u\,dx dt. \tag{4.32}$$

As $u(x,0) \le M < 0$ for $0 \le x \le 1$, by continuity $u(x,t)$ stays negative, at least for $t > 0$ small enough. If, by contradiction, u did not remain negative everywhere on T, there would exist a point $C = (x_0, t_0)$ at which it vanished for the first time, i.e.

$$u(x_0, t_0) = 0 \quad \text{and} \quad u(x,t) < 0 \text{ for } t < t_0.$$

By (4.32), since $u(A) \le M$, $u(B) \le M$ and $u < 0$ on T_C, we would have

$$0 = u(C) \le M < 0,$$

a contradiction. Therefore $u < 0$ on the whole T.

Solution 4.3.6. Call $u(x,t)$ the string shape at time t. The model is then

$$\begin{cases} u_{tt} - c^2 u_{xx} = 0 & x > 0,\ t > 0 \\ u(x,0) = g(x),\ u_t(x,0) = h(x) & x \ge 0 \\ u_x(0,t) = 0 & t \ge 0. \end{cases}$$

We shall mimic Problem 4.2.6 (page 229) and seek functions $\tilde{g}$ and $\tilde{h}$ defined on $\mathbb{R}$, coinciding with g and h on $x > 0$, and also such that the function

$$u(x,t) = \frac{1}{2}[\tilde{g}(x+ct) + \tilde{g}(x-ct)] + \frac{1}{2c}\int_{x-ct}^{x+ct}\tilde{h}(s)\,ds$$

(from d'Alembert's formula) satisfies $u_x(0,t) = 0$ for any $t > 0$. We compute

$$u_x(x,t) = \frac{1}{2}\left[\tilde{g}'(x+ct) + \tilde{g}'(x-ct)\right] + \frac{1}{2c}\left[\tilde{h}(x+ct) - \tilde{h}(x-ct)\right],$$

and the boundary condition reads

$$\frac{1}{2}\left[\tilde{g}'(ct) + \tilde{g}'(-ct)\right] + \frac{1}{2c}\left[\tilde{h}(ct) - \tilde{h}(-ct)\right] = 0.$$

It suffices to extend h as an even function and g' as an odd one (i.e g to become even).

Solution 4.3.7. From (4.17) we get

$$u(x,t) = -g\int_0^t\left[\tau - \left(\tau - \frac{x}{c}\right)\mathcal{H}\left(\tau - \frac{x}{c}\right)\right]d\tau = -\frac{1}{2}gt^2 + g\int_0^t\left(\tau - \frac{x}{c}\right)\mathcal{H}\left(\tau - \frac{x}{c}\right)d\tau.$$

The last integral is zero if $t \le x/c$, and equals

$$\frac{1}{2}\left(t - \frac{x}{c}\right)^2, \qquad \text{if } t > \frac{x}{c}.$$

Hence

$$u(x,t) = \begin{cases} -\frac{g}{2}\left[t^2 - \left(t - \frac{x}{c}\right)^2\right] & 0 \le x \le ct \\ -\frac{g}{2}t^2 & x \ge ct. \end{cases}$$

The interpretation is rather easy. The portion of string placed on $x \ge ct$ is only subject to gravity. The fixed end affects only the portion between $x = 0$ and $x = ct$ (and the wave propagates with speed c).

Solution 4.3.8. We resort to what we did for Problem 4.2.7 (page 230), and arrive at the ODE

$$-c^2 U_{xx}(x,s) + s^2 u(x,s) = F(s) + 1$$

which, once integrated, gives

$$U(x,s) = (1 + F(s))\frac{1 - e^{-xs/c}}{s^2}.$$

Now we can anti-transform and obtain

$$u(x,t) = t - \left(t - \frac{x}{c}\right)\mathcal{H}\left(t - \frac{x}{c}\right) - \frac{g}{2}\left[t^2 - \left(t - \frac{x}{c}\right)^2\mathcal{H}\left(t - \frac{x}{c}\right)\right]$$

$$= \begin{cases} \frac{x}{c} - \frac{g}{2}\left[t^2 - \left(t - \frac{x}{c}\right)^2\right] & 0 \le x \le ct \\ t - \frac{g}{2}t^2 & x \ge ct. \end{cases}$$

Solution 4.3.9. a) From Problem 4.2.9 on page 234 the fundamental frequencies are given by

$$\nu_j = \frac{\alpha_j}{4\pi} \sqrt{\frac{g}{L}}$$

where the α_j are the zeroes of the Bessel function J_0. These are, approximately,

$$\alpha_1 = 2.40483\ldots, \quad \alpha_2 = 5.52008\ldots, \quad \alpha_3 = 8.65373\ldots$$

so that

$$\nu_1 = 0.5990\ldots, \quad \nu_2 = 1.3750\ldots, \quad \nu_3 = 2.1558\ldots.$$

b) Again by Problem 4.2.9, the first fundamental modes are

$$u_j(x,t) = A_j J_0\left(\alpha_j \sqrt{x}\right) \cos\left(\frac{\sqrt{9.8}\alpha_j}{2} t\right) \qquad j = 1, 2, 3$$

where

$$A_j = \frac{1}{J_1^2(\alpha_j)} \left\{ \int_0^{1/2} \frac{x}{100} J_0\left(\alpha_j \sqrt{x}\right) dx + \int_{1/2}^1 \frac{(1-x)}{100} J_0\left(\alpha_j \sqrt{x}\right) dx \right\}.$$

Inserting the values obtained in part a) we get

$$A_1 = 0.003874\ldots, \quad A_2 = -0.005787\ldots, \quad A_3 = 0.002371\ldots.$$

The amplitudes are much smaller than 1 (the rest length).

c) See Fig. 4.13.

Solution 4.3.10. The equation is hyperbolic, with characteristics $x = $ constant and $x + 2y = $ constant. The general solution is

$$u(x,y) = 2e^x + e^{(x+2y)/2}[F(x) + G(x+2y)]$$

where F and G are two arbitrary functions.

Solution 4.3.11. The equation is hyperbolic, with characteristics defined by $x = $ constant and $y = $ constant (the equation is already in normal form). Set $v = u_y$ and differentiate in y:

$$v_{xy} + y v_y + v - v = 0,$$

i.e. $(v_y)_x + y v_y = 0$. Integrating in x gives $v_y = f(y)e^{-xy}$, and then integrating in y twice,

$$u(x,y) = yG(x) + G'(x) + \int_0^y (y - \eta)e^{-x\eta} f(\eta)\, d\eta,$$

with f, G arbitrary.

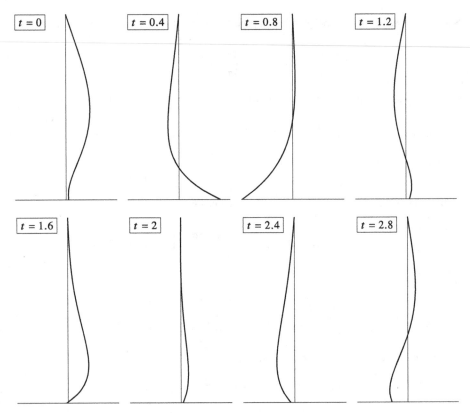

Fig. 4.13 Positions of the hanging chain (Exercise 4.3.9) at various instants

Solution 4.3.12.   **a)** Integrating on R we have:

$$\int_R u_{xy}\,(\xi,\eta)\,d\xi d\eta = \int_0^y d\eta \int_{\varphi(y)}^x u_{xy}\,(\xi,\eta)\,d\xi = \int_0^y \left[u_y\,(x,\eta) - u_y\,(\varphi\,(y),\eta)\right]d\eta$$
$$= u\,(x,y) - u\,(x,0) - u\,(\varphi\,(y),y) + u\,(\varphi\,(y),0).$$

Using the data and the equation we can write the following formula for u:

$$u\,(x,y) = h\,(y) + g\,(x) - g\,(\varphi\,(y)) + \int_0^y \int_{\varphi(y)}^x F\,(\xi,\eta)\,d\xi d\eta. \qquad (4.33)$$

b) It is easy to verify that (4.33) defines a solution $u \in C^2\,(Q) \cap C\,(\overline{Q})$ precisely when $h\,(0) = g\,(0)$.

c) Suppose u, v are solutions in $C^2\,(Q) \cap C\,(\overline{Q})$ of the same problem. Their difference $w = u - v$ solves $w_{xy} = 0$ with *vanishing* data. If we integrate on the rectangle defined by points $(a,b),(x,b),(a,y),(x,y)$, with $\varphi\,(y) < a < x$ and $0 < b < y$, we find, with similar computations,

$$0 = \int_R w_{xy}(\xi, \eta)\, d\xi\, d\eta = w(x, y) - w(a, y) - w(x, b) + w(a, b).$$

Passing to the limit $a \to \varphi(y)$ and $b \to 0$, we see that $w(a, y)$, $w(x, b)$ and $w(a, b)$ converge to the origin, and therefore $w(x, y) = 0$, proving uniqueness.

Solution 4.3.13. If $g, h \in C^2(0, \infty)$ are continuous at zero, there exists a unique solution continuous on $\overline{S}$ with second derivatives continuous on S, given by

$$u(x, t) = g\left(\frac{x+t}{2}\right) + h\left(\frac{x-t}{2}\right) - g(0).$$

Solution 4.3.14. a) The solution has the form $u(x, t) = F(x + t) + G(x - t)$. We impose the Cauchy data on the straight line $x = t$:

$$u(x, x) = F(2x) + G(0) = f(x), \tag{4.34}$$

which implies

$$F(z) = f\left(\frac{z}{2}\right) - G(0),$$

and

$$\frac{1}{\sqrt{2}}[u_x(x, x) - u_t(x, x)] = \sqrt{2} G'(0) = g(x).$$

The problem can be solved only if the Neumann datum g is constant: $g(x) = k$. In this case there are infinitely many solutions

$$u(x, t) = f\left(\frac{x+t}{2}\right) + \frac{k}{\sqrt{2}}(x - t) + G(x - t)$$

where G is any function such that $G(0) = G'(0) = 0$. The problem is, therefore, *ill posed*.

b) This problem, too, is ill posed. Every function

$$u(x, t) = F(x + t) - F(x - t)$$

with F even ($F(-z) = F(z)$) will solve the problem with zero data.

Solution 4.3.15. a) Let us differentiate the integrand function[9]:

$$E'(t) = \int_0^L [\rho u_t u_{tt} + \tau u_x u_{xt}]\, dx.$$

The boundary conditions force $u_t(0, t) = u_t(0, L) = 0$, whence

$$\int_0^L u_x u_{xt}\, dx = [u_x u_t]_0^L - \int_0^L u_{xx} u_t\, dx = -\int_0^L u_{xx} u_t\, dx,$$

---

[9] This is allowed if, for instance, u has continuous second derivatives on $0 \le x \le L, t \ge 0$.

and from the ODE we compute

$$E'(t) = \int_0^L u_t \left[\rho u_{tt} - \tau u_{xx} \right] dx = \gamma \int_0^L u_t u_{xxt} dx$$

$$= \gamma \left[u_t u_{xx} \right]_0^L - \gamma \int_0^L u_{xt}^2 dx = -\gamma \int_0^L u_{xt}^2 dx \le 0.$$

Interpretation: $E(t)$ is the total mechanical energy at time t. The formula

$$E'(t) = -\gamma \int_0^L u_{xt}^2 (x,t) dx$$

shows that the string dissipates energy at rate $-\gamma \int_0^L u_{xt}^2 dx$. In the model of a string small vibrations, u_x represents the relative displacement of the string particles, and u_{xt} controls the variation of u_x. The term $-\int_0^L u_{xt}^2 dx$ is therefore the kinetic energy per unit time that is dissipated because of the internal friction of the string particles.

b) If u, v solve (4.29), the quantity $w = u - v$ also solves (4.29), with data $f(x) = g(x) = 0$. Call $E(t)$ the energy associated to w. From part a) we have $E'(t) \le 0$ and $E(0) = 0$, since $w_x(x,0) = f'(x) = 0$ and $w_t(x,0) = g(x) = 0$. Consequently $E(t) = 0$ for any $t > 0$, which implies

$$w_x = w_t = 0$$

for any $0 < x < L$ and every $t > 0$. But then w is constant. Being initially zero, it must be identically zero, so $u = v$.

c) We seek solutions of the type $u(x,t) = v(x) w(t)$. Substituting into the equation, we obtain

$$\rho v(x) w''(t) - \tau v''(x) w(t) = \gamma v''(x) w'(t).$$

Now we separate variables and set $c^2 = \tau/\rho$ and $\varepsilon^2 = \gamma/\rho$:

$$\frac{v''(x)}{v(x)} = \frac{w''(t)}{c^2 w(t) + \varepsilon^2 w'(t)} = -\lambda^2.$$

The eigenvalue problem for v is $v''(x) + \lambda^2 v(x) = 0$, $v(0) = v(L) = 0$, solved by

$$v_n(x) = \sin \lambda_n x, \qquad \lambda_n = \frac{n\pi}{L}, \ n = 1, 2, \ldots \ .$$

For w we therefore have

$$w''(t) + \varepsilon^2 \lambda_n^2 w'(t) + c^2 \lambda_n^2 w(t) = 0.$$

The general integral depends on the sign of

$$\delta_n = \varepsilon^4 \lambda_n^2 - 4c^2.$$

When $\delta_n < 0$, i.e. $1 \le n < 2cL/(\pi \varepsilon^2)$, we have

$$w_n(t) = \exp\left(-\frac{\varepsilon^2 \lambda_n^2}{2} t \right) \left[a_n \sin\left(\frac{\lambda_n \sqrt{|\delta_n|}}{2} t \right) + b_n \cos\left(\frac{\lambda_n \sqrt{|\delta_n|}}{2} t \right) \right].$$

If there is an n such that $\delta_n = 0$, then

$$w_n(t) = (a_n + b_n t) \exp\left(-\frac{\varepsilon^2 \lambda_n^2}{2} t\right).$$

When $\delta_n > 0$

$$w_n(t) = a_n \exp\left(\frac{-\varepsilon^2 \lambda_n^2 + \lambda_n \sqrt{\delta_n}}{2} t\right) + b_n \exp\left(\frac{-\varepsilon^2 \lambda_n^2 - \lambda_n \sqrt{\delta_n}}{2} t\right).$$

The solution to (4.29) will be

$$u(x,t) = \sum_{n=1}^{\infty} w_n(t) \sin \lambda_n x.$$

Now assume that f and g can be expanded in sine series on $[0, L]$; the Fourier coefficients a_n and b_n will be determined by the data. To see if $u(x,t) \to 0$ as $t \to \infty$ there is no need to compute the coefficients explicitly. Suppose, in fact, that f, g are $C^1(\mathbb{R})$, so that

$$\sum_{n=1}^{\infty} (|a_n| + |b_n|) < \infty.$$

Then

$$\delta_n < 0 \implies |w_n(t)| \le (|a_n| + |b_n|) \exp\left(-\frac{\varepsilon^2 \lambda_n^2}{2} t\right) \le (|a_n| + |b_n|) \exp\left(-\frac{\varepsilon^2 \pi^2}{2L^2} t\right),$$

$$\delta_n = 0 \implies |w_n(t)| \le (|a_n| + t\,|b_n|) \exp\left(-\frac{\varepsilon^2 \lambda_n^2}{2} t\right) \le (|a_n| + t\,|b_n|) \exp\left(-\frac{\varepsilon^2 \pi^2}{2L^2} t\right),$$

$$\delta_n > 0 \implies -\varepsilon^2 \lambda_n^2 + \lambda_n \sqrt{\delta_n} = \varepsilon^2 \lambda_n^2 \left(-1 + \sqrt{1 - \frac{4c^2}{\varepsilon^4 \lambda_n^2}}\right) \le -\frac{2c^2}{\varepsilon^2} < 0$$

so

$$|w_n(t)| \le |a_n| \exp\left(-\frac{c^2}{\varepsilon^2} t\right) + |b_n| \exp\left(-\frac{\varepsilon^2 \pi^2}{2L^2} t\right).$$

It is easy to conclude $u(x,t) \to 0$ as $t \to \infty$.

Solution 4.3.16. For better clarity we denote by Δ_2, Δ_3 the Laplacians in two, three dimensions respectively.

a) We have

$$v_{tt} - \Delta_3 v = \left[u_{tt} - \Delta_2 u - k^2 u\right] e^{kz} = 0 \qquad \text{if } k = \sqrt{\lambda}.$$

b) The function v introduced in a) solves the Cauchy problem

$$\begin{cases} v_{tt} - \Delta_3 v = 0 & (x,y,z) \in \mathbb{R}^3,\ t > 0 \\ v(x,y,z,0) = f(x,y)e^{kz}, \quad v_t(x,y,z,0) = g(x,y)e^{kz} & (x,y,z) \in \mathbb{R}^3 \end{cases}$$

$(k = \sqrt{\lambda})$. Kirchhoff's formula gives

$$v(x, y, z, t) = \frac{\partial}{\partial t}\left[\frac{1}{4\pi t}\int_{\partial B_t(x,y,z)} f(\xi, \eta)e^{k\xi}\, d\sigma\right] + \frac{1}{4\pi t}\int_{\partial B_t(x,y,z)} g(\xi, \eta)e^{k\xi}\, d\sigma$$

and then

$$u(x, y, t) = v(x, y, 0, t) =$$
$$= \frac{\partial}{\partial t}\left[\frac{1}{4\pi t}\int_{\partial B_t(x,y,0)} f(\xi, \eta)e^{k\xi}\, d\sigma\right] + \frac{1}{4\pi t}\int_{\partial B_t(x,y,0)} g(\xi, \eta)e^{k\xi}\, d\sigma.$$

The integrals on the surface $\partial B_t(x, y, 0)$ can be reduced to the disc $K_t(x, y)$, as in Problem 4.2.18 on page 249.

Solution 4.3.17. It is clear that there can be waves with wave vector parallel to the (horizontal) z-axis; their potential is

$$\phi(z, t) = A \exp\{i(k_3 z - \omega t)\}.$$

But we may also imagine waves moving in different directions and being reflected on the walls of the waveguide. They are polarised on the xz-plane, so we write them as sums of an incoming wave and a reflected one: equivalently, we represent them by means of a potential of the form (Fig. 4.14)

$$\phi(x, z, t) = A \exp\left\{i\omega\left(\frac{x}{c}\cos\theta_I + \frac{z}{c}\sin\theta_I - t\right)\right\}$$
$$+ A \exp\left\{i\omega\left(-\frac{x}{c}\cos\theta_I + \frac{z}{c}\sin\theta_I - t\right)\right\}$$

where we used the result of Problem 4.2.22 on page 254.

The wave will get reflected if the homogeneous Neumann conditions on the walls

$$\phi_x(0, z, t) = \phi_x(d, z, t) = 0$$

are fulfilled. While the first one clearly holds, the other one gives

$$\exp\left\{i\frac{\omega d}{c}\cos\theta_I\right\} = \exp\left\{-i\frac{\omega d}{c}\cos\theta_I\right\}$$

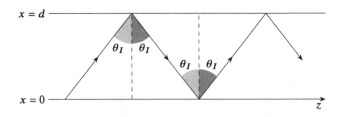

Fig. 4.14 Reflections along the waveguide

which implies

$$\frac{\omega d}{c} \cos \theta_I = -\frac{\omega d}{c} \cos \theta_I + 2n\pi, \quad n = 0, 1, 2, \ldots$$

and so

$$\cos \theta_I = \frac{n\pi c}{\omega d}. \tag{4.35}$$

As $0 < \cos \theta_I \leq 1$, there are *real* solutions only for integer numbers $n \leq N$, where N is the maximum integer such that

$$n \leq \frac{\omega d}{\pi c}.$$

Consequently *there is only a finite number N of incidence angles θ_I at which a plane harmonic can propagate along the waveguide.* Each angle corresponds actually to one *mode* of the waveguide. Given n, we have

$$\sin \theta_I = \sqrt{1 - \frac{n^2 \pi^2 c^2}{\omega^2 d^2}} \quad \text{and} \quad \frac{\omega}{c} \sin \theta_I = \sqrt{\frac{\omega^2}{c^2} - \frac{n^2 \pi^2}{d^2}}$$

and so

$$k_3 = \sqrt{\frac{\omega^2}{c^2} - \frac{n^2 \pi^2}{d^2}} \tag{4.36}$$

is the wavenumber along z. By Euler's relation $2 \cos \alpha = e^{i\alpha} + e^{-i\alpha}$, $\alpha = n\pi/d$, we can write the potential for the nth mode in a more significant way:

$$\phi_n (x, z, t) = 2A \cos \frac{n\pi x}{d} \exp \{i (k_3 z - \omega t)\}. \tag{4.37}$$

When $n = 0$ we recover the wave moving along z at speed c. When $0 < n \leq N$ the wave behaves as a *standing* wave along x and a *travelling* one along z.

Complex solutions to (4.35) correspond to *purely imaginary* wavenumbers $k_3 = i\kappa$, $\kappa > 0$. The potential (4.37) refers to waves that *decay exponentially* as $z \to +\infty$ (called *evanescent waves*).

Remark. An alternative way to look at this is the following. Fix the wavenumber k_3 and try to find the possible travelling frequencies. By (4.36) there exists a *cut-off frequency*

$$\omega_c = \frac{\pi c}{d}$$

and a sequence of *resonant frequencies*

$$\omega_n = c \sqrt{k_3^2 + \frac{n^2 \pi^2}{d^2}}$$

corresponding to the harmonics that can propagate.

Solution 4.3.18. We use cylindrical coordinates $r = \sqrt{x^2 + y^2}, \theta, z$ with $0 \leq \theta \leq 2\pi, z \in \mathbb{R}$. Waves of angular frequency ω moving along the pipe can be described by the potential

$$\Phi (r, \theta, z, t) = \varphi (r, \theta, z) e^{-i\omega t},$$

solution to

$$\Phi_{tt} - c^2 \left\{ \Phi_{rr} + \frac{1}{r} \Phi_r + \frac{1}{r^2} \Phi_{\theta\theta} + \Phi_{zz} \right\} = 0 \tag{4.38}$$

inside the cylinder. Moreover, the zero normal velocity on the walls enforces the homogeneous Neumann condition

$$\varphi_r(a, \theta, z) = 0.$$

Substituting $\varphi(r, \theta, z) e^{-i\omega t}$ in (4.38) we find for φ the equation

$$\varphi_{rr} + \frac{1}{r}\varphi_r + \frac{1}{r^2}\varphi_{\theta\theta} + \varphi_{zz} + \frac{\omega^2}{c^2}\varphi = 0. \tag{4.39}$$

Ler us determine the modes by analysing solutions with separate variables:

$$\varphi(r, \theta, z) = u(r) v(\theta) w(z)$$

with u bounded, $u'(a) = 0$ and $v(\theta)$ 2π−periodic. Substituting in (4.39) and dividing by φ, we get

$$\frac{u''(r) + \frac{1}{r}u'(r)}{u(r)} + \frac{1}{r^2}\frac{v''(\theta)}{v(\theta)} + \frac{w''(z)}{w(z)} + \frac{\omega^2}{c^2} = 0.$$

We set

$$\frac{u''(r) + \frac{1}{r}u'(r)}{u(r)} + \frac{1}{r^2}\frac{v''(\theta)}{v(\theta)} = -\frac{w''(z)}{w(z)} - \frac{\omega^2}{c^2} = \alpha$$

and define

$$\gamma = \frac{\omega^2}{c^2} - \alpha,$$

so that

$$w''(z) + \gamma w(z) = 0. \tag{4.40}$$

From

$$\frac{u''(r) + \frac{1}{r}u'(r)}{u(r)} + \frac{1}{r^2}\frac{v''(\theta)}{v(\theta)} = \alpha$$

we deduce

$$\frac{r^2u''(r) + ru'(r) - r^2\alpha u(r)}{u(r)} = -\frac{v''(\theta)}{v(\theta)} = \beta.$$

As v is 2π-periodic, by

$$v''(\theta) + \beta v(\theta) = 0$$

we have

$$\beta = m^2 \quad \text{with } m \geq 0 \text{ integer}$$

and

$$v(\theta) = v_0 \cos[m(\theta - \theta_0)].$$

Then the equation for u is

$$u''(r) + \frac{1}{r}u'(r) + \left(-\alpha - \frac{m^2}{r^2}\right)u(r) = 0. \tag{4.41}$$

If $\alpha = -\lambda^2$ (we shall consider only this case, for simplicity), equation (4.41) is a parametric Bessel equation[10] of order m. The only bounded solutions are

$$u_m(r) = u_0 J_m(\lambda r)$$

---

[10] Appendix A.

where J_m is the Bessel function of the first kind of order m. The constraint $u'(a) = 0$ reads

$$J'_m (\lambda a) = 0.$$

There are infinitely many zeroes $\lambda_{m1} < \lambda_{m2} < \ldots < \lambda_{mp} < \ldots$ of J'_m, all found in standard textbooks. The first one is zero. Here are a few other positive roots, approximated to two digits:

$$\lambda_{01} = 3.83 \quad \lambda_{02} = 7.02 \quad \lambda_{03} = 10.17$$
$$\lambda_{11} = 1.84 \quad \lambda_{12} = 5.33 \quad \lambda_{13} = 8.54$$
$$\lambda_{21} = 3.05 \quad \lambda_{22} = 6.71 \quad \lambda_{23} = 9.97.$$

So we have infinitely many solutions to (4.41) of the form

$$u_{mp}(r) = u_0 J_m (\lambda_{mp} r / a).$$

If the angular frequency is given, the values of γ are determined by

$$\gamma_{mp} = \frac{\omega^2}{c^2} - \frac{\lambda_{mp}^2}{a^2}. \tag{4.42}$$

When the right-hand side in (4.42) is positive,

$$\frac{\omega^2}{c^2} - \frac{\lambda_{mk}^2}{a^2} = k^2, \qquad k > 0,$$

the wave

$$\Phi_{mk}(r, \theta, z, t) = A_{mk} J_m \left(\frac{\lambda_{mk}}{a} r \right) \cos [m(\theta - \theta_0)] \exp \{i(kz - \omega t)\}$$

travels *undamped* along the pipe. If, instead,

$$\frac{\omega^2}{c^2} - \frac{\lambda_{mk}^2}{a^2} = -k^2,$$

the corresponding mode (bounded for $z > 0$)

$$\Phi_{mk}(r, \theta, z, t) = A_{mk} J_m \left(\frac{\lambda_{mk}}{a} r \right) \cos [m(\theta - \theta_0)] \exp \{-kz - i\omega t\}$$

becomes weaker as it propagates in the pipe.

Equation (4.42) shows that as the pipe radius decreases, more and more modes are damped, to the point that if the radius is small enough only one mode survives. In case $m = 0$, in fact, $\lambda_{00} = 0$ and the corresponding mode is the plane harmonic

$$\Phi_{00}(r, \theta, z, t) = A \exp \left\{ i\omega(\frac{z}{c} - t) \right\}$$

which propagates irrespective of the pipe' size. Therefore, as soon as

$$\omega < \lambda_{11} \frac{c}{a} = 1.84 \frac{c}{a},$$

only the plane wave will travel, while all others are "absorbed" by the pipe.

5

Functional Analysis

5.1 Backgrounds

We shall recall some basic facts on Hilbert spaces, distributions and Sobolev spaces.

Banach and Hilbert spaces

• A *Banach space* is a complete normed vector space[1] X. 'Normed' means that there is a positive function defined on X, called the *norm*,

$$\|\cdot\| = \|\cdot\|_X : X \to \mathbb{R}_+$$

with the following three properties: *positivity* ($\|x\| = 0 \iff x = 0$), *homogeneity* ($\|\lambda x\| = |\lambda| \|x\|$) and the *triangle inequality*

$$\|x + y\| \le \|x\| + \|y\|.$$

'Complete' means that *every Cauchy sequence[2] converges in X*. Given a norm, the distance between two vectorx x and y is given by $\|x - y\|$.

Typical examples of Banach spaces are the spaces $L^p(\Omega)$, $1 \le p \le \infty$, of Lebesgue-measurable functions f defined on[3] $\Omega \subset \mathbb{R}^n$ that are Lebesgue p-integrable, i.e.

$$\|f\|_{L^p(\Omega)} = \left(\int_\Omega |f|^p \right)^{1/p} < \infty \quad \text{for } 1 \le p < \infty,$$

or

$$\|f\|_{L^\infty(\Omega)} = \text{ess sup}_\Omega |f| < \infty \quad \text{if } p = \infty.$$

---

[1] We shall only deal with vector spaces over the reals.
[2] $\{x_n\} \subset X$ is called a Cauchy sequence if and only if $\|x_n - x_m\| \to 0$ as $m, n \to \infty$.
[3] Most of times Ω will be a *domain*, that is a *connected open subset* of $\mathbb{R}^n$.

© Springer International Publishing Switzerland 2015
S. Salsa, G. Verzini, *Partial Differential Equations in Action. Complements and Exercises*,
UNITEXT – La Matematica per il 3+2 87, DOI 10.1007/978-3-319-15416-9_5

• *Hölder's inequality.* Let $f \in L^p(\Omega)$, $g \in L^q(\Omega)$, with $1 \le p, q \le +\infty$, and $p^{-1} + q^{-1} = 1$ (p and q are called conjugate). Then the following inequality holds

$$\int_\Omega |uv| \, d\mathbf{x} \le \left(\int_\Omega |u|^p \, d\mathbf{x} \right)^{1/p} \left(\int_\Omega |v|^q \, d\mathbf{x} \right)^{1/q}.$$

In case $p = q = 2$ this goes under the name of the *Cauchy-Schwarz inequality.*

• A *pre-Hilbert* space is a vector space H equipped with an *inner product* (also *dot product*, especially in the case dim $H < \infty$)

$$(\cdot, \cdot) = (\cdot, \cdot)_H : H \times H \to \mathbb{R}$$

with the properties of *positivity* ($(x, x) \ge 0$, and $(x, x) = 0$ if and only if $x = 0$), *bilinearity*[4], *symmetry* ($(x, y) = (y, x)$). The inner product induces a norm

$$\|x\| = (x, x)^{1/2}.$$

If H is complete with respect to the induced norm it is called a *Hilbert space.*

Most prominently, $L^2(\Omega)$ is a Hilbert space with inner product

$$(u, v)_{L^2(\Omega)} = \int_\Omega uv \, d\mathbf{x}.$$

In particular, by positivity, if $\int_\Omega uv \, d\mathbf{x} = 0$ for any $v \in L^2(\Omega)$ then $u \equiv 0$ almost everywhere in Ω (to prove it just take $v = u$). By density, actually, it suffices that the integral vanishes for any $v \in C_0^\infty(\Omega)$.

• *Parallelogram rule.* In a Hilbert space H one always has

$$2\|u\|^2 + 2\|v\|^2 = \|u + v\|^2 + \|u - v\|^2.$$

This relation *characterises those norms that come from an inner product.*

• *Orthogonal decomposition theorem.* Let H be a Hilbert space and V a closed subspace in H. For every $x \in H$ there exists a unique $P_V x$ in V (the *projection* of x onto V) such that

$$\|P_V x - x\| = \inf_{v \in V} \|v - x\|.$$

Moreover, setting $Q_V x = x - P_V x$, we have $(v, Q_V x) = 0$ for all $v \in V$ and

$$\|x\|^2 = \|P_V x\|^2 + \|Q_V x\|^2 \quad \text{(Pythagoras' theorem)}.$$

• *Linear maps.* If F is a linear map between two Hilbert spaces H_1, H_2, the continuity of F is equivalent to its *boundedness: there exists a constant C such that*

$$\|Fu\|_{H_2} \le C \|u\|_{H_1}$$

for any $u \in H_1$. The smallest of such constants defines the *norm* of F.

---

[4] That is, linearity in both arguments.

• *Dual space and Riesz's theorem.* The set of real linear, continuous functionals on a Hilbert space H is called the *dual space* of H, and is denoted by H^*.

Riesz's representation theorem states that, for any $F \in H^*$ there exists a unique element $u \in H$ such that

$$(u, v) = Fv \qquad \text{for any } v \in H.$$

Moreover, $\|u\|_H = \|F\|_{H^*}$.

• *Weak convergence.* Let H be a Hilbert space. A sequence $\{x_n\} \subset H$ is said to *converge weakly*, or in the weak sense, to $\bar{x}$ (written $x_n \rightharpoonup \bar{x}$), if $(x_n, v) \to (\bar{x}, v)$ for any $v \in H$ as $n \to \infty$. If a sequence converges strongly it converges weakly, but the converse is not true.

If $x_n \rightharpoonup \bar{x}$ then $\{x_n\}$ is bounded and $\{\|x_n\|\}$ is lower semi-continuous; that is:

$$\|\bar{x}\| \leq \liminf \|x_n\|.$$

• *Compactness and weak compactness.* A subset K in a normed space X is said to be *compact* if any open covering of K admits a finite sub-covering. In a normed space, K is compact if and only if is *sequentially compact*, i.e. if every sequence in K admits a subsequence converging in K. The set K is called *relatively compact* (or *precompact*) if its closure is compact. This is equivalent to the following: K is relatively compact if and only if, for any $\varepsilon > 0$, K is contained in the union of finitely many balls of radius ε.

An operator $T : H_1 \to H_2$ between two Hilbert spaces is called *compact* if it maps bounded sets to relatively compact subsets, i.e. $\overline{T(B)} \subset H_2$ is compact for any $B \subset H_1$ bounded. In particular, if T is linear and compact then it is bounded and therefore continuous; if T is not linear it may be compact but not continuous. We shall call *completely continuous* a compact and continuous operator.

Every bounded set is *sequentially weakly relatively compact*: any bounded sequence admits a weakly convergent subsequence.

Distributions

Let $\Omega \subset \mathbb{R}^n$ be a domain. One denotes by $\mathcal{D}(\Omega)$ the vector space of *test functions*. These are smooth (infinitely differentiable) functions that have compact support[5]. Test functions come equipped with the following notion of convergence: $v_k \to v$ in $\mathcal{D}(\Omega)$ if the support of any v_k is contained in one single compact set, and all derivatives $D^\alpha v_k$ converge uniformly in Ω to $D^\alpha v$.

The *space of distributions on* Ω, denoted by $\mathcal{D}'(\Omega)$, is the set of continuous linear functionals on $\mathcal{D}(\Omega)$. If $F \in \mathcal{D}'(\Omega)$ we shall write

$$\langle F, v \rangle$$

---

[5] The support of a continuous map is the closure of the set of its non-zero values.

to indicate how F acts on a test function v. In particular, F is continuous for

$$\langle F, v_k \rangle \to \langle F, v \rangle$$

if $v_k \to v$ in $\mathcal{D}(\Omega)$. By using strings of m test functions one defines, in a completely similar manner, vector-valued distributions $\mathcal{D}'(\Omega; \mathbb{R}^m)$.

A function u in $L^1_{\text{loc}}(\Omega)$ may be seen as a distribution, under the canonical identification

$$\langle u, v \rangle = \int_\Omega uv \, d\mathbf{x}.$$

• *Derivative of a distribution.* Take $F \in \mathcal{D}'(\Omega)$. The *derivative* $\partial_{x_i} F$ is the distribution defined by the formula

$$\langle \partial_{x_i} F, v \rangle = -\langle F, \partial_{x_i} v \rangle, \qquad \forall v \in \mathcal{D}(\Omega).$$

A remarkable result is: suppose $F \in \mathcal{D}'(\Omega)$ and ∇F is the zero distribution; then F is a constant function (Ω is a domain, hence connected).

• *Rapidly-decreasing functions and tempered distributions.* Denote by $\mathcal{S}(\mathbb{R}^n)$ the space of functions $v \in C^\infty(\mathbb{R}^n)$ that *decrease 'very quickly'* when $|\mathbf{x}| \to \infty$, i.e. such that, for any $m \in \mathbb{N}$ and every multi-index $\boldsymbol{\alpha} = (\alpha_1, \ldots, \alpha_n)$,

$$D^{\boldsymbol{\alpha}} v(\mathbf{x}) = o\left(|\mathbf{x}|^{-m}\right), \qquad |\mathbf{x}| \to \infty.$$

(We use the customary shortcut $D^{\boldsymbol{\alpha}} = \partial_{x_1}^{\alpha_1} \ldots \partial_{x_n}^{\alpha_n}$.)

If $\{v_k\} \subset \mathcal{S}(\mathbb{R}^n)$ and $v \in \mathcal{S}(\mathbb{R}^n)$ we write $v_k \to v$ in $\mathcal{S}(\mathbb{R}^n)$ if, for every multi-indices $\boldsymbol{\alpha}, \boldsymbol{\beta}$,

$$\mathbf{x}^{\boldsymbol{\beta}} D^{\boldsymbol{\alpha}} v_k \to \mathbf{x}^{\boldsymbol{\beta}} D^{\boldsymbol{\alpha}} v \quad \text{uniformly in } \mathbb{R}^n.$$

A distribution $T \in \mathcal{D}'(\mathbb{R}^n)$ is called *tempered* if

$$\langle T, v_k \rangle \to 0$$

for any $\{v_k\} \subset \mathcal{D}(\mathbb{R}^n)$ such that $v_k \to 0$ in $\mathcal{S}(\mathbb{R}^n)$. The set of tempered distributions is denoted by $\mathcal{S}'(\mathbb{R}^n)$.

• *Fourier transforms of tempered distributions.* Let $T \in \mathcal{S}'(\mathbb{R}^n)$. The Fourier transform $\widehat{T} = \mathcal{F}[T]$ is the tempered distribution defined by

$$\langle \widehat{T}, v \rangle = \langle T, \widehat{v} \rangle, \qquad \forall v \in \mathcal{S}(\mathbb{R}^n).$$

Sobolev spaces

Let $\Omega \subset \mathbb{R}^n$ be a domain. The Sobolev spaces we will use most are: the Hilbert space

$$H^1(\Omega) = \left\{ u \in L^2(\Omega) : \nabla u \in L^2(\Omega; \mathbb{R}^n) \right\}$$

(all derivatives are meant in distributional sense) with the inner product

$$(u, v)_{H^1(\Omega)} = \int_\Omega [\nabla u \cdot \nabla v + uv] \, d\mathbf{x},$$

and its closed subspace $H_0^1(\Omega) = \overline{\mathcal{D}(\Omega)}^{H^1(\Omega)}$. Elements of $H_0^1(\Omega)$ have *zero trace* on $\partial\Omega$. In a similar way one may define the Hilbert spaces $H^m(\Omega)$ of functions whose derivatives up to order m, included, belong to $L^2(\Omega)$.

• *Sobolev spaces in $\mathbb{R}$.* We have the inclusion

$$H^1(a, b) \subset C([a, b])$$

(this is false in dimension $n \geq 2$). Furthermore, for any $u \in H^1(a, b)$ the Fundamental Theorem of Calculus holds:

$$u(y) - u(x) = \int_x^y u'(s) \, ds \qquad a \leq x \leq y \leq b.$$

• *Poincaré inequality and equivalent norms.* If Ω is bounded there exists a constant C_P, only depending on Ω and n, such that

$$\|u\|_{L^2(\Omega)} \leq C_P \|\nabla u\|_{L^2(\Omega)}$$

for any $u \in H_0^1(\Omega)$. This fact allows to use in $H_0^1(\Omega)$ the equivalent norm $\|\nabla u\|_{L^2(\Omega)}$. This will be always our choice, unless otherwise stated.

Poincaré's inequality holds on other subspaces of $H^1(\Omega)$ (such as the space of functions with zero average).

• *The dual of H_0^1.* The dual of $H_0^1(\Omega)$, denoted by $H^{-1}(\Omega)$, consists of elements that can be written (albeit not uniquely) as $f + \operatorname{div} \mathbf{f}$, where $f \in L^2(\Omega; \mathbb{R})$ and $\mathbf{f} \in L^2(\Omega; \mathbb{R}^n)$.

• *Traces.* Let Ω be a bounded and Lipschitz domain (or half-space) and set $\Gamma = \partial\Omega$. Then there is a well-defined *trace operator*

$$\gamma_0 : H^1(\Omega) \to L^2(\Gamma),$$

which is linear, continuous, and such that

1. $\gamma_0 u = u_{|\Gamma}$, if $u \in C(\overline{\Omega})$.
2. $\|\gamma_0 u\|_{L^2(\Gamma)} \leq C_* \|u\|_{H^1(\Omega)}$, with C_* independent of u.

(Abusing notations we shall write $u_{|\Gamma}$ instead of $\gamma_0 u$ even when $u \notin C(\overline{\Omega})$.) The space of traces of functions in $H^1(\Omega)$, i.e. the image of γ_0, is denoted by

$$H^{1/2}(\partial\Omega) = \{u_{|\Gamma} : u \in H^1(\Omega)\}.$$

It is a Hilbert space with norm

$$\|w\|_{H^{1/2}(\Gamma)} = \inf\left\{\|u\|_{H^1(\Omega)} : u \in H^1(\Omega),\ u_{|\Gamma} = w\right\}.$$

For any $u \in H^1(\Omega)$, we have

$$\|u|_\Gamma\|_{L^2(\Gamma)} \le C_*\|u|_\Gamma\|_{H^{1/2}(\Gamma)} \le C_*\|u\|_{H^1(\Gamma)}.$$

• *Rellich's theorem*. Let Ω be a bounded Lipschitz domain. Then the embedding $H^1(\Omega) \hookrightarrow L^2(\Omega)$ is *compact*.

5.2 Solved Problems

- **5.2.1 – 5.2.11** : Hilbert spaces.
- **5.2.12 – 5.2.20** : Distributions.
- **5.2.21 – 5.2.30** : Sobolev spaces.

5.2.1 Hilbert spaces

Problem 5.2.1 (The space of continuous functions). *a) Let $C([-1, 1])$ denote the space of real, continuous functions on the interval* $[-1, 1]$, *with norm*

$$\|f\| = \|f\|_{C([-1,1])} = \max_{t\in[-1,1]} |f(t)|.$$

Check that $C([-1, 1])$ is a Banach space, but that its norm cannot be induced by an inner product (thus $C([-1, 1])$ cannot be a Hilbert space). Hint. If the norm is induced by an inner product the parallelogram rule must hold.

b) Let $C^\star([-1, 1])$ be the space of real, continuous functions on $[-1, 1]$ with norm

$$\|f\|_\star = \|f\|_{L^2(-1,1)} = \left(\int_{-1}^1 |f(t)|^2\, dt\right)^{1/2}.$$

Check that although this norm comes from an inner product, it does not make the space complete (so that neither $C^\star([-1, 1])$ is a Hilbert space). Hint. Show that

$$f_n(t) = \begin{cases} 0 & -1 \le t \le 0 \\ nt & 0 \le t \le \frac{1}{n} \\ 1 & \frac{1}{n} \le t \le 1, \end{cases}$$

is a Cauchy sequence in $C^\star([-1, 1])$ but it converges to the Heaviside function $\mathscr{H}$ ($\mathscr{H}(x) = 0$ for $x < 0$, $\mathscr{H}(x) = 1$ for $x \ge 0$).

Solution. a) It is enough to show that $C\left([-1,1]\right)$ is complete; in fact, let $\{f_n\} \subset C\left([-1,1]\right)$ be a Cauchy sequence for the sup norm. Given $\varepsilon > 0$, then, for $n > m \geq N\left(\varepsilon\right)$ we have

$$|f_n\left(x\right) - f_m\left(x\right)| < \varepsilon, \quad \text{for any } x \in [-1,1]. \tag{5.1}$$

Equation (5.1) indicates that for any $x \in [-1,1]$ the sequence of real numbers $\{f_n\left(x\right)\}$ is a Cauchy sequence, so

$$\lim_{n \to \infty} f_n\left(x\right) = f\left(x\right)$$

exists and is finite. Passing to the limit in (5.1) as $n \to \infty$, we obtain

$$|f\left(x\right) - f_m\left(x\right)| < \varepsilon, \quad \text{for any } x \in [-1,1], \tag{5.2}$$

i.e. $f_n \to f$ *uniformly in* $[-1,1]$. As uniform convergence of a sequence of continuous functions implies the continuity of the limit, we deduce $f \in C\left([-1,1]\right)$. Every Cauchy sequence in $C\left([-1,1]\right)$ therefore converges to some element of $C\left([-1,1]\right)$, which is thus complete.

In order to show that the norm can not be induced by an inner product, we show that the parallelogram rule fails. For this we need to find two functions f, g for which

$$\|f + g\|^2 + \|f - g\|^2 \neq 2\|f\|^2 + 2\|g\|^2.$$

For example, let

$$f = 1 + x \quad \text{and} \quad g = 1 - x.$$

Then

$$\|1 + x\| = \|1 - x\| = 2 \quad \text{and} \quad \|(1 + x) + (1 - x)\| = \|(1 + x) - (1 - x)\| = 2,$$

thus the parallelogram rule does not hold and the space is not Hilbert.

b) The given norm is the norm of $L^2(-1,1)$, induced by the inner product

$$(f,g)_{L^2(-1,1)} = \int_{-1}^{1} f(t)g(t)\,dt,$$

as is easy to see. Let us consider the given sequence. To fix ideas suppose $m > n$. Then

$$\|f_n - f_m\|_\star^2 = \int_0^{1/m} ((m-n)t)^2\,dt + \int_{1/m}^{1/n} (1 - nt)^2\,dt =$$

$$= \frac{(m-n)^2}{3m^3} + \frac{1}{3n}\left(1 - \frac{n}{m}\right)^3 \leq \frac{1}{3m} + \frac{1}{3n}.$$

Consequently, for $n, m \to +\infty$, $\|f_n - f_m\|_\star \to 0$ and the sequence is a Cauchy sequence for this norm. Analogously, if $\mathcal{H}$ is Heaviside's step function,

$$\|\mathcal{H} - f_n\|_\star^2 = \int_0^{1/n} (1 - nt)^2\,dt = \frac{1}{3n},$$

so f_n converges to $\mathcal{H}$ in L^2. As $\mathcal{H} \notin C^*([-1,1])$ (it is discontinuous at $x = 0$), f_n is a Cauchy sequence in $C^*([-1,1])$ that does not converge in $C^*([-1,1])$, so the space is not complete.

Problem 5.2.2 (Projections in a Hilbert space). *Set $H = L^2(-1, 1)$ and consider the subspace V of odd functions:*

$$V = \{u \in H : u(-t) = -u(t) \text{ for (almost) every } t \in (-1, 1)\}.$$

a) Verify that the orthogonal decomposition theorem holds.

b) Determine $V^{\perp}$ and then write the expressions for $P_V f$, $Q_V f$ for an arbitrary $f \in H$.

Solution. a) As H is a Hilbert space, we only need to check that V is a closed subspace in H. Since a linear combination of odd functions is odd, V is a linear space. To prove it is closed we must check that the limit (in $L^2(-1, 1)$-norm) of a sequence of odd functions is still odd:

$$\{u_n\} \subset V, \, u_n \to \bar{u} \qquad \text{implies } \bar{u} \in V.$$

For that we can, for instance, use the fact that if a sequence converges in L^2 it admits a subsequence that converges almost everywhere. So there exists $\{u_{n_k}\}$ such that

$$u_{n_k}(t) \to \bar{u}(t)$$

for almost every $t \in (-1, 1)$; then

$$0 = u_{n_k}(t) + u_{n_k}(-t) \to \bar{u}(t) + \bar{u}(-t) \qquad \text{for almost every } t \in (-1, 1),$$

whence $\bar{u} \in V$.

b) To identify $V^{\perp}$ we seek all $u \in L^2(-1, 1)$ such that $\int_{-1}^{1} uv \, dt = 0$ for any $v \in V$. Take $u \in V^{\perp}$; then:

$$0 = \int_{-1}^{1} u(t)v(t) \, dt = \int_{-1}^{0} u(t)v(t) \, dt + \int_{0}^{1} u(t)v(t) \, dt =$$

$$= -\int_{0}^{1} u(-t)v(t) \, dt + \int_{0}^{1} u(t)v(t) \, dt = \int_{0}^{1} [u(t) - u(-t)]v(t) \, dt.$$

Again, by the freedom in choosing v, we deduce

$$u(t) - u(-t) = 0 \quad \text{a.e. in } (0, 1)$$

and so $V^{\perp}$ is *the subspace of even functions*. To find the projections onto V and $V^{\perp}$ of a function f it is enough to decompose f as the sum of its even and odd parts

$$f(t) = \frac{f(t) + f(-t)}{2} + \frac{f(t) - f(-t)}{2}.$$

Since an even function and an odd one are always orthogonal with respect to the $L^2(-1, 1)$ scalar product, we obtain that

$$P_V f(t) = \frac{f(t) - f(-t)}{2}, \qquad Q_V f(t) = \frac{f(t) + f(-t)}{2}.$$

Problem 5.2.3 (Orthogonal splitting). *Set* $Q = (0, 1) \times (0, 1)$, $H = L^2(Q)$, *and consider the subspace*

$$V = \{u \in H : u(x, y) = v(x) \text{ a.e., with } v \in L^2(0, 1)\}.$$

Determine the projection operators P_V, $P_{V^\perp}$. *Decompose the function* $f(x, y) = xy$.

Solution. Note that, given $u \in H$, $P_V u$ is the best approximation (in L^2 sense) of u by means of a function depending only on x.

By the orthogonal decomposition theorem, the decomposition exists if V is closed. It is easy to see that V is a linear space, so we just check its closure. Setting $u_n(x, y) = v_n(x)$, we suppose u_n converges to $\bar{u}$ in $L^2(Q)$. The claim is that $\bar{u}$ depends only on x and belongs to $L^2(0, 1)$. Since

$$\int_0^1 \int_0^1 [v_n(x) - v_m(x)]^2 dx dy = \int_0^1 [v_n(x) - v_m(x)]^2 dx,$$

the sequence $\{v_n\}$ is Cauchy in $L^2(0, 1)$ and hence converges to $v = v(x)$, $v \in L^2(0, 1)$. But

$$\int_0^1 [v_n(x) - v(x)]^2 dx = \int_0^1 \int_0^1 [v_n(x) - v(x)]^2 dx dy,$$

and the uniqueness of limits implies $v = \bar{u}$, making V closed.

Let us find $V^\perp$. This subspace consists of functions $u \in H$ such that for any v in $L^2(0, 1)$

$$0 = \int_Q u(x, y)v(x) \, dx dy = \int_0^1 \left(\int_0^1 u(x, y) \, dy \right) v(x) \, dx.$$

We deduce

$$V^\perp = \left\{ u \in H : \int_0^1 u(x, y) \, dy = 0, \text{ for almost every } x \in (0, 1) \right\}.$$

Now it is immediate to see that for any $f \in H$

$$P_V f(\cdot) = \int_0^1 f(\cdot, y) \, dy,$$

$$Q_V f(\cdot) = f(\cdot, y) - \int_0^1 f(\cdot, y) \, dy.$$

In particular, taking $f(x, y) = xy$,

$$xy = \frac{1}{2}x + \left(xy - \frac{1}{2}x\right),$$

where the first summand belongs to V and the one in brackets lives in $V^\perp$.

Problem 5.2.4 (Norms of functionals and Riesz's theorem). *Given $H = L^2(0, 1)$ and $u \in H$, consider the functional*

$$Fu = \int_0^{1/2} u(t)\, dt.$$

a) Check that F is well defined and belongs to H^*.

b) Compute $\|F\|_{H^*}$ in two ways, first using the definition and then Riesz's theorem.

Solution. a) Clearly, if $u \in L^2(0, 1)$ then $u \in L^1(0, 1/2)$. Moreover, by Schwarz's inequality

$$\left|\int_0^{1/2} u(t)\, dt\right| \leq \int_0^{1/2} |u(t)|\, dt \leq \left(\int_0^{1/2} 1\, dt\right)^{1/2} \cdot \left(\int_0^{1/2} |u(t)|^2\, dt\right)^{1/2}$$

$$\leq \frac{\sqrt{2}}{2}\left(\int_0^1 |u(t)|^2\, dt\right)^{1/2} = \frac{\sqrt{2}}{2}\|u\|_H. \tag{5.3}$$

Therefore F is a well-defined functional on H and is bounded. Being linear it is also continuous, and belongs to H^*.

b) We recall that by definition

$$\|F\|_{H^*} = \sup_{\|u\|_H \leq 1} |Fu|.$$

From (5.3) we immediately deduce $\|F\|_{H^*} \leq \sqrt{2}/2$. Let us show $\|F\|_{H^*} = \sqrt{2}/2$, by exhibiting a function with unitary norm such that $Fu = \sqrt{2}/2$. It is reasonable to choose u so to concentrate its norm on the interval $[0, 1/2]$. In particular, if we pick

$$u(t) = \begin{cases} \sqrt{2} & 0 < t < \dfrac{1}{2} \\ 0 & \dfrac{1}{2} < t < 1, \end{cases}$$

we have $\|u\|_H = 1$ and $Fu = \sqrt{2}/2$, whence the claim.

An alternative method is to invoke Riesz's theorem. Since $F \in H^*$, the theorem yields the existence (and uniqueness) of an $f \in H$ such that

$$Fu = \int_0^1 f(t)u(t)\,dt \qquad \text{and} \qquad \|F\|_{H^*} = \|f\|_H.$$

Therefore

$$f(t) = \begin{cases} 1 & 0 < t < \dfrac{1}{2} \\ 0 & \dfrac{1}{2} < t < 1, \end{cases}$$

and then

$$\|F\|_{H^*} = \frac{\sqrt{2}}{2}.$$

Problem 5.2.5 (Operators: norm, invertibility, eigenvalues). *Let $H = L^2(0, +\infty)$ and consider the linear operator*

$$(Lu)(t) = u(2t)$$

from H to itself.

a) Prove that L is continuous and compute its norm.

b) Compute the (real) eigenvalues of L, if any.

Solution. a) L is linear, so we have to prove its boundedness. As

$$\|Lu\|_H = \left(\int_0^{+\infty} u^2(2t)\,dt \right)^{1/2} = \left(\frac{1}{2} \int_0^{+\infty} u^2(s)\,ds \right)^{1/2} = \frac{\sqrt{2}}{2} \|u\|_H, \qquad (5.4)$$

L is bounded (hence continuous), and has norm $\sqrt{2}/2$.

b) We must find non-zero solutions to

$$Lu = \lambda u, \quad \text{i.e.} \quad u(2t) = \lambda u(t), \quad \text{a.e.} \quad t > 0. \qquad (5.5)$$

We will show that (5.5) implies that $u \equiv 0$, thus proving that L has no real eigenvalue. Computing the norms of both sides and exploiting (5.4) we find

$$\frac{\sqrt{2}}{2} \|u\| = |\lambda| \|u\|.$$

Consequently if u is not identically zero, $\lambda = \pm\sqrt{2}/2$. Observe that applying L to both sides of (5.5) gives

$$u(4t) = L(Lu) = L(\lambda u) = \lambda^2 u.$$

Applying L n times, we find that the eigenfunction u satisfies

$$u(2^n t) = \lambda^n u(t) \qquad (5.6)$$

for almost every $t > 0$ and any integer $n \geq 0$. But

$$u(t) = u\left(2^n \frac{t}{2^n}\right) = \lambda^n u(2^{-n}t),$$

so

$$u(2^{-n}t) = \lambda^{-n}u(t)$$

and (5.6) holds for any $n \in \mathbb{Z}$. Therefore to find u we only need to know its value on, say, $[1, 2)$. In particular

$$\int_{2^n}^{2^{n+1}} u^2(s)\,ds = 2^n \int_1^2 u^2(2^n t)\,dt = 2^n \int_1^2 \lambda^{2n} u^2(t)\,dt = \int_1^2 u^2(t)\,dt.$$

By writing

$$(0, +\infty) = \bigcup_{n \in \mathbb{Z}} [2^n, 2^{n+1}),$$

we have

$$\|u\|_H^2 = \int_0^{+\infty} u^2(t)\,dt = \sum_{n \in \mathbb{Z}} \int_{2^n}^{2^{n+1}} u^2(t)\,dt = \sum_{n \in \mathbb{Z}} \int_1^2 u^2(t)\,dt.$$

If u is an eigenfunction, then

$$0 < \|u\|_H < +\infty.$$

On the other hand the above series is the sum of infinitely many identical terms, and converges if and only if its (only) term vanishes identically, i.e. $u = 0$ almost everywhere on $(1, 2)$. By property (5.6), we have $u = 0$ almost everywhere on $(0, +\infty)$. To sum up, not even for $\lambda = \pm\sqrt{2}/2$ the equation $Lu = \lambda u$ has non-trivial solutions. Consequently L does not have real eigenvalues.

**Problem 5.2.6 (Compactness). *a) Write $H = l^2$, i.e. the Hilbert space*

$$l^2 = \left\{ x = \{x_n\}_{n \geq 1} : \sum_{n \geq 1} x_n^2 < \infty \right\}$$

with inner product $(x, y)_{l^2} = \sum_{n \geq 1} x_n y_n$. Given $M > 0$, define the set

$$K = \left\{ x \in l^2 : \sum_{n \geq 1} x_n^2 < M, \ \sum_{n \geq 1} n^2 x_n^2 < M \right\}.$$

Prove that for any given $\varepsilon > 0$ there exists a finite-dimensional subspace $V_\varepsilon \subset H$ such that

$$\mathrm{dist}\,(x, V_\varepsilon) = \|x - P_{V_\varepsilon} x\| < \varepsilon \qquad \text{for any } x \in K.$$

b) Deduce K is relatively compact in H (use the definition of compactness involving coverings by ε-spheres).

Solution. a) We can interpret H as an infinite-dimensional generalisation of $\mathbb{R}^n$: sequences, in fact, can be seen as vectors with an infinite number of coordinates, and the definition of norm is nothing else but the infinite-dimensional version of Pythagoras' theorem! It can be proved that a Hilbert basis (a complete orthonormal system) for H is given by the infinitely many vectors $\mathbf{e}_i = (x_n)$, with $x_i = 1$, $x_n = 0$ $\forall n \neq i$. The finite-dimensional subspaces of H easiest to describe are those whose elements have only a finite number of non-zero coordinates. Said better, set

$$V_k = \{\mathbf{x} \in H : x_n = 0 \quad \text{for any } n \geq k\}.$$

We invite the reader to check V_k is a vector subspace of dimension $k - 1$ (hence closed), and that, for any $\mathbf{x} \in H$, its distance from V_k is

$$\|\mathbf{x} - P_{V_k}\mathbf{x}\| = \left(\sum_{n=k}^{+\infty} x_n^2\right)^{1/2}. \tag{5.7}$$

Making the distance between an element of H and V_k arbitrarily small is the same as making small the kth remainder of the corresponding series. By definition of K, for any $\mathbf{x} \in K$ and any natural number k, we have

$$M > \sum_{n=1}^{+\infty} n^2 x_n^2 \geq \sum_{n=k}^{+\infty} n^2 x_n^2 \geq k^2 \sum_{n=k}^{+\infty} x_n^2.$$

Recalling (5.7) we have

$$\mathbf{x} \in K \quad \Rightarrow \quad \|\mathbf{x} - P_{V_k}\mathbf{x}\| \leq \frac{\sqrt{M}}{k}.$$

Now it suffices to choose $k > \sqrt{M}/\varepsilon$ to conclude.

b) We recall that a set is called relatively compact if its closure is compact. In particular a set $A \subset H$ is relatively compact when for any $\varepsilon > 0$ there is a finite number of points $\mathbf{y}_1, \ldots, \mathbf{y}_N$ such that

$$A \subset \bigcup_{i=1}^{N} B_\varepsilon(\mathbf{y}_i)$$

where $B_\varepsilon(\mathbf{y}_i)$ is the ball of radius ε centred at $\mathbf{y}_i$. Fix $\varepsilon > 0$ and

$$k > 2\frac{\sqrt{M}}{\varepsilon}.$$

By part a) every point in K has distance less than $\varepsilon/2$ apart from V_k. Consider

$$F = K \cap V_k,$$

a bounded subset (since $\mathbf{x} \in F$ implies $\|\mathbf{x}\|_{V_k}^2 \le M$) of the finite-dimensional space V_k, and hence relatively compact. By the characterisation mentioned earlier there are N points $\mathbf{y}_1, \dots, \mathbf{y}_N$ such that

$$F \subset \bigcup_{i=1}^{N} B_{\varepsilon/2}(\mathbf{y}_i).$$

Take $\mathbf{x} \in K$. Then $P_{V_k} \mathbf{x} \in F$ and there is an index i for which $P_{V_k} \mathbf{x} \in B_{\varepsilon/2}(\mathbf{y}_i)$. Thus

$$\|\mathbf{x} - \mathbf{y}_i\| \le \|\mathbf{x} - P_{V_k} \mathbf{x}\| + \|P_{V_k} \mathbf{x} - \mathbf{y}_i\| \le \frac{\varepsilon}{2} + \frac{\varepsilon}{2} = \varepsilon$$

i.e. $\mathbf{x} \in B_\varepsilon(\mathbf{y}_i)$. Since the argument holds for any $\mathbf{x} \in K$, we have proved

$$K \subset \bigcup_{i=1}^{N} B_\varepsilon(\mathbf{y}_i), \qquad \text{i.e. } K \text{ is relatively compact.}$$

Problem 5.2.7 (Weak convergence). *Let H be a Hilbert space and $\{u_n\} \subset H$ a sequence such that*

$$u_n \rightharpoonup \bar{u}$$

and

$$\text{either} \quad \|u_n\| \to \|\bar{u}\| \qquad \text{or} \qquad \|u_n\| \le \|\bar{u}\|.$$

Show that u_n converges (strongly) to $\bar{u}$.

Solution. We recall that u_n converges weakly to $\bar{u}$ if $(u_n, v) \to (\bar{u}, v)$ for any $v \in H$. In particular, choosing $v = \bar{u}$ gives $(u_n, \bar{u}) \to (\bar{u}, \bar{u}) = \|\bar{u}\|^2$. We obtain

$$\|u_n - \bar{u}\|^2 = (u_n, u_n) - 2(u_n, \bar{u}) + (\bar{u}, \bar{u}) = \|u_n\|^2 - \|\bar{u}\|^2 + 2\underbrace{\left[\|\bar{u}\|^2 - (u_n, \bar{u})\right]}_{\to 0}.$$

In both cases we deduce $\|u_n - \bar{u}\| \to 0$, i.e. $u_n \to \bar{u}$ strongly.

Problem 5.2.8 (Nonlinear compact operator). *Write $H = L^2(0, 1)$ and consider the nonlinear operator*

$$T[f](t) = \int_0^1 (t + f(s))^2 \, ds.$$

a) Show T is well defined and continuous from H to H.

b) Using the criterium for subsets of $L^2(0, 1)$ (see [18, Chap. 6]), prove that T is compact.

Solution. a) If $f \in H$ then

$$|T[f](t)| = \int_0^1 (t + f(s))^2 \, ds \le 2 \int_0^1 (t^2 + f^2(s)) \, ds \le 2 + 2\|f\|_2^2, \qquad (5.8)$$

and T is well defined on H. Let us check that the range of T is contained in H. By (5.8)

$$\|T[f]\|_2^2 = \int_0^1 \left[\int_0^1 (t + f(s))^2 \, ds \right]^2 dt \leq [2 + 2\|f\|_2^2]^2 ,$$

so $T : H \to H$. To prove continuity (T is not linear, so we really need a proof) let f, g be in H. Then

$$\|T[f] - T[g]\|_2^2 = \int_0^1 \left[\int_0^1 (t + f(s))^2 - (t + g(s))^2 \, ds \right]^2 dt =$$

$$= \int_0^1 \left[\int_0^1 (2t + f(s) + g(s)) \, (f(s) - g(s)) \, ds \right]^2 dt \leq$$

$$\leq \int_0^1 \left[\int_0^1 (2t + f(s) + g(s))^2 \, ds \right] \cdot \left[\int_0^1 (f(s) - g(s))^2 \, ds \right] dt \leq$$

$$\leq [4 + 2\|f + g\|_2^2]^2 \cdot \|f - g\|_2^2$$

(we have used Schwarz's inequality and, in the last step, (5.8)). Hence, if $f \to g$ in H we have $T[f] \to T[g]$ in H, so T is continuous.

b) Write

$$\mathcal{F} = \{f \in H : \|f\|_2 \leq M\}.$$

In order to prove that $T[\mathcal{F}]$ is relatively compact in H we shall use the criterion for L^2-compactness, and check whether $T[\mathcal{F}]$ is bounded and there exist C, α such that

$$\|T[f](\cdot + h) - T[f](\cdot)\|_2 \leq C|h|^\alpha \tag{5.9}$$

for any $f \in \mathcal{F}$ vanishing outside $(0, 1)$. As we saw in part a), from (5.8) we have, for any $f \in \mathcal{F}$,

$$\|T[f]\|_2^2 \leq [2 + 2\|f\|_2^2]^2 \leq [2 + 2M^2]^2 ,$$

and so $T[\mathcal{F}]$ is bounded. But we also have

$$\|T[f](\cdot + h) - T[f](\cdot)\|_2^2 \leq \int_0^1 \left[\int_0^1 (t + h + f(s))^2 - (t + f(s))^2 \, ds \right]^2 dt =$$

$$= \int_0^1 \left[\int_0^1 (2h \, (t + f(s)) + h^2) \, ds \right]^2 dt \leq h^2 \int_0^1 \left[\int_0^1 (h + 2t + 2f(s)) \, ds \right]^2 dt$$

and Schwarz's inequality plus (5.8) give (5.9) with $\alpha = 2$.

Problem 5.2.9 (Iterated projections[a]). *Let V and W be closed subspaces in a Hilbert space H. Define the sequence $\{x_n\}$ of projections as follows: $x_0 \in H$ is given, and*

$$x_{2n+1} = P_W x_{2n}, \quad x_{2n+2} = P_V x_{2n+1} \qquad \text{when } n \geq 0.$$

Prove the following assertions:

a) $V \cap W = \{0\}$ implies $x_n \to 0$.

b) $V \cap W \neq \{0\}$ implies $x_n \to P_{V \cap W} x_0$.

Hint. *a) Show, in this order: $\|x_n\|$ decreases, $x_n \rightharpoonup 0$ and $\|x_n\|^2 = (x_{2n-1}, x_0)$; b) Reduce to to previous case by subtracting $P_{V^\perp \cap W} x_0$.*

---

[a] This problem is related to *Schwarz's alternating method*, see Problem 6.2.15, Chap. 6 (page 361).

Solution. a) The idea – as we are looking at projections – is that the sequence of norms should decrease. Therefore $\{x_n\}$ will be bounded, with finite limit, in particular equal to zero. Let us check that $\|x_n\|$ decreases. Since

$$(x_{n+1}, x_n) = (x_{n+1}, x_{n+1}) = \|x_{n+1}\|^2 ,$$

for every $n \geq 0$, we have

$$\begin{aligned} \|x_{n+1} - x_n\|^2 &= \|x_{n+1}\|^2 - 2(x_{n+1}, x_n) + \|x_n\|^2 \\ &= \|x_{n+1}\|^2 - 2\|x_{n+1}\|^2 + \|x_n\|^2 \\ &= -\|x_{n+1}\|^2 + \|x_n\|^2 \end{aligned}$$

so $\|x_n\| \geq \|x_{n+1}\|$ and $\|x_n\| \downarrow l \geq 0$. Moreover

$$\|x_{n+1} - x_n\| \to 0.$$

We will show that $x_n \rightharpoonup 0$ weakly and then $l = 0$ (see Problem 5.2.7 on page 286).

Let $\{x_{2n_k}\}$ be any weakly convergent subsequence of $\{x_{2n}\}$: $x_{2n_k} \rightharpoonup x$, with x in V. As also $x_{2n_k+1} \rightharpoonup x$ (in fact the distance between x_{2n_k} and x_{2n_k+1} tends to 0) one has $x \in W$, and so $x = 0$. But the subsequence is arbitrary, so $x_{2n} \rightharpoonup 0$. Similarly for $\{x_{2n+1}\}$, and altogether $x_n \rightharpoonup 0$.

Now, to fix ideas let us assume $x_n \in V$, and therefore $x_{n-1} \in W$. As orthogonal projections are symmetric operators,

$$\|x_n\|^2 = (x_n, x_{n-1}) = (x_n, P_W x_{n-2}) = (P_W x_n, x_{n-2}) = (x_{n+1}, x_{n-2}).$$

Iterating the argument gives

$$\|x_n\|^2 = (x_{n+1}, x_{n-2}) = (x_{n+2}, x_{n-3}) = \cdots = (x_{2n-1}, x_0)$$

and since $x_{2n-1} \rightharpoonup 0$ we see that $\|x_n\| \to 0$.

b) If $V \cap W \neq \{0\}$ we set

$$z_0 = x_0 - P_{V \cap W} x_0.$$

The sequence starting with z_0 and generated by projecting on V and W as before, is given by

$$z_n = x_n - P_{V \cap W} x_0.$$

It is easy to see that $P_{V\cap W}z_n = 0$, and then z_n belongs to $V \setminus (V^\perp \cap W)$ or to $W \setminus (V \cap W^\perp)$, whose intersection reduces to $\{0\}$. In this way one falls back to the previous situation, whence $z_n \to 0$, that is $x_n \to P_{V\cap W}x_0$.

Problem 5.2.10 (Projection onto a convex set). *Let H be a Hilbert space and $K \subset H$ a (non-empty) closed, convex subset. Prove, along the lines of the Projection Theorem [18, Chap. 6, Sect. 4], that for any $x \in H$ there is a unique $P_K x \in K$ such that*

$$\|P_K x - x\| = \inf_{v \in K} \|v - x\|.$$

Verify that $P_K x$ is uniquely determined by the variational inequality

$$(x - P_K x, v - P_K x) \le 0 \qquad \text{for any } v \in K. \tag{5.10}$$

Solution. Define

$$d = \inf_{v \in K} \|v - x\|$$

and let $\{v_n\} \subset K$ be a minimising sequence, i.e. for any n

$$v_n \in K, \qquad d^2 \le \|v_n - x\|^2 \le d^2 + \frac{1}{n}. \tag{5.11}$$

We shall prove that v_n is a Cauchy sequence. The parallelogram rule for $x - v_n$, $x - v_m$ gives

$$2\|x - v_n\|^2 + 2\|x - v_m\|^2 = \|v_n - v_m\|^2 + 4\left\|x - \frac{v_n + v_m}{2}\right\|^2.$$

Since K is convex and v_n and v_m belong to K, also $(v_n + v_m)/2$ belongs to K and its distance from x is greater than or equal to d. So we can write

$$\|v_n - v_m\|^2 = 2\|x - v_n\|^2 + 2\|x - v_m\|^2 - 4\left\|x - \frac{v_n + v_m}{2}\right\|^2 \le$$

$$\le 2d^2 + \frac{2}{n} + 2d^2 + \frac{2}{m} - 4d^2 = \frac{2}{n} + \frac{2}{m}.$$

Then $\|v_n - v_m\| \to 0$ as m, n tend to infinity, and the sequence $\{v_n\}$ is Cauchy sequence. But H is complete, so there exists $w \in H$ with $v_n \to w$ in H. Keeping into account the closure of K (and the norm continuity) we have, from (5.11),

$$w \in K, \qquad \|x - w\| = d.$$

To show uniqueness for w consider $w' \in K$, with $\|x - w'\| = d$. The parallelogram rule (for $x - w$, $x - w'$) gives (again: K convex implies $(w + w')/2 \in K$)

$$\|w - w'\|^2 = 2\|x - w\|^2 + 2\|x - w'\|^2 - 4\left\|x - \frac{w + w'}{2}\right\|^2 \le 2d^2 + 2d^2 - 4d^2 = 0$$

i.e. $w = w'$. Hence, to any $x \in H$ we may associate a unique element $w = P_K x \in K$, the projection of x to K, that achieves the shortest distance.

Let us now prove that $P_K x$ satisfies (5.10). Fix $v \in K$ and set, for $0 \le t \le 1$,

$$u = t P_K x + (1 - t)v.$$

The convexity of K forces $u \in K$ and also

$$\|x - P_K x\|^2 \le \|x - u\|^2 = \|x - P_K x - (1-t)(v - P_K x)\|^2 =$$
$$= \|x - P_K x\|^2 + (1-t)^2 \|v - P_K x\|^2 - 2(1-t)(x - P_K x, v - P_K x).$$

That is, for $t < 1$,

$$0 \le (1-t)\|v - P_K x\|^2 - 2(x - P_K x, v - P_K x).$$

Passing to the limit as $t \to 1^-$ we find (5.10).

Vice versa, let $y \in K$ be such that

$$(x - y, v - y) \le 0, \qquad \text{for any } v \in K.$$

In particular, with $v = P_K x$, we have

$$0 \ge (x - y, P_K x - y) = (x - P_K x + P_K x - y, P_K x - y) = (x - P_K x, P_K x - y) + \|P_K x - y\|^2.$$

We have just shown $(x - P_K x, P_K x - y) \ge 0$, so now $\|P_K x - y\|^2 \le 0$ and then $y = P_K x$. Therefore $P_K x$ is characterised by (5.10).

Problem 5.2.11 (Projection onto a convex set in L^2). *Let Ω be a domain in $\mathbb{R}^n$, $H = L^2(\Omega)$ and*

$$K = \{v \in L^2(\Omega) ; \ a \le v(x) \le b \ \text{a.e. in } \Omega\}$$

with a, b given constants.

a) Verify that K is closed and convex. Using Problem 5.2.10 deduce that the projection $\hat{u} = P_K(u)$ of any $u \in L^2(\Omega)$ is characterised by the variational inequality

$$\int_\Omega [\hat{u}(x) - u(x)] [v(x) - \hat{u}(x)] \, dx \ge 0, \qquad \forall v \in K. \tag{5.12}$$

b) Define $\Omega = E^- \cup E_0 \cup E^+$, where

$$E^- = \{x : \hat{u}(x) < u(x)\}, E_0 = \{x : \hat{u}(x) = u(x)\}, E^+ = \{x : \hat{u}(x) > u(x)\}.$$

Prove that, a.e. in Ω,

$$\hat{u}(x) = \begin{cases} a & \text{in } E^+ \\ u(x) & \text{in } E_0 \\ b & \text{in } E^-. \end{cases} \tag{5.13}$$

c) Deduce that (5.12) is equivalent to the pointwise inequality:

$$[\hat{u}(x) - u(x)] [v(x) - \hat{u}(x)] \ge 0 \quad \text{a.e. in } \Omega, \forall v \in K \tag{5.14}$$

and thus find the formula, valid a.e. in Ω,

$$\hat{u}(x) = P_{[a,b]}(u(x)) = \min\{b, \max\{a, u(x)\}\} = \max\{a, \min\{b, u(x)\}\}. \tag{5.15}$$

Solution. a) To prove that K is closed we have to show that if $\{u_n\}$ is a sequence in L^2 such that $a \le u_n \le b$ and $u_n \to \bar{u}$ in L^2, then $a \le \bar{u} \le b$. For that it is enough to notice that L^2 convergence implies pointwise convergence almost everywhere of at least one subsequence: hence $a \leq \bar{u}(x) \leq b$ almost everywhere, and then the claim follows.

As for convexity, fix u_1, u_2 in K. For $0 \leq \lambda \leq 1$

$$a = \lambda a + (1 - \lambda)a \leq \lambda u_1(x) + (1 - \lambda)u_2(x) \leq \lambda b + (1 - \lambda)b = b.$$

Therefore any convex combination of elements of K belongs to K, and we can apply Problem 5.2.10. In particular (5.12) is another way of writing (5.10) in the present context.

b) First, observe that all three sets are measurable and defined up to zero-measure sets. By contradiction, suppose there exists an $A \subset E^+$, of positive measure, with $\hat{u}(x) > a$. Define

$$v(x) = \begin{cases} a & \text{in } A \\ \hat{u}(x) & \text{in } \Omega \backslash A. \end{cases}$$

Then $v \in K$ and

$$\int_\Omega [\hat{u}(x) - u(x)][v(x) - \hat{u}(x)]\,dx = \int_A [\hat{u}(x) - u(x)][a - \hat{u}(x)]\,dx < 0,$$

which contradicts (5.12). The same can be done for E^-.

c) By (5.13) we have, for any $v \in K$:

$$[a - u(x)][v(x) - a] \geq 0 \quad \text{a.e. in } E^+,$$
$$[b - u(x)][v(x) - b] \geq 0 \quad \text{a.e. in } E^-$$

and then (5.14) follows. The latter, in turn, indicates that

$$\hat{u}(x) = P_{[a,b]}u(x)$$

a.e. in Ω, and therefore (5.15) is a consequence of the following one-dimensional formula, which is not hard to prove:

$$\forall z \in \mathbb{R}, \ P_{[a,b]}(z) = \min\{b, \max\{a, z\}\} = \max\{b, \min\{a, z\}\}.$$

5.2.2 Distributions

Problem 5.2.12 (Distributions and Fourier series). *Prove that*

$$\sum_{k=1}^{+\infty} c_k \sin kx$$

converges in $\mathcal{D}'(\mathbb{R})$ if the numerical sequence $\{c_k\}$ grows 'slowly', meaning that there exists $p \in \mathbb{R}$, $C > 0$ such that $|c_k| \leq Ck^p$ for every $k \geq 1$.

Solution. We need to prove that for any $v \in \mathcal{D}(\mathbb{R})$, the series

$$\sum_{k=1}^{\infty} c_k \int_{\mathbb{R}} v(x) \sin kx \, dx \qquad (5.16)$$

converges. Let N be an integer greater than $p + 2$; we can assume that N is even and set $N = 2n$. Take any test function v, whose support is contained in some interval, say $[a, b]$. Notice that v and all its derivatives vanish at a and b. If we integrate N times by parts we find

$$\int_{\mathbb{R}} v(x) \sin kx \, dx = \int_a^b v(x) \sin kx \, dx = \frac{1}{k} \int_a^b v'(x) \cos kx \, dx$$

$$= -\frac{1}{k^2} \int_a^b v''(x) \sin kx \, dx = \frac{(-1)^n}{k^N} \int_a^b v^{(N)}(x) \sin kx \, dx.$$

Hence

$$\left| \int_{\mathbb{R}} v(x) \sin kx \, dx \right| \leq \frac{(b-a)}{k^N} \max \left| v^{(N)} \right|$$

and

$$\left| c_k \int_{\mathbb{R}} v(x) \sin kx \, dx \right| \leq C(b-a) \max \left| v^{(N)} \right| \frac{1}{k^{N-p}} \leq C(b-a) \max \left| v^{(N)} \right| \frac{1}{k^2}$$

for $N > p + 2$. The series (5.16) is therefore convergent.

Problem 5.2.13 (Support of a distribution). *a) Prove that if $F \in \mathcal{D}'(\mathbb{R}^n)$, $v \in \mathcal{D}(\mathbb{R}^n)$ and v vanishes on an open set containing $\mathrm{supp}(F)$, then $\langle F, v \rangle = 0$.*
b) Is it true that $\langle F, v \rangle = 0$ if v vanishes only on the support of F?

Solution. a) Call K the support of F; K is the complement of the largest open set Ω such that, if v is a test function with support in Ω, then $\langle F, v \rangle = 0$. But if v is zero on an open set containing K then $\mathrm{supp}(v)$ must be contained in Ω and therefore $\langle F, v \rangle = 0$.

b) This is false. It suffices to take $F = \delta'$ and v a test function such that

$$v(0) = 0, v'(0) \neq 0.$$

The support of F is $K = \{0\}$ and v vanishes on K, but

$$\langle F, v \rangle = \langle \delta', v \rangle = -\langle \delta, v' \rangle = -v'(0) \neq 0.$$

Problem 5.2.14 (Differential equation in $\mathcal{D}'$). *Given $a \in C^\infty(\mathbb{R})$, solve the equation*

$$G' + aG = \delta' \qquad \text{in } \mathcal{D}'(\mathbb{R}).$$

Solution. First observe that, if $v \in \mathcal{D}(\mathbb{R})$ and $g \in C^\infty(\mathbb{R})$, then $gv \in \mathcal{D}(\mathbb{R})$. Hence for any $F \in \mathcal{D}'(\mathbb{R})$ we can define the distribution $gF \in \mathcal{D}'(R)$ by

$$\langle gF, v \rangle = \langle F, gv \rangle, \qquad \forall v \in \mathcal{D}(\mathbb{R}).$$

As $a \in C^\infty(\mathbb{R})$, also $h(x) = \exp \int_0^x a(s)\, ds$ belongs to $C^\infty(\mathbb{R})$, and we can multiply both sides of the equation by h, obtaining

$$\frac{d}{dx}[G(x)h(x)] = h\delta'.$$

But

$$\langle h\delta', v \rangle = \langle \delta', hv \rangle = -\langle \delta, h'v + hv' \rangle = -h'(0)v(0) - h(0)v'(0)$$
$$= -a(0)v(0) - v'(0) = \langle -a(0)\delta + \delta', v \rangle,$$

so that

$$\frac{d}{dx}[G(x)h(x)] = -a(0)\delta + \delta'.$$

The primitives of $-a(0)\delta + \delta'$ are given by

$$-a(0)\mathcal{H} + \delta + c,$$

where $\mathcal{H}$ is the Heaviside step function and $c \in \mathbb{R}$, so we have

$$G(x) \exp \int_0^x a(s)\, ds = -a(0)\mathcal{H} + \delta + c.$$

The general solution of the differential equation is then

$$G(x) = (-a(0)\mathcal{H} + \delta + c)\, e^{-\int_0^x a(s)ds} = \delta + (-a(0)\mathcal{H} + c)\, e^{-\int_0^x a(s)ds}.$$

Problem 5.2.15 (Rapidly decreasing functions). *Verify that:*

a) The function $v(\mathbf{x}) = e^{-|\mathbf{x}|^2}$ *belongs to* $\mathcal{S}(\mathbb{R}^n)$.

b) The function $v(\mathbf{x}) = e^{-|\mathbf{x}|^2} \sin\left(e^{|\mathbf{x}|^2}\right)$ *does not belong to* $\mathcal{S}(\mathbb{R}^n)$.

Solution. a) Take $v(\mathbf{x}) = e^{-|\mathbf{x}|^2}$. Each derivative $D^\alpha v$ of order k has the form

$$(\text{polynomial of degree } k \text{ in } x_1, \ldots, x_n)\, e^{-|\mathbf{x}|^2}.$$

Therefore whichever $m \in \mathbb{N}$ we consider, we have

$$\lim_{|\mathbf{x}| \to \infty} |\mathbf{x}|^m D^\alpha v(\mathbf{x}) = 0$$

and $D^\alpha v(\mathbf{x}) = o\left(|\mathbf{x}|^{-m}\right)$, so that v decreases rapidly at infinity.

b) Let $v(\mathbf{x}) = e^{-|\mathbf{x}|^2} \sin\left(e^{|\mathbf{x}|^2}\right)$. Then

$$D_{x_j} v(\mathbf{x}) = 2x_j \left[-e^{-|\mathbf{x}|^2} \sin\left(e^{|\mathbf{x}|^2}\right) + \cos\left(e^{|\mathbf{x}|^2}\right)\right]$$

and so

$$\lim_{|\mathbf{x}| \to \infty} D_{x_j} v(\mathbf{x}) \text{ does not exist.}$$

We conclude that $v \notin \mathcal{S}(\mathbb{R}^n)$.

Problem 5.2.16 (Distributions and tempered distributions). *Verify that if $F \in \mathcal{D}'(\mathbb{R}^n)$ has compact support then $F \in \mathcal{S}'(\mathbb{R}^n)$.*

Solution. Let F have compact support $K \subset \mathbb{R}^n$ and take $\{v_k\} \subset \mathcal{D}(\mathbb{R}^n)$ such that $v_k \to 0$ in $\mathcal{S}(\mathbb{R}^n)$. We have to show that $\langle F, v_k \rangle \to 0$. Choose open sets A and B so that $K \subset A \subset B$. If w is a test function with support in B and $w \equiv 1$ on A, then for any other test function v

$$\langle F, v \rangle = \langle F, vw \rangle.$$

In fact, the test function $z = v - vw$ vanishes on A and hence has support in the complement of K; consequently $\langle F, z \rangle = \langle F, v \rangle - \langle F, vw \rangle = 0$.

Since (check this fact) $v_k w \to 0$ in $\mathcal{D}(\mathbb{R}^n)$, we conclude

$$\langle F, v_k \rangle = \langle F, v_k w \rangle \to 0$$

and therefore F is tempered.

Problem 5.2.17 (L^p functions and tempered distributions). *a) Verify that, for any $1 \le p \le +\infty$, $L^p(\mathbb{R}^n) \subset \mathcal{S}'(\mathbb{R}^n)$.*

b) Show that if $u_k \to u$ in $L^p(\mathbb{R}^n)$ then $u_k \to u$ in $\mathcal{S}'(\mathbb{R}^n)$.

Solution. Take $f \in L^p(\mathbb{R}^n)$, $1 \le p \le \infty$ and $\{v_k\} \subset \mathcal{D}(\mathbb{R}^n)$ such that $v_k \to 0$ in $\mathcal{S}(\mathbb{R}^n)$. If $q = p/(p-1)$ is the conjugate exponent to p, we have

$$\langle f, v_k \rangle = \int_{\mathbb{R}^n} \frac{f(\mathbf{x})}{(1 + |\mathbf{x}|)^{(n+1)/q}} (1 + |\mathbf{x}|)^{(n+1)/q} v_k(\mathbf{x}) \, d\mathbf{x},$$

and by Hölder's inequality

$$|\langle f, v_k \rangle| \le \|f\|_{L^p} \sup_{\mathbb{R}^n} \left[(1 + |\mathbf{x}|)^{(n+1)/q} |v_k(\mathbf{x})|\right] \left(\int_{\mathbb{R}^n} \frac{1}{(1 + |\mathbf{x}|)^{(n+1)}} d\mathbf{x}\right)^{1/q}.$$

But

$$\int_{\mathbb{R}^n} \frac{1}{(1 + |\mathbf{x}|)^{(n+1)}} d\mathbf{x} = \omega_n \int_0^{+\infty} \frac{\rho^{n-1}}{(1 + \rho)^{(n+1)}} d\rho < \infty,$$

and moreover, $v_k \to 0$ in $\mathcal{S}(\mathbb{R}^n)$ gives

$$\sup_{\mathbb{R}^n} \left[(1 + |\mathbf{x}|)^{(n+1)/q} |v_k(\mathbf{x})| \right] \to 0.$$

Thus, overall $\langle f, v_k \rangle \to 0$. This says that f is tempered.

b) Take $u_k \to 0$ in $L^p(\mathbb{R}^n)$, $v \in \mathcal{S}(\mathbb{R}^n)$, $1 \le p \le \infty$ and $q = p/(p-1)$. Arguing as above we obtain

$$|\langle u_k, v \rangle| \le \underbrace{\|u_k\|_{L^p}}_{\to 0} \underbrace{\sup_{\mathbb{R}^n} \left[(1 + |\mathbf{x}|)^{(n+1)/q} |v(\mathbf{x})| \right]}_{<\infty} \underbrace{\left(\int_{\mathbb{R}^n} \frac{1}{(1 + |\mathbf{x}|)^{(n+1)}} d\mathbf{x} \right)^{1/q}}_{<\infty},$$

so $\langle u_k, v \rangle \to 0$, and then $u_k \to 0$ in $\mathcal{S}'(\mathbb{R}^n)$.

Problem 5.2.18 (Fourier transform of a periodic distribution). *Let $u \in \mathcal{D}'(\mathbb{R})$ be periodic of period T. Compute the Fourier transform of u, then anti-transform the formula obtained and discuss.*

(We remind that F is said periodic of period T if

$$\langle F(x+T), v \rangle := \langle F, v(x-T) \rangle = \langle F, v \rangle$$

for any test function v, and that every periodic distribution is tempered.)

Solution. As u is periodic,

$$u(x+T) - u(x) = 0.$$

Taking the Fourier transform we find

$$(e^{iT\xi} - 1)\widehat{u}(\xi) = 0. \tag{5.17}$$

The zeroes of $e^{iT\xi} - 1$ are $\xi_n = 2\pi n/T$, $n \in \mathbb{Z}$, all simple (multiplicity one). The general solution of (5.17) is given by the following formula[6]:

$$\widehat{u}(\xi) = \sum_{n \in \mathbb{Z}} c_n \delta \left(\xi - \frac{2n\pi}{T} \right) \tag{5.18}$$

where $c_n \in \mathbb{C}$ are arbitrary constants, and the series converges[7] in $\mathcal{D}'(\mathbb{R})$. As the inverse Fourier transform of $\delta(\xi - a)$ is $e^{-iax}/2\pi$, when we transform back, we find the Fourier series

$$u(x) = \frac{1}{2\pi} \sum_{n \in \mathbb{Z}} c_n \exp \left(-i \frac{2n\pi x}{T} \right),$$

which holds in $\mathcal{D}'(\mathbb{R})$ (and also in $\mathcal{S}'(\mathbb{R})$). The numbers $c_n/2\pi$ are therefore the Fourier coefficients of the expansion of u.

---

[6] [18, Chap. 7, Sect. 5].

[7] It can be proved that it also converges in $\mathcal{S}'(\mathbb{R})$ as well.

Remark. Computing the Fourier coefficients $\widehat{u}_n = c_n/2\pi$ is a rather delicate matter. A relatively easy case is the following. Suppose we can decompose

$$u = u_1 + u_2$$

using T-periodic distributions u_1, u_2 such that $u_1 \in L^1_{\mathrm{loc}}(\mathbb{R})$. If x_0 *does not belong to the support of u_2, and we write $\widetilde{u}_2$ for the restriction of u_2 to $(x_0, x_0 + T)$, we have*

$$\widehat{u}_n = \frac{1}{T} \int_{x_0}^{x_0+T} u_1(x) \, e^{-i2n\pi x/T} \, dx + \langle \widetilde{u}_2, e^{-i2n\pi x/T} \rangle. \tag{5.19}$$

Note that the choice $u = u_1$ gives the usual formula. An explicit example can be found in Exercise 5.3.12 on page 311.

Problem 5.2.19 (Rapidly decreasing functions and tempered distributions). *Suppose $u \in L^1_{\mathrm{loc}}(\mathbb{R}^n)$ and let $P = P(\mathbf{x})$ be a polynomial with*

$$\frac{u}{P} \in L^1(\mathbb{R}^n).$$

Prove that $u \in \mathcal{S}'(\mathbb{R}^n)$.

Solution. Take $u \in L^1_{\mathrm{loc}}(\mathbb{R}^n)$ and P a polynomial such that $\dfrac{u}{P} \in L^1(\mathbb{R}^n)$. Let $\{v_k\} \subset \mathcal{D}(\mathbb{R}^n)$ be such that $v_k \to 0$ in $\mathcal{S}(\mathbb{R}^n)$. In particular $Pv_k \to 0$ uniformly in $\mathbb{R}^n$. Then

$$|\langle u, v_k \rangle| = \left| \int_{\mathbb{R}^n} u(\mathbf{x}) v_k(\mathbf{x}) \, d\mathbf{x} \right| = \left| \int_{\mathbb{R}^n} \frac{u(\mathbf{x})}{P(\mathbf{x})} P(\mathbf{x}) v_k(\mathbf{x}) \, d\mathbf{x} \right|$$

$$\leq \sup_{\mathbb{R}^n} |Pv_k| \int_{\mathbb{R}^n} \left| \frac{u(\mathbf{x})}{P(\mathbf{x})} \right| d\mathbf{x} \to 0,$$

because $u/P \in L^1(\mathbb{R}^n)$. Consequently u is tempered.

Problem 5.2.20 (Global Cauchy problem for the wave equation). *Let $g, h \in \mathcal{D}(\mathbb{R}^n)$.*

a) Show that $u \in C^2(\mathbb{R}^n \times [0, +\infty)$ is a solution of the global Cauchy problem ($n = 1, 2, 3$)

$$\begin{cases} u_{tt} - c^2 \Delta u = 0 & \mathbf{x} \in \mathbb{R}^n, t > 0 \\ u(\mathbf{x}, 0) = g(\mathbf{x}), \ u_t(\mathbf{x}, 0) = h(\mathbf{x}) & \mathbf{x} \in \mathbb{R}^n \end{cases} \tag{5.20}$$

if and only if $u^0(\mathbf{x}, t) = \mathcal{H}(t) u(\mathbf{x}, t)$ (the extension of u to zero for $t < 0$) is a solution in $\mathcal{D}'(\mathbb{R}^n \times \mathbb{R})$ of the equation

$$u^0_{tt} - c^2 \Delta u^0 = g(\mathbf{x}) \otimes \delta'(t) + h(\mathbf{x}) \otimes \delta(t). \tag{5.21}$$

b) Let $K = K(\mathbf{x}, t)$ be the fundamental solution of the wave equation in dimension $n = 1, 2, 3$. Deduce that $K^0(\mathbf{x}, t) = \mathcal{H}(t) K(\mathbf{x}, t)$ satisfies

$$K^0_{tt} - c^2 \Delta K^0 = \delta(\mathbf{x}) \otimes \delta(t) \qquad \text{in } \mathcal{D}'(\mathbb{R}^n \times \mathbb{R}).$$

Solution. a) Recall that for or every $\varphi \in \mathcal{D}\left(\mathbb{R}^n \times \mathbb{R}\right)$ we have, by definition of tensor product,

$$\langle g\left(\mathbf{x}\right) \otimes \delta'\left(t\right), \varphi\left(x, t\right)\rangle = \langle g\left(\mathbf{x}\right), \langle \delta'\left(t\right), \varphi\left(\mathbf{x}, t\right)\rangle\rangle = -\langle g\left(\mathbf{x}\right), \langle \delta\left(t\right), \varphi_t\left(\mathbf{x}, t\right)\rangle\rangle$$

$$= -\langle g\left(\mathbf{x}\right), \varphi_t\left(\mathbf{x}, 0\right)\rangle = -\int_{\mathbb{R}} g\left(\mathbf{x}\right) \varphi_t\left(\mathbf{x}, 0\right) d\mathbf{x},$$

$$\langle h\left(\mathbf{x}\right) \otimes \delta\left(t\right), \varphi\left(x, t\right)\rangle = \langle h\left(\mathbf{x}\right), \langle \delta\left(t\right), \varphi\left(\mathbf{x}, t\right)\rangle\rangle$$

$$= \langle h\left(\mathbf{x}\right), \varphi\left(x, 0\right)\rangle = \int_{\mathbb{R}} h\left(\mathbf{x}\right) \varphi\left(\mathbf{x}, 0\right) d\mathbf{x}.$$

Let now u^0 satisfy (5.21). This means that,

$$\int_{\mathbb{R}}\int_{\mathbb{R}^n} u^0\left(\varphi_{tt} - c^2\Delta\varphi\right) dt d\mathbf{x} = -\int_{\mathbb{R}^n} g\left(\mathbf{x}\right)\varphi_t\left(\mathbf{x}, 0\right) d\mathbf{x} + \int_{\mathbb{R}^n} h\left(\mathbf{x}\right) \varphi\left(\mathbf{x}, 0\right) d\mathbf{x}.$$
(5.22)

Integrating twice by parts the first integral, we get, taking into account that $u^0 = u^0\left(\mathbf{x}, t\right) = \mathcal{H}\left(t\right) u\left(\mathbf{x}, t\right)$,

$$\int_0^{+\infty}\int_{\mathbb{R}^n} u\left(\varphi_{tt} - c^2\Delta\varphi\right) d\mathbf{x} dt = \int_0^{+\infty}\int_{\mathbb{R}^n}\left(u_{tt} - c^2\Delta u\right)\varphi d\mathbf{x} dt$$

$$-\int_{\mathbb{R}^n} u\left(\mathbf{x}, 0\right)\varphi_t\left(\mathbf{x}, 0\right) d\mathbf{x} + \int_{\mathbb{R}^n} u_t\left(\mathbf{x}, 0\right)\varphi\left(\mathbf{x}, 0\right) d\mathbf{x}. \quad (5.23)$$

Comparing (5.22) and (5.23) we infer that

$$\int_0^{+\infty}\int_{\mathbb{R}^n}\left(u_{tt} - c^2\Delta u\right)\varphi d\mathbf{x} dt - \left[\int_{\mathbb{R}^n} u\left(\mathbf{x}, 0\right) - g\left(\mathbf{x}\right)\right]\varphi_t\left(\mathbf{x}, 0\right) d\mathbf{x}$$

$$+ \int_{\mathbb{R}^n}\left[u_t\left(\mathbf{x}, 0\right) - h\left(\mathbf{x}\right)\right]\varphi\left(\mathbf{x}, 0\right) d\mathbf{x} = 0 \quad (5.24)$$

for every $\varphi \in \mathcal{D}\left(\mathbb{R}^n \times \mathbb{R}\right)$. Take $\varphi \in \mathcal{D}\left(\mathbb{R}^n \times (0, +\infty)\right)$. Then $\varphi_t\left(\mathbf{x}, 0\right) = \varphi\left(\mathbf{x}, 0\right) = 0$ and (5.24) gives

$$\int_0^{+\infty}\int_{\mathbb{R}^n}\left(u_{tt} - c^2\Delta u\right)\varphi d\mathbf{x} dt = 0.$$

The arbitrariness of φ implies that $u_{tt} - c^2\Delta u = 0$ in $\mathbb{R}^n \times (0, +\infty)$. Now we can go back to $\varphi \in \mathcal{D}\left(\mathbb{R}^n \times \mathbb{R}\right)$. Since $u_{tt} - c^2\Delta u = 0$ we get from (5.24)

$$-\left[\int_{\mathbb{R}^n} u\left(\mathbf{x}, 0\right) - g\left(\mathbf{x}\right)\right]\varphi_t\left(\mathbf{x}, 0\right) d\mathbf{x} + \int_{\mathbb{R}^n}\left[u_t\left(\mathbf{x}, 0\right) - h\left(\mathbf{x}\right)\right]\varphi\left(\mathbf{x}, 0\right) d\mathbf{x} = 0. \quad (5.25)$$

Choose now $\psi_0, \psi_1 \in \mathcal{D}\left(\mathbb{R}^n\right)$ and $b\left(t\right) \in \mathcal{D}\left(\mathbb{R}\right)$ such that $b\left(0\right) = 1$ and $b'\left(0\right) = 0$. The function

$$\psi\left(\mathbf{x}, t\right) = \left(\psi_0\left(\mathbf{x}\right) + t\psi_1\left(\mathbf{x}\right)\right) b\left(t\right)$$

belongs to $\mathcal{D}\,(\mathbb{R}^n \times \mathbb{R})$ and $\psi\,(\mathbf{x},0) = \psi_0\,(\mathbf{x})$, $\psi_t\,(\mathbf{x},0) = \psi_1\,(\mathbf{x})$. Inserting ψ into (5.25) we obtain

$$-\int_{\mathbb{R}^n} [u\,(\mathbf{x},0) - g\,(\mathbf{x})]\,\psi_1\,(\mathbf{x})\,d\mathbf{x} + \int_{\mathbb{R}^n} [u_t\,(\mathbf{x},0) - h\,(\mathbf{x})]\,\psi_0\,(\mathbf{x})\,d\mathbf{x} = 0.$$

The arbitrariness of ψ_0,ψ_1 implies $u\,(\mathbf{x},0) = g\,(\mathbf{x})$ and $u_t\,(\mathbf{x},0) = h\,(\mathbf{x})$.

Conversely, let u satisfy (5.20). A double integration by parts, using the initial conditions, gives

$$0 = \int_0^{+\infty}\!\!\int_{\mathbb{R}^n} u\left(\varphi_{tt} - c^2\Delta\varphi\right) d\mathbf{x}dt + \int_{\mathbb{R}^n} g\,(\mathbf{x})\,\varphi_t\,(\mathbf{x},0)\,d\mathbf{x} - \int_{\mathbb{R}^n} h\,(\mathbf{x})\,\varphi\,(\mathbf{x},0)\,d\mathbf{x}$$

which is equivalent to (5.22) since $u^0 = 0$ for $t < 0$.

b) It is enough to remember that K is a solution of the wave equations with initial data $K\,(\mathbf{x},0) = 0$, $K_t\,(\mathbf{x},0) = \delta\,(\mathbf{x})$.

5.2.3 Sobolev spaces

Problem 5.2.21 (Singularities in Sobolev spaces). *Consider the function*

$$u(x,y) = \left(\sqrt{x^2 + y^2}\right)^\alpha$$

and let $D = \{(x,y) \in \mathbb{R}^2 : x^2 + y^2 < 1\}$ be the unit disc. Determine the values $\alpha \in \mathbb{R}$ for which:

    a) $u \in L^2(D)$.   *b) $u \in H^1(D)$.*   *c) $u \in H^{-1}(D)$.*

Solution. a) It is enough to compute the integral of u^2 and check whether it is finite. In polar coordinates

$$\|u\|_2^2 = \int_D (x^2 + y^2)^\alpha\,dxdy = \int_0^{2\pi} d\theta \int_0^1 r^{2\alpha+1}\,dr = 2\pi \int_0^1 r^{2\alpha+1}\,dr.$$

The integral is finite if and only if $2\alpha + 1 > -1$. Hence

$$u \in L^2(D) \qquad \Longleftrightarrow \qquad \alpha > -1.$$

b) First of all u must be in $L^2(D)$, i.e. $\alpha > -1$. Next, the gradient of u (in the sense of distributions) must belong to $L^2(D;\mathbb{R}^2)$. When $\alpha \leq 0$ we have $u \notin C^1\,(\overline{D})$, so, *a priori*, distribution derivatives may not coincide with classical ones. However, for $\alpha > -1$, this can be proved (see Exercise 5.3.15 on page 312). Thus, in such case,

$$u_x(x,y) = \alpha x(x^2 + y^2)^{(\alpha-2)/2}, \qquad u_y(x,y) = \alpha y(x^2 + y^2)^{(\alpha-2)/2}.$$

Hence

$$\|\nabla u\|_2^2 = \int_D \alpha^2 (x^2 + y^2)^{\alpha-1} \, dxdy = 2\pi\alpha^2 \int_0^1 r^{2\alpha-1} \, dr.$$

Therefore $2\alpha - 1 > -1$ or $\alpha = 0$, and so

$$u \in H^1(D) \qquad \text{if and only if} \qquad \alpha \geq 0.$$

c) For u to belong to the dual of $H_0^1(D)$ we must find a constant C satisfying

$$\left| \int_D uv \, dxdy \right| \leq C \left(\int_D |\nabla v|^2 \, dxdy \right)^{1/2} \qquad \text{for any } v \in H_0^1(D).$$

As we are in dimension 2, for any $1 \leq p < +\infty$ we have[8] $H^1(D) \subset L^p(D)$. From Poincaré's inequality we infer the existence of a constant C_P (depending on D and p) such that any $v \in H_0^1(D)$ fulfils

$$\|v\|_{L^p(D)} \leq C_P \|\nabla v\|_{L^2(D;\mathbb{R}^2)}.$$

If $q > 1$ is the conjugate to p, the above inequality combines with Hölder's inequality to give

$$\left| \int_D uv \, dxdy \right| \leq \left(\int_D u^q \, dxdy \right)^{1/q} \left(\int_D v^p \, dxdy \right)^{1/p} \leq$$

$$\leq C_P \left(\int_D u^q \, dxdy \right)^{1/q} \left(\int_D |\nabla v|^2 \, dxdy \right)^{1/2}.$$

This inequality shows that if $u \in L^q(D)$, with $q > 1$, then $u \in H^{-1}(D)$. So let us see for which numbers α, u belongs to $L^q(D)$ (for some $q > 1$). As

$$\int_D (x^2 + y^2)^{q\alpha/2} \, dxdy = 2\pi \int_0^1 r^{q\alpha+1} \, dr$$

we must have $q\alpha + 1 > -1$. This is true (for some $q > 1$) as soon as $\alpha > -2$. The converse statement can be proved, too: when $\alpha \leq -2$ the function u does not belong to the dual of $H_0^1(D)$. It suffices to pick $v \in \mathcal{D}(D) \subset H_0^1(D)$ satisfying

$$v \geq 0 \text{ in } D, \qquad v \geq 1 \text{ in } D' = \left\{ (x, y) \in \mathbb{R}^2 : x^2 + y^2 < \frac{1}{2} \right\},$$

to obtain (note that $u \geq 0$ in D)

$$\int_D uv \, dxdy \geq \int_{D'} u \, dxdy = 2\pi \int_0^{1/2} r^\alpha \cdot r \, dr = +\infty \quad \text{if } \alpha \leq -2.$$

---

[8] By Sobolev's embedding theorem, see [18, Chap. 7, Sect. 10].

To sum up,

$$u \in H^{-1}(D) \qquad \text{if and only if} \qquad \alpha > -2.$$

Problem 5.2.22 (Singularities in trace spaces). *Consider the function*

$$u(x, y) = \left(\sqrt{x^2 + y^2} \right)^{\alpha}$$

and let $D_+ = \{(x, y) \in \mathbb{R}^2 : x > 0, y > 0, x^2 + y^2 < 1\}$. *Find all* $\alpha \in \mathbb{R}$ *for which:*

a) $u \in L^2(\partial D_+)$.   *b)* $u \in H^1(\partial D_+)$.   *c)* $u \in H^{1/2}(\partial D_+)$.

Solution. a) The function u has a singularity only at the origin and when $\alpha < 0$; in particular, it is smooth along the arc

$$\{(x, y) : x > 0, y > 0, x^2 + y^2 = 1\} \subset \partial D_+.$$

On the segment

$$S = \{(x, 0) : 0 < x < 1\} \subset \partial D_+$$

we have

$$\|u\|^2_{L^2(S)} = \int_0^1 u^2(x, 0)\, dx = \int_0^1 x^{2\alpha}\, dx,$$

and the integral converges if $2\alpha > -1$. By symmetry this holds also on $\{(0, y) : 0 < y < 1\}$, therefore

$$u \in L^2(\partial D_+) \qquad \text{if and only if} \qquad \alpha > -\frac{1}{2}.$$

b) As in the previous exercise, for $\alpha > -1/2$, the distributional derivatives of u are classical. Hence

$$\|u_x\|^2_{L^2(S)} = \int_0^1 \alpha^2 x^{2\alpha-2}\, dx.$$

The integral is finite if $\alpha = 0$ or $2\alpha - 2 > -1$. Then

$$u \in H^1(\partial D_+) \qquad \text{if and only if} \qquad \alpha = 0 \text{ or } \alpha > \frac{1}{2}.$$

c) The domain is Lipschitz so we can resort to the theory of traces. We saw in the previous problem that the extension of u to D_+, defined by the same algebraic expression, is in $H^1(D)$ (whence in $H^1(D_+)$) for $\alpha \geq 0$. Immediately, then,

$$u \in H^{1/2}(\partial D_+) \qquad \text{provided} \qquad \alpha \geq 0.$$

Problem 5.2.23 (Norm estimates in $L^\infty(\mathbb{R})$). *Prove that* $u \in H^1(\mathbb{R})$ *implies* $u \in L^\infty(\mathbb{R})$, *and*

$$\|u\|_{L^\infty(\mathbb{R})} \leq \|u\|_{H^1(\mathbb{R})}.$$

Solution. Let us prove the inequality for C^∞ functions with compact support. By density the result will hold in general (recall that $H^1(\mathbb{R}) = H_0^1(\mathbb{R})$). In fact, take $u \in \mathcal{D}(\mathbb{R})$, and observe that

$$u^2(t) = \int_{-\infty}^t 2u(s)u'(s)\,ds \le 2\left(\int_{-\infty}^t u^2(s)\,ds\right)^{1/2}\left(\int_{-\infty}^t u'^2(s)\,ds\right)^{1/2}$$

$$\le \int_{-\infty}^t u^2(s)\,ds + \int_{-\infty}^t u'^2(s)\,ds$$

(in the last line we used the elementary inequality $2ab \le a^2 + b^2$). We can now take the supremum over t and write

$$u \in \mathcal{D}(\mathbb{R}) \qquad \Rightarrow \qquad \|u\|_{L^\infty(\mathbb{R})} \le \|u\|_{H^1(\mathbb{R})}. \tag{5.26}$$

Let now $\{u_n\} \subset \mathcal{D}(\mathbb{R})$ and $u_n \to u$ in $H^1(\mathbb{R})$. Equation (5.26) implies

$$\|u_n - u_m\|_{L^\infty(\mathbb{R})} \le \|u_n - u_m\|_{H^1(\mathbb{R})},$$

so $\{u_n\}$ is a Cauchy sequence in $L^\infty(\mathbb{R})$. Therefore u_n converges to u also in $L^\infty(\mathbb{R})$, and (5.26) holds for $u \in H^1(\mathbb{R})$ as well.

Problem 5.2.24 (Norm estimates in $L^\infty(a, b)$). *Prove that $u \in H^1(a, b)$ implies $u \in L^\infty(a, b)$, and*

$$\|u\|_{L^\infty(a,b)} \le C\|u\|_{H^1(a,b)}$$

with $C = \max\{(b - a)^{-1/2}, (b - a)^{1/2}\}$.

Solution. Since u is continuous on the compact set $[a, b]$, it is bounded. We must show the existence of a constant C such that

$$|u(t)| \le C(\|u\|_2 + \|u'\|_2)$$

for any $t \in [a, b]$ and any $u \in H^1(a, b)$. To this end, let us remark that by the mean-value theorem and the continuity of u we may fix a point $\xi \in [a, b]$ with

$$u^2(\xi) = \frac{1}{b - a}\int_a^b u^2(s)\,ds = \frac{1}{b - a}\|u\|_2^2.$$

Hence

$$|u(t)| = \left|u(\xi) + \int_\xi^t u'(s)\,ds\right| \le \frac{1}{\sqrt{b - a}}\|u\|_2 + \int_a^b |u'(s)|\,ds \le$$

$$\le \frac{1}{\sqrt{b - a}}\|u\|_2 + \sqrt{b - a}\,\|u'\|_2 \le C(\|u\|_2 + \|u'\|_2),$$

where

$$C = \max\left\{\frac{1}{\sqrt{b-a}}, \sqrt{b-a}\right\}.$$

Problem 5.2.25 (Riesz's theorem and the Dirac δ measure). *Let δ denote the one-dimensional Dirac delta measure at 0. Prove that δ belongs to the dual of $H^1(-1, 1)$. Determine the element of $H^1(-1, 1)$ (which exists by Riesz's theorem) representing δ with respect to the inner product of $H^1(-1, 1)$.*

Solution. Clearly δ is a linear functional that is well defined on $H^1(-1, 1)$, since the functions in that space are continuous and it makes sense to take their value at any point. The continuity of δ follows from Problem 5.2.24, where it is proved that

$$|v(0)| \leq \sqrt{2}(\|v\|_2 + \|v'\|_2) \qquad \text{for any } v \in H^1(-1, 1).$$

Thus δ belongs to dual of H^1. By Riesz's representation theorem there is a function $u \in H^1(-1, 1)$ such that

$$\int_{-1}^{1} \left[u'(t)v'(t) + u(t)v(t)\right] dt = v(0), \qquad \text{for any } v \in H^1(-1, 1).$$

To identify this function we suppose u is regular on $[-1, 0]$ and $[0, 1]$, separately. Integrating by parts, we find

$$\int_{-1}^{0} \left[u'(t)v'(t) + u(t)v(t)\right] dt = u'(0^-)v(0) - u'(-1)v(-1) - \int_{-1}^{0} \left[u''(t) - u(t)\right] v(t)\, dt$$

and

$$\int_{0}^{1} \left[u'(t)v'(t) + u(t)v(t)\right] dt = u'(1)v(1) - u'(0^+)v(0) - \int_{0}^{1} \left[u''(t) - u(t)\right] v(t)\, dt.$$

If we add the two relations we find, after rearranging the terms,

$$-\int_{-1}^{1} \left[u''(t) - u(t)\right] v(t)\, dt + v(0)\left[u'(0^-) - u'(0^+)\right] - u'(-1)v(-1) + u'(1)v(1) = v(0).$$

By choosing v with support in $(-1, 0)$, we get:

$$\int_{-1}^{0} \left[u''(t) - u(t)\right] v(t)\, dt = 0,$$

forcing $u''(t) - u(t) = 0$ in $(-1, 0)$, as v is arbitrary. Analogously, $u''(t) - u(t) = 0$ in $(0, 1)$. If we now take a general v, we are left with

$$v(0)\left[u'(0^-) - u'(0^+)\right] - u'(-1)v(-1) + u'(1)v(1) = v(0).$$

Since the values $v(0)$, $v(1)$ and $v(-1)$ are arbitrary, we deduce the "transmission" condition

$$u'(0^-) - u'(0^+) = 1,$$

and the Neumann conditions

$$u'(-1) = u'(1) = 0.$$

Summarising, we have reduced to the following problem:

$$\begin{cases} u''(t) - u'(t) = 0 & \text{when } -1 < t < 0,\ 0 < t < 1 \\ u'(-1) = u'(1) = 0, & u'(0^-) - u'(0^+) = 1. \end{cases}$$

The general integral of the ODE is a linear combination of exponentials, so

$$u(t) = \begin{cases} A_1 e^t + A_2 e^{-t} & -1 \le t \le 0 \\ B_1 e^t + B_2 e^{-t} & 0 \le t \le 1. \end{cases}$$

The boundary conditions and the continuity at 0 give

$$\begin{cases} A_1 e^{-1} - A_2 e = 0 \\ B_1 e - B_2 e^{-1} = 0 \\ A_1 - A_2 - B_1 + B_2 = 1 \\ A_1 + A_2 = B_1 + B_2 \end{cases}$$

whence

$$A_2 = B_1 = \frac{1}{2(e^2 - 1)}, \qquad A_1 = B_2 = \frac{e^2}{2(e^2 - 1)},$$

and finally

$$u(t) = \begin{cases} \dfrac{1}{2(e^2 - 1)} \left[e^{2+t} + e^{-t} \right] & -1 \le t \le 0 \\ \dfrac{1}{2(e^2 - 1)} \left[e^t + e^{2-t} \right] & 0 \le t \le 1. \end{cases}$$

It is not hard to see that $u \in H^1(-1, 1)$ (it is piecewise C^1), and integrating by parts one proves (do it) that this is indeed the required function. (Uniqueness is a consequence of Riesz's theorem.)

Problem 5.2.26 (Riesz's theorem and integral average). *Prove that the functional*

$$Lu = \int_0^1 u(t)\, dt$$

belongs to $H^{-1}(0, 1)$. Determine the element that represents it in $H_0^1(0, 1)$.

Solution. L is a linear and continuous functional on $L^2(0, 1)$ and therefore, *a fortiori*, on $H_0^1(0, 1)$. In particular, continuity follows from the Schwarz and Poincaré inequalities:

$$\left| \int_0^1 u(t)\, dt \right| \le \left(\int_0^1 dt \right)^{1/2} \left(\int_0^1 |u(t)|^2\, dt \right)^{1/2} \le \|u\|_{L^2(0,1)} \le C_P \|u'\|_{L^2(0,1)}.$$

By Riesz's theorem there is a unique $u \in H_0^1(0, 1)$ such that

$$\int_0^1 u'(t)v'(t)\, dt = \int_0^1 v(t)\, dt, \qquad \text{for all } v \in H_0^1(0, 1).$$

To find u explicitly we proceed formally, supposing u regular (with at least two continuous derivatives in $[0, 1]$), so to be able to integrate by parts. We obtain

$$u'(1)v(1) - u'(0)\, v(0) - \int_0^1 u''(t)v(t)\, dt = \int_0^1 v(t)\, dt, \qquad \text{for any } v \in H_0^1(0, 1)$$

whence

$$\int_0^1 [u''(t) + 1]v(t)\, dt = 0 \qquad \text{for any } v \in H_0^1(0, 1).$$

As v was arbitrary, we infer

$$u''(t) = -1 \quad \text{in } (0, 1)$$

with $u(1) = u(0) = 0$. Then

$$u(t) = \frac{1}{2}(t - t^2).$$

By reading the argument backwards we see easily that u represents L for the inner product of $H_0^1(0, 1)$.

Problem 5.2.27 (Riesz's theorem). *Let f and g be regular functions and consider the functional*

$$Lv = \int_0^1 \big(f(t)v'(t) + g(t)v(t) \big)\, dt.$$

Prove that L belongs to the dual of $H^1(0, 1)$.

Calling u the element of $H^1(0, 1)$ representing L, write a boundary-value problem that determines u uniquely.

Solve the problem explicitly in case $f(t) = t(t-1)$ and $g(t) = 2t$.

Solution. The functional L is linear, so we just have to prove its boundedness. By Schwarz's inequality

$$|Lv| \le \int_0^1 \big(|f(t)v'(t)| + |g(t)v(t)| \big)\, dt \le \|f\|_{L^2(0,1)} \|v'\|_{L^2(0,1)} + \|g\|_{L^2(0,1)} \|v\|_{L^2(0,1)},$$

and therefore, as long as f and g are square-integrable, L is continuous. By Riesz's theorem there exists a unique $u \in H^1(0, 1)$ such that

$$\int_0^1 \left(u'(t)v'(t) + u(t)v(t) \right) dt = \int_0^1 \left(f(t)v'(t) + g(t)v(t) \right) dt \tag{5.27}$$

for any $v \in H^1(0, 1)$. Mimicking the previous exercises we assume u is sufficiently regular so to allow for integration by parts. Then

$$\int_0^1 u'(t)v'(t)\, dt = u'(1)v(1) - u'(0)v(0) - \int_0^1 u''(t)v(t)\, dt$$

and

$$\int_0^1 f(t)v'(t)\, dt = f(1)v(1) - f(0)v(0) - \int_0^1 f'(t)v(t)\, dt.$$

Substituting into (5.27) gives

$$\int_0^1 (-u'' + u + f' - g)v\, dt + [u'(1) - f(1)]v(1) - [u'(0) - f(0)]v(0) = 0$$

for any $v \in H^1(0, 1)$. Choosing v vanishing at the endpoints, we obtain

$$\int_0^1 (-u'' + u + f' - g)v\, dt = 0,$$

forcing

$$u'' - u = f' - g \quad \text{in } (0, 1).$$

Then, for general $v \in H^1(0, 1)$, we are left with

$$[u'(1) - f(1)]v(1) - [u'(0) - f(0)]v(0) = 0$$

which gives

$$u'(1) = f(1) \quad \text{and} \quad u'(0) = f(0).$$

In conclusion, u is solves

$$\begin{cases} u''(t) - u(t) = f'(t) - g(t) \\ u'(0) = f(0),\ u'(1) = f(1). \end{cases}$$

In case we take $f(t) = t^2 - t$ and $g(t) = 2t$, we find

$$u'' - u = -1,\ u'(0) = u'(1) = 0.$$

The solution is $u(t) \equiv 1$.

Problem 5.2.28 (Projections). *Define $H = H^1(-1, 1)$ and*

$$V = \{u \in H : u(0) = 0\}.$$

After proving V is a closed subspace in H, compute the projection onto V of f, where $f(t) = 1$ on $[-1, 1]$.

Solution. That V is a subspace is straightforward (recall that functions in H are continuous, so pointwise values make sense). We have to prove that V is closed, i.e. that if $u_n(0) = 0$ and $u_n \to \bar{u}$ in H, then $\bar{u}(0) = 0$. For this we observe that convergence in H implies uniform convergence. As seen in Problem 5.2.24 (page 301), in fact,

$$\|u\|_{L^\infty(-1,1)} \leq \sqrt{2}\|u\|_H.$$

In turn, uniform convergence implies pointwise convergence *everywhere* on $[-1, 1]$. In particular if $u_n \to \bar{u}$ in H then $u_n(0) \to \bar{u}(0)$ and V is closed. The projection theorem guarantees the existence (and uniqueness) of the projection $P_V f$. To find the latter explicitly, we may minimise over V the quadratic functional

$$E(u) = \|f - u\|_H^2 = \int_{-1}^1 \left[(u'(t))^2 + (1 - u(t))^2\right] dt.$$

We know that the minimum $u = P_V f$ satisfies the necessary condition

$$(u - f, v)_H = \int_{-1}^1 \left[u'(t)v'(t) - (1 - u(t))v(t)\right] dt = 0, \qquad \text{for any } v \in H. \quad (5.28)$$

Assuming u regular and integrating by parts, we find

$$\int_{-1}^1 u'(t)v'(t)\, dt = u'(t)v(t)\big|_{-1}^1 - \int_{-1}^1 u''(t)v(t)\, dt,$$

and so

$$u'(1)v(1) - u'(-1)v(-1) - \int_{-1}^1 \left[u'' + 1 - u\right] v\, dt = 0, \qquad (5.29)$$

for any $v \in H^1(-1, 1)$. In particular, the equality must hold for every $v \in H_0^1(-1, 1)$, so that

$$\int_{-1}^1 \left[u'' + 1 - u\right] v\, dt = 0, \qquad \text{for any } v \in H_0^1(-1, 1),$$

which implies

$$u'' + 1 - u = 0, \qquad \text{a.e. in } (-1, 1).$$

But then the integral in (5.29) is always zero, and we are left with

$$u'(1)v(1) - u'(-1)v(-1) = 0, \qquad \text{for any } v \in H^1(-1, 1),$$

whence

$$u'(-1) = u'(1) = 0.$$

As $u(0) = 0$ (in fact $u \in V$) we find that u solves

$$\begin{cases} u''(t) - u(t) = -1 & -1 < t < 0, \, 0 < t < 1 \\ u'(-1) = u'(1) = 0, & u(0) = 0. \end{cases}$$

Then

$$u(t) = \begin{cases} 1 + A_1 e^t + A_2 e^{-t} & -1 \le t \le 0 \\ 1 + B_1 e^t + B_2 e^{-t} & 0 \le t \le 1, \end{cases}$$

with

$$\begin{cases} A_1 e^{-1} - A_2 e = 0 \\ B_1 e - B_2 e^{-1} = 0 \\ 1 + A_1 + A_2 = 1 + B_1 + B_2 = 0. \end{cases}$$

We find

$$A_2 = B_1 = -\frac{1}{e^2 + 1}, \qquad A_1 = B_2 = -\frac{e^2}{e^2 + 1}$$

and finally

$$u(t) = \begin{cases} 1 - \dfrac{1}{e^2 + 1} \left[e^{2+t} + e^{-t} \right] & -1 \le t \le 0 \\ 1 - \dfrac{1}{e^2 + 1} \left[e^t + e^{2-t} \right] & 0 \le t \le 1. \end{cases}$$

Since there is a unique element in $H^1(-1, 1)$ satisfying (5.28), it follows that u coincides with $P_V f$.

Problem 5.2.29 (Piecewise defined function). *Define*

$$Q^+ = \{(x, y) \in \mathbb{R}^2; 0 < x < 1, 0 < y < 1\}$$

$$Q^- = \{(x, y) \in \mathbb{R}^2; -1 < x < 0, 0 < y < 1\}$$

and $Q = Q^+ \cup Q^- \cup \{x = 0, 0 < y < 1\}$, then consider

$$u(x, y) = \begin{cases} xy^2 & \text{in } Q^+ \\ 1 - x^2 y & \text{in } Q^-. \end{cases}$$

Compute the gradient of u in the sense of distributions. Decide whether $u \in H^1(Q)$.

Solution. On Q^+, and separately on Q^-, u is a polynomial, and thus it defines a distribution; note that u jumps by -1 horizontally (for example) across the vertical segment $\{x = 0, 0 < y < 1\}$. Let v be a test function and let us compute the distributional gradient of u. By definition

$$\langle u_x, v \rangle = -\langle u, v_x \rangle = - \int_Q u(x,y) \, v_x(x,y) \, dxdy =$$

$$= - \int_0^1 dy \int_0^1 xy^2 v_x(x,y) \, dx - \int_0^1 dy \int_{-1}^0 (1 - x^2 y) \, v_x(x,y) \, dx.$$

Integrating by parts in x, we find, as $v(1,y) = 0$:

$$\langle u_x, v \rangle = \int_0^1 dy \int_0^1 y^2 v(x,y) \, dx - \int_0^1 v(0,y) \, dy - \int_0^1 dy \int_{-1}^0 2xy \, v(x,y) \, dx.$$

Set

$$w_1(x,y) = \begin{cases} y^2 & \text{in } Q^+ \\ -2xy & \text{in } Q^- \end{cases}$$

and let $F = \delta \otimes 1$ be the distribution[9]

$$\langle F, v \rangle = \int_0^1 v(0,y) \, dy.$$

Then

$$u_x = w_1 - \delta \otimes 1.$$

The distribution $-\delta \otimes 1$ has support on the discontinuity curve of u, and signals the presence of a -1 jump across the y-axis in the direction $x > 0$. Let us compute u_y. By definition

$$\langle u_y, v \rangle = -\langle u, v_y \rangle = - \int_Q u(x,y) \, v_y(x,y) \, dxdy =$$

$$= - \int_0^1 dx \int_0^1 xy^2 v_y(x,y) \, dy - \int_{-1}^0 dx \int_0^1 (1 - x^2 y) \, v_y(x,y) \, dy.$$

Integrating by parts in y, since $v(x,0) = v(x,1) = 0$, we have

$$\langle u_x, v \rangle = \int_0^1 dx \int_0^1 2xy \, v(x,y) \, dy - \int_{-1}^0 dx \int_0^1 x^2 v(x,y) \, dy.$$

Defining

$$w_2(x,y) = \begin{cases} 2xy & \text{in } Q^+ \\ -x^2 & \text{in } Q^-, \end{cases}$$

we then obtain

$$u_y = w_2.$$

There are no distributions concentrated on the discontinuity, because along the vertical direction u is not discontinuous. Whereas $u_y \in L^2(Q)$, the distribution representing u_x cannot be identified with a function of $L^2(Q)$. Therefore $u \notin H^1(Q)$.

---

[9] This is a Dirac distribution along the segment $x = 0, 0 < y < 1$.

Problem 5.2.30 (Gluing functions in H^1). *Let $\Omega \subset \mathbb{R}^n$ be a regular bounded domain, divided into two regular domains Ω_1, Ω_2 with a regular interface Γ (hypersurface):*

$$\Omega = \Omega_1 \cup \Omega_2 \cup \Gamma, \qquad \Omega_1 \cap \Omega_2 = \Omega_1 \cap \Gamma = \Omega_2 \cap \Gamma = \emptyset.$$

Take $u_1 \in H^1(\Omega), u_2 \in H^1(\Omega)$. Under which conditions does the function

$$u(\mathbf{x}) = \begin{cases} u_1(\mathbf{x}) & \mathbf{x} \in \Omega_1 \\ u_2(\mathbf{x}) & \mathbf{x} \in \Omega_2 \end{cases}$$

belong to $H^1(\Omega)$?

Solution. As $u_i \in L^2(\Omega_i)$, $i = 1, 2$ we immediately have

$$u \in L^2(\Omega)$$

and so $u \in \mathcal{D}'(\Omega)$. If we want $u \in H^1(\Omega)$, ∇u must by represented (as a distribution) by some vector $\mathbf{w} \in L^2(\Omega; \mathbb{R}^n)$, i.e.

$$\langle \nabla u, \mathbf{F} \rangle = \int_\Omega \mathbf{w} \cdot \mathbf{F} \, d\mathbf{x}$$

for any vector field $\mathbf{F} \in \mathcal{D}(\Omega; \mathbb{R}^n)$. Since ∇u_1 and ∇u_2 must be elements of $L^2(\Omega_1; \mathbb{R}^n)$ and $L^2(\Omega_2; \mathbb{R}^n)$ respectively, the candidate $\mathbf{w}$ should be defined by

$$\mathbf{w} = \begin{cases} \nabla u_1 & \text{in } \Omega_1 \\ \nabla u_2 & \text{in } \Omega_2 \end{cases}$$

with the proviso that u should not "jump" when crossing Γ (see Problem 5.2.29). A natural condition is to demand that the *traces* of u_1 and u_2 on Γ coincide. Let us check this fact. By definition of distribution gradient

$$\langle \nabla u, \mathbf{F} \rangle = -\langle u, \text{div } \mathbf{F} \rangle = -\int_\Omega u \, \text{div } \mathbf{F} \, d\mathbf{x} = -\int_{\Omega_1} u_1 \, \text{div } \mathbf{F} \, d\mathbf{x} - \int_{\Omega_2} u_2 \, \text{div } \mathbf{F} \, d\mathbf{x} \tag{5.30}$$

(Γ has measure zero, being regular). Let $\boldsymbol{\nu}$ denote the unit normal to Γ, pointing outward with respect to Ω_1 (so that $-\boldsymbol{\nu}$ points outward with respect to Ω_2). As $\mathbf{F} = 0$ on $\partial\Omega$, the divergence theorem in H^1 tells that

$$\int_{\Omega_1} u_1 \, \text{div } \mathbf{F} \, d\mathbf{x} = \int_\Gamma u_1 \mathbf{F} \cdot \boldsymbol{\nu} \, d\sigma - \int_{\Omega_1} \nabla u_1 \cdot \mathbf{F} \, d\mathbf{x} \tag{5.31}$$

and

$$\int_{\Omega_2} u_2 \, \text{div } \mathbf{F} \, d\mathbf{x} = -\int_\Gamma u_2 \mathbf{F} \cdot \boldsymbol{\nu} \, d\sigma - \int_{\Omega_2} \nabla u_2 \cdot \mathbf{F} \, d\mathbf{x}. \tag{5.32}$$

If $u_1 = u_2$ on Γ, we find, from (5.30), (5.31) and (5.32),

$$\langle \nabla u, \mathbf{F} \rangle = \int_{\Omega} \mathbf{w} \cdot \mathbf{F} \, d\mathbf{x}.$$

The condition $u_1 = u_2$ is therefore sufficient (and also necessary[10]) for $u \in H^1(\Omega)$.

5.3 Further Exercises

5.3.1. *Extend the functions of $H = L^2(0, 2\pi)$ to the entire $\mathbb{R}$ in a periodic way. Consider the subspace of $2\pi/3$-periodic functions*

$$V = \left\{ u \in H : u\left(t + \frac{2}{3}\pi\right) = u(t) \quad \text{for almost every } t \in \mathbb{R} \right\}.$$

Check that V is closed in H, then determine $V^{\perp}$ and the projections P_V, $P_{V^{\perp}}$.

5.3.2. *In $H = L^2(0, 1)$ consider the closed subspace V of quadratic polynomials. Compute $P_V f$ when $f(t) = t^3$.*

5.3.3. *Set $Q = (0, 1) \times (0, 1)$, $H = L^2(Q)$, and consider subspaces*

$$K = \{u \in H : u(x, y) = \text{constant}\},$$

$$V = \left\{ u \in H : u(x, y) = v(x), v \in L^2(0, 1), \int_0^1 v(x)dx = 0 \right\},$$

$$W = \left\{ u \in H : u(x, y) = w(y), w \in L^2(0, 1), \int_0^1 w(y)dy = 0 \right\}.$$

Verify that

$$S = K \oplus V \oplus W$$

is closed and determine the projections P_S, $P_{S^{\perp}}$. Decompose the function $f(x, y) = xy$ accordingly.

5.3.4. *Take $H = L^2(B_1)$ (B_1 being the unit disc in $\mathbb{R}^2$) and consider the linear operator on H:*

$$(Lu)(x, y) = u(y, -x).$$

a) Verify L is continuous and compute its norm.
b) Find the possible (real) eigenvalues.

5.3.5. *Let $\{x_k\} \subset \mathbb{R}$ be a sequence going to $+\infty$. Prove that*

$$\sum_{k=1}^{+\infty} c_k \delta(x - x_k)$$

converges in $\mathcal{D}'(\mathbb{R})$ irrespective of the behaviour of the real sequence $\{c_k\}$.

---

[10] We leave this for the reader to check.

5.3.6. Let $f : \mathbb{R} \to \mathbb{R}$ denote the 2π-periodic extension of

$$f_0(x) = \frac{\pi - x}{2} \qquad 0 < x < 2\pi.$$

a) Write the Fourier series of f and check that it converges in $\mathcal{D}'(\mathbb{R})$ to f.

b) Deduce the (remarkable) formula

$$\sum_{n=1}^{\infty} \cos nx = -\frac{1}{2} + \pi \sum_{n=-\infty}^{\infty} \delta(x - 2\pi n), \qquad \text{valid in } \mathcal{D}'(\mathbb{R}). \qquad (5.33)$$

5.3.7. Give an example of sequence of test functions $\{v_k\} \subset \mathcal{D}(\mathbb{R})$ that converges to 0 in $\mathcal{S}(\mathbb{R})$, but not in $\mathcal{D}(\mathbb{R})$.

5.3.8. Prove that any polynomial is a tempered distribution, whilst $e^x \notin \mathcal{S}'(\mathbb{R})$.

5.3.9. Show that the following distributions belong to $\mathcal{S}'(\mathbb{R})$:

a) p. v.$(1/x)$.

b) $\mathrm{comb}(x)$.

5.3.10. *a)* Let $B_r \subset \mathbb{R}^3$ be the ball with radius r and centre at the origin, and $T \in \mathcal{D}'(\mathbb{R}^3)$ be defined by

$$\langle T, v \rangle = \int_{\partial B_r} v \, d\sigma$$

(a sort of 'delta' function spread over the spherical surface ∂B_r; it coincides with $\delta(|\mathbf{x}| - r)$). Prove that T is a tempered distribution and compute its Fourier transform.

b) Use part a) to find the fundamental solution to the equation

$$u_{tt} - \Delta u = 0 \qquad \text{in } \mathbb{R}^3.$$

5.3.11. Prove the following result: if there exist m and C such that

$$|c_n| < C \left(1 + |n|^m\right)$$

for any $n \in \mathbb{Z}$, then the distribution

$$F = \sum_{n=-\infty}^{+\infty} c_n \delta(x - n)$$

belongs to $\mathcal{S}'(\mathbb{R})$ [11].

5.3.12. Let comb_T be the distribution

$$\mathrm{comb}_T(x) = \sum_{n \in \mathbb{Z}} \delta(x - nT),$$

known as the Dirac comb.

---

[11] The condition is also necessary: *if F is tempered, there exist m and C such that $|c_n| \leq C\left(1 + |n|^m\right)$, for any integer n.*

a) Prove that comb_T is T-periodic; using Problem 5.2.18 on page 295 (and the remark therein!) compute its Fourier series.

b) Deduce the formulas:

$$\widehat{\mathrm{comb}_T} = \frac{2\pi}{T}\,\mathrm{comb}_{2\pi/T},$$

and (Poisson's formula)

$$\sum_{n\in\mathbb{Z}} \widehat{v}\,(nT) = \frac{2\pi}{T}\sum_{n\in\mathbb{Z}} v\left(\frac{2\pi n}{T}\right)$$

for any $v \in \mathcal{S}'\,(\mathbb{R})$.

5.3.13. *a)* Take $f \in H^1\,(0,\pi)$. Discuss the possibility of expanding f in a series of cosines, or sines, only.

b) Compute the best Poincaré constant C_P, for which

$$\|u\|_{L^2(0,L)} \le C_P \|u'\|_{L^2(0,L)} \qquad \text{for any } u \in V,$$

in the cases $V = H_0^1(0, L)$ and $V = \{u \in H^1(0, L) : u(0) = 0\}$.

5.3.14. Relying on Problem 5.2.6 (page 284) and the previous exercise, prove that the embedding $H^1(0, \pi) \hookrightarrow L^2(0, \pi)$ is compact.

5.3.15. Let $\alpha > -1$ and consider the map

$$u(x, y) = (x^2 + y^2)^{\alpha/2},$$

defined (almost everywhere) on the unit disc in $\mathbb{R}^2$. Prove that even if u is not C^1, its distribution gradient coincides with the usual gradient.

5.3.16. Let B_1 be the unit disc centred at the origin in $\mathbb{R}^2$. Prove that

$$v(x, y) = \log\left(\log\left(1 + \frac{1}{\sqrt{x^2 + y^2}}\right)\right) - \log\,(\log 2)$$

belongs to $H_0^1(B_1)$.

5.3.17. (Heisenberg's Uncertainty Principle) Let $\psi \in H^1(\mathbb{R})$ be such that $x[\psi(x)]^2 \to 0$ as $|x| \to \infty$, and $\int_{\mathbb{R}} [\psi\,(x)]^2\,dx = 1$. Prove that

$$1 \le 4 \int_{\mathbb{R}} x^2\,|\psi\,(x)|^2\,dx \int_{\mathbb{R}} |\psi'\,(x)|^2\,dx.$$

(Interpretation: if ψ is Schrödinger's wavefunction, the first integral above measures the total density of a particle, the second one is the total momentum).

5.3.18. True of false?

a) If $u \in H^1\,(\mathbb{R})$, then $u\,(x) \to 0$ as $x \to \pm\infty$.

b) If $u \in H^1\,(\mathbb{R}^n)$, $n \ge 2$, then $u\,(\mathbf{x}) \to 0$ as $|\mathbf{x}| \to +\infty$.

c) If $\Omega \subset \mathbb{R}^n$ is an unbounded domain, Poincaré's inequality

$$\int_\Omega v^2 \le C_P \int_\Omega |\nabla v|^2 \qquad \forall v \in H_0^1(\Omega)$$

does not hold.

5.3.19. Let Ω be bounded and Lipschitz. Prove that if $a \in L^\infty(\Omega)$, $a \geq 0$ a.e. and $\int_\Omega a(\mathbf{x})\, d\mathbf{x} > 0$, then

$$\|u\|_\star = \left(\int_\Omega (|\nabla u|^2 + a(\mathbf{x})\, u^2)\, d\mathbf{x} \right)^{1/2}$$

is equivalent to $\|u\|_{H^1(\Omega)}$.

5.3.20. Suppose Ω is bounded and Lipschitz, and $f : \partial\Omega \to \mathbb{R}$ is Lipschitz on $\partial\Omega$ with Lipschitz constant L.

a) Prove that

$$\tilde{f}(\mathbf{x}) = \min_{\partial\Omega} \{ f(\mathbf{y}) + L\, |\mathbf{x} - \mathbf{y}| \}$$

is Lipschitz in $\mathbb{R}^n$ and coincides with f on $\partial\Omega$.

b) Deduce $f \in H^{1/2}(\partial\Omega)$.

5.3.21. Solve Problem 5.2.25 on page 302 with $H^1(\mathbb{R})$ replacing $H^1(-1, 1)$.

5.3.22. Let δ be the Dirac delta function at 0. Prove that δ belongs to $H^{-1}(-1, 1)$. Determine the representative element in $H_0^1(-1, 1)$ according to Riesz's theorem.

5.3.23. Prove that the functional

$$Lu = \int_0^1 u'(t)\, dt$$

belongs to the dual of $H^1(0, 1)$. What is the Riesz's element representing it in $H^1(0, 1)$?

5.3.24. *a)* Given f, g regular, consider the functional

$$Lv = \int_0^1 \left(f(t)v'(t) + g(t)v(t) \right) dt.$$

Prove that $L \in H^{-1}(0, 1)$. Letting $u \in H_0^1(0, 1)$ be the element representing it, write a boundary-value problem that determines u uniquely.

b) Compute u explicitly if $f = g = 1$.

5.3.25. Set $H = H_0^1(-1, 1)$ and

$$V = \{u \in H : u(0) = 0\}.$$

Prove that V is closed in H, and compute the projection onto V of f, where $f(t) = 1 - t^2$ on $[-1, 1]$.

5.3.26. (Subspaces of $H_0^1(\Omega)$) Let $\mathbb{R}^n \supset \Omega = \Omega_1 \cup \Omega_2$ be a domain, and take $V = H_0^1(\Omega)$ with inner product $(u, v)_{H_0^1(\Omega)} = \int_\Omega \nabla u \cdot \nabla v$. Define

$$V_1 = \left\{ u \in H_0^1(\Omega_1), u = 0 \text{ in } \Omega\backslash\Omega_1 \right\} \quad \text{and} \quad V_2 = \left\{ u \in H_0^1(\Omega_2), u = 0 \text{ in } \Omega\backslash\Omega_2 \right\}.$$

Prove that V_1 and V_2 are closed subspaces in V, and

$$V_1^\perp \cap V_2^\perp = \{0\}.$$

5.3.1 Solutions

Solution 5.3.1. V is a subspace and is closed, since L^2 convergence implies convergence almost everywhere of one subsequence. Take $u \in V^\perp$. Then for any $v \in V$, we can write

$$0 = \int_0^{2\pi} u(t)v(t)\,dt = \int_0^{2\pi/3} u(t)v(t)\,dt + \int_{2\pi/3}^{4\pi/3} u(t)v(t)\,dt + \int_{4\pi/3}^{2\pi} u(t)v(t)\,dt.$$

Changing variable in the last two integrals and using the periodicity of v, we find

$$0 = \int_0^{2\pi/3} \left[u(t) + u\left(t + \frac{2}{3}\pi\right) + u\left(t + \frac{4}{3}\pi\right) \right] v(t)\,dt.$$

This equation holds for any $v \in L^2(0, 2\pi/3)$, so we deduce that

$$u(t) + u\left(t + \frac{2}{3}\pi\right) + u\left(t + \frac{4}{3}\pi\right) = 0 \quad \text{a.e. in } (0, 2\pi/3).$$

Integrating on $(2\pi/3, 4\pi/3)$ or $(4\pi/3, 2\pi)$, rather than on $(0, 2\pi/3)$, and repeating the above argument, we finally deduce that

$$V^\perp = \left\{ u \in L^2(0, 2\pi) : u(t) + u\left(t + \frac{2}{3}\pi\right) + u\left(t + \frac{4}{3}\pi\right) = 0 \quad \text{a.e. in } (0, 2\pi) \right\}.$$

Hence

$$P_V f(t) = \frac{1}{3}\left[f(t) + f\left(t + \frac{2}{3}\pi\right) + f\left(t + \frac{4}{3}\pi\right) \right]$$

and

$$P_{V^\perp} f(t) = \frac{1}{3}\left[2f(t) - f\left(t + \frac{2}{3}\pi\right) - f\left(t + \frac{4}{3}\pi\right) \right].$$

Solution 5.3.2. We have to minimise the distance in $L^2(0, 1)$ between the cubic function $p(t) = t^3$ and some polynomial of order two to be determined. The space

$$V = \left\{ at^2 + bt + c : a, b, c \in \mathbb{R} \right\}$$

is closed as finite-dimensional. So we have to minimise

$$\int_0^1 \left(t^3 - at^2 - bt - c \right)^2 dt$$

as a, b, c vary[12]. By the orthogonal decomposition theorem, setting $P_V t^3 = At^2 + Bt + C$ and $g(t) = t^3 - P_V t^3$, we have

$$\int_0^1 g(t)(at^2 + bt + c)\,dt = 0 \qquad \text{for any } a, b, c.$$

---

[12] Clearly, one could compute the integral directly and then minimise the resulting function of three variables.

In particular, the above equation holds when *two* of the coefficients are zero and the third one is 1. This gives three equations

$$0 = \int_0^1 g(t)\, dt = \int_0^1 (t^3 - At^2 - Bt - C)\, dt = \frac{1}{4} - \frac{A}{3} - \frac{B}{2} - C$$

$$0 = \int_0^1 g(t)t\, dt = \int_0^1 (t^4 - At^3 - Bt^2 - Ct)\, dt = \frac{1}{5} - \frac{A}{4} - \frac{B}{3} - \frac{C}{2}$$

$$0 = \int_0^1 g(t)t^2\, dt = \int_0^1 (t^5 - At^4 - Bt^3 - Ct^2)\, dt = \frac{1}{6} - \frac{A}{5} - \frac{B}{4} - \frac{C}{3},$$

i.e. the system

$$\begin{cases} 4A + 6B + 12C = 3 \\ 15A + 20B + 30C = 12 \\ 12A + 15B + 20C = 10. \end{cases}$$

With a little work we find $A = 3/2$, $B = -3/5$ and $C = 1/20$, whence

$$P_V f = \frac{3}{2}t^2 - \frac{3}{5}t + \frac{1}{20}.$$

Solution 5.3.3. To prove S is closed we may argue as in Problem 5.2.3 (page 281) for inspiration. We wish to characterise $S^\perp$. If $h \in S^\perp$ then

$$\int_Q h(x, y)(v(x) + w(y) + k)\, dx dy = 0$$

for any v, w and k. In particular, choosing $w \equiv 0$, we deduce that

$$\int_0^1 (v(x) + k)\left[\int_0^1 h(x, y) dy \right] dx = 0.$$

Since every function $u \in L^2(0, 1)$ can be written in the form $u(x) = v(x) + k$ with $v \in V$ and $k = \int_0^1 u(x)\, dx$, we infer that

$$\int_0^1 h(x, y)\, dy = 0 \quad \text{a.e. in } (0, 1).$$

Analgously, we deduce that

$$\int_0^1 h(x, y)\, dx = 0 \quad \text{a.e. in } (0, 1).$$

We conclude that

$$S^\perp = \left\{ h \in L^2(Q) : \int_0^1 h(x, y)\, dx = \int_0^1 h(x, y)\, dy = 0 \ \text{a.e. in } (0, 1) \right\}.$$

Set

$$P_S f(x, y) = v_f(x) + w_f(y) + k_f.$$

Let us determine v_f. For any $g \in L^2(0, 1)$ we have, since $f - P_S f \in S^{\perp}$,

$$0 = \int_Q [f(x, y) - (v_f(x) + w_f(y) + k_f)] g(x) \, dx \, dy =$$

$$= \int_0^1 \left[\int_0^1 f(x, y) \, dy - v_f(x) - k_f \right] g(x) \, dx$$

(recall $\int_0^1 w_f(y) \, dy = 0$), so that

$$\int_0^1 f(x, y) \, dy - v_f(x) - k_f = 0$$

and

$$v_f(x) = \int_0^1 f(x, y) \, dy - k_f$$

almost everywhere. In the same way, for any $g \in L^2(0, 1)$ we have

$$0 = \int_Q [f(x, y) - (v_f(x) + w_f(y) + k_f)] g(y) \, dx \, dy =$$

$$= \int_0^1 \left[\int_0^1 f(x, y) \, dx - w_f(y) - k_f \right] g(y) \, dy,$$

and therefore

$$w_f(y) = \int_0^1 f(x, y) \, dx - k_f.$$

Finally,

$$0 = \int_Q [f(x, y) - (v_f(x) + w_f(y) + k_f)] \, dx \, dy = \int_Q f(x, y) \, dx \, dy - k_f,$$

and

$$k_f = \int_Q f(x, y) \, dx \, dy.$$

To sum up,

$$P_S f(x, y) = \int_0^1 f(x, y) \, dy + \int_0^1 f(x, y) \, dx - \int_Q f(x, y) \, dx \, dy,$$

and $P_{S^{\perp}} f = f - P_S f$. In particular,

$$xy = \underbrace{\frac{2x + 2y - 1}{4}}_{\in S} + \underbrace{\frac{4xy - 2x - 2y + 1}{4}}_{\in S^{\perp}}.$$

Solution 5.3.4. **a)** L is linear, so we have to prove it is bounded. As

$$\|Lu\|_H^2 = \int_{B_1} (Lu)^2(x, y) \, dx \, dy = \int_{B_1} u^2(y, -x) \, dx \, dy = \|u\|_H^2,$$

L is continuous and $\|L\| = 1$.

b) Let $\lambda \in \mathbb{R}$, $v \neq 0$ be such that

$$Lv = \lambda v.$$

Taking norms and using part a) we obtain $|\lambda| = 1$, so the only possible eigenvalues are $\lambda = \pm 1$. Now we have to find non-zero functions v_+ and v_- satisfying

$$v_+(y, -x) = v_+(x, y) \qquad \text{or} \qquad v_-(y, -x) = -v_-(x, y).$$

For instance: $v_+(x, y) = x^2 + y^2$ and $v_-(x, y) = xy$ show that ± 1 do occur as eigenvalues.

Solution 5.3.5.   Take $\{c_k\} \subset \mathbb{R}$. We have to show that for any $v \in \mathcal{D}(\mathbb{R})$, the series

$$\sum_{k=1}^{\infty} c_k \langle \delta(x - x_k), v \rangle \tag{5.34}$$

converges. Notice that the support of v is compact (bounded, in particular) and $x_k \to +\infty$. Hence, if we take a large enough k_0, for any $k > k_0$ the points x_k will not belong to the support of v, and therefore $v(x_k) = 0$ for $k > k_0$. But then the series in practice reduces to a finite number of summands, and its sum is obviously finite.

Solution 5.3.6.   **a)** As f is odd we can expand in sine series

$$f \sim \sum_{n=1}^{\infty} \frac{\sin nx}{n}.$$

The equality is understood in $L_{\text{loc}}^2(\mathbb{R})$, and therefore in $\mathcal{D}'(\mathbb{R})$.
   b) Let us denote by χ_A the characteristic function of the set A:

$$\chi_A(x) = \begin{cases} 1 & \text{if } x \in A \\ 0 & \text{if } x \notin A. \end{cases}$$

Due to the periodicity, f can be written as

$$f(x) = \sum_{n=-\infty}^{\infty} f(x) \chi_{[2n\pi, 2(n+1)\pi)}(x) = \sum_{n=-\infty}^{\infty} f(x - 2n\pi) \chi_{[2n\pi, 2(n+1)\pi)}(x)$$

$$= \sum_{n=-\infty}^{\infty} \frac{\pi - (x - 2n\pi)}{2} \chi_{[2n\pi, 2(n+1)\pi)}(x)$$

and so

$$\sum_{n=1}^{\infty} \frac{\sin nx}{n} = \sum_{n=-\infty}^{\infty} \frac{\pi - (x - 2n\pi)}{2} \chi_{[2n\pi, 2(n+1)\pi)}(x), \tag{5.35}$$

a relation valid in $\mathcal{D}'(\mathbb{R})$. Now, series that converge in $\mathcal{D}'(\mathbb{R})$ can be differentiated term by term, to give another convergent series in $\mathcal{D}'(\mathbb{R})$. Then (5.33) follows from (5.35), by differentiating each term and making use of the relation

$$\frac{d}{dx} g(x) \chi_{[a,b)}(x) = -g(b^-) \delta(x - b) + g(a^+) \delta(x - a) + g'(x) \chi_{[a,b)}(x),$$

valid for any $g \in C^1(a, b)$ with finite limits for $x \to a^+$ and $x \to b^-$ (prove this last fact).

Solution 5.3.7. By comparing the two definitions of convergence one sees that convergence in $\mathcal{S}$ does not require any assumption on $\mathrm{supp}(v_k)$. One example is the following: fix a test function v equal zero outside $[0, 1]$, and set

$$v_k(x) = v(x - k)e^{-x}.$$

For any $k \geq 1$, v_k is a test function that vanishes outside $[k, k + 1]$. In particular, the support of v_k is not contained in any given compact set, and v_k does not converge in $\mathcal{D}(\mathbb{R})$. On the other hand for any $m, p \geq 0$

$$\sup_{\mathbb{R}} |x^m D^p v_k(x)| \to 0 \quad \text{as } k \to +\infty$$

(prove it), so $v_k \to 0$ in $\mathcal{S}(\mathbb{R})$.

Solution 5.3.8. It suffices to show that any monomial that has the form $P(x) = x_1^{\alpha_1} x_2^{\alpha} \cdots x_n^{\alpha_n}$ is a tempered distribution. Let $m \geq 0$ denote its degree and $\{v_k\} \subset \mathcal{D}(\mathbb{R}^n)$ be such that $v_k \to 0$ in $\mathcal{S}(\mathbb{R}^n)$. Notice that

$$|P(x)| = |x_1^{\alpha_1} x_2^{\alpha} \cdots x_n^{\alpha_n}| \leq |x|^m$$

and that the function

$$h(x) = |x|^m (1 + |x|)^{-(m+n+1)}$$

is integrable in $\mathbb{R}^n$. In fact if ω_n denotes the measure of the surface of the unit sphere in $\mathbb{R}^n$,

$$\int_{\mathbb{R}^n} \frac{|x|^m}{(1 + |x|)^{m+n+1}} dx = \omega_n \int_0^{+\infty} \frac{\rho^{m+n-1}}{(1 + \rho)^{m+n+1}} d\rho = M < \infty.$$

By definition of convergence in $\mathcal{S}(\mathbb{R}^n)$,

$$\sup_{\mathbb{R}^n} \left[(1 + |x|)^{m+n+1} |v_k(x)| \right] \to 0 \quad \text{as } k \to +\infty.$$

Therefore

$$|\langle P, v_k \rangle| = \left| \int_{\mathbb{R}^n} P(x) v_k(x) \, dx \right| \leq \int_{\mathbb{R}^n} \frac{|x|^m}{(1 + |x|)^{m+n+1}} (1 + |x|)^{m+n+1} |v_k(x)| \, dx$$

$$\leq M \sup_{\mathbb{R}^n} \left[(1 + |x|)^{m+n+1} |v_k(x)| \right] \to 0,$$

so P defines a tempered distribution.

Now set $u(x) = e^x$ and take $v \in \mathcal{D}(\mathbb{R})$ *non-negative*, equal to e^{-x} on $[-1, 1]$ and zero outside $[-2, 2]$. Consider

$$v_k(x) = v(x - k)e^{-x}.$$

We argue as in Exercise 5.3.7 to find that $v_k \to 0$ in $\mathcal{S}(\mathbb{R})$. Yet

$$\langle u, v_k \rangle = \int_{k-2}^{k+2} v(x - k) \, dx \geq \int_{k-1}^{k+1} dx = 2 \not\to 0.$$

In conclusion, u is not a tempered distribution.

Solution 5.3.9. **a)** Let us show that $F \in \mathcal{S}'(\mathbb{R}^n)$ implies $F' \in \mathcal{S}'(\mathbb{R}^n)$. If $\{v_k\} \subset \mathcal{D}(\mathbb{R}^n)$ and $v_k \to 0$ in $\mathcal{S}(\mathbb{R}^n)$, then also $v_k' \to 0$ in $\mathcal{S}(\mathbb{R}^n)$. Therefore

$$\langle F', v_k \rangle = -\langle F, v_k' \rangle \to 0$$

because F is tempered. By virtue of Problem 5.2.19 on page 296, $u(x) = \log|x|$ is tempered, and hence its derivative is given by p. v.$(1/x)$.

b) The Dirac comb is a tempered distribution. To see it, let $\{v_k\} \subset \mathcal{D}(\mathbb{R})$ be such that $v_k \to 0$ in $\mathcal{S}(\mathbb{R})$. Then

$$\langle \text{comb}, v_k \rangle = \sum_{n=-\infty}^{+\infty} v_k(n).$$

Convergence in $\mathcal{S}(\mathbb{R})$ guarantees that $n^2 v_k(n) \to 0$ uniformly in n when $k \to \infty$. Given $\varepsilon > 0$ we can then write

$$n^2 |v_k(n)| < \varepsilon$$

for $k \geq k_0$ large enough. Therefore when $k \geq k_0$

$$|\langle \text{comb}, v_k \rangle| \leq \varepsilon \sum_{n=-\infty}^{+\infty} \frac{1}{n^2} \leq C\varepsilon$$

and so $\langle \text{comb}, v_k \rangle \to 0$.

Solution 5.3.10. **a)** T is tempered because its support ∂B_r is compact. By definition, given any $v \in \mathcal{S}(\mathbb{R}^3)$,

$$\langle \widehat{T}, v \rangle = \langle T, \widehat{v} \rangle = \int_{\partial B_r} \widehat{v}\, d\sigma = \int_{\partial B_r} d\sigma \int_{\mathbb{R}^3} e^{-i x \cdot \sigma} v(x)\, dx =$$

$$= \int_{\mathbb{R}^3} v(x) \left(\int_{\partial B_r} e^{-i x \cdot \sigma}\, d\sigma \right) dx.$$

Hence

$$\widehat{T}(\xi) = \int_{\partial B_r} e^{-i \xi \cdot \sigma}\, d\sigma.$$

To deal with the integral we pass to spherical coordinates (r, θ, ψ) with vertical axis coinciding with ξ, $0 < \theta < 2\pi$, $0 < \psi < \pi$. Set $|\xi| = \rho$ and observe that on ∂B_r

$$\xi \cdot \sigma = r\rho \cos\psi$$

and $d\sigma = r^2 \sin\psi\, d\psi$, so

$$\int_{\partial B_r} e^{-i \xi \cdot \sigma}\, d\sigma = 2\pi r^2 \int_0^\pi e^{-ir\rho \cos\psi} \sin\psi\, d\psi = 4\pi r \frac{r}{\rho} \sin r\rho.$$

In conclusion

$$\widehat{T}(\xi) = 4\pi r \frac{\sin r |\xi|}{|\xi|}.$$

b) The fundamental solution of the wave equation in dimension $n = 3$ is the solution of the problem

$$\begin{cases} u_{tt} - \Delta u = 0 & \text{in } \mathbb{R}^3 \times (0, +\infty) \\ u(x, 0) = 0, \ u_t(x, 0) = \delta(x - y) & \text{in } \mathbb{R}^3. \end{cases}$$

Since the equation is translation-invariant we can assume $y = 0$. We set

$$\hat{u}(\xi, t) = \int_{\mathbb{R}^3} e^{-i \xi \cdot x} u(x, t)\, dx,$$

the Fourier transform of u in the spatial variables. We find that $\hat{u}$ solves

$$\begin{cases} \hat{u}_{tt} + |\boldsymbol{\xi}|^2 \hat{u} = 0 & t > 0 \\ \hat{u}(\boldsymbol{\xi}, 0) = 0, \ \hat{u}_t(\boldsymbol{\xi}, 0) = 1, \end{cases}$$

for any $\boldsymbol{\xi} \in \mathbb{R}^3$. The general solution is $\hat{u}(\boldsymbol{\xi}, t) = C_1 e^{i|\boldsymbol{\xi}|t} + C_2 e^{-i|\boldsymbol{\xi}|t}$. The initial condition impose

$$C_1 + C_2 = 0 \quad \text{and} \quad C_1 i|\boldsymbol{\xi}| - C_2 i|\boldsymbol{\xi}| = 1,$$

so

$$\hat{u}(\boldsymbol{\xi}, t) = \frac{1}{2i|\boldsymbol{\xi}|} \left(e^{i|\boldsymbol{\xi}|t} - e^{-i|\boldsymbol{\xi}|t} \right) = \frac{\sin(|\boldsymbol{\xi}|t)}{|\boldsymbol{\xi}|}.$$

At this stage we can use part a) (with $r = t$) and get

$$u(\mathbf{x}, t) = \frac{\delta(|\mathbf{x}| - t)}{4\pi t}$$

so, in the end,

$$K(\mathbf{x}, \mathbf{y}, t) = \frac{\delta(|\mathbf{x} - \mathbf{y}| - t)}{4\pi t}.$$

Solution 5.3.11. Consider

$$F = \sum c_n \delta(x - n), \qquad \text{where } c_n < (1 + |n|^m),$$

and a sequence $\{v_k\} \subset \mathcal{D}(\mathbb{R})$ such that $v_k \to 0$ in $\mathcal{S}(\mathbb{R})$. Then

$$\langle F, v_k \rangle = \sum_{n=-\infty}^{+\infty} c_n v_k(n).$$

By definition of convergence in $\mathcal{S}(\mathbb{R})$ we have

$$n^{m+2} v_k(n) \to 0$$

uniformly in n, as $k \to \infty$. Given $\varepsilon > 0$, we write

$$n^{m+2} |v_k(n)| < \varepsilon$$

for some large $k \geq k_0$. When $k \geq k_0$, then,

$$|\langle F, v_k \rangle| \leq \varepsilon \sum_{n=-\infty}^{+\infty} \frac{(1 + n^m)}{n^{m+2}} \leq C\varepsilon$$

and $\langle F, v_k \rangle \to 0$.

Solution 5.3.12. **a)** Let us check comb_T is T-periodic. Take a test function v and compute:

$$\langle \mathrm{comb}_T(x + T), v \rangle = \langle \mathrm{comb}_T, v(x - T) \rangle = \sum_{n \in \mathbb{Z}} v(nT - T) = \sum_{n \in \mathbb{Z}} v(nT) = \langle \mathrm{comb}_T, v \rangle.$$

By Problem 5.2.18 on page 295, the Fourier series of $u = \text{comb}_T$ is given in $\mathcal{D}'(\mathbb{R})$ by the formula:

$$\text{comb}_T(x) = \sum_{n \in \mathbb{Z}} \widehat{u}_n \exp\left(-i\frac{2n\pi x}{T}\right).$$

To compute $\widehat{u}_n$ we use the remark on page 296: the point $x_0 = -T/2$ does not belong to the support of comb_T, and the restriction of comb_T to

$$(x_0, x_0 + T) = (-T/2, T/2)$$

coincides with $\delta(x)$. By (5.19), with $u_1 = 0$,

$$\widehat{u}_n = \frac{1}{T}\langle \delta, e^{-i2n\pi x/T}\rangle = \frac{1}{T}$$

whence the Fourier series reads

$$\text{comb}_T(x) = \frac{1}{T}\sum_{n \in \mathbb{Z}} \exp\left(-i\frac{2n\pi x}{T}\right).$$

b) From (5.18), since $\widehat{u}_n = c_n/2\pi$, we have

$$\widehat{\text{comb}_T}(\xi) = \frac{2\pi}{T}\sum_{n \in \mathbb{Z}} \delta\left(\xi - \frac{2n\pi}{T}\right) = \frac{2\pi}{T}\,\text{comb}_{2\pi/T}(\xi).$$

If now $v \in \mathcal{S}'(\mathbb{R})$:

$$\langle \widehat{\text{comb}_T}, v\rangle = \langle \text{comb}_T, \widehat{v}\rangle = \sum_{n \in \mathbb{Z}} \widehat{v}(nT)$$

while

$$\frac{2\pi}{T}\langle \text{comb}_{2\pi/T}, v\rangle = \frac{2\pi}{T}\sum_{n \in \mathbb{Z}} v\left(\frac{2\pi n}{T}\right),$$

giving the Poisson formula.

Solution 5.3.13. As $f' \in L^2(0, \pi)$, by considering the even and odd extensions to $(-\pi, \pi)$ we can write, in the two cases

$$f'(x) = \sum_{n=0}^{\infty} a_n \cos nx \quad \text{and} \quad f'(x) = \sum_{n=1}^{\infty} b_n \sin nx$$

respectively, where convergence is in $L^2(0, \pi)$. Note[13]

$$a_0 = \frac{1}{\pi}\int_0^{\pi} f' = \frac{f(\pi) - f(0)}{\pi}.$$

Since Fourier series can be integrated term by term, in the first case we have:

$$f(x) = f(0) + \sum_{n=0}^{\infty}\int_0^x a_n \cos nx \, dx = f(0) + \frac{f(\pi) - f(0)}{\pi}x + \sum_{n=1}^{\infty} A_n \sin nx$$

---

[13] The functions of $H^1(a, b)$ are *absolutely continuous*, so the Fundamental Theorem of Calculus holds.

with $A_n = a_n/n$. Assuming

$$f(\pi) = f(0) = 0,$$

it follows

$$f(x) = \sum_{n=1}^{\infty} A_n \sin nx$$

with the series *converging uniformly on the entire* $\mathbb{R}$.

In the other case, term-by-term integration shows that

$$f(x) = f(0) + \beta + \sum_{n=1}^{\infty} B_n \cos nx$$

where

$$B_n = -\frac{b_n}{n} \quad \text{and} \quad \beta = \sum_{n=1}^{\infty} \frac{b_n}{n}.$$

Each function in $H^1(0, \pi)$ can therefore be expanded in cosines, and the series *converges uniformly on the whole* $\mathbb{R}$.

b) Take $u \in H_0^1(0, L)$. Repeating the steps of part a) for $2L$-periodic functions we obtain

$$u = \sum_{n=1}^{\infty} A_n \sin\left(\frac{n\pi}{L}x\right), \qquad u' = \sum_{n=1}^{\infty} a_n \cos\left(\frac{n\pi}{L}x\right),$$

with $a_n = \dfrac{n\pi}{L} A_n$. Parseval's identity gives

$$\|u\|_{L^2(0,L)}^2 = \sum_{n=1}^{\infty} A_n^2 \quad \text{and} \quad \|u'\|_{L^2(0,L)}^2 = \sum_{n=1}^{\infty} a_n^2 = \sum_{n=1}^{\infty} \left(\frac{n\pi}{L}\right)^2 A_n^2 \geq \left(\frac{\pi}{L}\right)^2 \sum_{n=1}^{\infty} A_n^2.$$

Hence

$$\|u\|_{L^2(0,L)} \leq \frac{L}{\pi} \|u'\|_{L^2(0,L)} \qquad \text{for any } u \in H_0^1(0, L).$$

Note that if $A_1 = 1$ and $A_n = 0$ when $n \geq 2$, that is $u = \sin\frac{\pi}{L}x$, the above inequalities become equalities, so the constant $C_P = L/\pi$ is optimal.

Let now

$$V = \{u \in H^1(0, L) : u(0) = 0\}.$$

It is easy to see[14] that

$$\tilde{u}(x) = \begin{cases} u(x) & 0 \leq x \leq L \\ u(L - x) & L \leq x \leq 2L \end{cases}$$

belongs to $H_0^1(0, 2L)$, and that

$$\|\tilde{u}\|_{L^2(0,L)} = \sqrt{2}\|u\|_{L^2(0,L)}, \quad \|\tilde{u}'\|_{L^2(0,2L)} = \sqrt{2}\|u'\|_{L^2(0,2L)}.$$

Previous arguments show

$$\|\tilde{u}\|_{L^2(0,2L)} \leq \frac{2L}{\pi} \|\tilde{u}'\|_{L^2(0,2L)},$$

---

[14] This fact is left to the reader.

so that

$$\|u\|_{L^2(0,L)} \le \frac{2L}{\pi} \|u'\|_{L^2(0,L)} \qquad \text{for any } u \in V.$$

Solution 5.3.14. We have to prove that the unit ball in $H^1(0, \pi)$,

$$B = \left\{ u \in H^1(0, \pi) : \|u'\|_2^2 + \|u\|_2^2 \le 1 \right\},$$

is relatively compact in $L^2(0, \pi)$. To do that we recall, from Exercise 5.3.13, that we can expand u in cosine Fourier series:

$$u = \sum_{n=0}^{+\infty} B_0 \cos nt$$

where

$$B_0 = \frac{1}{\pi} \int_0^\pi u(t) \cos nt \, dt \quad \text{and} \quad B_n = \frac{2}{\pi} \int_0^\pi u(t) \cos nt \, dt,$$

and the series *converges uniformly* in $\mathbb{R}$. Parseval's identity gives

$$\|u\|_2^2 = 2B_0^2 + \sum_{n=0}^{+\infty} B_n^2.$$

The map that associates to a function u the sequence of its Fourier coefficients B_n is an isomorphism between the vector spaces $L^2(0, 2\pi)$ and l^2; it is also an isometry of the underlying metric structures. Therefore the claim follows provided we prove that B can be identified with a subspace of l^2, in analogy to the space K of Problem 5.2.6 (page 284).

From Exercise 5.3.13 we know that $B_n = -b_n/n$, where the b_n, $n \ge 1$, are the Fourier coefficients of the *sine* series of u'. What is more, Parseval's identity gives

$$\|u'\|_2^2 = \sum_{n=1}^{+\infty} n^2 B_n^2.$$

In terms of Fourier series, then,

$$\|u'\|_2^2 + \|u\|_2^2 \le 1 \quad \Longleftrightarrow \quad 2B_0^2 + \sum_{n=1}^{+\infty} B_n^2 + \sum_{n=1}^{+\infty} n^2 B_n^2 \le 1.$$

The ball $B \subset H^1(0, \pi)$ can be identified with the subset of l^2

$$K = \left\{ (B_n) : \sum_{n=0}^{+\infty} B_n^2 + \sum_{n=1}^{+\infty} n^2 B_n^2 \le 1 \right\}.$$

(the factor 2 in front of B_0 is irrelevant). By Problem 5.2.6 the proof is complete.

Solution 5.3.15. As $\alpha > -1$ we have already seen (Problem 5.2.21 on page 298) that $u \in L^2(B_1)$, so $u \in \mathcal{D}'(B_1)$. Let us call v the partial derivative of u with respect to x in distributional sense, and keep the notation u_x for the ordinary derivative (defined outside the origin). We compute v. Taking

$\varphi \in \mathcal{D}(B_1)$, we have

$$\langle v, \varphi \rangle = -\langle u, \varphi_x \rangle = -\int_{B_1} u(x, y)\varphi_x(x, y) \, dxdy = -\int_{B_\varepsilon} u\varphi_x \, dxdy - \int_{B_1 \setminus B_\varepsilon} u\varphi_x \, dxdy =$$

$$= -\int_{B_\varepsilon} u\varphi_x \, dxdy - \int_{\partial B_\varepsilon} u\varphi\nu_x \, ds + \int_{B_1 \setminus B_\varepsilon} u_x\varphi \, dxdy,$$

where ν_x denotes the h-component of the unit normal to ∂B_ε, pointing away from $B_1 \setminus B_\varepsilon$ (recall $\varphi = 0$ on ∂B_1). Now,

$$\left| \int_{B_\varepsilon} u\varphi_x \, dxdy \right| \le C \int_{B_\varepsilon} |u| \, dxdy = C' \int_0^\varepsilon r^{\alpha+1} \, dr = C'' \varepsilon^{\alpha+2} \to 0,$$

and

$$\left| \int_{\partial B_\varepsilon} u\varphi\nu_x \, ds \right| \le C \int_{\partial B_\varepsilon} |u| \, dxdy = C' \int_0^{2\pi} \varepsilon^\alpha \, d\theta = C'' \varepsilon^{\alpha+1} \to 0$$

as $\varepsilon \to 0$, since $\alpha > -1$. Overall,

$$\langle v, \varphi \rangle = \int_{B_1} u_x\varphi \, dxdy.$$

The same argument for u_y will then lead to the claim.

Solution 5.3.16.   **a)** First remember that

$$\int_2^\infty t^a \log^b t \, dt$$

is finite if and only if either $a < -1$, or $a = 1$ and $b < -1$.

Exactly as in the previous exercise one sees that the distributional derivatives of v coincide with the classical ones, so that

$$\nabla v(x, y) = \left(\log\left(1 + \frac{1}{\sqrt{x^2 + y^2}}\right) \right)^{-1} \cdot \left(1 + \frac{1}{\sqrt{x^2 + y^2}}\right)^{-1} \cdot \frac{(2x, 2y)}{(x^2 + y^2)^{3/2}}.$$

In polar coordinates

$$\int_{B_1} v^2(x, y) \, dxdy = 2\pi \int_0^1 \left[\log\log(1 + r^{-1}) \right]^2 r \, dr < +\infty$$

(the integrand is finite on a neighbourhood of the origin) and

$$\int_{B_1} |\nabla v(x, y)|^2 \, dxdy = 8\pi \int_0^1 \frac{1}{\log^2(1 + r^{-1})} \cdot \frac{1}{r(1 + r)^2} dr.$$

Setting $t = 1 + r^{-1}$, we find:

$$\frac{1}{8\pi} \int_{B_1} |\nabla v(x, y)|^2 \, dxdy = \int_2^\infty \frac{1}{\log^2 t} \cdot \frac{t - 1}{t^2} dt < \infty.$$

But $v(x, y) = 0$ on ∂B_1, so $v \in H_0^1(B_1)$.

Solution 5.3.17. We can write

$$0 = x\psi^2(x)\Big|_{-\infty}^{+\infty} = \int_{\mathbb{R}} \frac{d}{dx}[x\psi^2(x)]\,dx = \int_{\mathbb{R}} |\psi(x)|^2\,dx + 2\int_{\mathbb{R}} x\psi(x)\psi'(x)\,dx,$$

from which

$$1 = -2\int_{\mathbb{R}} x\psi(x)\psi'(x)\,dx \le 2\left(\int_{\mathbb{R}} x^2|\psi(x)|^2\,dx\right)^{1/2} \cdot \left(\int_{\mathbb{R}} |\psi'(x)|^2\,dx\right)^{1/2}.$$

Squaring the latter gives the required inequality.

Solution 5.3.18. **a)** True. First, we prove that the limit exists. In fact

$$u^2(x) - u^2(0) = 2\int_0^x u(s)\,u'(s)\,ds,$$

and the limit exists since $u \cdot u' \in L^1(\mathbb{R})$. Then

$$\lim_{x\to+\infty} u^2(x) = l \ge 0.$$

We claim that $l = 0$. If, on the contrary, l were positive, given $\varepsilon > 0$ such that $l - \varepsilon > 0$ we would have

$$u^2(x) > l - \varepsilon$$

for $x > N = N(\varepsilon)$, and then

$$\int_{\mathbb{R}} u^2(x)\,dx \ge \int_N^{+\infty} u^2(x)\,dx > \int_N^{+\infty} (l-\varepsilon)dx = +\infty,$$

contradicting the fact that $u \in L^2(\mathbb{R})$.

b) False, in general, for any $n > 1$. For instance, let $n = 3$. In the ball $B_R(x_0)$ consider the radial function

$$u_R(x) = \begin{cases} 1 & \text{if } |x - x_0| < \dfrac{R}{2} \\[2mm] \dfrac{R}{|x - x_0|} - 1 & \text{if } \dfrac{R}{2} \le |x - x_0| \le R. \end{cases}$$

Notice that $u_R(x) = 0$ if $|x - x_0| = R$, while $u_R(x) = 1$ for $|x - x_0| = R/2$. Therefore u_R is continuous in $B_R(x_0)$ and vanishes on the boundary. Moreover $0 \le u_R \le 1$, and for $R/2 \le |x - x_0| \le R$,

$$|\nabla u_R(x)| = \frac{R}{|x - x_0|^2} \le \frac{2}{R}.$$

Now we will construct a function $u \in H^1(\mathbb{R}^3)$ without limit for $|x| \to +\infty$. Let us choose, for any integer $k \ge 1$,

$$x_k = (k, 0, 0) \quad \text{and} \quad R_k = \frac{1}{k^2}$$

and define

$$u(x) = \begin{cases} u_{R_k}(x) & \text{in } B_{R_k}(x_k),\, k \ge 1 \\ 0 & \text{if } x \notin \cup_{k\ge 1} B_{R_k}(x_k) \end{cases}$$

where

$$u_{R_k}(\mathbf{x}) = \begin{cases} 1 & \text{if } |\mathbf{x} - \mathbf{x}_k| < \dfrac{R_k}{2} \\ \dfrac{R_k}{|\mathbf{x} - \mathbf{x}_k|} - 1 & \text{if } \dfrac{R_k}{2} \le |\mathbf{x} - \mathbf{x}_k| \le R_k. \end{cases}$$

As $0 \le u_{R_k} \le 1$,

$$\int_{\mathbb{R}^3} u^2 = \sum_{k \ge 1} \int_{B_{R_k}(\mathbf{x}_k)} u_{R_k}^2 < \frac{4\pi}{3} \sum_{k \ge 1} R_k^3 = \frac{4\pi}{3} \sum_{k \ge 1} \frac{1}{k^6} < \infty.$$

Additionally,

$$|\nabla u_{R_k}| \le \frac{2}{R_k}, \qquad \int_{\mathbb{R}^3} |\nabla u|^2 = \sum_{k \ge 1} \int_{B_{R_k}(\mathbf{x}_k)} |\nabla u_{R_k}|^2 < \frac{8\pi}{3} \sum_{k \ge 1} R_k = 2 \sum_{k \ge 1} \frac{1}{k^2} < \infty.$$

Hence u lives in $H^1(\mathbb{R}^3)$ and is bounded; yet $|\mathbf{x}_k| = k \to +\infty$ and $u(\mathbf{x}_k) = 1 \nrightarrow 0$, while if $\mathbf{y}_k = (0, 0, k)$ then $|\mathbf{y}_k| = k \to +\infty$ and $u(\mathbf{y}_k) = 0$. The limit of $u(\mathbf{x})$ when $|\mathbf{x}| \to \infty$ does not exist.

c) False. To have an inequality of Poincaré type it is enough that the domain is bounded in one direction only. When $n > 1$, for example, we may consider the strip

$$\Omega = \left\{ (\mathbf{x}', x_n) \in \mathbb{R}^n : \mathbf{x}' \in \mathbb{R}^{n-1}, \, 0 < x_n < d \right\}.$$

Take $v \in \mathcal{D}(\Omega)$. As $v(\mathbf{x}', 0) = 0$,

$$v^2(\mathbf{x}', x_n) = \left(\int_0^{x_n} v_{x_n}(\mathbf{x}', s)\, ds \right)^2 \le x_n \int_0^{x_n} v_{x_n}^2(\mathbf{x}', s)\, ds \le x_n \int_0^d v_{x_n}^2(\mathbf{x}', s)\, ds$$

(using the Cauchy-Schwarz inequality). Integrating in x_n over $(0, d)$ and in $\mathbf{x}'$ over $\mathbb{R}^{n-1}$ gives

$$\int_\Omega v^2 \le \frac{d^2}{2} \int_\Omega v_{x_n}^2 \le \frac{d^2}{2} \int_\Omega |\nabla v|^2.$$

A density argument allows to extend the inequality to $v \in H_0^1(\Omega)$.

Solution 5.3.19. We have to prove there are positive constants $C_1 < C_2$ such that

$$C_1 \|u\|_{H^1(\Omega)} \le \|u\|_\star \le C_2 \|u\|_{H^1(\Omega)}. \tag{5.36}$$

On the one hand

$$\int_\Omega (|\nabla u|^2 + a(\mathbf{x}) u^2)\, d\mathbf{x} \le \int_\Omega |\nabla u|^2\, d\mathbf{x} + \|a\|_{L^\infty(\Omega)} \int_\Omega u^2\, d\mathbf{x},$$

and the second inequality in (5.36) holds with $C_2 = \sqrt{\max(1, \|a\|_{L^\infty(\Omega)})}$.

To prove the other inequality, it is enough to find a constant C such that

$$\int_\Omega u^2\, d\mathbf{x} \le C \int_\Omega (|\nabla u|^2 + a(\mathbf{x}) u^2)\, d\mathbf{x}.$$

Suppose not. Then for every $n \in \mathbb{N}$ there exists $u_n \in H^1(\Omega)$, $u_n \neq 0$, such that

$$\frac{1}{n} \int_\Omega u_n^2 \, d\mathbf{x} \geq \int_\Omega (|\nabla u_n|^2 + a(\mathbf{x}) u_n^2) \, d\mathbf{x}.$$

We may suppose $\|u_n\|_{L^2(\Omega)} = 1$. Therefore, since $a \geq 0$,

$$\int_\Omega |\nabla u_n|^2 \to 0, \qquad \int_\Omega a(\mathbf{x}) u_n^2 \, d\mathbf{x} \to 0, \qquad \int_\Omega u_n^2 \, d\mathbf{x} \to 1. \qquad (5.37)$$

On the other hand, u_n is bounded in $H^1(\Omega)$, thus $u_n \rightharpoonup \bar{u}$ in $H^1(\Omega)$ and, by Rellich's theorem, $u_n \to \bar{u}$ in $L^2(\Omega)$ strongly (up to extracting a subsequence). Using the weak semicontinuity of the norm, we infer

$$\int_\Omega |\nabla \bar{u}|^2 \leq \liminf \int_\Omega |\nabla u_n|^2 = 0, \qquad \int_\Omega \bar{u}^2 = \lim \int_\Omega u_n^2 = 1,$$

and $\bar{u}$ is a non-zero constant. We claim that

$$\int_\Omega a(\mathbf{x}) u_n^2 \, d\mathbf{x} \to \int_\Omega a(\mathbf{x}) \bar{u}^2 \, d\mathbf{x} = \bar{u}^2 \int_\Omega a(\mathbf{x}) \, d\mathbf{x} > 0.$$

This would contradict (5.37), hence concluding the proof. The claim follows from the Cauchy-Schwarz inequality, which implies

$$\left| \int_\Omega a(\mathbf{x}) u_n^2 \, d\mathbf{x} - \int_\Omega a(\mathbf{x}) \bar{u}^2 \, d\mathbf{x} \right| \leq \int_\Omega a(\mathbf{x}) \left| u_n^2 - \bar{u}^2 \right| d\mathbf{x}$$

$$\leq \|a\|_{L^\infty(\Omega)} \|u_n + \bar{u}\|_{L^2(\Omega)} \|u_n - \bar{u}\|_{L^2(\Omega)}$$

$$\leq 2\|a\|_{L^\infty(\Omega)} \|u_n - \bar{u}\|_{L^2(\Omega)} \to 0,$$

since $u_n \to \bar{u}$ in $L^2(\Omega)$.

Solution 5.3.20.   a) Let $\mathbf{x}_1, \mathbf{x}_2 \in \mathbb{R}^n$, and $\mathbf{y}_2 \in \partial\Omega$ be such that

$$\tilde{f}(\mathbf{x}_2) = f(\mathbf{y}_2) + L|\mathbf{x}_2 - \mathbf{y}_2|.$$

By definition, $\tilde{f}(\mathbf{x}_1) \leq f(\mathbf{y}_2) + L|\mathbf{x}_1 - \mathbf{y}_2|$, therefore

$$\tilde{f}(\mathbf{x}_1) - \tilde{f}(\mathbf{x}_2) \leq L(|\mathbf{x}_1 - \mathbf{y}_2| - |\mathbf{x}_2 - \mathbf{y}_2|) \leq L|\mathbf{x}_1 - \mathbf{x}_2|;$$

changing the role of $\mathbf{x}_1$ and $\mathbf{x}_2$ one obtains that $\tilde{f}$ is Lipschitz on $\mathbb{R}^n$.
   Now take $\mathbf{x} \in \partial\Omega$. As f is Lipschitz, for $\mathbf{y} \in \partial\Omega$

$$f(\mathbf{x}) \leq f(\mathbf{y}) + L|\mathbf{x} - \mathbf{y}|,$$

and so $f(\mathbf{x}) \leq \tilde{f}(\mathbf{x})$. At the same time, choosing $\mathbf{y} = \mathbf{x}$ in $\tilde{f}$ gives $\tilde{f}(\mathbf{x}) \leq f(\mathbf{x})$ right away. Hence $f \equiv \tilde{f}|_{\partial\Omega}$.
   b) $\tilde{f}$ is Lipschitz, so $\tilde{f}|_\Omega \in H^1(\Omega)$. But then its trace on $\partial\Omega$, i.e. f, belongs to $H^{1/2}(\Omega)$.

Solution 5.3.21.   If we follow Problem 5.2.25 (page 302) we must seek $u \in H^1(\mathbb{R})$ such that

$$\int_\mathbb{R} [u'(t)v'(t) + u(t)v(t)] \, dt = v(0) \qquad \text{for any } v \in H^1(\mathbb{R}).$$

Supposing u is regular enough to integrate by parts on $(-\infty, 0)$ and $(0, \infty)$, we may exploit Exercise 5.3.18, a), to rewrite the above expression as

$$-\int_{\mathbb{R}} \left[u''(t) - u(t)\right] v(t)\, dt + v(0) \left[u'(0^-) - u'(0^+)\right] = v(0).$$

Let us choose first v zero at 0, and then an arbitrary v; we find that u must solve

$$\begin{cases} u''(t) - u'(t) = 0 & t \neq 0 \\ \lim_{|x| \to \infty} u(x) = 0, & u'(0^-) - u'(0^+) = 1 \end{cases}$$

(we have used Exercise 5.3.18, a) again). The solution has the form

$$u(t) = \begin{cases} A_1 e^t + A_2 e^{-t} & t \leq 0 \\ B_1 e^t + B_2 e^{-t} & t \geq 0, \end{cases}$$

with

$$\begin{cases} A_2 = B_1 = 0 \\ A_1 - A_2 - B_1 + B_2 = 1 \\ A_1 + A_2 = B_1 + B_2. \end{cases}$$

Hence the candidate for representing δ in $H^1(\mathbb{R})$ is

$$u(t) = \begin{cases} \dfrac{1}{2} e^{-t} & t \leq 0 \\ \dfrac{1}{2} e^t & t \geq 0. \end{cases}$$

The reader should check that the above function is indeed the function required.

Solution 5.3.22. The fact that δ is a linear, continuous functional on $H_0^1(-1, 1)$ follows from arguments akin to those of Problem 5.2.25 on page 302. By Riesz's theorem there exists $u \in H_0^1(-1, 1)$ such that

$$\int_{-1}^{1} u'(t) v'(t)\, dt = v(0) \qquad \text{for any } v \in H_0^1(-1, 1).$$

To be able to integrate by parts we suppose u is regular on $[-1, 0]$ and $[0, 1]$. Then

$$\int_{-1}^{1} u'(t) v'(t)\, dt = \int_{-1}^{0} u'(t) v'(t)\, dt + \int_{0}^{1} u'(t) v'(t)\, dt =$$

$$= (u'(0^-) - u'(0^+)) v(0) - \int_{-1}^{0} u''(t) v(t)\, dt - \int_{0}^{1} u''(t) v(t)\, dt.$$

As v is arbitrary, we have for u the following:

$$\begin{cases} u''(t) = 0 & \text{when } -1 < t < 0 \text{ and } 0 < t < 1, \\ u(-1) = u(1) = 0, & u'(0^-) - u'(0^+) = 1. \end{cases}$$

Then

$$u(t) = \begin{cases} A_1 + A_2 t & -1 \leq t \leq 0 \\ B_1 + B_2 t & 0 \leq t \leq 1. \end{cases}$$

The boundary conditions together with the continuity at 0 force

$$\begin{cases} A_1 - A_2 = 0 \\ B_1 + B_2 = 0 \\ A_2 - B_2 = 1 \\ A_1 = B_1, \end{cases}$$

so $A_1 = A_2 = B_1 = 1/2$, $B_2 = -1/2$ and ultimately

$$u(t) = \frac{1}{2}(1 - |t|).$$

Solution 5.3.23. We are in the same situation of Problem 5.2.27 (page 304), with $f \equiv 1$ and $g \equiv 0$. The function u that should represent L will solve

$$\begin{cases} u''(t) - u(t) = 0 & 0 < x < 1 \\ u'(0) = u'(1) = 1, \end{cases} \qquad \text{that is,} \qquad u(t) = \frac{1}{1+e}\left(e^t + e^{1-t}\right).$$

Solution 5.3.24. a) As in Problem 5.2.27 on page 304, it is easy to prove L is linear and continuous. We seek $u \in H_0^1(0,1)$ so that

$$\int_0^1 u'(t)v'(t)\, dt = \int_0^1 \left(f(t)v'(t) + g(t)v(t)\right) dt, \quad \text{for any } v \in H_0^1(0,1).$$

Assume u regular enough. Then

$$\int_0^1 u'(t)v'(t)\, dt = u'(1)v(1) - u'(0)v(0) - \int_0^1 u''(t)v(t)\, dt = -\int_0^1 u''(t)v(t)\, dt$$

and

$$\int_0^1 f(t)v'(t)\, dt = f(1)v(1) - f(0)v(0) - \int_0^1 f'(t)v(t)\, dt = -\int_0^1 f'(t)v(t)\, dt.$$

Therefore

$$\int_0^1 \left(-u''(t) + f'(t) - g(t)\right) v(t)\, dt = 0 \qquad \text{for any } v \in H_0^1(0,1),$$

which implies that the Riesz's element u is the solution of the problem

$$\begin{cases} u''(t) = f'(t) - g(t) \\ u(0) = u(1) = 0. \end{cases}$$

b) The problem becomes

$$\begin{cases} u''(t) = -1 \\ u(0) = u(1) = 0, \end{cases} \qquad \text{with solution} \qquad u(t) = \frac{1}{2}(t - t^2).$$

Solution 5.3.25. According to Problem 5.2.28 on page 306 the space V is closed in H. To find $P_V f$ we must minimise over V the quadratic functional

$$E(u) = \|f - u\|_H^2 = \int_{-1}^1 (2t - u'(t))^2\, dt.$$

The minimum $u = P_V f$ must satisfy the necessary condition

$$(f - u, v)_{H_0^1(-1,1)} = \int_{-1}^1 (2t - u'(t))v'(t)\, dt = 0 \qquad \text{for any } v \in H_0^1(-1, 1).$$

Assume we can integrate by parts:

$$\int_{-1}^1 (2t - u'(t))v'(t)\, dt = -\int_{-1}^1 (2 - u''(t))v(t)\, dt = 0$$

for any $v \in H$. The candidate projection u solves

$$\begin{cases} u''(t) = 2 & \text{for } -1 < t < 0,\, 0 < t < 1 \\ u(-1) = u(1) = 0, & u(0) = 0. \end{cases}$$

Therefore

$$u(t) = \begin{cases} t^2 + A_1 t + A_2 & -1 \le t \le 0 \\ t^2 + B_1 t + B_2 & 0 \le t \le 1, \end{cases}$$

where

$$\begin{cases} 1 - A_1 + A_2 = 0 \\ 1 + B_1 + B_2 = 0 \\ A_2 = B_2 = 0. \end{cases}$$

A direct computation yields

$$u(t) = t^2 - |t|.$$

Solution 5.3.26. If $u \in H_0^1(\Omega_1)$, the zero extension of u on $\Omega \setminus \Omega_1$ belongs to $H_0^1(\Omega)$ (Problem 5.2.30 on page 309), making V_1 a subspace of V. To check its closure take $v_n \in V_1$ converging to $v_0 \in V$. In particular, $\{v_n\}$ is a Cauchy sequence in $H_0^1(\Omega_1)$ and hence $v_n \to v_0$ in $H_0^1(\Omega_1)$. Let $\bar{v}_0$ be the zero extension of v_0 on $\Omega \setminus \Omega_1$. Then $\bar{v}_0 \in V_1$ and

$$v_n \to \bar{v}_0$$

in V. Consequently $v_0 = \bar{v}_0 \in V_1$, so that V_1 is closed. The same reasoning goes through for V_2.

Proving $V_1^\perp \cap V_2^\perp = \{0\}$ amounts to showing[15] that $V_1 + V_2$ is dense in V, which in turn is a consequence of

$$\mathcal{D}(\Omega_1) + \mathcal{D}(\Omega_2) = \mathcal{D}(\Omega).$$

So we shall prove the latter. Take any $v \in \mathcal{D}(\Omega)$; we aim to construct $v_1 \in \mathcal{D}(\Omega_1)$ and $v_2 \in \mathcal{D}(\Omega_2)$ such that $v = v_1 + v_2$. Call K the support of v. Then

$$\text{dist}(K, \partial\Omega) = d > 0.$$

For $j = 1, 2$ define

$$A_j^{\delta/2} = \left\{ \mathbf{x} \in \Omega_j : \text{dist}(\mathbf{x}, \partial\Omega_j) > \frac{\delta}{2} \right\},$$

in such a way that

$$K \subset A_1 \cup A_2.$$

The functions

$$v_1 = v\frac{w_1}{w_1 + w_2} \quad \text{and} \quad v_2 = v\frac{w_2}{w_1 + w_2}$$

---

[15] Prove it.

will do the job, provided we find functions $w_1 \in \mathcal{D}(\Omega_1)$ and $w_2 \in \mathcal{D}(\Omega_2)$ such that $w_1 = w_2$ and $w_1 + w_2 > 0$ in K. Let us concentrate on w_1; the argument is completely similar for w_2. Call $z = \chi(A_1)$ the characteristic function of $A_1^{d/2}$ and take the mollifier[16]

$$\eta(\mathbf{x}) = \begin{cases} c \exp\left(\dfrac{1}{|\mathbf{x}|^2 - 1}\right) & 0 \le |\mathbf{x}| < 1 \\ 0 & |\mathbf{x}| \ge 1 \end{cases}$$

where $c = \left(\int_{\mathbb{R}^n} \eta\right)^{-1}$. Set

$$\eta_\varepsilon(\mathbf{x}) = \varepsilon^{-n} \eta\left(\frac{\mathbf{x}}{\varepsilon}\right)$$

and then take

$$w_1 = \eta_\varepsilon * z.$$

For $\varepsilon = d/2$ the function w_1 has the desired properties.

---

[16] [18, Chap. 7, Sect. 2].

6

Variational Formulations

6.1 Backgrounds

We start by recalling the basic facts about variational problems.

• *Bilinear forms and coercivity.* Let V be a Hilbert space. A map

$$B : V \times V \to \mathbb{R}$$

is called a *bilinear form* if it is linear in either argument. The form

$$B^*(u, v) = B(v, u)$$

is called *adjoint* form of B. B is:

$\to$ *Self-adjoint* (or *symmetric*) if

$$B(v, u) = B(u, v).$$

$\to$ *Continuous* if and only if it is *bounded*, i.e. there is a constant $M > 0$ such that

$$|B(u, v)| \leq M \|u\|_V \|v\|_V.$$

$\to$ *Coercive* if there is a constant $a > 0$ such that

$$B(u, u) \geq a \|u\|_V^2,$$

for every $u, v \in V$.

Let

$$V \hookrightarrow H = H^* \hookrightarrow V^*$$

be a *Hilbert triplet*: V is continuously embedded in H, dense in H; H is identified with its dual H^*, and it is dense and continuously embedded in V^*,

© Springer International Publishing Switzerland 2015
S. Salsa, G. Verzini, *Partial Differential Equations in Action. Complements and Exercises*,
UNITEXT – La Matematica per il 3+2 87, DOI 10.1007/978-3-319-15416-9_6

The form B is *weakly coercive* with respect to the Hilbert triplet $\{V, H, V^*\}$ if there exist $\lambda_0 \in \mathbb{R}$ and $a > 0$ such that

$$B(u, u) + \lambda_0 \|u\|_H^2 \geq a \|u\|_V^2$$

for any $u \in V$.

• *Abstract variational problem (stationary case)*: determine $u \in V$ such that

$$B(u, v) = Fv, \qquad \forall v \in V \tag{6.1}$$

where B is a bilinear form on V and $F \in V^*$.

Theorem 6.1 (Lax-Milgram). *If B is continuous and coercive with coercivity constant a, there exists a unique solution $\bar{u}$ to problem (6.1), and the following stability estimate holds:*

$$\|\bar{u}\|_V \leq \frac{1}{a} \|F\|_{V^*}.$$

Moreover, if B is self-adjoint, then $\bar{u}$ is the unique minimiser of the 'energy' functional

$$E(v) = \frac{1}{2} B(v, v) - Fv, \qquad \text{for } v \in V.$$

Theorem 6.2 (Fredholm alternative). *Let B be a weakly coercive bilinear form on V w.r.t. the Hilbert triplet $\{V, H, V^*\}$, with V, H separable and V compactly embedded in H, and let $F \in V^*$.*

Suppose $\mathcal{N}(B)$, $\mathcal{N}(B^*)$ denote the solution subspaces (eigenspaces) of the homogeneous problems

$$B(u, v) = 0, \qquad \forall v \in V \qquad \text{and} \qquad B^*(w, v) = 0, \qquad \forall v \in V.$$

Then

a) $\dim \mathcal{N}(B) = \dim \mathcal{N}(B^*) = d < \infty.$
b) Problem (6.1) has a solution if and only if $Fw = 0$ for any $w \in \mathcal{N}(B^*)$.

• *Dirichlet eigenvalues of the operator $-\Delta$.* Let $\Omega \subset \mathbb{R}^n$ be a bounded Lipschitz domain and consider the eigenvalue problem

$$-\Delta u = \lambda u, \qquad u \in H_0^1(\Omega).$$

There exists a sequence of numbers

$$0 < \lambda_1 < \lambda_2 \leq \lambda_3 \leq \dots,$$

$\lambda_k \to +\infty$, for which the problem has non-trivial solutions if and only if $\lambda = \lambda_k$. The corresponding eigenfunctions (suitably normalized) form an orthonormal system in $L^2(\Omega)$ and an orthogonal system in $H_0^1(\Omega)$. In particular, the first eigenvalue λ_1 is *simple*, i.e. the first eigenspace has the form

$$\{t\varphi_1 : t \in \mathbb{R}\}$$

and φ_1 can be chosen strictly positive on Ω.

Similar results hold for the other boundary conditions.

• *Abstract variational problem (first-order evolution case)*: let $\{V, H, V^*\}$ be a Hilbert triplet with V, H separable and $B(\cdot, \cdot, t)$ a bilinear form on V (no assumption on the t-dependence is needed, except for measurability). Find a function $t \mapsto \mathbf{u}(t)$ such that

$$\mathbf{u} \in L^2(0, T; V), \qquad \dot{\mathbf{u}} \in L^2(0, T; V^*)$$

and

$$\begin{cases} \dfrac{d}{dt}(\mathbf{u}(t), v)_H + B(\mathbf{u}(t), v, t) = (\mathbf{f}(t), v)_H & \forall v \in V, \\ \mathbf{u}(0) = g \end{cases}$$

for almost every t in $(0, T)$, in distributional sense on $(0, T)$.

Theorem 6.3. *If* $\mathbf{f} \in L^2(0, T; H)$, $g \in H$ *and* B *is continuous and weakly coercive, uniformly*[1] *in* t, *the variational problem has a unique solution.*

• *Abstract variational problem (second-order evolution case)*: given a Hilbert triplet $\{V, H, V^*\}$ with V, H separable and $B(\cdot, \cdot)$ bilinear, symmetric on V, find a function $t \mapsto \mathbf{u}(t)$ such that

$$\mathbf{u} \in L^2(0, T; V), \qquad \dot{\mathbf{u}} \in L^2(0, T; H), \qquad \ddot{\mathbf{u}} \in L^2(0, T; V^*)$$

and

$$\begin{cases} \dfrac{d^2}{dt^2}(\mathbf{u}(t), v)_H + B(\mathbf{u}(t), v) = (\mathbf{f}(t), v)_H & \forall v \in V, \\ \mathbf{u}(0) = g \\ \dot{\mathbf{u}}(0) = h \end{cases}$$

for almost every $t \in (0, T)$, in distributional sense on $(0, T)$.

Theorem 6.4. *If* $\mathbf{f} \in L^2(0, T; H)$, $g \in V$, $h \in H$ *and* B *is continuous, self-adjoint and weakly coercive, the variational problem has a unique solution.*

A typical Hilbert triplet is given by $\{V, H, V^*\}$ where

$$H_0^1(\Omega) \subset V \subset H^1(\Omega),$$

and $H = L^2(\Omega)$.

---

[1] The continuity constant M and the coercivity constant a do not depend on t.

6.2 Solved Problems

6.2.1 One-dimensional problems

Problem 6.2.1 (Dirichlet conditions). *Write the variational formulation of the problem:*

$$\begin{cases} (x^2 + 1)u'' - xu' = \sin 2\pi x & 0 < x < 1 \\ u(0) = u(1) = 0. \end{cases}$$

Show that it has a unique solution $u \in H_0^1(0, 1)$ and find a constant C for which

$$\|u'\|_{L^2(0,1)} \le C.$$

Solution. As the Dirichlet conditions are homogeneous, we choose test functions in $H_0^1(0, 1)$ (which is the closure, for the usual norm, of $C_0^1(0, 1)$). Multiply the equation by $v \in H_0^1(0, 1)$ and integrate. We find:

$$\int_0^1 \left[(x^2 + 1)u''(x) - xu'(x) \right] v(x)\, dx = \int_0^1 \sin(2\pi x) v(x)\, dx. \qquad (6.2)$$

Integrating by parts we may write

$$\int_0^1 (x^2 + 1)u''(x)v(x)\, dx$$

$$= \left[(x^2 + 1)u'(x)v(x) \right]_0^1 - \int_0^1 u'(x)\frac{d}{dx}\left[(x^2 + 1)v(x) \right] dx =$$

$$= -\int_0^1 \left[(x^2 + 1)u'(x)v'(x) + 2xu'(x)v(x) \right] dx$$

and then substituting into (6.2),

$$\int_0^1 \left[(x^2 + 1)u'(x)v'(x) + 3xu'(x)v(x) \right] dx = -\int_0^1 \sin(2\pi x)v(x)\, dx.$$

Setting

$$B(u, v) = \int_0^1 \left[(x^2 + 1)u'(x)v'(x) + 3xu'(x)v(x) \right] dx,$$

$$Fv = -\int_0^1 \sin(2\pi x)v(x)\, dx,$$

we obtain the following variational formulation: *find $u \in H_0^1(0, 1)$ such that*

$$B(u, v) = Fv \quad \text{for any } v \in H_0^1(0, 1). \tag{6.3}$$

Note that if $u \in C^2$, $v \in C^1$ satisfy (6.3), we can repeat the integration the other way around to obtain

$$\int_0^1 \left[(x^2 + 1)u''(x) - xu'(x) - \sin(2\pi x)\right] v(x)\, dx \quad \text{for all } v \in C_0^1(0, 1).$$

As v is arbitrary, we deduce

$$(x^2 + 1)u''(x) - xu'(x) - \sin(2\pi x) = 0 \quad \text{in } (0, 1)$$

so that u is the classical solution to the initial problem. This indicates that the variational formulation is coherent with the classical one for regular solutions. For the analysis we need the Lax-Milgram theorem. Clearly, B is bilinear and F is linear. As for the continuity, by the Schwarz and Poincaré inequalities (the latter holds here with $C_P = 1/\pi$, see Exercise 5.3.13, Chap. 5, page 312), we get

$$|B(u, v)| \le \|x^2 + 1\|_{L^\infty} \|u'\|_{L^2} \|v'\|_{L^2} + \|3x\|_{L^\infty} \|u'\|_{L^2} \|v\|_{L^2} \le$$
$$\left(2 + \frac{3}{\pi}\right) \|u'\|_{L^2} \|v'\|_{L^2}$$

(where $L^p = L^p(0, 1)$) and

$$|Fv| \le \|\sin 2\pi x\|_{L^2} \|v\|_{L^2} \le \frac{\sqrt{2}}{2\pi} \|v'\|_{L^2}.$$

Concerning the coercivity of B we have to estimate (from below)

$$B(u, u) = \int_0^1 \left[(x^2 + 1)(u')^2 + 3xu'u\right] dx.$$

Noting that

$$uu' = (u^2)'/2,$$

and integrating by parts, using Poincaré's inequality, we can write

$$\int_0^1 3xuu'\, dx = \int_0^1 \frac{3}{2}x(u^2)'\, dx = -\int_0^1 \frac{3}{2}u^2\, dx$$
$$= -\frac{3}{2}\|u\|_{L^2}^2 \ge -\frac{3}{2\pi^2}\|u'\|_{L^2}^2.$$

Since $x^2 + 1 \ge 1$,

$$B(u, u) \ge a\|u'\|_{L^2}^2,$$

where $a = 1 - \frac{3}{2\pi^2} > 0$. Hence B is coercive and we can apply Lax–Milgram to obtain the existence of a unique solution u to (6.3). Moreover,

$$\|u'\|_{L^2} \leq \frac{1}{a}\|F\|_{H^{-1}} \leq \left(1 - \frac{3}{2\pi^2}\right)\frac{\sqrt{2}}{2\pi},$$

which is the required inequality.

Problem 6.2.2 (Neumann conditions). *Write the variational formulation of the problem:*

$$\begin{cases} e^x u'' + e^x u' - cu = 1 - 2x & 0 < x < 1 \\ u'(0) = u'(1) = 0. \end{cases}$$

Discuss the existence of solutions as the real parameter c varies.

Solution. The equation can be written as

$$-(e^x u')' + cu = 2x - 1.$$

Given the Neumann boundary conditions, we choose $V = H^1(0, 1)$ as test functions. Multiply the equation by a generic test function v and integrate over $(0, 1)$:

$$\int_0^1 \left[-(e^x u')' + cu\right] v \, dx = \int_0^1 (2x - 1)v \, dx.$$

Integrating by parts now gives

$$\int_0^1 \left[-(e^x u')' + cu\right] v \, dx = \left[-e^x u' v\right]_0^1 + \int_0^1 \left[e^x u' v' + cuv\right] dx.$$

Setting

$$B(u, v) = \int_0^1 \left[e^x u' v' + cuv\right] dx, \quad Fv = \int_0^1 (2x - 1)v \, dx,$$

gives the following variational formulation: *determine $u \in V$ such that*

$$B(u, v) = Fv \quad \text{for any } v \in V. \tag{6.4}$$

To check the coherence of this formulation with the classical one, observe that if $u \in C^2$ is a weak solution, integrating by parts gives

$$\int_0^1 \left[e^x(u'' + u') - cu + 2x - 1\right] v \, dx - \left[e^x u' v\right]_0^1 = 0 \quad \text{for every } v \in C^1(0, 1). \tag{6.5}$$

In particular, the relation (6.5) must hold for every v that vanishes at the boundary: that is

$$\int_0^1 \left[e^x u'' + e^x u' - cu + 2x - 1\right] v \, dx = 0 \quad \text{for every } v \in C_0^1(0, 1),$$

which forces
$$e^x u'' + e^x u' - cu + 2x - 1 = 0 \quad \text{on } (0, 1),$$

i.e. u solves the starting equation. Substituting in (6.5) produces

$$-eu'(1)v(1) + u'(0)v(0) = 0 \quad \text{for every } v \in C^1(0, 1),$$

hence $u'(0) = u'(1) = 0$. In other words a regular weak solution to (6.4) is indeed a classical solution to the initial problem, so the variational formulation is correct.

For the analysis of the problem let us try to apply the Lax-Milgram theorem, or, if not possible, Fredholm's alternative. That F is linear and B bilinear is clear. As for the continuity, write $L^2 = L^2(0, 1)$, so

$$|Fv| \leq \|2x - 1\|_{L^2} \|v\|_{L^2} \leq \frac{\sqrt{3}}{3} \left(\|v'\|_{L^2} + \|v\|_{L^2} \right)$$

and

$$|B(u, v)| \leq \|e^x\|_{L^\infty} \|u'\|_{L^2} \|v'\|_{L^2} + |c| \|u\|_{L^2} \|v\|_{L^2} \leq \max\{e, |c|\} \|u\|_V \|v\|_V,$$

hence F belongs to V^* and B is continuous. Coercivity: as $e^x \geq 1$ on $[0, 1]$, we have

$$B(u, u) = \int_0^1 \left[e^x (u')^2 + cu^2 \right] dx \geq \|u'\|_{L^2}^2 + c \|u\|_{L^2}^2.$$

If $c > 0$, B is coercive, with constant $\min\{1, c\}$. By the Lax-Milgram theorem we deduce there is a unique solution u to (6.3). Vice versa, if $c \leq 0$,

$$B(u, u) + (1 - c) \|u\|_{L^2}^2 \geq \|u'\|_{L^2}^2 + \|u\|_{L^2}^2$$

and B is only weakly coercive for the triplet $\{V, H = L^2(0, 1), V^*\}$. In fact the embedding of V in H is compact, and we can use Fredholm's alternative, by which (6.4) has solutions if and only if

$$Fw = \int_0^1 (1 - 2x)w \, dx = 0$$

for every solution w of the adjoint homogeneous problem:

$$B(v, w) = 0, \qquad \text{for any } v \in V.$$

Note that $B(u, v) = B(v, u)$, making the problem self-adjoint. In formulas, if w solves the adjoint homogeneous problem, for every $v \in V$ we have

$$\int_0^1 \left[e^x v' w' + cvw \right] dx = 0,$$

which is the weak formulation (prove this fact) of the Sturm-Liouville problem

$$\begin{cases} -(e^x u')' = -cu & 0 < x < 1 \\ u'(0) = u'(1) = 0. \end{cases}$$

We know[2] that such problem admits a sequence of simple eigenvalues $\{-c_k\} = \{\lambda_k\}$, with $\lambda_1 = 0$ and $\lambda_k > 0$ for $k \geq 1$. In particular, each eigenspace has dimension 1 and is generated by $\psi_k \in V$, $\|\psi_k\| = 1$. When $-c \neq \lambda_k$, the problem has only the zero solution. In summary,

if $c \neq -\lambda_k$ *for every* k there exists a unique solution $u \in V$ to (6.4).

Conversely,

if $c = -\lambda_k$ then (6.4) can be solved if and only if $\displaystyle\int_0^1 x \psi_k(x)\, dx = 0$

(in fact $\int_0^1 \psi_k(x)\, dx = 0$ for every $k \geq 1$; why is this?), and then there are infinitely many solutions $u = \bar{u} + C\psi_k$, where $\bar{u}$ is a particular solution.

In general the explicit determination of ψ_k, and so the verification of the compatibility conditions, is not elementary. The case $c = -\lambda_1 = 0$ is easy and one can see that the solutions of the adjoint homogeneous problem are only the constants (i.e. $\psi_1 \equiv 1$). Therefore the problem has solutions, for

$$\int_0^1 (1 - 2x) \cdot 1\, dx = 0.$$

Integrating we get the infinitely many solutions

$$u(t) = (x^2 + x + 1)e^{-x} + C,$$

with C an arbitrary constant.

Problem 6.2.3 (Robin-Dirichlet conditions). *Write the variational formulation of the problem*
$$\begin{cases} \cos x\, u'' - \sin x\, u' - xu = 1 & 0 < x < \pi/6 \\ u'(0) = -u(0),\ u(\pi/6) = 0 \end{cases}$$
and discuss existence and uniqueness. Determine a stability estimate for the solution.

Solution. The equation can be written in divergence form

$$-(\cos x\, u')' + xu = -1.$$

Given the homogeneous Dirichlet condition at $x = \pi/6$, we choose as functional space

$$V = \left\{ v \in H^1(0, \pi/6) : v(\pi/6) = 0 \right\}.$$

Multiplying by $v \in V$ and integrating we have

$$\int_0^{\pi/6} \left[-(\cos xu')' + xu \right] v\, dx = \int_0^{\pi/6} -v\, dx, \qquad \text{for every } v \in V.$$

---

[2] Appendix A.

An integration by parts gives

$$\int_0^{\pi/6} -(\cos x\, u')' v\, dx = \left[-\cos x\, u'v\right]_0^{\pi/6} + \int_0^{\pi/6} \cos x\, u'v'\, dx =$$

$$= u'(0)v(0) + \int_0^{\pi/6} \cos x\, u'v'\, dx =$$

$$= -u(0)v(0) + \int_0^{\pi/6} \cos x\, u'v'\, dx,$$

where, in the last equality, we used Robin's condition at $x = 0$. The weak formulation thus reads: *determine $u \in V$ such that*

$$B(u, v) = Fv \quad \text{for every } v \in V, \tag{6.6}$$

where

$$B(u, v) = \int_0^{\pi/6} \left[\cos x\, u'v' + xuv\right] dx - u(0)v(0), \qquad Fv = \int_0^{\pi/6} -v\, dx.$$

As in previous problems, it is possible to check (and we invite the reader to do so) that if the variational solution u is regular, the classical formulation follows from the weak formulation after integration.

The functional F is linear and B is a bilinear form. Let us check continuity. For every $v \in V$ we may write

$$v(\pi/6) - v(0) = \int_0^{\pi/6} v'(t)\, dt,$$

whence ($L^p = L^p(0, \pi/6)$)

$$|v(0)| \le \int_0^{\pi/6} |v'(x)|\, dx \le \sqrt{\frac{\pi}{6}}\, \|v'\|_{L^2}$$

(in the last step we used Schwarz's inequality on the product $|v'| \cdot 1$). The functions in V satisfy Poincaré's inequality

$$\|v\|_{L^2} \le C_P \|v'\|_{L^2},$$

hence we can employ the equivalent norm $\|v\|_V = \|v'\|_{L^2}$. Keeping previous inequalities into account, we deduce

$$|B(u, v)| \le \|u'\|_{L^2}\|v'\|_{L^2} + \|x\|_{L^\infty}\|u\|_{L^2}\|v\|_{L^2} + |u(0)||v(0)| \le$$

$$\le \left(1 + \frac{\pi}{6}C_P^2 + \frac{\pi}{6}\right) \|u'\|_{L^2}\|v'\|_{L^2},$$

and

$$|Fv| \le \|1\|_{L^2}\|v\|_{L^2} \le C_P \sqrt{\frac{\pi}{6}}\, \|v'\|_{L^2}.$$

Both B and F are continuous. Let us examine the coercivity of B. Recalling that $\cos x \geq \sqrt{3}/2$ and $x \geq 0$ on $[0, \pi/6]$ we can write

$$B(u,u) = \int_0^{\pi/6} \left[\cos x \, (u')^2 + x u^2 \right] dx - u^2(0) \geq$$

$$\geq \frac{\sqrt{3}}{2} \|u'\|_{L^2}^2 - \frac{\pi}{6} \|u'\|_{L^2}^2 = a \|u'\|_{L^2}^2$$

where $a = \frac{\sqrt{3}}{2} - \frac{\pi}{6} > 0$. Then we can apply the Lax-Milgram theorem, and obtain the existence of a unique solution u to (6.6). By the same theorem, moreover,

$$\|u'\|_{L^2} \leq \frac{1}{a} \|F\|_{V^*} \leq C_P \left(\frac{\sqrt{3}}{2} - \frac{\pi}{6} \right)^{-1} \sqrt{\frac{\pi}{6}}.$$

Thanks to Exercise 5.3.13, Chap. 5 (on page 312), we know that the best Poincaré constant in V is $C_P = 1/3$. In conclusion,

$$\|u'\|_{L^2} \leq \frac{\sqrt{2\pi}}{9 - \pi\sqrt{3}} \approx 0.7043\ldots .$$

Problem 6.2.4 (Problems on unbounded domains). *Write the variational formulation of*

$$\begin{cases} -u'' + \dfrac{2+x^2}{1+x^2}u = \dfrac{1}{1+x^2} & x \in \mathbb{R} \\ u(x) \to 0 & x \to \pm\infty \end{cases}$$

and prove existence and uniqueness. Determine an $L^\infty(\mathbb{R})$-estimate for the solution.

Solution. Problems on unbounded domains are usually more delicate to treat. At present, however, the space $V = H^1(\mathbb{R})$, that coincides with the closure of $C_0^\infty(\mathbb{R})$ in the usual norm, is particularly suited as it incorporates the condition of vanishing at infinity (Exercise 5.3.18 a), on page 312 in Chap. 5. What is more, a function $v \in C_0^\infty(\mathbb{R})$ has compact support and so we can multiply by v and integrate on $\mathbb{R}$, without worrying about the integrals convergence. Then

$$\int_{-\infty}^{+\infty} \left[-u'' + \frac{2+x^2}{1+x^2}u \right] v \, dx = \int_{-\infty}^{+\infty} \frac{1}{1+x^2} v \, dx.$$

Integrating by parts we get

$$\int_{-\infty}^{+\infty} -u'' v \, dx = -\left[u'v \right]_{-\infty}^{+\infty} + \int_{-\infty}^{+\infty} u'v' \, dx,$$

giving the variational formulation: *determine $u \in V$ such that:*

$$B(u,v) = Fv \quad \text{for every } v \in V, \tag{6.7}$$

where

$$B(u, v) = \int_{-\infty}^{+\infty} \left[u'v' + \frac{2 + x^2}{1 + x^2} uv \right] dx, \qquad Fv = \int_{-\infty}^{+\infty} \frac{1}{1 + x^2} v \, dx.$$

The reader can check that if u is regular and solves (6.7), then it is a classical solution.

To apply Lax-Milgram we need to prove the continuity of F, B and the coercivity of B. As

$$|Fv| \le \left\| \frac{1}{1 + x^2} \right\|_{L^2(\mathbb{R})} \|v\|_{L^2(\mathbb{R})} = \sqrt{\pi} \|v\|_{L^2(\mathbb{R})}, \tag{6.8}$$

and

$$|B(u, v)| \le \|u'\|_{L^2(\mathbb{R})} \|v'\|_{L^2(\mathbb{R})} + \left\| \frac{2 + x^2}{1 + x^2} \right\|_{L^\infty(\mathbb{R})} \|u\|_{L^2(\mathbb{R})} \|v\|_{L^2(\mathbb{R})} \le 2\|u\|_V \|v\|_V,$$

F and B are continuous. Since $\inf_{\mathbb{R}} \left(\frac{2+x^2}{1+x^2} \right) = 1$, we have

$$B(u, u) = \int_{-\infty}^{+\infty} \left[(u')^2 + \frac{2 + x^2}{1 + x^2} u^2 \right] dx \ge \|u\|_V^2,$$

showing B is coercive, with coercivity constant 1. Consequently the Lax-Milgram theorem guarantees existence and uniqueness of the solution.

Concerning the estimate, from[3] $\|u\|_{L^\infty(\mathbb{R})} \le \|u\|_V$ and (6.8) we obtain $\|u\|_{L^\infty(\mathbb{R})} \le \|F\|_{V^*} \le \pi$.

Problem 6.2.5 (Legendre equation). *Let*

$$V = \left\{ v \in L^2(-1, 1) : (1 - x^2)^{1/2} v' \in L^2(-1, 1) \right\}.$$

a) Verify that the formula

$$(u, v)_V = \int_{-1}^{1} [uv + (1 - x^2) u'v'] \, dx \tag{6.9}$$

defines an inner product making V into a Hilbert space.

b) Study the variational problem

$$(u, v)_V = Fv \equiv \int_{-1}^{1} fv \, dx, \qquad \text{for every } v \in V, \tag{6.10}$$

where $f \in L^2(-1, 1)$, and interpret it in the strong sense.

c) Study the eigenvalue problem

$$(u, v)_V = \lambda \int_{-1}^{1} uv \, dx, \qquad \text{for every } v \in V.$$

---

[3] Problem 5.2.23, Chap. 5 (page 300).

Solution. a) Let us verify the properties of the inner product. That $(u, u)_V$ is non-negative is clear, while

$$(u, u)_V = \int_{-1}^{1} [u^2 + (1 - x^2)(u')^2] \, dx = 0$$

implies $u = 0$ a.e.[4] on $(-1, 1)$. Hence $(u, v)_V$ is an inner product, so we just have to prove V is complete with respect to it. Take $\{u_n\}$ a Cauchy sequence in V. Then $\{u_n\}$ is a Cauchy sequence in $L^2(-1, 1)$, thus $u_n \to u_0 \in L^2(-1, 1)$. Analogously, $\{(1 - x^2)^{1/2} u'_n\}$ is a Cauchy sequence in $L^2(-1, 1)$ and therefore

$$(1 - x^2)^{1/2} u'_n \to w \in L^2(-1, 1).$$

On the other hand, if $\varphi \in \mathcal{D}(-1, 1)$, then $\left[(1 - x^2)^{1/2} \varphi\right]' \in \mathcal{D}(-1, 1)$, and

$$\langle (1 - x^2)^{1/2} u'_n, \varphi \rangle = -\int_{-1}^{1} u_n \left[(1 - x^2)^{1/2} \varphi\right]' \, dx$$

$$\to -\int_{-1}^{1} u_0 \left[(1 - x^2)^{1/2} \varphi\right]' \, dx = \langle (1 - x^2)^{1/2} u'_0, \varphi \rangle$$

Therefore $(1 - x^2)^{1/2} u'_0 = w \in L^2(-1, 1)$, which means V is complete.

b) Denote by $\|\cdot\|$ and $\|\cdot\|_V$ the $L^2(-1, 1)$-norm and the norm induced by (6.9). As

$$|Fv| = \left| \int_{-1}^{1} fv \, dx \right| \leq \|f\| \|v\| \leq \|f\| \|v\|_V,$$

the functional F defines an element in V^*. Riesz's theorem implies there exists a unique u solving (6.10) and

$$\|u\|_V \leq \|f\|.$$

For the strong interpretation let us proceed formally and integrate by parts the second term of the inner product; (6.10) reads

$$\left[(1 - x^2) u'v\right]_{-1}^{1} + \int_{-1}^{1} \{u - [(1 - x^2) u']'\} v \, dx = \int_{-1}^{1} fv \, dx \qquad \text{for every } v \in V,$$

and as v is arbitrary we have

$$-[(1 - x^2) u']' + u = f \qquad \text{in } (-1, 1) \tag{6.11}$$

and also

$$\lim_{x \to \pm 1 \mp} (1 - x^2) u'(x) = 0.$$

The boundary conditions are thus the 'natural' homogeneous Neumann conditions. Note that the limit is necessary because functions in V may be unbounded in a neighbourhood of ± 1.

---

[4] In fact, $u \in V$ implies $u \in H^1_{\text{loc}}(-1, 1)$, hence $u \in C(-1, 1)$.

c) From part b) the eigenvalue problem

$$(u, v)_V = \lambda \int_{-1}^{1} uv \, dx \qquad \text{for every } v \in V$$

is equivalent to the (*Legendre*) equation

$$-[(1 - x^2) u']' = (\lambda - 1) u \qquad \text{on } (-1, 1),$$

with boundary conditions

$$\lim_{x \to \pm 1} (1 - x^2) u' (x) = 0.$$

The only solutions[5] are *Legendre's polynomials* P_n, defined recursively by the (Rodrigues) formula

$$P_n (x) = \frac{1}{2^n n!} \frac{d^n}{dx^n} (x^2 - 1)^n, \qquad n \in \mathbb{N},$$

with eigenvalues $\lambda_n = 1 + n (n + 1)$. For example

$$P_0 (x) = 1, \ P_1 (x) = x, \ P_2 (x) = \frac{1}{3} (3x^2 - 1), \ P_4 (x) = \frac{1}{2} (5x^3 - 3x).$$

Problem 6.2.6 (*Variational problem in* H^1). *Consider the following weak problem: determine* $u \in H^1(0, 1)$ *such that*

$$\int_0^1 u'v' \, dt = v(1) - v(0) - \int_0^1 (\log t)v' \, dt, \qquad \forall v \in H^1(0, 1).$$

Study when it can be solved. Show that there exist infinitely many solutions and determine them explicitely.

Solution. Let

$$B(u, v) = \int_0^1 u'v' \, dt, \qquad Fv = v(1) - v(0) - \int_0^1 (\log t)v' \, dt.$$

First of all, F is a continuous linear functional on $V = H^1(0, 1)$: indeed, observing that

$$\int_0^1 \log^2 t \, dt = \int_{-\infty}^0 s^2 e^s \, ds = 2$$

we have[6]

$$|Fv| \leq 2\|v\|_{L^\infty} + \|\log t\|_{L^2} \|v'\|_{L^2} \leq \left(2 + \sqrt{2}\right) \|v\|_{H^1}.$$

At the same time, it is easy to check that B is continuous but only weakly coercive on V (see for instance Exercise 6.3.1 on page 381.c)). The homogeneous problem is self-adjoint

---

[5] Appendix A.

[6] For pointwise estimates of functions in $H^1(a, b)$ we always use Problem 5.2.24 (page 301), Chap. 5.

and has only the constants as solutions. We can apply Fredholm's alternative, and obtain that the problem can be solved if and only if $Fv = 0$ for every v constant. But this compatibility condition is automatic, so the problem has infinitely many solutions that differ by an additive constant.

In order to write the associated boundary-value problem let us rearrange terms:

$$\int_0^1 \left[u' - \log t \right] v' \, dt = v(1) - v(0), \quad \forall v \in V.$$

By direct inspection we infer

$$u' - \log t = 1,$$

from which the solutions are

$$u(t) = t \log t + C.$$

Notice that $u \notin C^1([0, 1])$ ($u'(0)$ does not exist), so that this problem does not have a corresponding classical version.

6.2.2 Elliptic problems

Problem 6.2.7 (Dirichlet conditions). *Write the variational formulation of the problem*

$$\begin{cases} -\Delta u + c(\mathbf{x})u = f(\mathbf{x}) & \mathbf{x} \in \Omega \\ u = 0 & \mathbf{x} \in \partial\Omega, \end{cases}$$

where Ω is a regular bounded domain in $\mathbb{R}^n$, $f \in L^2(\Omega)$. What are the constraints on c that ensure the Lax-Milgram theorem can be applied? Write the corresponding stability estimate.

Solution. The homogeneous Dirichlet conditions suggest to place the weak formulation in $H_0^1(\Omega)$. So let us multiply by some $v \in H_0^1(\Omega)$ and integrate[7]. Since (by Gauss's formula)

$$\int_\Omega -\Delta u \, v \, d\mathbf{x} = -\int_{\partial\Omega} \partial_\nu u \, v \, d\sigma + \int_\Omega \nabla u \cdot \nabla v \, d\mathbf{x} = \int_\Omega \nabla u \cdot \nabla v \, d\mathbf{x},$$

we obtain

$$\int_\Omega [\nabla u \cdot \nabla v + c(\mathbf{x})uv] \, d\mathbf{x} = \int_\Omega fv \, d\mathbf{x} \quad \text{for every } v \in H_0^1(\Omega).$$

But u, v are in L^2, so to make sense of the integral let us assume[8] $c \in L^\infty(\Omega)$. So set

$$B(u, v) = \int_\Omega [\nabla u \cdot \nabla v + c(\mathbf{x})uv] \, d\mathbf{x}, \quad Fv = \int_\Omega fv \, d\mathbf{x},$$

---

[7] More precisely, we use $v \in \mathcal{D}(\Omega)$ and extend the resulting equation to $H_0^1(\Omega)$ by density.

[8] Or, more sophisticatedly, note that by Sobolev's embedding (see [18, Chap. 7, Sect. 10]) u and v belong to $L^{p^*}(\Omega)$, $p^* = 2n/(n-2)$, so it suffices to demand $c \in L^q(\Omega)$, with $q = n/2$ for $n \geq 3$ and any q for $n = 2$.

and then we have the variational formulation: *determine $u \in H_0^1(\Omega)$ such that*

$$B(u, v) = Fv \quad \text{for every } v \in H_0^1(\Omega). \tag{6.12}$$

Vice versa, if $u \in C_0^2(\Omega)$ solves (6.12), we can use Gauss's formula in the opposite way

$$\int_\Omega [-\Delta u + c(\mathbf{x})u - f] \, v \, d\mathbf{x} \quad \text{for every } v \in C_0^1(\Omega),$$

so that $-\Delta u + c(\mathbf{x})u - f = 0$ a.e. on Ω. In particular, even if the coefficients c and f are just continuous, the equality holds everywhere and u is the classical solution.

Now, to use Lax-Milgram theorem we need to understand the continuity of F and B and the coercivity of B (that F and B are linear is straightforward). To this end let us recall that for every $u \in H_0^1(\Omega)$ we have Poincaré's inequality

$$\|u\|_{L^2(\Omega)} \leq C_P \|\nabla u\|_{L^2(\Omega)},$$

which entitles us to use in $H_0^1(\Omega)$ the equivalent norm $\|u\|_{H_0^1(\Omega)} = \|\nabla u\|_{L^2(\Omega)}$. As for the continuity, from Schwarz's and Poincaré's inequalities ($L^p = L^p\Omega$),

$$|B(u, v)| \leq \|\nabla u\|_{L^2} \|\nabla v\|_{L^2} + \|c\|_{L^\infty} \|u\|_{L^2} \|v\|_{L^2} \leq \left(1 + \|c\|_{L^\infty} C_P^2\right) \|\nabla u\|_{L^2} \|\nabla v\|_{L^2}$$

and

$$|Fv| \leq \|f\|_{L^2} \|v\|_{L^2} \leq C_P \|f\|_{L^2} \|\nabla v\|_{L^2}.$$

Coercivity: we have to bound (from below)

$$B(u, u) = \int_\Omega \left[|\nabla u|^2 + c(\mathbf{x})u^2 \right] d\mathbf{x}.$$

One possibility is imposing $c(\mathbf{x}) \geq 0$ a.e. onà Ω. Thus B is coercive, with constant $a = 1$. There are less restrictive conditions on c, though, provided we take a smaller coercivity constant. In fact, suppose that

$$c(\mathbf{x}) \geq -\gamma, \qquad \text{for almost every } \mathbf{x} \in \Omega.$$

By Poincaré's inequality

$$\int_\Omega c(\mathbf{x})u^2 \, d\mathbf{x} \geq -\gamma \|u\|_{L^2}^2 \geq -\gamma C_P^2 \|\nabla u\|_{L^2}^2$$

and therefore

$$B(u, u) \geq a \|\nabla u\|_{L^2}^2,$$

where

$$a = 1 - \gamma C_P^2.$$

Thus to have $a > 0$ (coercivity) it is enough to take c not 'too' negative[9], i.e.

$$\gamma < \frac{1}{C_P^2}.$$

Using Lax-Milgram theorem implies, with the previous hypotheses, that there is a unique solution u to (6.12), which satisfies the estimate

$$\|\nabla u\|_{L^2} \leq \frac{1}{a}\|F\|_{H^{-1}} \leq \frac{C_P}{1 - \gamma C_P^2}\|f\|_{L^2}.$$

Problem 6.2.8 (A minimum problem). *Let $Q = (0, 1) \times (0, 1) \subset \mathbb{R}^2$. Minimise, as $v \in H_0^1(Q)$ varies, the functional*

$$E(v) = \int_Q \left\{ \frac{1}{2}|\nabla v|^2 - xv \right\} dxdy.$$

Write the Euler equation and prove there exists a unique minimiser $u \in H_0^1(Q)$. Find an explicit formula for u.

Solution. With $v, w \in H_0^1(Q)$ we set

$$B(v, w) = \int_Q \nabla v \cdot \nabla w \, dxdy \qquad \text{and} \qquad Fv = \int_Q xv \, dxdy,$$

and write $E(v) = \frac{1}{2}B(v, v) - Fv$. As B is self-adjoint, u minimises E if and only if it solves the *Euler equation*

$$\int_Q \nabla u \cdot \nabla v \, dxdy = \int_Q xv \, dxdy, \qquad \text{i.e. } B(u, v) = Fv, \qquad \forall v \in H_0^1(Q). \quad (6.13)$$

We have

$$|Fv| \leq \|x\|_{L^2}\|v\|_{L^2} \leq \frac{C_P}{\sqrt{2}}\|\nabla v\|_{L^2}$$

where C_P is the Poincaré constant for $H_0^1(Q)$. Hence F is linear, continuous, while B is bilinear, continuous and coercive (it is the scalar product of $H_0^1(Q)$). Using the Lax-Milgram theorem (or better, Riesz's theorem), we obtain existence and uniqueness of the solution u minimising E.

For an explicit expression, note that (6.13) is the variational formulation of the following Dirichlet problem:

$$\begin{cases} -\Delta u = x & \text{in } Q \\ u = 0 & \text{on } \partial Q. \end{cases}$$

---

[9] Using Sobolev's embedding theorem, in line with the previous footnote argument, the role of γ is taken by $\|c^-\|_{L^q}$, where q is as above and $c^-(x) = \inf\{0, -c(x)\}$ is the *negative part* of c.

Thus, we can use the methods of Chap. 2, in particular separation of variables, to compute it. Let us expand the source term in a sines-Fourier series in y, that is

$$x = \sum_{n=1}^{+\infty} x b_n \sin (n \pi y)$$

where

$$b_n = 2 \int_0^1 \sin (n \pi y) \, dy = \frac{2}{n \pi} [-\cos n \pi + 1] = \begin{cases} 0 & \text{for } n \text{ even} \\ 4/n \pi & \text{for } n \text{ odd,} \end{cases}$$

and let us seek solutions of the form

$$u(x, y) = \sum_{n=1}^{+\infty} u_n(x) \sin (n \pi y).$$

Notice that u satisfies $u(x, 0) = u(x, 1) = 0$ automatically. Differentiating (formally), we obtain

$$u_{xx}(x, y) = \sum_{n=1}^{+\infty} u_n'' \sin (n \pi y), \qquad u_{yy}(x, y) = \sum_{n=1}^{+\infty} -n^2 \pi^2 u_n(x) \sin (n \pi y)$$

and therefore

$$\Delta u(x, y) = \sum_{n=1}^{+\infty} \left[u_n''(x) - n^2 \pi^2 u_n(x) \right] \sin (n \pi y) = \sum_{n=1}^{+\infty} x b_n \sin (n \pi y). \qquad (6.14)$$

Overall, the u_n solve the boundary-value problems

$$\begin{cases} u_n''(x) - n^2 \pi^2 u_n(x) = b_n x \\ u_n(0) = u_n(1) = 0. \end{cases}$$

When n is even $b_n = 0$, and $u_n \equiv 0$. When n is odd the equation reads

$$u_n''(x) - n^2 \pi^2 u_n(x) = \frac{4}{n \pi} x,$$

whose general integral is

$$u_n(x) = C_1 e^{n \pi x} + C_2 e^{-n \pi x} - \frac{4}{n^3 \pi^3} x.$$

The boundary conditions force

$$u_n(x) = \frac{4}{n^3 \pi^3} \left[\frac{e^{n \pi x} - e^{-n \pi x}}{e^{n \pi} - e^{-n \pi}} - x \right].$$

Finally,

$$u(x, y) = \frac{4}{\pi^3} \sum_{k=0}^{+\infty} \frac{1}{(2k+1)^3} \left[\frac{\sinh((2k+1)\pi x)}{\sinh((2k+1)\pi)} - x \right] \sin((2k+1)\pi y). \quad (6.15)$$

By Weierstrass's criterion, since

$$\frac{\sinh((2k+1)\pi x)}{\sinh((2k+1)\pi)} \le 1 \qquad \text{for } 0 \le x \le 1,\, k \ge 0,$$

both (6.15) and the series of the partial derivatives (whose terms have order $1/(2k+1)^2$) converge uniformly on $\overline{Q}$, so $u \in C_0^1(\overline{Q}) \subset H_0^1(Q)$. Moreover, also the second-derivatives series converges uniformly on every compact set contained in Q, so we can differentiate term by term and check that u is the (at this point classical) solution.

The series of second derivatives does not converge uniformly on $\overline{Q}$: from (6.14), Δu is discontinuous on $\overline{Q}$, because it equals x on Q and 0 on $y = 1$; despite the coefficients being regular, u is not $C^2(\overline{Q})$. This should come as no surprise, because the domain is only Lipschitz.

Problem 6.2.9 (Riesz's theorem and Laplacian). *Given a bounded regular domain $\Omega \subset \mathbb{R}^n$, consider $H_0^1(\Omega)$ with product*

$$(u, v) = \int_\Omega \nabla u \cdot \nabla v \, d\mathbf{x},$$

and let $F \in H^{-1}(\Omega)$.

a) Apply Riesz's theorem to F and deduce a suitable variational problem.

b) Write the corresponding classical problem and interpret it in view of Riesz's theorem.

Solution. a) Using $(\nabla u, \nabla v)$ as scalar product in $H_0^1(\Omega)$, Riesz's theorem says that *there is a unique $u \in H_0^1(\Omega)$ such that*

$$\int_\Omega \nabla u \cdot \nabla v \, d\mathbf{x} = Fv \qquad \text{for every } v \in H_0^1(\Omega), \quad (6.16)$$

and

$$\|u\|_{H_0^1(\Omega)} = \|F\|_{H^{-1}(\Omega)}.$$

The variational problem (6.16) is thus well posed on $H_0^1(\Omega)$.

b) The space of test functions $\mathcal{D}(\Omega)$ is dense in $H_0^1(\Omega)$, so the variational formulation of a) is equivalent to

$$\int_\Omega \nabla u \cdot \nabla \varphi \, d\mathbf{x} = \langle F, \varphi \rangle \qquad \text{for every } v \in \mathcal{D}(\Omega),$$

where $\langle F, \varphi \rangle$ is the duality pairing between $\mathcal{D}(\Omega)$ and $\mathcal{D}'(\Omega)$. Using the language of distributions we may write, for any $\varphi \in \mathcal{D}(\Omega)$,

$$0 = \langle \nabla u, \nabla \varphi \rangle - \langle F, \varphi \rangle = \langle -\Delta u - F, \varphi \rangle.$$

Hence

$$-\Delta u = F$$

in distributional sense. Therefore for every $F \in H^{-1}(\Omega)$, $u = (-\Delta)^{-1} F$ is the element of $H_0^1(\Omega)$ for which

$$Fv = ((-\Delta)^{-1} F, v)_{H_0^1(\Omega)} \quad \text{for every } v \in H_0^1(\Omega),$$

and moreover

$$\|F\|_{H^{-1}(\Omega)} = \|(-\Delta)^{-1} F\|_{H_0^1(\Omega)}.$$

This amounts to say that the *canonical isomorphism* in Riesz's theorem is actually the operator

$$(-\Delta)^{-1} : H^{-1}(\Omega) \longrightarrow H_0^1(\Omega).$$

Problem 6.2.10 (Bilinear forms on subspaces). *Given a bounded and regular domain $\Omega \subset \mathbb{R}^n$, consider the subspace*

$$V = \left\{ u \in H^1(\Omega) : \int_\Omega u \, d\mathbf{x} = 0 \right\}$$

in $H^1(\Omega)$ (with usual scalar product).

a) Say to which problem the following variational formulation is associated: find $u \in V$ such that

$$\int_\Omega \nabla u \cdot \nabla v \, d\mathbf{x} = 0, \quad \text{for every } v \in V.$$

b) Use the Lax-Milgram theorem and deduce that the problem has the unique solution $u = 0$.

c) Deduce that two solutions of

$$\begin{cases} -\Delta u = f & \mathbf{x} \in \Omega \\ \partial_\nu u = g & \mathbf{x} \in \partial\Omega, \end{cases}$$

with $f \in L^2(\Omega)$ and $g \in L^2(\partial\Omega)$, differ by a constant.

Solution. a) First observe that V is closed in $H^1(\Omega)$, so we have the decomposition

$$H^1(\Omega) = V \oplus V^\perp$$

where $V^{\perp}$ is the space of constants[10]. Consequently, for every $v \in H^1(\Omega)$, we can write

$$v = \tilde{v} + C_v \qquad \text{where } C_v = \int_{\Omega} v \, d\mathbf{x} \quad \text{and } \tilde{v} \in V.$$

Let $u \in V$ be a solution to the given problem and $v \in H^1(\Omega)$. Then

$$\int_{\Omega} \nabla u \cdot \nabla v \, d\mathbf{x} = \int_{\Omega} \nabla u \cdot \nabla (\tilde{v} + C_v) \, d\mathbf{x} = \int_{\Omega} \nabla u \cdot \nabla \tilde{v} \, d\mathbf{x} = 0,$$

so u solves the equivalent problem: *determine $u \in V$ such that*

$$\int_{\Omega} \nabla u \cdot \nabla v \, d\mathbf{x} = 0, \qquad \textit{for every } v \in H^1(\Omega).$$

Assuming u regular and integrating by parts with Gauss's formula gives

$$0 = \int_{\Omega} \nabla u \cdot \nabla v \, d\mathbf{x} = \int_{\partial\Omega} \partial_{\nu} u \, v \, d\sigma - \int_{\Omega} -\Delta u \, v \, d\mathbf{x} \qquad (6.17)$$

for every $v \in H^1(\Omega)$. In particular,

$$\int_{\Omega} -\Delta u \, v \, d\mathbf{x} = 0 \qquad \text{for every } v \in H^1_0(\Omega),$$

which implies $\Delta u = 0$ in Ω. Substituting in (6.17) we find

$$\int_{\partial\Omega} \partial_{\nu} u \, v \, d\sigma = 0 \qquad \text{for every } v \in H^1(\Omega),$$

hence $\partial_{\nu} u = 0$ on $\partial\Omega$. To sum up, the weak formulation corresponds to the boundary-value problem

$$\begin{cases} -\Delta u = 0 & \mathbf{x} \in \Omega \\ \partial_{\nu} u = 0 & \mathbf{x} \in \partial\Omega \\ \int_{\Omega} u \, d\mathbf{x} = 0. \end{cases}$$

b) Recall that Poincaré's inequality

$$\|u\|_{L^2(\Omega)} \leq C_P \|\nabla u\|_{L^2(\Omega)}$$

holds on the subspace V of $H^1(\Omega)$ of functions with zero average on a regular bounded domain, thus we have the equivalence of the norms. We want to prove that the bilinear form

$$B(u, v) = \int_{\Omega} \nabla u \cdot \nabla v \, d\mathbf{x}$$

---

[10] The proof is left to the reader.

is continuous and coercive on V. From Lax-Milgram, we will deduce that the variational problem has only the zero solution. Now,

$$|B(u, v)| \leq \|\nabla u\|_{L^2(\Omega)} \|\nabla v\|_{L^2(\Omega)} \leq \|u\|_{H^1(\Omega)} \|v\|_{H^1(\Omega)}$$

and by Poincaré's inequality, for every λ,

$$B(u, u) = \|\nabla u\|_{L^2(\Omega)}^2 = \lambda \|\nabla u\|_{L^2(\Omega)}^2 + (1 - \lambda)\|\nabla u\|_{L^2(\Omega)}^2$$

$$\geq \lambda \|\nabla u\|_{L^2(\Omega)}^2 + \frac{1 - \lambda}{C_P^2} \|u\|_{L^2(\Omega)}^2 = \frac{1}{C_P^2 + 1}\|u\|_{H^1(\Omega)}^2$$

(in the end we chose $\lambda = (1 - \lambda)/C_P^2 = 1/(C_P^2 + 1)$). From these inequalities the claim follows.

c) Let u_1, u_2 be solutions of the given problem. Then $w = u_1 - u_2$ solves

$$\int_\Omega \nabla w \cdot \nabla v \, d\mathbf{x} = 0 \qquad \text{for every } v \in H^1(\Omega). \tag{6.18}$$

Using the orthogonal decomposition of $H^1(\Omega)$ seen in a), we may write $w = \tilde{w} + C_w$ and $v = \tilde{v} + C_v$. Then (6.18) reads

$$\int_\Omega \nabla \tilde{w} \cdot \nabla \tilde{v} \, d\mathbf{x} = 0 \qquad \text{for every } \tilde{v} \in V.$$

But by b) we obtain $\tilde{w} = 0$, i.e. $w = C_w$, proving the claim.

Problem 6.2.11 (Robin conditions). *Let Ω be a bounded regular domain in $\mathbb{R}^n$.*

a) Show that the formula

$$\|u\|_*^2 = \int_\Omega |\nabla u|^2 \, d\mathbf{x} + \int_{\partial\Omega} u^2 \, d\sigma$$

defines on $H^1(\Omega)$ an equivalent norm to the standard one.

b) Write the variational formulation of the Robin problem

$$\begin{cases} -\Delta u = f & \mathbf{x} \in \Omega \\ \partial_\nu u + \gamma u = 0 & \mathbf{x} \in \partial\Omega, \end{cases}$$

with $\gamma > 0$, and discuss whether it is well posed.

Solution. a) Let us prove the equivalence between $\|u\|_{H^1(\Omega)}$ and $\|u\|_*$ for functions in $C^\infty(\overline{\Omega})$. As Ω bounded and regular (Lipschitz would be enough) we get the equivalence in $H^1(\Omega)$ by density. So take $u \in C^\infty(\overline{\Omega})$. One way to proceed is the following. Consider the vector field (on $\mathbb{R}^n$)

$$\mathbf{F} = \frac{\mathbf{x}}{n}.$$

Clearly div $\mathbf{F} = 1$, and since Ω is bounded, $|\mathbf{F}| \leq M$ on $\overline{\Omega}$. We can use Gauss's formula on u^2 and $\mathbf{F}$ and write

$$\int_\Omega u^2 \, \mathrm{div} \, \mathbf{F} \, d\mathbf{x} = \int_{\partial\Omega} u^2 \mathbf{F} \cdot \boldsymbol{v} \, d\sigma - \int_\Omega \nabla(u^2) \cdot \mathbf{F} \, d\mathbf{x}. \tag{6.19}$$

Therefore

$$\int_\Omega u^2 \, \mathrm{div} \, \mathbf{F} \, d\mathbf{x} = \int_\Omega u^2 \, d\mathbf{x},$$

$$\left| \int_{\partial\Omega} u^2 \mathbf{F} \cdot \boldsymbol{v} \, d\sigma \right| \leq \int_{\partial\Omega} u^2 |\mathbf{F}| \, d\sigma \leq M \int_{\partial\Omega} u^2 \, d\sigma$$

and

$$\left| \int_\Omega \nabla(u^2) \cdot \mathbf{F} \, d\mathbf{x} \right| \leq \int_\Omega 2|u||\nabla u||\mathbf{F}| \, d\mathbf{x} \leq \frac{1}{4} \int_\Omega u^2 \, d\mathbf{x} + 4M^2 \int_\Omega |\nabla u|^2 \, d\mathbf{x}$$

(we used the elementary fact $2ab \leq a^2 + b^2$, with $a = |u|/2$ and $b = 2M|\nabla u|$). Substituting in (6.19) gives

$$\frac{3}{4} \int_\Omega u^2 \, d\mathbf{x} \leq M \int_{\partial\Omega} u^2 \, d\sigma + 4M^2 \int_\Omega |\nabla u|^2 \, d\mathbf{x}$$

i.e.

$$\|u\|_{H^1}^2 = \int_\Omega |\nabla u|^2 \, d\mathbf{x} + \int_\Omega u^2 \, d\mathbf{x}$$
$$\leq \frac{19}{3} M^2 \int_\Omega |\nabla u|^2 \, d\mathbf{x} + \frac{4}{3} M \int_{\partial\Omega} u^2 \, d\sigma \leq C \|u\|_*^2,$$

where $C = M \max \{19M, 4\}/3$. As already mentioned, this holds on $H^1(\Omega)$ by a density argument.

Conversely, the theory of traces implies $u|_{\partial\Omega} \in H^{1/2}(\partial\Omega)$, and the latter is embedded in $L^2(\Omega)$. Put in formulas, there exists a constant C_* such that, for every $u \in H^1(\Omega)$,

$$\|u|_{\partial\Omega}\|_{L^2(\partial\Omega)} \leq C_* \|u\|_{H^1(\Omega)}$$

so

$$\|u\|_* \leq \sqrt{1 + C_*^2} \, \|u\|_{H^1(\Omega)}.$$

Overall, $C_1 \| \cdot \|_{H^1(\Omega)} \leq \| \cdot \|_* \leq C_2 \| \cdot \|_{H^1(\Omega)}$, making the norms equivalent. Eventually, the norm $\| \cdot \|_*$ is induced by the scalar product

$$(u, v)_* = \int_\Omega \nabla u \cdot \nabla v \, d\mathbf{x} + \int_{\partial\Omega} uv \, d\sigma.$$

b) By the previous result, let us introduce the Hilbert space $H_*^1(\Omega)$ of H^1 functions equipped with norm $\| \cdot \|_*$ (and the relative scalar product). Multiplying by $v \in H_*^1(\Omega)$ and integrating:

$$\int_\Omega -\Delta u \, v \, d\mathbf{x} = \int_\Omega f v \, d\mathbf{x}.$$

Gauss's formula and the Robin condition give

$$\int_\Omega -\Delta u\, v\, d\mathbf{x} = \int_\Omega \nabla u \cdot \nabla v\, dx - \int_{\partial\Omega} \partial_\nu u\, v\, d\sigma$$
$$= \int_\Omega \nabla u \cdot \nabla v\, dx + \gamma \int_{\partial\Omega} u\, v\, d\sigma.$$

So let us set

$$B(u,v) = \int_\Omega \nabla u \cdot \nabla v\, dx + \gamma \int_{\partial\Omega} u\, v\, d\mathbf{x}, \quad Fv = \int_\Omega f v\, d\mathbf{x},$$

and write a variational formulation: *determine* $u \in H^1_*(\Omega)$ *such that*

$$B(u,v) = Fv \quad \text{for every } v \in H^1_*(\Omega).$$

The bilinear form B satisfies

$$|B(u,v)| \leq \max\{1,\gamma\}|(u,v)_*| \leq \max\{1,\gamma\}\|u\|_*\|v\|_*,$$
$$B(u,u) = \int_\Omega |\nabla u|^2\, dx + \gamma \int_{\partial\Omega} u^2\, d\sigma \geq \min\{1,\gamma\}\|u\|_*^2$$

(in the former we used Schwarz's inequality for the scalar product in $H^1_*(\Omega)$). Hence B is continuous and coercive (as $\gamma > 0$) and to use Lax-Milgram we just need to check the continuity on $H^1_*(\Omega)$ of the linear functional. As norms are equivalent, this holds for $f \in L^2(\Omega)$, for example (it suffices that it belong to the dual of $H^1(\Omega)$). In that case the problem is always well posed.

Problem 6.2.12 (Neumann conditions). *Let* $\Omega \subset \mathbb{R}^n$ *be a bounded and regular domain,* $\mathbf{b} \in L^\infty(\Omega;\mathbb{R}^n)$ *and* $f \in L^2(\Omega)$. *Write the variational formulation of*

$$\begin{cases} -\Delta u + \mathbf{b}(\mathbf{x}) \cdot \nabla u = f & \mathbf{x} \in \Omega \\ \partial_\nu u = 0 & \mathbf{x} \in \partial\Omega, \end{cases}$$

and discuss when it can be solved.

Solution. Due to the Neumann boundary conditions we choose $V = H^1(\Omega)$ and formulate the problem in weak form as follows: *determine* $u \in V$ *such that*

$$B(u,v) = \int_\Omega [\nabla u \cdot \nabla v + v\mathbf{b}(\mathbf{x}) \cdot \nabla u]\, d\mathbf{x} = \int_\Omega f v\, d\mathbf{x} = Fv \quad \text{for every } v \in V.$$

As $f \in L^2(\Omega)$ the linear functional F is bounded. Similarly, setting $\|\mathbf{b}\|_{L^\infty(\Omega)} = b$, we have

$$|B(u,v)| \leq \|\nabla u\|_{L^2(\Omega)}\|\nabla v\|_{L^2(\Omega)} + b\|\nabla u\|_{L^2(\Omega)}\|v\|_{L^2(\Omega)} \leq (1+b)\|u\|_V\|v\|_V$$

and also the bilinear form B is continuous.

The form B *cannot* be coercive: if $u = k \neq 0$, constant, $\|u\|_{L^2(\Omega)} = k\sqrt{|\Omega|} > 0$ while $B(u, u) = 0$. Yet B is weakly coercive for the Hilbert triple $\{V, H = L^2(\Omega), V^*\}$. In fact by Schwarz's inequality (and $2ab \leq a^2 + b^2$):

$$\int_\Omega u\mathbf{b} \cdot \nabla u \, d\mathbf{x} \geq -b\|\nabla u\|_{L^2(\Omega)}\|u\|_{L^2(\Omega)} \geq -\frac{1}{4}\|\nabla u\|^2_{L^2(\Omega)} - b^2\|u\|^2_{L^2(\Omega)} \quad (6.20)$$

so

$$B(u, u) = \int_\Omega |\nabla u|^2 \, d\mathbf{x} + \int_\Omega u\mathbf{b} \cdot \nabla u \, d\mathbf{x} \geq \frac{3}{4}\|\nabla u\|^2_{L^2(\Omega)} - b^2\|u\|^2_{L^2(\Omega)}$$

i.e.

$$B(u, u) + \left(\frac{3}{4} + b^2\right)\|u\|_{L^2(\Omega)} \geq \frac{3}{4}\|u\|^2_V$$

which shows the weak coercivity of B. As V is compactly embedded in H we can use Fredholm's alternative. Therefore the problem has solutions if and only if

$$\int_\Omega fw \, d\mathbf{x} = 0 \quad (6.21)$$

for every w that solves the homogeneous problem

$$\int_\Omega [\nabla w \cdot \nabla v + w\mathbf{b} \cdot \nabla v] \, d\mathbf{x} = 0 \quad \text{for every } v \in V.$$

This condition cannot be expressed in an elementary manner, except in special cases like $\mathbf{b} = \mathbf{0}$. In that case, in fact, the only homogeneous solutions are the constants, and (6.21) reduces to

$$\int_\Omega f \, d\mathbf{x} = 0.$$

Remark. Find the mistake in the following argument. If div $\mathbf{b} = 0$, instead of (6.20) we can write

$$\int_\Omega u\mathbf{b} \cdot \nabla u \, d\mathbf{x} = \frac{1}{2}\int_\Omega \mathbf{b} \cdot \nabla(u^2) \, d\mathbf{x} = \int_{\partial\Omega} u^2 \mathbf{b} \cdot \mathbf{v} \, d\sigma$$

whence, if $\mathbf{b} \cdot \mathbf{v} \geq b_0 > 0$ (the flow of $\mathbf{b}$ across $\partial\Omega$ is outgoing),

$$B(u, u) = \int_\Omega |\nabla u|^2 \, d\mathbf{x} + \int_\Omega u\mathbf{b} \cdot \nabla u \, d\mathbf{x} \geq \|\nabla u\|^2_{L^2(\Omega)} + b_0\|u\|^2_{L^2(\partial\Omega)}.$$

Recalling Problem 6.2.11, the form B is coercive!!

Problem 6.2.13 (Fredholm alternative). *Let $Q = (0, \pi) \times (0, \pi)$. Discuss the Dirichlet problem*

$$\begin{cases} \Delta u + 2u = f & \text{in } Q \\ u = 0 & \text{on } \partial Q. \end{cases}$$

Examine, in particular, the cases

$$f(x, y) = 1 \quad \text{and} \quad f(x, y) = x - \frac{\pi}{2}.$$

Solution. Set $V = H_0^1(Q)$ with the gradient norm in $L^2(Q)$, so the weak formulation is

$$B(u, v) = \int_Q [\nabla u \cdot \nabla v - 2uv]\, dxdy = \int_Q -fv\, dxdy \equiv Fv, \quad \text{for every } v \in V.$$

By Schwarz's and Poincaré's inequalities B is bilinear and continuous on V; in fact

$$|B(u, v)| \le (1 + 2C_P^2)\|u\|_V\|v\|_V.$$

If $f \in L^2(Q)$ (as in the given cases) also F is continuous on V. The form B *cannot* be coercive, since the homogeneous problem

$$B(u, v) = \int_Q (\nabla u \cdot \nabla v - 2uv)\, dxdy = 0, \quad \text{for every } v \in V, \quad (6.22)$$

which is the variational form of

$$\begin{cases} \Delta u + 2u = 0 & \text{in } Q \\ u = 0 & \text{on } \partial Q, \end{cases} \quad (6.23)$$

has nonzero solutions on V (found by separating variables as explained in Chap. 2). Easily we find

$$\bar{u}(x, y) = c \sin x \sin y \qquad c \in \mathbb{R} \quad (6.24)$$

In other words, $\lambda = 2$ is a Dirichlet eigenvalue of $-\Delta$ and (6.24) are the eigenfunctions. Inserting $u = v = \bar{u}$ in (6.22) produces

$$B(\bar{u}, \bar{u}) = 0$$

so the bilinear form is not coercive. Accidentally, the eigenfunction $\sin x \sin y$ is *positive* on Q. By general principles we deduce that 2 is the *first Dirichlet eigenvalue* of the operator $-\Delta$ (the smallest one). The other eigenvalues can be computed by variable separation. Let us write $u(x, y) = p(x)q(y)$ and substitute into $-\Delta u = \lambda u$. A few computations give

$$\frac{p''(x)}{p(x)} = -\frac{q''(y)}{q(y)} - \lambda$$

which breaks into two boundary-value problems

$$p'' + \mu^2 p = 0, \quad p(0) = p(\pi) = 0$$

and

$$q'' + (\lambda - \mu^2)q = 0, \quad q(0) = q(\pi) = 0.$$

This gives the eigenvalue sequence

$$\lambda_{nm} = n^2 + m^2, \quad \varphi_{nm}(x, y) = \sin nx \sin my, \quad n, m \ge 1, \text{ integer.}$$

Back to the variational problem, we cannot use the Lax-Milgram theorem. Let us try with Fredholm's alternative. The bilinear form B is *weakly coercive*; in fact,

$$B(u,u) + 2 \int_Q u^2 dx dy = \|u\|_V^2$$

and the Hilbert triple $\{V, L^2(Q), V^*\}$ satisfies the assumptions. Hence the problem has solutions if and only if F vanishes on every solution of the adjoint homogeneous problem. As B is symmetric, the adjoint homogeneous problem coincides with (6.22), so the problem has a solution if and only if

$$\int_Q f(x,y) \sin x \sin y \, dx dy = 0.$$

When $f(x,y) = 1$, $\int_Q \sin x \sin y \, dx dy = 4$, so *no* solution exists. Vice versa, when $f(x,y) = x - \pi/2$ we have

$$\int_Q \left(x - \frac{\pi}{2} \right) \sin x \sin y \, dx dy = 0$$

and there are infinitely many solutions of the form

$$u(x,y) = U(x,y) + c \sin x \sin y.$$

In order to find U we write it as superposition of eigenfunctions φ_{nm}, that is[11]

$$U(x,y) = \sum_{n,m \geq 1} u_{nm} \sin nx \sin my$$

and we require that it satisfies the equation

$$\Delta U + 2U = \sum_{n,m \geq 1} \left(-n^2 - m^2 + 2 \right) u_{nm} \sin nx \sin my = x - \frac{\pi}{2}.$$

Expand f in sines-Fourier series. The coefficients f_{nm} are

$$f_{nm} = \frac{4}{\pi^2} \int_0^\pi \int_0^\pi \left(x - \frac{\pi}{2} \right) \sin nx \sin my \, dx dy, \qquad n, m \geq 1.$$

Then

$$f_{nm} = \begin{cases} 0 & n \text{ odd and any } m, \text{ or } m \text{ even and any } n, \\ -\dfrac{8}{\pi} \dfrac{1}{2h(2k+1)} & n = 2h \text{ and } m = 2k+1, h \geq 1, k \geq 0 \end{cases}$$

---

[11] Chap. 2.

so

$$u_{nm} = \begin{cases} 0 & n \text{ odd, or } m \text{ even,} \\ \dfrac{-4}{\pi h \, (2k+1) \, [2 - 4h^2 - (2k+1)^2]} & n = 2h \text{ and } m = 2k+1, h \geq 1, k \geq 0. \end{cases}$$

Finally,

$$U(x,y) = \sum_{h \geq 1, k \geq 0} \frac{-4}{\pi h \, (2k+1) \, [2 - 4h^2 - (2k+1)^2]} \sin 2hx \sin (2k+1) \, y.$$

Problem 6.2.14 (Fredholm alternative). *Set* $Q = (0, \pi) \times (0, \pi)$. *Discuss the problem*

$$\begin{cases} \Delta u + \lambda u = 1 & \text{in } Q \\ u = 0 & \text{on } \partial Q \setminus \{y = 0\} \\ -u_y(x, 0) = x & \text{for } 0 \leq x \leq \pi \end{cases}$$

where λ *is a real parameter.*

Solution. For the variational formulation it is natural to choose the subspace in $H^1(Q)$ adapted to the Dirichlet conditions, that is:

$$V = \left\{ v \in H^1(Q) : u = 0 \text{ on } \partial Q \setminus \{y = 0\} \right\}.$$

Since the Poincaré inequality $\|v\|_{L^2(Q)} \leq C_P \|\nabla v\|_{L^2(Q)}$ holds, we can choose an equivalent norm $\|v\|_V = \|\nabla v\|_{L^2(Q)}$. Multiply by $v \in V$ and integrate. The left-hand side becomes

$$\int_Q [\Delta u + \lambda u] \, v \, dxdy = \int_{\partial Q} \partial_\nu u \, v \, ds - \int_Q [\nabla u \cdot \nabla v - \lambda uv] \, dxdy$$

$$= -\int_0^\pi u_y(x, 0) v(x, 0) \, dx - \int_Q [\nabla u \cdot \nabla v - \lambda uv] \, dxdy$$

$$= \int_0^\pi x v(x, 0) \, dx - \int_Q [\nabla u \cdot \nabla v - \lambda uv] \, dxdy.$$

The weak formulation reads: *determine* $u \in V$ *such that*

$$B(u, v) = Fv \quad \text{for every } v \in V,$$

where

$$B(u, v) = \int_Q [\nabla u \cdot \nabla v - \lambda uv] \, dxdy$$

and

$$Fu = -\int_Q v \, dxdy + \int_0^\pi x v(x, 0) \, dx.$$

From the trace inequality we have $\|v\|_{L^2(\partial Q)} \le C_* \|v\|_V$. Hence

$$|Fv| \le \pi \|v\|_{L^2(Q)} + \pi \sqrt{\frac{\pi}{3}} \|v(\cdot,0)\|_{L^2(0,\pi)} \le \pi \left(C_P + \sqrt{\frac{\pi}{3}} C_* \right) \|v\|_V,$$

so the linear functional F is continuous. Similarly, since

$$|B(u,v)| \le \|\nabla u\|_{L^2(Q)} \|\nabla v\|_{L^2(Q)} + |\lambda| \|u\|_{L^2(Q)} \|v\|_{L^2(Q)} \le (1 + |\lambda| C_P^2) \|u\|_V \|v\|_V,$$

B is continuous.

Take $\lambda \le 0$. Then

$$B(u,u) \ge \|u\|_V^2$$

so B is coercive, too, and the Lax-Milgram theorem guarantees existence and uniqueness of the weak solution.

Conversely, take $\lambda > 0$. Now we can only guarantee weak coercivity: in fact

$$B(u,u) + \lambda \|u\|_{L^2(Q)}^2 \ge \|u\|_V^2.$$

Hence we can use Fredholm's alternative with respect to the Hilbert triple $\{V, L^2(Q), V^*\}$. By Rellich's theorem, in fact, the embedding of V in $L^2(Q)$ is compact. As B is symmetric, the problem admits solutions if and only if

$$\int_Q w(x,y)\,dxdy - \int_0^\pi xw(x,0)\,dx = 0 \tag{6.25}$$

for every w that solves the adjoint homogeneous problem $B(w,v) = 0$ for every $v \in V$, which is the weak formulation of

$$\begin{cases} -\Delta w = \lambda w & \text{in } Q \\ w = 0 & \text{on } \partial Q \setminus \{y = 0\} \\ -w_y(x,0) = 0 & 0 \le x \le \pi. \end{cases} \tag{6.26}$$

From the general theory we know that (6.26) has a sequence of eigenvalues $0 < \lambda_1 < \lambda_2 \le \lambda_3 \le \ldots$, while it only has the zero solution for $\lambda \ne \lambda_k$. Therefore when $\lambda \ne \lambda_k$ the solution exists and is unique. Vice versa, if $\lambda = \lambda_k$, the problem has solutions if and only if the compatibility condition (6.25) holds, and then there are infinitely many solutions. In this case the eigenvalues can be found by separating the variables, as was done in the previous problem; we find

$$\lambda = \lambda_{nm} = n^2 + \frac{(2m+1)^2}{4} \qquad (n \ge 1, m \ge 0)$$

with eigenfunctions

$$\varphi_{nm}(x,y) = \sin nx \cos\left(\frac{2m+1}{2} y\right).$$

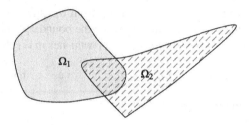

Fig. 6.1 Domain for Problem 6.2.15

If $\lambda \neq \lambda_{nm}$ the problem has one solution only. If λ is one of the λ_{nm} the problem can be solved if and only if

$$\int_Q \sin nx \cos \left(\frac{2m+1}{2}y\right)\, dxdy - \int_0^\pi x \sin nx \, dx = 0.$$

A simple computation shows that the sum on the left is zero for no value of $n \geq 1$ and $m \geq 0$, thus the compatibility conditions do not hold and so if λ is one of the λ_{nm}, there is no solution.

Problem 6.2.15 (Schwarz's alternating method[a]). *Referring to Fig. 6.1, let $\Omega \subset \mathbb{R}^2$ be a domain such that $\Omega = \Omega_1 \cup \Omega_2$ where Ω_1 and Ω_2 have regular boundary (Lipschitz is enough), with $\Omega_1 \cap \Omega_2 \neq \emptyset$. Set $V = H_0^1(\Omega)$ with inner product $(u, v) = \int_\Omega \nabla u \cdot \nabla v$ and define (see Exercise 5.3.26 on page 313, Chap. 5)*

$$V_1 = \left\{u \in H_0^1(\Omega_1), u = 0 \text{ in } \Omega \setminus \Omega_1\right\} \quad \text{and} \quad V_2 = \left\{u \in H_0^1(\Omega_2), u = 0 \text{ in } \Omega \setminus \Omega_2\right\}.$$

Given $f \in L^2(\Omega)$ and an arbitrary element $u_0 \in H_0^1(\Omega)$, define u_{2n+1} $(n \geq 0)$ and u_{2n} $(n \geq 1)$ by the following recursive procedure:

$$\begin{cases} -\Delta u_{2n+1} = f & \text{in } \Omega_1 \\ u_{2n+1} = u_{2n} & \text{in } (\Omega \setminus \Omega_1) \cup \partial\Omega_1 \end{cases} \quad \text{and} \quad \begin{cases} -\Delta u_{2n} = f & \text{in } \Omega_2 \\ u_{2n} = u_{2n-1} & \text{in } (\Omega \setminus \Omega_2) \cup \partial\Omega_2. \end{cases}$$

Prove that u_n converges in $H_0^1(\Omega)$ to the solution $u \in H_0^1(\Omega)$ of

$$-\Delta u = f \quad \text{in } \Omega,$$

by showing that

$$u_{2n+1} - u = P_{V_1^\perp}(u_{2n} - u) \quad \text{and} \quad u - u_{2n} = P_{V_2^\perp}(u - u_{2n-1})$$

and using Problem 5.2.9, Chap. 5 (page 288).

---
[a] Useful in numerical approximation methods for boundary-value problems.

Solution. The sequence u_n is built starting from u_0, solving the Dirichlet problem one time on Ω_1 and the next on Ω_2 and updating the boundary data with u_{n-1}. Note that $u_{2n+1} - u$ is harmonic in Ω_1 and every $v_1 \in V_1$ vanishes in Ω_1. So we can write

$$(u_{2n+1} - u, v_1) = 0 \quad \forall v_1 \in V_1, \qquad\qquad \text{i.e.} \qquad u_{2n+1} - u \in V_1^{\perp}.$$

Adding and subtracting u_{2n} gives, for every $n \geq 0$,

$$(u_{2n+1} - u_{2n}, v_1) = (u - u_{2n}, v_1), \qquad \forall v_1 \in V_1.$$

As $u_{2n+1} - u_{2n} \in V_1$, the last equation amounts to say that

$$u_{2n+1} - u_{2n} = P_{V_1}(u - u_{2n})$$

or (as $u - u_{2n} = (u - u_{2n+1}) + (u_{2n+1} - u_{2n})$)

$$u - u_{2n+1} = P_{V_1^{\perp}}(u - u_{2n}).$$

Arguing in a similar manner we find

$$u - u_{2n} = P_{V_2^{\perp}}(u - u_{2n-1}).$$

Set $x_0 = u_0$, $x_{2n+1} = u - u_{2n+1}$ and $x_n = u - u_{2n}$. The sequence (x_n) fits into the iterative pattern of Problem 5.2.9, Chap. 5. Now, $u_n \to u$ in $H_0^1(\Omega)$ is the same as $x_n \to 0$ in V. To use Problem 5.2.9, Chap. 5, we need V_1 and V_2 to be closed in V and $V_1^{\perp} \cap V_2^{\perp} = \{0\}$. But this follows from Exercise 5.3.26, Chap. 5.

Problem 6.2.16 (Oblique derivative). *Let $\Omega \subset \mathbb{R}^2$ be a bounded domain with C^1 boundary. For $b \in C^1(\overline{\Omega})$, consider the bilinear form on $H^1(\Omega)$:*

$$a(u, v) = \int_{\Omega} \left[u_x v_x + u_y v_y + b u_x v_y - b u_y v_x + b_y u_x v - b_x u_y v \right] dx dy.$$

a) Given $f \in L^2(\Omega), \mathbf{f} = (f_1, f_2) \in L^2(\Omega; \mathbb{R}^2)$ and $g \in L^2(\partial\Omega)$, interpret in classical sense the variational problem

$$a(u, v) = Fv, \qquad\qquad \text{for any } v \in H^1(\Omega) \tag{6.27}$$

where $Fv = \int_{\Omega}(fv + \mathbf{f} \cdot \nabla v) + \int_{\partial\Omega} gv\, ds$.

b) Tell under which conditions the problem has solutions. Discuss the case $b \equiv$ constant.

Solution. a) First of all let us write the bilinear form in a clearer way. Setting

$$A(x, y) = \begin{pmatrix} 1 & b(x, y) \\ -b(x, y) & 1 \end{pmatrix} \qquad \text{and} \qquad \mathbf{b}(x, y) = (b_y, -b_x)$$

we have

$$a(u, v) = \int_\Omega [A(x, y) \nabla u \cdot \nabla v + (\mathbf{b}(x, y) \cdot \nabla u) v] \, dxdy.$$

Thus (6.27) is the variational formulation of

$$\begin{cases} \mathcal{L}u = -\operatorname{div}(A(x, y) \nabla u) + \mathbf{b}(x, y) \cdot \nabla u = f - \operatorname{div} \mathbf{f} & \text{in } \Omega, \\ \partial_\nu^A u \equiv A(x, y) \nabla u \cdot \nu = g & \text{on } \partial\Omega, \end{cases} \tag{6.28}$$

where $\nu = (\nu_1, \nu_2)$ is the outward normal to $\partial\Omega$. Let us make (6.28) more explicit. As

$$A(x, y) \nabla u = \left(u_x + b(x, y) u_y, -b(x, y) u_x + u_y\right)$$

we have

$$\operatorname{div}(A(x, y) \nabla u) = u_{xx} + bu_{yx} + b_x u_y - b_y u_x - bu_{xy} + u_{yy}$$

so

$$-\operatorname{div}(A(x, y) \nabla u) + \mathbf{b}(x, y) \cdot \nabla u = -\Delta u.$$

Introducing the unit vector $\boldsymbol\tau = (-\nu_2, \nu_1)$, tangent to $\partial\Omega$, we obtain:

$$A(x, y) \nabla u \cdot \nu = (u_x + bu_y)\nu_1 + (-bu_x + u_y)\nu_2 =$$
$$= \nabla u \cdot \nu + b\nabla u \cdot \boldsymbol\tau = \nabla u \cdot (\nu + b\boldsymbol\tau)$$
$$= \partial_\sigma u$$

where $\sigma = \nu + b\boldsymbol\tau$. Eventually, we have

$$\begin{cases} \Delta u = \operatorname{div} \mathbf{f} - f & \text{in } \Omega, \\ \partial_\sigma u = g & \text{on } \partial\Omega. \end{cases} \tag{6.29}$$

Note that

$$\nu \cdot \frac{\sigma}{|\sigma|} = \nu \cdot \frac{\nu + b\boldsymbol\tau}{\sqrt{1 + b^2}} = \frac{1}{\sqrt{1 + b^2}}, \tag{6.30}$$

so the vector σ *is never tangent* to $\partial\Omega$; for this reason (6.29) is called an *oblique-derivative problem*.

b) If u is solution to problem (6.29) and $c \in \mathbb{R}$, $u + c$ solves the same problem; the Lax-Milgram lemma does not apply, for the bilinear form, albeit continuous on $H^1(\Omega)$, is *not* coercive. Yet it is *weakly coercive*. In fact, set $M = \max |\mathbf{b}|$, so that

$$\left| \int_\Omega (\mathbf{b}(x, y) \cdot \nabla v) v \, dxdy \right| \le M \int_\Omega |\nabla v| \, |v| \, dxdy \le M \, \|\nabla v\|_{L^2(\Omega)} \|v\|_{L^2(\Omega)}$$

$$\le \frac{1}{2} \|\nabla v\|_{L^2(\Omega)} + \frac{M^2}{2} \|v\|_{L^2(\Omega)}^2$$

whence

$$a\left(v,v\right)=\int_{\Omega}\left[A\left(x,y\right)\nabla v\cdot\nabla v+\left(\mathbf{b}\left(x,y\right)\cdot\nabla v\right)v\right]dxdy$$

$$\geq\frac{1}{2}\left\|\nabla v\right\|_{L^{2}(\Omega)}-\frac{M^{2}}{2}\left\|v\right\|_{L^{2}(\Omega)}^{2}$$

and then

$$a\left(v,v\right)+M^{2}\left\|v\right\|_{L^{2}(\Omega)}^{2}\geq\min\left\{\frac{1}{2},\frac{M^{2}}{2}\right\}\left\|v\right\|_{H^{1}(\Omega)}^{2}.$$

By Rellich's theorem the immersion of $V=H^{1}\left(\Omega\right)$ in $H=L^{2}\left(\Omega\right)$ is compact.

By Fredholm's alternative, problem (6.27) can be solved if and only if F is orthogonal, in $H^{1}\left(\Omega\right)$, to the solution of the *adjoint homogeneous* problem

$$a_{*}\left(w,v\right)=0\quad\text{for every }v\in H^{1}\left(\Omega\right)\tag{6.31}$$

where

$$a_{*}\left(w,v\right)=a\left(v,w\right)=\int_{\Omega}\left[A\left(x,y\right)\nabla v\cdot\nabla w+\left(\mathbf{b}\left(x,y\right)\cdot\nabla v\right)w\right]dxdy=$$

$$=\int_{\Omega}\left[A^{\top}\left(x,y\right)\nabla w\cdot\nabla v+\left(\mathbf{b}\left(x,y\right)\cdot\nabla v\right)w\right]dxdy.$$

The adjoint equation (6.31) is the variational formulation of

$$\begin{cases}\mathcal{L}^{*}w=-\operatorname{div}\left(A^{\top}\left(x,y\right)\nabla w\right)-\operatorname{div}(\mathbf{b}\left(x,y\right)w)=0 & \text{in }\Omega,\\ \partial_{\boldsymbol{\nu}}^{A^{\top}}w\equiv A^{\top}\left(x,y\right)\nabla w\cdot\boldsymbol{\nu}=0 & \text{on }\partial\Omega\end{cases}\tag{6.32}$$

which reduces, with computations of the previous kind, to the following (Robin-like) problem:

$$\begin{cases}\Delta w=0 & \text{in }\Omega,\\ \partial_{\boldsymbol{\sigma}^{*}}w+\left(\partial_{\boldsymbol{\tau}}b\right)w=0 & \text{on }\partial\Omega,\end{cases}\tag{6.33}$$

where

$$\boldsymbol{\sigma}^{*}=\boldsymbol{\nu}-b\boldsymbol{\tau}.$$

Problem (6.27) can be solved if and only if

$$\int_{\Omega}\left(fw+\mathbf{f}\cdot\nabla w\right)+\int_{\partial\Omega}gw\,ds=0,$$

for every solution w of (6.31).

If b is constant, the solutions to the adjoint homogeneous problem are only the constants. In fact, if $w\in H^{1}\left(\Omega\right)$ solves (6.31), choosing $v=w$ gives

$$0=a_{*}\left(w,w\right)=\int_{\Omega}A^{\top}\left(x,y\right)\nabla w\cdot\nabla w\,dxdy=\int_{\Omega}\left|\nabla w\right|^{2}dxdy$$

whence $w =$ is constant. Problem (6.27) can be solved, therefore, if and only if

$$\int_\Omega f + \int_{\partial\Omega} g \, ds = 0.$$

Problem 6.2.17 (Transmission conditions). *Let Ω_1 and Ω be bounded Lipschitz domains in $\mathbb{R}^n$ such that $\overline{\Omega}_1 \subset \Omega$ and set $\Omega_2 = \Omega \setminus \overline{\Omega}_1$. On Ω_1 and Ω_2 consider the bilinear forms*

$$a_k (u, v) = \int_{\Omega_k} A^k (\mathbf{x}) \nabla u \cdot \nabla v \, d\mathbf{x} \qquad (k = 1, 2)$$

where A^k are uniformly elliptic matrices. Suppose the entries of A^k are continuous on $\overline{\Omega}_k$, while we allow

$$A(\mathbf{x}) = \begin{cases} A^1 (\mathbf{x}) & \text{in } \Omega_1 \\ A^2 (\mathbf{x}) & \text{in } \Omega_2 \end{cases}$$

to be discontinuous across $\Gamma = \partial\Omega_1$. Let $u \in H_0^1 (\Omega)$ be the variational solution of

$$a (u, v) = a_1 (u, v) + a_2 (u, v) = \int_\Omega f v \, d\mathbf{x} \qquad \text{for every } v \in H_0^1 (\Omega) \qquad (6.34)$$

with $f \in L^2 (\Omega)$.

Calling u_k the restriction of u to Ω_k, determine (formally) what type of problem the pair u_1, u_2 satisfies, and in particular which (two) conditions on Γ express the coupling of u_1 and u_2.

Solution. If we restrict to $v \in C_0^\infty (\Omega_k)$, (6.34) reduces to

$$a_k (u, v) = \int_\Omega f_k v \, d\mathbf{x},$$

where $f_k = f_{|\Omega_k}$, and then

$$\mathscr{L}_k u = -\operatorname{div}\left(A^k (\mathbf{x}) \nabla u_k \right) = f_k \quad \text{in } \Omega_k \qquad (k = 1, 2). \qquad (6.35)$$

On the other hand, as $u \in H_0^1 (\Omega)$, it follows $u_2 = 0$ on $\partial\Omega$, and on Γ the traces of u_1 and u_2 must coincide (Problem 5.2.30 on page 309, Chap. 5):

$$u_1 = u_2 \quad \text{on } \Gamma$$

which gives us the first constraint.

To find the second condition we proceed formally, multiplying (6.35) by $v \in C_0^\infty (\Omega)$, integrating on Ω_k and using Gauss's formula; if ν_k is the unit normal of Γ, pointing away from Ω_k, and

$$v_k^* = (A^k)^T (\mathbf{x}) \, \nu_k \qquad\qquad \text{is the } \textit{outward unit conormal,}$$

then

$$-\int_{\partial\Omega_1} \frac{\partial u_k}{\partial v_k^*} \, v d\sigma + a_k \, (u_k, v) = \int_{\Omega_k} f_k v \, d\mathbf{x};$$

adding these gives

$$\int_\Gamma v \left(\frac{\partial u_1}{\partial v_1^*} + \frac{\partial u_2}{\partial v_2^*} \right) d\sigma = 0.$$

As v was arbitrary we obtain the (*transmission condition*)

$$\frac{\partial u_1}{\partial v_1^*} + \frac{\partial u_2}{\partial v_2^*} = 0. \tag{6.36}$$

Remark. As $u \in H_0^1 (\Omega)$, its trace on Γ belongs to $H^{1/2} (\Gamma)$ but the trace of the normal and conormal derivatives is, *a priori*, not well defined. Notwithstanding, for functions such that $\mathcal{L}_k u \in L^2 (\Omega_k), k = 1, 2$, as in the present case, it is possible to define $\partial_{v_1^*} u_1 + \partial_{v_2^*} u_2$ as an element of the dual of $H^{1/2} (\Gamma)$ and to give a meaining to (6.36).

6.2.3 Evolution problems

Problem 6.2.18 (Drift-diffusion-reaction). *Consider the problem*

$$\begin{cases} u_t - (a \, (x) \, u_x)_x + b \, (x) \, u_x + c \, (x) \, u = f \, (x, t) & 0 < x < 1, 0 < t < T, \\ u \, (x, 0) = u_0 \, (x) & 0 \le x \le 1, \\ u \, (0, t) = 0, \; u \, (1, t) = k \, (t) & 0 \le t \le T. \end{cases}$$

a) Modify the unknown function so to reduce to homogeneous Dirichlet conditions.

b) Suitably choosing the functional space, write a weak formulation.

c) Discuss whether the problem is well posed, clarifying the conditions on the coefficients and data, and giving an energy estimate of the solution.

Solution. a) Let us proceed formally. We reduce first to vanishing boundary conditions. We interpolate the Dirichlet data on the interval $[0, 1]$ using

$$g \, (x, t) = xk \, (t)$$

and set $w = u - g$. Note

$$g \, (0, t) = 0 \quad \text{and} \quad g \, (1, t) = k \, (t).$$

Then

$$u_t \, (x, t) = w_t \, (x, t) + x\dot{k} \, (t) \quad \text{and} \quad u_x \, (x, t) = w_x \, (x, t) + k \, (t).$$

Substituting into the differential equation gives the following problem for w:

$$\begin{cases} w_t - (a\,(x)\,w_x)_x + b\,(x)\,w_x + c\,(x)\,w = F\,(x,t) + k\,(t)\,a'(x) & 0 < x < 1, 0 < t < T, \\ w\,(x,0) = w_0\,(x) & 0 \le x \le 1, \\ w\,(0,t) = 0,\ w\,(1,t) = 0 & 0 \le t \le T, \end{cases}$$

where

$$F\,(x,t) = f\,(x,t) - [b\,(x) + xc\,(x)]k\,(t) - x\dot{k}\,(t)$$

and

$$w_0\,(x) = u_0\,(x) - k\,(0)\,x.$$

b) Let $(\cdot,\cdot)$ be the inner product on $L^2\,(0,1)$ and set

$$B(u,v) = \int_0^1 \left[a\,(x)\,u'\,(x)\,v'\,(x) + b\,(x)\,u'\,(x)\,v\,(x) + c\,(x)\,u(x)v\,(x)\right] dx$$

for every pair $u, v \in V = H_0^1\,(0,1)$. On V we choose $\|u'\|_{L^2(0,1)}$ as norm.

Now set $\mathbf{w}\,(t) = w\,(\cdot,t)$, interpreting $w\,(x,t)$ as function of t with values in V. A weak formulation is: find $\mathbf{w} \in L^2\,(0,T;V)$ such that $\dot{\mathbf{w}} \in L^2\,(0,T;V^*)$ and:

i) $\dfrac{d}{dt}(\mathbf{w}\,(t),v) + B\,(\mathbf{w}\,(t),v) = (\mathbf{F}\,(t),v) - k(t)(a,v')$

for every $v \in V$, in distributional sense in $(0,T)$.

ii) $\|\mathbf{w}\,(t) - w_0\|_{L^2(0,1)} \to 0$ as $t \to 0^+$.

c) For the problem to be well posed we require the coefficients a, b and c be bounded, and the diffusion coefficient a positive[12]; precisely:

$$|a\,(x)|, |b\,(x)|, |c\,(x)| \le M, \qquad a\,(x) \ge a_0 > 0 \qquad \text{a.e. on } (0,1).$$

We also need the functional

$$v \mapsto (\mathbf{F}\,(t),v) - k(t)(a,v')$$

to define, for almost every t, an element in V^*. As a, b and c are bounded, it suffices to ask

$$\mathbf{f} \in L^2\,(0,T;L^2\,(0,1)) \quad \text{and} \quad k \in H^1\,(0,T).$$

Finally, we assume w_0, or equivalently u_0, belongs to $L^2\,(0,1)$.

There remains to examine the bilinear form B. To use the Faedo-Galerkin theory we have to check if B is *continuous* and *weakly coercive on V*.

---

[12] Then the equation is *uniformly parabolic*.

Using the boundedness of the coefficients and Schwarz's and Poincaré's inequalities ($L^2 = L^2(0, 1)$), we have:

$$|B(u, v)| \leq M \int_0^1 |u'v' + u'v + uv| \, dx$$

$$\leq M \left(\|u'\|_{L^2} \|v'\|_{L^2} + \|u'\|_{L^2} \|v\|_{L^2} + \|u\|_{L^2} \|u\|_{L^2} \right)$$

$$\leq M \left(1 + C_P + C_P^2 \right) \|u'\|_{L^2} \|v'\|_{L^2}$$

so that B is continuous on V.

Let us check the weak coercivity:

$$B(u, u) = \int_0^1 \left[a(x)(u')^2 + b(x)u'u + c(x)u^2 \right] dx.$$

Concerning the middle integral term, we may write, for every $\varepsilon > 0$,

$$\left| \int_0^1 b(x) u'u \, dx \right| \leq M \|u'\|_{L^2} \|u\|_{L^2} \leq \frac{M\varepsilon}{2} \|u'\|_{L^2}^2 + \frac{M}{2\varepsilon} \|u\|_{L^2}^2$$

so that

$$B(u, u) \geq \left(a_0 - \frac{M\varepsilon}{2} \right) \|u'\|_{L^2}^2 - \left(\frac{M}{2\varepsilon} + M \right) \|u\|_{L^2}^2.$$

Choose $\varepsilon = \frac{a_0}{M}$. Then, setting $\lambda_0 = \frac{M^2}{2a_0} + M$ we have

$$B(u, u) \geq \frac{a_0}{2} \|u'\|_{L^2}^2 - \lambda_0 \|u\|_{L^2}^2$$

which entails that the bilinear form

$$B_0(u, v) = B(u, v) + \lambda_0(u, v)$$

is coercive on V or, equivalently, B is weakly coercive (with coercivity constant $a_0/2$). Setting $\mathbf{z}(t) = e^{-\lambda_0 t} \mathbf{w}(t)$, the function $\mathbf{z}$ is a solution to:

i')

$$\frac{d}{dt}(\mathbf{z}(t), v) + B_0(\mathbf{z}(t), v) = (e^{-\lambda_0 t} \mathbf{F}(t), v)$$

for every $v \in V$, in distributional sense in $(0, T)$.

ii') $\|\mathbf{z}(t) - w_0\|_{L^2} \to 0$ as $t \to 0^+$.

Under the mentioned hypotheses, the problem has a unique weak solution **z**. From the theory we then obtain an energy estimate, written in terms of $z\,(x,t)$:

$$\max_{t\in[0,T]} \int_0^1 z^2\,(x,t)^2\,dx + a_0 \int_0^T \int_0^1 z_x^2\,(x,t)\,dxdt$$

$$\leq e^{(\lambda_0+1)T} \left\{ \int_0^T \int_0^1 e^{-\lambda_0 t} F^2\,(x,t)\,dxdt + \int_0^1 z_0^2\,(x)\,dx \right\}.$$

At this point it is not hard to transfer the conclusions back to the original problem.

Problem 6.2.19 (Asymptotic stability). *Consider*

$$\begin{cases} u_t - \Delta u = 0 & \mathbf{x} \in \Omega, \, t \in (0,T) \\ u(\mathbf{x},0) = g(\mathbf{x}) & \mathbf{x} \in \Omega \\ u(\sigma,t) = 0 & \sigma \in \partial\Omega, t \in (0,T), \end{cases}$$

where Ω is bounded and regular, and $g \in L^2(\Omega)$.

a) Write a weak formulation. Using the Faedo-Galerkin method with the subspaces spanned by the Dirichlet eigenfunctions of the operator $-\Delta$ ([18, Chap. 8, Sect. 3]), deduce the existence of a solution for every $t > 0$, and find its expression.

b) If u denotes the solution, prove that

$$\|\nabla u\|_{L^2(\Omega)} \leq \frac{1}{\sqrt{2et}} \|g\|_{L^2(\Omega)} \qquad \text{for every } t > 0.$$

c) If $g \in H_0^1(\Omega)$, prove that

$$\|\nabla u\|_{L^2(\Omega)} \leq e^{-\lambda_1 t} \|\nabla g\|_{L^2(\Omega)} \qquad \text{for every } t > 0,$$

where λ_1 is the first eigenvalue of $-\Delta$.

Solution. a) Let us set $\mathbf{u}\,(t) = u\,(\cdot,t)$, by interpreting $u\,(x,t)$ as a function in t with values in $V = H_0^1(\Omega)$ (this choice is dictated by the boundary conditions). Call $(\cdot,\cdot)$ the usual scalar product on $H = L^2(\Omega)$ and define

$$B(u,v) = \int_\Omega \nabla u \cdot \nabla v \, d\mathbf{x}.$$

A weak formulation is: seek $\mathbf{u} \in L^2\,(0,T;V)$ such that $\dot{\mathbf{u}} \in L^2\,(0,T;V^*)$ and

i) $\dfrac{d}{dt}(\mathbf{u}\,(t),v) + B\,(\mathbf{u}\,(t),v) = 0$

for every $v \in V$, in distributional sense on $(0,T)$.

ii) $\|\mathbf{u}\,(t) - g\|_H \to 0$ for $t \to 0^+$.

The conditions for being well posed are fulfilled (the bilinear form is the scalar product on V, and hence continuous and uniformly coercive in t).

Let $0 < \lambda_1 < \lambda_2 \leq \lambda_3 \leq \ldots$ be the eigenvalues of

$$u \in V : -\Delta u = \lambda u$$

and $\varphi_1, \varphi_2, \ldots$ the corresponding normalised eigenfunctions in $L^2(\Omega)$. These form a complete orthonormal system in $L^2(\Omega)$ and a complete orthogonal system in V. Recall that for every v

$$B(\varphi_k, v) = \lambda_k(\varphi_k, v) \tag{6.37}$$

whence

$$\|\nabla \varphi_k\|_{L^2}^2 = \lambda_k \|\varphi_k\|_{L^2}^2 = \lambda_k.$$

Let V_m be the space spanned by the first m eigenfunctions. Set

$$\mathbf{u}_m = \sum_{k=1}^{m} c_{m,k}(t)\varphi_k \qquad \text{and} \qquad g = \sum_{k=1}^{\infty} g_k \varphi_k.$$

By (6.37) the Faedo-Galerkin approximating problem is

$$0 = \sum_{k=1}^{m} (\dot{c}_{m,k}(t)\varphi_k, v) + B\left(c_{m,k}(t)\varphi_k, v\right) = \sum_{k=1}^{m} \left((\dot{c}_{m,k}(t) + \lambda_k c_{m,k}(t))\varphi_k, v\right)$$

and the coefficient $c_{m,k}$ solves $\dot{c}_{m,k}(t) + \lambda_k c_{m,k}(t) = 0$ with initial condition $c_{m,k}(0) = g_k$. Therefore

$$c_{m,k}(t) = g_k e^{-\lambda_k t}.$$

The approximate solutions are then

$$\mathbf{u}_m(t) = \sum_{k=1}^{m} g_k e^{-\lambda_k t} \varphi_k$$

which converge in $L^2(0, T; V)$ to the solution

$$\mathbf{u}(t) = \sum_{k=1}^{\infty} g_k e^{-\lambda_k t} \varphi_k(\mathbf{x})$$

as $m \to \infty$. In practice we have used variable separation.

b) Since φ_k form an orthogonal basis in V, we can write, back in the original notation,

$$\nabla u(\mathbf{x}, t) = \sum_{k=1}^{\infty} g_k e^{-\lambda_k t} \nabla \varphi_k(\mathbf{x})$$

and from (6.37) we have

$$\|\nabla u(\mathbf{x}, t)\|_{L^2(\Omega)}^2 = \sum_{k=1}^{\infty} |g_k|^2 \lambda_k e^{-2\lambda_k t}.$$

The function $f(z) = ze^{-az}$ has a maximum at $z = 1/a$, with $f(1/a) = 1/ae$, so for every $k \geq 1$

$$\lambda_k e^{-2\lambda_k t} \leq \frac{1}{2et}.$$

Hence

$$\|\nabla u(\mathbf{x}, t)\|_{L^2(\Omega)} \leq \frac{1}{\sqrt{2et}} \|g\|_{L^2(\Omega)}, \quad t > 0.$$

c) If g is in $H_0^1(\Omega)$ we can get a better estimate, since

$$\nabla g = \sum_{k=1}^{\infty} g_k \nabla \varphi_k$$

and so

$$\|\nabla g\|_{L^2(\Omega)}^2 = \sum_{k=1}^{\infty} g_k^2 \|\nabla \varphi_k\|_{L^2(\Omega)}^2 = \sum_{k=1}^{\infty} |g_k|^2 \lambda_k.$$

Then

$$\|\nabla u(\mathbf{x}, t)\|_{L^2(\Omega)}^2 = \sum_{k=1}^{\infty} |g_k|^2 \lambda_k e^{-2\lambda_k t} \leq e^{-2\lambda_1 t} \sum_{k=1}^{\infty} |g_k|^2 \lambda_k = e^{-\lambda_1 t} \|\nabla g\|_{L^2(\Omega)}^2$$

as required.

Remark. From b) and c) we also see that if $f = f(\mathbf{x})$ belongs, for instance, to $L^2(\Omega)$, the solution to the evolution problem

$$\begin{cases} u_t - \Delta u = f & \text{in } \Omega \times (0, T) \\ u(\mathbf{x}, 0) = g(\mathbf{x}) & \text{in } \Omega \\ u(\sigma, t) = 0 & \text{on } \partial\Omega \times (0, T), \end{cases}$$

converges in $H_0^1(\Omega)$-norm to the solution u_∞ of the stationary problem

$$\begin{cases} -\Delta u_\infty = f & \text{in } \Omega \\ u_\infty(\sigma) = 0 & \text{on } \partial\Omega \end{cases}$$

at a rate depending on the regularity of the initial datum. In such a case we say that u_∞ is asymptotically stable.

Problem 6.2.20 (Wave equation, Neumann conditions). *Consider*

$$\begin{cases} u_{tt} - u_{xx} = f(x, t) & \text{in } Q_T = (0, 1) \times (0, T), \\ u(x, 0) = u_0(x), \, u_t(x, 0) = u_1(x) & \text{in } [0, 1] \\ u_x(0, t) = u_x(1, t) = 0 & \text{for } 0 \leq t \leq T. \end{cases}$$

After finding a weak formulation, discuss whether the problem is well posed and give an energy estimate of the solution.

Solution. Set $H = L^2(0,1)$ with inner product (v,w) and $V = H^1(0,1)$. Suppose $f \in L^2(Q_T)$, $u_0 \in H^1(0,1)$, $u_1 \in L^2(0,1)$. Finally, let

$$B(u,v) = \int_0^1 u_x v_x \, dx.$$

A possible weak formulation is: find a function $\mathbf{u}(t) = u(\cdot,t)$ such that

$$\mathbf{u} \in L^2(0,T;V), \qquad \dot{\mathbf{u}} \in L^2(0,T;H), \qquad \ddot{\mathbf{u}} \in L^2(0,T;V^*),$$

together with:

i) For every $v \in V$,

$$\frac{d^2}{dt^2}(\mathbf{u},v) + B(\mathbf{u},v) = (\mathbf{f}(t),v)$$

in $\mathscr{D}'(0,1)$ and for almost every $t \in (0,T)$.

ii) $\|\mathbf{u}(t) - u_0\|_V \to 0$ and $\|\dot{\mathbf{u}}(t) - u_1\|_H \to 0$ as $t \to 0^+$.

To check that the problem is well posed we can use the Faedo-Galerkin theory. In fact the bilinear form $B(u,v)$ is continuous and weakly coercive on V, whilst the functional

$$Fv = (\mathbf{f}(t),v)$$

is continuous on V. So there exists a unique weak solution u, and

$$\max_{[0,T]} \|\mathbf{u}(t)\|_V^2 \le e^T \left\{ \|u_0\|_V + \|u_1\|_H + \int_0^T \|\mathbf{f}(s)\|_H^2 \, ds \right\}.$$

Problem 6.2.21 (Wave equation with concentrated reaction). *Consider the problem*

$$\begin{cases} u_{tt} - u_{xx} + u(x,t)\delta(x) = f(x,t) & \text{in } Q_T = (a,b) \times (0,T), \\ u(x,0) = u_0(x), \, u_t(x,0) = u_1(x) & \text{in } [a,b] \\ u(a,t) = u(b,t) = 0 & \text{for } 0 \le t \le T \end{cases}$$

where $\delta(x)$ is the Dirac distribution centred at the origin.

a) Write a weak formulation.

b) Say if the problem is well posed, and in that case give an energy estimate for the solution.

Solution. a) If $0 \notin (a,b)$ the problem reduces to the standard wave equation, so we assume $0 \in (a,b)$. Suppose $f \in L^2(Q_T)$, $u_0 \in H_0^1(a,b)$, $u_1 \in L^2(a,b)$. Let $H = L^2(a,b)$, with scalar product $(\cdot,\cdot)$, and $V = H_0^1(a,b)$ with inner product $(v,w)_V = (u_x,v_x)$. Recall that $V \subset C([a,b])$, so the differential equation can be written as

$$u_{tt} - u_{xx} + u(0,t) = f(x,t).$$

A possible weak formulation is: find $\mathbf{u}(t) = u(\cdot, t)$ such that

$$\mathbf{u} \in L^2(0, T; V), \qquad \dot{\mathbf{u}} \in L^2(0, T; H), \qquad \ddot{\mathbf{u}} \in L^2(0, T; V^*),$$

and satisfying:

i) For every $v \in V$,

$$\frac{d^2}{dt^2}(\mathbf{u}, v) + (u_x, v_x) + u(0, t) v(0) = (f, v)$$

in $\mathscr{D}'(a, b)$ for almost every $t \in (a, b)$.

ii) $\|\mathbf{u}(t) - u_0\|_V \to 0$ and $\|\dot{\mathbf{u}}(t) - u_1\|_H \to 0$ as $t \to 0^+$.

b) The bilinear form $a(w, v) = (w_x, v_x) + u(0, t) v(0)$ is continuous and coercive on V. In fact, by Problem 5.2.24 on page 301 (Chap. 5), we have

$$|a(w, v)| \le (1 + (b - a)) \|w_x\|_H \|v_x\|_H$$

and

$$a(w, w) = (w_x, w_x) + w^2(0) \ge \|w_x\|_H^2.$$

Hence *there exists a unique weak solution u*, and moreover

$$\max_{[0,T]} \|u_x(\cdot, t)\|_H^2 \le e^T \left\{ \|u_0\|_V + \|u_1\|_H + \int_0^T \|f(\cdot, s)\|_H^2 \, ds \right\}. \tag{6.38}$$

This estimate can be deduced from the general theory, but it can also be found directly: multiply the PDE by u_t and integrate on (a, b):

$$\int_a^b (u_{tt} u_t - u_{xx} u_t + u^2(0, t)) \, dx = \int_a^b f(x, t) u_t(x, t) \, dx.$$

Integrating the second term by parts, since $u_t(a, t) = u_t(b, t) = 0$, we have

$$\int_a^b u_{xx} u_t \, dx = \int_a^b u_x u_{xt} \, dx = \frac{1}{2} \frac{d}{dt} \int_a^b u_x^2(x, t) \, dx.$$

Furthermore,

$$\left| \int_a^b fv \, dx \right| \le \|f(\cdot, t)\|_H \|u_t(\cdot, t)\|_H \le \frac{1}{2} \|f(\cdot, t)\|_H^2 + \frac{1}{2} \|u_t(\cdot, t)\|_H^2.$$

Hence

$$\frac{d}{dt} \int_a^b [u_t^2(x, t) + u_x^2(x, t)] \, dx + 2 \int_a^b u^2(0, t) \, dx \le \|f(\cdot, t)\|_H^2 + \|u_t(\cdot, t)\|_H^2.$$

Integrating on $(0, t), t \leq T$, we find

$$\|u_t\,(\cdot,t)\|_H^2 + \|u_x\,(\cdot,t)\|_H^2$$
$$\leq \|u_1\|_H^2 + \|u_0\|_V^2 + \int_0^t \|f\,(\cdot,s)\|_H^2\,ds + \int_0^t \|u_t\,(\cdot,s)\|_H^2\,ds \quad (6.39)$$

and in particular

$$\|u_t\,(\cdot,t)\|_H^2 \leq \|u_1\|_H^2 + \|u_0\|_V^2 + \int_0^t \|f\,(\cdot,s)\|_H^2\,ds + \int_0^t \|u_t\,(\cdot,s)\|_H^2\,ds.$$

Gronwall's lemma[13] now gives

$$\|u_t\,(\cdot,t)\|_H^2 \leq e^t \left\{ \|u_1\|_H^2 + \|u_0\|_V^2 + \int_0^t \|f\,(\cdot,s)\|_H^2\,ds \right\}$$

and then

$$\int_0^T \|u_t\,(\cdot,t)\|_H^2 \leq (e^T - 1) \left\{ \|u_1\|_H^2 + \|u_0\|_V^2 + \int_0^T \|f\,(\cdot,s)\|_H^2\,ds \right\}.$$

Substituting in the right-hand side of (6.39) gives back (6.38).

To be more rigorous, (6.38) can be proved for the Galerkin approximations, and extended to the solution by a limiting procedure.

Problem 6.2.22 (Oblique-derivative conditions for the heat equation). *Consider the problem*

$$\begin{cases} u_t - \Delta u = f\,(\mathbf{x},t) & \text{in } Q_T = \Omega \times (0,T) \\ u\,(\mathbf{x},0) = g\,(\mathbf{x}) & \text{in } \Omega \\ \partial_\nu u + h\,(t)\,u_t = 0 & \text{on } S_T = \partial\Omega \times (0,T) \end{cases}$$

where Ω is a bounded Lipschitz domain in $\mathbb{R}^n$ and ν the outward unit normal to $\partial\Omega$. Take $f \in L^2\,(Q_T)$, $g \in H^1\,(\Omega)$ and $h \in C\,([0,T])$ such that $0 < h_0 \leq h\,(t) \leq h_1$.

a) Let $H = L^2\,(\Omega)$ and $V = H^1\,(\Omega)$; write a weak formulation.

b) If V_m and u_m denote the Galerkin approximations of V and u respectively, find energy estimates for u_m, $\partial_t u_m$ and $\partial_t\,(\widetilde{u}_m)$, where $\widetilde{u}_m$ is the trace[a] of u_m on $\partial\Omega$.

c) Deduce existence, uniqueness and an energy estimate for the solution to the original problem.

---

[a] To avoid confusion, here we will use a specific notation for the trace of a function on the boundary.

---

[13] [18, Chap. 10, Sect. 3].

Solution. This problem does not belong to the so-called standard theory, yet we can use the latter with a few modifications.

a) As usual we use the notation $\mathbf{u}(t) = u(\mathbf{x}, t)$, so that $\dot{\mathbf{u}} = \partial_t u$, and similarly for any function depending on t. Let us first proceed formally, multiplying by $v \in V$ and integrating by parts; keeping the boundary condition into account,

$$\int_\Omega \dot{\mathbf{u}} v \, d\mathbf{x} + \int_{\partial\Omega} h(t)\tilde{\dot{\mathbf{u}}}(t)\tilde{v} \, d\sigma + \int_\Omega \nabla u(t) \cdot \nabla v \, d\mathbf{x} = \int_\Omega \mathbf{f}(t) v \, d\mathbf{x}.$$

As $f \in L^2(Q_T)$ and $g \in H^1(\Omega)$, we seek solutions

$$\mathbf{u} \in L^2(0, T; V) \text{ with } \dot{\mathbf{u}} \in L^2(0, T; H).$$

In particular, $\mathbf{u} \in C([0, T]; H)$. Due to the presence of an integral involving the trace of $\dot{\mathbf{u}}(t)$ on $\partial\Omega$ we also require that

$$\tilde{\dot{\mathbf{u}}} \in L^2\left(0, T; L^2(\partial\Omega)\right).$$

Moreover:
i) For every $v \in V$ and a.e. on $(0, T)$,

$$(\dot{\mathbf{u}}(t), v) + h(t)\left(\tilde{\dot{\mathbf{u}}}(t), \tilde{v}\right)_{L^2(\partial\Omega)} + B(\mathbf{u}(t), v) = (\mathbf{f}(t), v),$$

where, as usual, $(\cdot, \cdot)$ is the inner product on $L^2(\Omega)$ and $B(w, v) = (\nabla w, \nabla v)$. To complete the weak formulation we require that
ii) $\mathbf{u}(t) \to \mathbf{g}$ in H as $t \to 0^+$.

b) The Galerkin approximate problem for $\mathbf{u}_m$ is then:

$$(\dot{\mathbf{u}}_m(t), v) + h(t)\left(\tilde{\dot{\mathbf{u}}}_m(t), \tilde{v}\right)_{L^2(\partial\Omega)} + B(\mathbf{u}_m(t), v) = (\mathbf{f}(t), v) \tag{6.40}$$

for every $v \in V_m$.

If $\{w_1, \ldots, w_m\}$ is an orthonormal basis of V_m in V and orthogonal in H, set

$$\mathbf{u}_m(t) = \sum_{k=1}^m c_{mk}(t) w_k \quad \text{and} \quad G_m = \sum_{k=1}^m g_k w_k, \text{ with } g_k = (g, w_k)_V.$$

As $g \in V$, then $G_m \to g$ in V. Substituting these expressions into (6.40) and choosing $v = w_s$, $s = 1, \ldots, m$, we find the following ODE system for the unknown coefficients $c_{mk}(t)$:

$$\sum_{k=1}^m M_{ks}(t) \dot{c}_{mk}(t) + c_{ms}(t) = f_s(t), \quad s = 1, \ldots, m \tag{6.41}$$

where $f_s(t) = (\mathbf{f}(t), w_s)$ and

$$M_{ks}(t) = \|w_k\|_H^2 \delta_{ks} + h(t)(w_k, w_s)_{L^2(\partial\Omega)}$$

with initial condition

$$c_{mk}(0) = g_k, \quad k = 1, \ldots, m.$$

The matrix $(M_{ks}(t))$ is positive definite: indeed

$$\sum_{k,s} M_{ks}\xi_k\xi_s = \sum_k \|w_k\|_H^2 |\xi^k|^2 + h \left\|\sum_k \xi_k w_k\right\|_{L^2(\partial\Omega)}.$$

As $f_s \in L^2(0,T)$ and $g \in V$, there is a unique solution

$$(c_{m1}(t),\ldots,c_{mm}(t)) \in H^1(0,T;\mathbb{R}^m)$$

of (6.41), corresponding to $\mathbf{u}_m \in H^1(0,T;V)$. Note that $\dot{\mathbf{u}}_m(t) \in V$, whose trace $\tilde{\dot{\mathbf{u}}}_m(t)$ is in $L^2(\partial\Omega)$ for almost every t. Setting $v = \dot{\mathbf{u}}_m(t)$ in (6.40) we may write

$$\|\dot{\mathbf{u}}_m(t)\|_H^2 + h(t)\left\|\tilde{\dot{\mathbf{u}}}_m(t)\right\|_{L^2(\partial\Omega)}^2 + \frac{1}{2}\frac{d}{dt}\|\nabla\mathbf{u}_m(t)\|_H^2 = (\mathbf{f}(t),\dot{\mathbf{u}}_m(t));$$

recalling that $h(t) \geq h_0 > 0$ and

$$|(\mathbf{f}(t),\dot{\mathbf{u}}_m(t))| \leq \frac{1}{2}\|\mathbf{f}(t)\|_H^2 + \frac{1}{2}\|\dot{\mathbf{u}}_m(t)\|_H^2,$$

it follows

$$\|\dot{\mathbf{u}}_m(t)\|_H^2 + 2h_0\left\|\tilde{\dot{\mathbf{u}}}_m(t)\right\|_{L^2(\partial\Omega)}^2 + \frac{d}{dt}\|\nabla\mathbf{u}_m(t)\|_H^2 = \|\mathbf{f}(t)\|_H^2.$$

Integrating from 0 to $t \leq T$, we obtain

$$\|\dot{\mathbf{u}}_m\|_{L^2(0,t;H)}^2 + 2h_0\left\|\tilde{\dot{\mathbf{u}}}_m\right\|_{L^2(0,t;L^2(\partial\Omega))}^2 + \|\nabla\mathbf{u}_m(t)\|_H^2 \leq \|\nabla g\|_H^2 + \|\mathbf{f}\|_{L^2(0,t;H)}^2.$$
$$(6.42)$$

This means, in particular, that

$$\begin{aligned}
\dot{\mathbf{u}}_m &\quad \text{is equi-bounded on } L^2(0,T;H)\\
\tilde{\dot{\mathbf{u}}}_m &\quad \text{is equi-bounded on } L^2(0,T;L^2(\partial\Omega))\\
\nabla\mathbf{u}_m &\quad \text{is equi-bounded on } L^\infty(0,T;H).
\end{aligned}$$

Only the norm of $\mathbf{u}_m$ in $L^2(0,T;H)$ is missing. Set $v = \mathbf{u}_m(t)$ in (6.40), so that

$$\frac{1}{2}\frac{d}{dt}\|\mathbf{u}_m(t)\|_H^2 + h(t)(\tilde{\dot{\mathbf{u}}}_m(t),\tilde{\mathbf{u}}_m(t))_{L^2(\partial\Omega)} + \|\nabla\mathbf{u}_m(t)\|_H^2 = (\mathbf{f}(t),\mathbf{u}_m(t)). \quad (6.43)$$

As

$$\|\tilde{\mathbf{u}}_m(t)\|_{L^2(\partial\Omega)}^2 \leq C_*\left(\|\mathbf{u}_m(t)\|_H^2 + \|\nabla\mathbf{u}_m(t)\|_H^2\right),$$

we have

$$\left|h(t)(\tilde{\dot{\mathbf{u}}}_m(t),\tilde{\mathbf{u}}_m(t))_{L^2(\partial\Omega)}\right| \leq h_1\left\|\tilde{\dot{\mathbf{u}}}_m(t)\right\|_{L^2(\partial\Omega)}\|\tilde{\mathbf{u}}_m(t)\|_{L^2(\partial\Omega)}$$

$$\leq C_1\left\|\tilde{\dot{\mathbf{u}}}_m(t)\right\|_{L^2(\partial\Omega)}^2 + \frac{1}{2}\left(\|\mathbf{u}_m(t)\|_H^2 + \|\nabla\mathbf{u}_m(t)\|_H^2\right),$$

where $C_1 = \dfrac{h_1^2 C_*^2}{2}$. Then (6.43) becomes

$$\frac{1}{2}\frac{d}{dt}\|\mathbf{u}_m(t)\|_H^2 + \|\nabla \mathbf{u}_m(t)\|_H^2 \le \frac{1}{2}\left(\|\mathbf{f}(t)\|_H^2 + \|\mathbf{u}_m(t)\|_H^2\right)$$

$$+ C_1 \left\|\tilde{\mathbf{u}}_m(t)\right\|_{L^2(\partial\Omega)}^2 + \frac{1}{2}\left(\|\mathbf{u}_m(t)\|_H^2 + \|\nabla \mathbf{u}_m(t)\|_H^2\right),$$

which implies

$$\frac{d}{dt}\|\mathbf{u}_m(t)\|_H^2 \le \|\mathbf{f}(t)\|_H^2 + 2\|\mathbf{u}_m(t)\|_H^2 + 2C_1\left\|\tilde{\mathbf{u}}_m(t)\right\|_{L^2(\partial\Omega)}^2.$$

Integrating between 0 and $t \le T$, and exploiting (6.42),

$$\|\mathbf{u}_m(t)\|_H^2 \le \|g\|_H^2 + \|\mathbf{f}\|_{L^2(0,t;H)}^2 + 2\int_0^t \|\mathbf{u}_m(s)\|_H^2\, ds + 2C_1\left\|\tilde{\mathbf{u}}_m\right\|_{L^2(0,t;L^2(\partial\Omega))}^2$$

$$\le C_2\left[\|g\|_V^2 + \|\mathbf{f}\|_{L^2(0,t;H)}^2\right] + 2\int_0^t \|\mathbf{u}_m(s)\|_H^2\, ds,$$

where $C_2 = 1 + \dfrac{C_1}{h_0}$. By Gronwall's lemma

$$\|\mathbf{u}_m(t)\|_H^2 \le e^{2t} C_2\left[\|g\|_V^2 + \|\mathbf{f}\|_{L^2(0,T;H)}^2\right],$$

and therefore

$$\mathbf{u}_m \text{ is bounded in } L^\infty(0,T;H).$$

c) By the results of b), there is a subsequence, still denoted by $\mathbf{u}_m$, such that

$$\mathbf{u}_m \rightharpoonup \mathbf{u} \quad \text{in } L^2(0,T;V)$$
$$\dot{\mathbf{u}}_m \rightharpoonup \dot{\mathbf{u}} \quad \text{in } L^2(0,T;H)$$
$$\tilde{\mathbf{u}}_m \rightharpoonup \tilde{\mathbf{u}} \quad \text{in } L^2(0,T;L^2(\partial\Omega)).$$

This is enough to pass to the limit in (6.40) and then conclude that $\mathbf{u}$ is the unique weak solution of the original equation.

Problem 6.2.23 (Wave equation, equipartition of energy). *Let* $\Omega \subset \mathbb{R}^3$ *be a bounded domain*, $Q_T = \Omega \times (0, T)$, $V = H_0^1(\Omega)$, $H = L^2(\Omega)$. *Given* $u_0 \in V$ *and* $u_1 \in H$, *call* $\mathbf{u} \in L^2(0, T; V)$ *the (unique) weak solution to*

$$\begin{cases} \ddot{\mathbf{u}}(t) - \Delta \mathbf{u} = 0 & \text{in } Q_T, \\ \mathbf{u}(0) = u_0, \dot{\mathbf{u}}(0) = u_1 \end{cases}$$

such that $\dot{\mathbf{u}} \in L^2(0, T; H)$ *and* $\ddot{\mathbf{u}} \in L^2(0, T; V^*)$.

a) Let $V_m = \text{span}\{w_1, \ldots, w_m\}$, *where the* $\{w_j\}$ *are the Dirichlet eigenfunctions of* $-\Delta$, *and* $\{\mathbf{u}_m\}$ *the Galerkin sequence ([18, Chap. 10, Sect. 7]) approximating* $\mathbf{u}$. *Prove that*

$$E_m(t) = \frac{1}{2} \|\dot{\mathbf{u}}_m(t)\|_H^2 + \frac{1}{2} \|\nabla \mathbf{u}_m(t)\|_H^2 = E_m(0), \qquad \text{for every } t \geq 0.$$

b) Setting

$$K_m(t) = \frac{1}{2t} \int_0^t \|\dot{\mathbf{u}}_m(s)\|_H^2 \, ds, \quad P_m(t) = \frac{1}{2t} \int_0^t \|\nabla \mathbf{u}_m(s)\|_H^2 \, ds,$$

deduce from a) that, for every $m \geq 1$,

$$K_m(t) \to \frac{E_m(0)}{2}, \quad P_m(t) \to \frac{E_m(0)}{2} \qquad \text{as } t \to +\infty. \tag{6.44}$$

Solution. a) Call $(\cdot, \cdot)$ the inner product of $L^2(\Omega)$. The general theory, in the present context, says that $\mathbf{u}_m \in H^2(0, T; V)$ satisfies

$$(\ddot{\mathbf{u}}_m(t), v) + (\nabla \mathbf{u}_m(t), \nabla v) = 0 \tag{6.45}$$

for every $v \in V$ and almost all $t \in (0, T)$. As $\dot{\mathbf{u}}_m \in L^2(0, T; V)$ for almost every given t, we can insert $v = \dot{\mathbf{u}}_m(t)$ into (6.45); then

$$(\ddot{\mathbf{u}}_m(t), \dot{\mathbf{u}}_m(t)) + (\nabla \mathbf{u}_m(t), \nabla \dot{\mathbf{u}}_m(t)) = \frac{1}{2} \frac{d}{dt} \|\dot{\mathbf{u}}_m(t)\|_H^2 + \frac{1}{2} \frac{d}{dt} \|\nabla \mathbf{u}_m(t)\|_H^2 = 0$$

whence

$$E_m(t) \equiv \frac{1}{2} \|\dot{\mathbf{u}}_m(t)\|_H^2 + \frac{1}{2} \|\nabla \mathbf{u}_m(t)\|_H^2 = E_m(0), \qquad t \geq 0. \tag{6.46}$$

b) From (6.46),

$$K_m(t) + P_m(t) = \frac{1}{2t} \int_0^t \left\{ \|\dot{\mathbf{u}}_m(s)\|_H^2 + \|\nabla \mathbf{u}_m(s)\|_H^2 \right\} ds = E_m(0). \tag{6.47}$$

As $\mathbf{u}_m \in L^2(0, T; V)$, for almost every given t, we set $v = \mathbf{u}_m(t)$ in (6.45); then

$$(\ddot{\mathbf{u}}_m(t), \mathbf{u}_m(t)) + \|\nabla \mathbf{u}_m(s)\|_H^2 = 0.$$

Integrating by parts,

$$\int_0^t (\ddot{\mathbf{u}}_m(s), \mathbf{u}_m(s))ds = -\int_0^t (\dot{\mathbf{u}}_m(s), \dot{\mathbf{u}}_m(s))ds + (\dot{\mathbf{u}}_m(t), \mathbf{u}_m(t)) - (u_1, u_0).$$

Hence

$$P_m(t) - K_m(t) = \frac{-(u_1, u_0) + (\dot{\mathbf{u}}_m(t), \mathbf{u}_m(t))}{2t}. \tag{6.48}$$

On the other hand, from (6.46) and Poincaré's inequality

$$|(\dot{\mathbf{u}}_m(t), \mathbf{u}_m(t))| \le \frac{1}{2}\|\dot{\mathbf{u}}_m(t)\|_H^2 + \frac{1}{2}\|\mathbf{u}_m(t)\|_H^2 \le \frac{1}{2}\|\dot{\mathbf{u}}_m(t)\|_H^2 + \frac{C_P^2}{2}\|\nabla\mathbf{u}_m(t)\|_H^2$$

$$\le \max\left\{1, C_P^2\right\} E_m(0).$$

Therefore $P_m(t) - K_m(t) \to 0$ as $t \to +\infty$, which together with (6.47) implies

$$K_m(t) \to \frac{E_m(0)}{2}, \quad P_m(t) \to \frac{E_m(0)}{2} \quad \text{as } t \to +\infty.$$

Remark. Condition (6.44) is also satisfied by the solution $\mathbf{u}$. The proof requires a little more work, especially for proving (6.46). The difficulty lies in the fact that $\dot{\mathbf{u}} \in L^2(0, T; H)$ and $\ddot{\mathbf{u}} \in L^2(0, T; V^*)$, so we cannot insert $\dot{\mathbf{u}}$ directly into the weak equation

$$\langle \ddot{\mathbf{u}}(t), v\rangle_* + (\nabla\mathbf{u}(t), \nabla v) = 0$$

where $\langle \ddot{\mathbf{u}}(t), v\rangle_*$ is the pairing of V and V^*. To do that we would need a better regularity for $\dot{\mathbf{u}}$, for instance $\dot{\mathbf{u}} \in L^2(0, T; V)$.

Problem 6.2.24 (Mixed boundary conditions). *Let*

$$\Omega = \left\{\mathbf{x} \in \mathbb{R}^2 : x_1^2 + 4x_2^2 < 4\right\}, \quad \Gamma_D = \partial\Omega \cap \{x_1 \ge 0\}, \quad \Gamma_N = \partial\Omega \cap \{x_1 < 0\}.$$

Consider the problem

$$\begin{cases} u_t - \text{div}(A_\alpha \nabla u) + \mathbf{b}\cdot\nabla u - \alpha u = x_2 & \text{in } \Omega \times (0, T) \\ u(\mathbf{x}, t) = 0 & \text{on } \Gamma_D \times (0, T) \\ -A_\alpha \nabla u \cdot \mathbf{n} = \cos x_1 & \text{on } \Gamma_N \times (0, T) \\ u(\mathbf{x}, 0) = \mathcal{H}(x_1) & \text{on } \Omega \end{cases} \tag{6.49}$$

where

$$A_\alpha = \begin{bmatrix} 1 & 0 \\ 0 & \alpha\, e^{x_1^2 + x_2^2} \end{bmatrix}, \quad \mathbf{b} = \begin{bmatrix} \sin(x_1 + x_2) \\ x_1^2 + x_2^2 \end{bmatrix}$$

and $\mathcal{H}$ is the Heaviside function. For which values of the parameter $\alpha \in \mathbb{R}$ is the problem parabolic? For such values, give a weak formulation of the problem and deduce existence and uniqueness of the solution.

Solution. For every $\boldsymbol{\xi} \in \mathbb{R}^2$, we have

$$A_\alpha(x_1, x_2)\boldsymbol{\xi} \cdot \boldsymbol{\xi} \geq \xi_1^2 + \alpha\xi_2^2 \geq \min(1, \alpha)(\xi_1^2 + \xi_2^2). \tag{6.50}$$

Therefore, the problem is parabolic for every $\alpha > 0$.

We have a mixed problem with homogeneous Dirichlet conditions on the boundary Γ_D and, then, we are allowed to use the Poincaré inequality; we consider

$$V = H^1_{0,\Gamma_D}(\Omega), \text{ with } \|v\|_V^2 = \|\nabla v\|^2_{L^2(\Omega)}.$$

Let us proceed formally multiplying the first equation of the problem (6.49) by $v \in V$ and integrating

$$\langle \dot{\mathbf{u}}, v \rangle_* + \int_\Omega A_\alpha \nabla \mathbf{u} \cdot \nabla v \, d\mathbf{x} - \int_{\Gamma_N} A_\alpha \nabla \mathbf{u} \cdot \mathbf{n} \, v \, d\sigma + \int_\Omega \mathbf{b} \cdot \nabla \mathbf{u} \, v \, d\mathbf{x}$$
$$- \alpha \int_\Omega \mathbf{u} \, v \, d\mathbf{x} = \int_\Omega x_2 v \, d\mathbf{x}.$$

On account of the third equation of problem (6.49), we find

$$\langle \dot{\mathbf{u}}, v \rangle_* + \int_\Omega A_\alpha \nabla \mathbf{u} \cdot \nabla v \, d\mathbf{x} + \int_\Omega \mathbf{b} \cdot \nabla \mathbf{u} \, v \, d\mathbf{x} - \alpha \int_\Omega \mathbf{u} \, v \, d\mathbf{x}$$
$$= \int_\Omega x_2 v \, d\mathbf{x} - \int_{\Gamma_N} \cos x_1 \, v \, d\sigma.$$

Define:

$$B(u, v) = \int_\Omega A_\alpha \nabla \mathbf{u} \cdot \nabla v \, d\mathbf{x} + \int_\Omega \mathbf{b} \cdot \nabla \mathbf{u} \, v \, d\mathbf{x} - \alpha \int_\Omega \mathbf{u} \, v \, d\mathbf{x}$$
$$Fv = \int_\Omega x_2 v \, d\mathbf{x} - \int_{\Gamma_N} \cos x_1 \, v \, d\sigma.$$

The weak formulation of problem (6.49) consists in finding $\mathbf{u} \in L^2(0, T; V)$ such that $\dot{\mathbf{u}} \in L^2(0, T; V')$, $\mathbf{u}(0) = u_0$ in Ω, and for every $v \in V$, a.e. t in $[0, T]$,

$$\langle \dot{\mathbf{u}}(t), v \rangle_* + B(\mathbf{u}(t), v) = Fv.$$

We prove the continuity of B and F and the weak coercivity of B.

We recall that for the functions in V a trace inequality $\|v\|_{\Gamma_N} \leq C_T \|v\|_V$ holds. Then, using also Schwarz's inequality, we can write:

$$|Fv| \leq \left(\int_\Omega x_2^2 d\mathbf{x} \right)^{1/2} \|v\|_{L^2(\Omega)} + \left(\int_{\Gamma_N} \cos^2 x_1 d\sigma \right)^{1/2} \|v\|_{\Gamma_N}$$
$$\leq \left(|\Omega|^{1/2} + |\Gamma_N|^{1/2} C_T \right) \|v\|_V = (\sqrt{\pi} + \sqrt{3}C_T)\|v\|_V$$

where $|\Omega|$ and $|\Gamma_N|$ denote the measures of the sets Ω and Γ_N.

Since in Ω we have $x_1^2 + x_2^2 \le 4$, using once more Schwarz's and Poincaré's inequalities we have

$$
\begin{aligned}
|B(u,v)| \le & \int |u_{x_1} v_{x_1} + \alpha e^{x_1^2 + x_2^2} u_{x_2} v_{x_2}| d\mathbf{x} + \alpha \int_\Omega |uv| \, d\mathbf{x} \\
& + \int_\Omega |\sin(x_1 + x_2) u_{x_1} + (x_1^2 + x_2^2) u_{x_2}||v| d\mathbf{x} \\
\le & \max\{1, \alpha e^4\} \int_\Omega |\nabla u \cdot \nabla v| d\mathbf{x} + \alpha \|u\|_{L^2(\Omega)} \|v\|_{L^2(\Omega)} + 5\|\nabla u\|_{L^2(\Omega)} \|v\|_{L^2(\Omega)} \\
\le & \left(\max\{1, \alpha e^4\} + \alpha C_P^2 + 5 C_P \right) \|u\|_V \|v\|_V.
\end{aligned}
$$

To estimate the drift term, we use Young's inequality:

$$
\begin{aligned}
& \left| \int_\Omega \boldsymbol{\beta} \cdot \nabla u \, v \, d\mathbf{x} \right| \\
& \le \int_\Omega (|\sin(x_1^2 + x_2^2)||u_{x_1}||v| + (x_1^2 + x_2^2)|u_y||v|) d\mathbf{x} \le \int_\Omega (|u_{x_1}||v| + 4|u_y||v|) d\mathbf{x} \\
& \le \varepsilon \|u_{x_1}\|_{L^2(\Omega)}^2 + \frac{1}{4\varepsilon} \|v\|_{L^2(\Omega)}^2 + \varepsilon \|u_{x_2}\|_{L^2(\Omega)}^2 + \frac{4}{\varepsilon} \|v\|_{L^2(\Omega)}^2 \le \varepsilon \|u\|_V^2 + \frac{17}{4\varepsilon} \|v\|_{L^2(\Omega)}^2.
\end{aligned}
$$

Thanks to the uniform ellipticity of A_α proved in (6.50), we deduce that

$$
B(u,u) + \lambda \|u\|_{L^2(\Omega)} \ge (\min(1,\alpha) - \epsilon) \|u\|_V^2 + \left(\lambda - \frac{17}{4\varepsilon} - \alpha \right) \|u\|_{L^2(\Omega)}^2.
$$

Now we choose ε and, consequently, λ in order that the two coefficients appearing in the right-hand side are positive. Hence, the weak coercivity of the bilinear form B is proved.

We conclude that problem (6.49) has a unique weak solution, satisfying the estimate

$$
\|\mathbf{u}\|_{C([0,T];H)}, \ \|\mathbf{u}\|_{L^2(0,T;V)}, \ \|\dot{\mathbf{u}}\|_{L^2(0,T;V^*)} \le C(\Omega).
$$

6.3 Further Exercises

6.3.1. *Establish whether the bilinear forms below are continuous on the given Hilbert spaces, and under which conditions they are coercive (or weakly coercive).*

a) $H = \mathbb{R}^n$, $a(\mathbf{x}, \mathbf{y}) = \displaystyle\sum_{i,j=1}^n a_{ij} x_i y_j$, where (a_{ij}) is a symmetric $n \times n$-matrix.

b) $H = H^1(0,1)$,

$$
a(u,v) = \int_0^1 A(x) u' v' dx + \int_0^1 B(x) u' v \, dx + \int_0^1 C(x) uv \, dx,
$$

A, B, C in $L^\infty(a,b)$.

c) $H = H_0^1(\Omega)$, Ω domain in $\mathbb{R}^n$,

$$a(u, v) = \int_\Omega \alpha(x)\nabla u \cdot \nabla v \, d\mathbf{x},$$

with $\alpha \in L^\infty(\Omega)$.

6.3.2. *Given the following boundary-value problems on* $(0, 1)$, *write the weak formulations and discuss the applicability of the Lax-Milgram theorem.*

a) $-u'' + (e^{-t}u)' = 4$, $u(0) = 1$, $u(1) = 2$.
b) $-(4 + t^2)u'' + 3u = \sin t$, $u'(0) = 1$, $u'(1) + u(1) = 0$.
c) $-u'' + u' + u = 0$, $u(0) = u(1)$, $u'(0) = u'(1)$.

6.3.3. *Discuss, in the two cases below, whether*

$$B(u, v) = Fv, \quad \text{for every } v \in V$$

is a well-posed problem. Then write the corresponding boundary-value problem in classical form.

a) $V = H^1(0, 1)$, $Fv = v(1)$, $B(u, v) = \int_0^1 (x + 1)u'v' \, dx + u(0)v(0)$.

b) $V = \{v \in H^1(0, 2) : v(2) = 0\}$,
$$Fv = \int_0^2 fv \, dx, \quad B(u, v) = \int_0^2 \left[u'v' - 2x^2 u'v - 4xuv\right] dx.$$

6.3.4. *Referring to Problem 6.2.3 (page 340), write a variational formulation of the Robin-Dirichlet problem:*

$$\begin{cases} \mathscr{L}u \equiv \cos x\, u'' - \sin x\, u' - xu = f & 0 < x < L \\ u'(0) = -u(0), \ u(L) = 0, \end{cases}$$

where $0 < L < \pi/2$ *and* $f \in L^2(0, L)$. *Discuss the solvability of the problem.*

6.3.5. *(Hermite equation) Let*

$$V = \left\{v : \mathbb{R} \to \mathbb{R} : e^{-x^2/2}v \in L^2(\mathbb{R}), e^{-x^2/2}v' \in L^2(\mathbb{R})\right\}.$$

a) *Verify that the formula*

$$(u, v)_V = \int_\mathbb{R} [uv + u'v']e^{-x^2} dx \qquad (6.51)$$

defines an inner product which makes V *into a Hilbert space.*
b) *Study the variational problem*

$$(u, v)_V = Fv \equiv \int_\mathbb{R} fv\, e^{-x^2} dx \qquad \text{for every } v \in V, \qquad (6.52)$$

with $f \in V$, *and interpret it in the classical sense.*

6.3.6. *a)* *Given* $\Omega \subset \mathbb{R}^n$, *a bounded regular domain,* $f \in L^2(\Omega)$ *and* $g \in L^2(\partial\Omega)$, *say to which problem the following variational formulation is associated*

$$\int_\Omega \nabla u \cdot \nabla v \, d\mathbf{x} + \int_\Omega uv \, d\mathbf{x} = \int_\Omega fv \, d\mathbf{x} + \int_{\partial\Omega} gv \, d\sigma, \qquad \text{for every } v \in H^1(\Omega). \quad (6.53)$$

b) *Is this well posed?*

6.3.7. Let B_1 denote the unit ball in $\mathbb{R}^n$. Consider the subspace of $H^1(B_1)$ given by

$$V = \left\{ u \in H^1(B_1) : \int_{\partial B_1} u \, d\sigma = 0 \right\}$$

and the following variational problem: determine $u \in V$ such that

$$\int_{B_1} (\nabla u \cdot \nabla v + uv) \, d\mathbf{x} = \int_{B_1} fv \, d\mathbf{x}, \qquad \text{for every } v \in V.$$

Can one apply the Lax-Milgram theorem? To which problem does the variational formulation correspond?

6.3.8. Given a bounded regular domain $\Omega \subset \mathbb{R}^n$, consider the system

$$\begin{cases} -\Delta u_1 + u_1 - u_2 = f_1 & \text{in } \Omega \\ -\Delta u_2 + u_1 + u_2 = f_2 & \text{in } \Omega \\ u_1 = u_2 = 0 & \text{on } \partial\Omega. \end{cases}$$

Write a weak formulation and apply the Lax-Milgram theorem to prove the existence and uniqueness of the solution, for every $\mathbf{f} = (f_1, f_2) \in L^2(\Omega) \times L^2(\Omega)$.

6.3.9. Let $\Omega = (0,1) \times (0,1)$, and set

$$\Gamma_N = \{(0,y) : 0 < y < 1\}$$

and $\Gamma_D = \partial\Omega \setminus \Gamma_N$. Study

$$\begin{cases} -\operatorname{div}(A(x,y)\nabla u + \mathbf{b}u) = f & \text{in } \Omega \\ u = 0 & \text{on } \Gamma_D \\ 3u_x(0,y) - 4u_y(0,y) + hu = 0 & 0 < y < 1 \end{cases}$$

where $f \in L^2(\Omega)$, $h > 0$ is constant and

$$A(x,y) = \begin{pmatrix} 2 + \sin xy & 3 \\ -3 & 4 - \sin xy \end{pmatrix}, \qquad \mathbf{b} = (-4x - y, -2x + y).$$

6.3.10. In relationship to Problem 6.2.12 on page 355, consider

$$\begin{cases} -\Delta u + \mathbf{b} \cdot \nabla u = f & \mathbf{x} \in \Omega \\ \partial_\nu u = 0 & \mathbf{x} \in \partial\Omega, \end{cases}$$

with $\mathbf{b} \in \mathbb{R}^n$ constant.

a) Multiply the equation by $e^{-\mathbf{b}\cdot\mathbf{x}}$, and write it in divergence form.

b) Provide a necessary and sufficient condition for the problem to admit solutions.

6.3.11. Let $Q = (0, \pi) \times (0, \pi)$. In relationship to Problem 6.2.13, study the Dirichlet problem

$$\begin{cases} \Delta u + 2u = \mathrm{div}\,\mathbf{f} & \text{in } Q \\ u = 0 & \text{on } \partial Q, \end{cases}$$

when

$$\mathbf{f} = (\log(\sin x), 0).$$

6.3.12. (Alternating method, Neumann conditions) *In relationship to Problem 6.2.15 let $\Omega = \Omega_1 \cup \Omega_2 \subset \mathbb{R}^2$ be a domain, where Ω_1, Ω_2 have regular boundary and $\Omega_1 \cap \Omega_2 \neq \emptyset$. Adapt Schwarz's alternating method to build a sequence $\{u_n\}$ converging in $H^1(\Omega)$ to the solution of the Neumann problem*

$$-\Delta u + u = f \quad \text{in } \Omega, \quad \partial_\nu u = 0 \quad \text{on } \partial\Omega, \tag{6.54}$$

where $f \in L^2(\Omega)$ and ν is the outward normal to $\partial\Omega$.

6.3.13. Let $B_1 = \{(x, y) : x^2 + y^2 < 1\}$. *Write the variational formulations of the following problems and discuss solvability:*

a)
$$\begin{cases} \Delta u = 0 & \text{in } B_1 \\ (x + y)u_x + (y - x)u_y = 1 + \alpha x^2 & \text{on } \partial B_1 \end{cases} \qquad \alpha \in \mathbb{R}.$$

b)
$$\begin{cases} -\Delta u = (x^2 - y^2)^n & \text{in } B_1 \\ (x + y)u_x + (y - x)u_y = 0 & \text{on } \partial B_1 \end{cases} \qquad n \in \mathbb{N}.$$

6.3.14. (Obstacle problem) *Let Ω be a bounded, convex domain in $\mathbb{R}^n$ and ψ a strictly concave function[14] on Ω such that $\max_\Omega \psi > 0$ and $\psi < 0$ on $\partial\Omega$. Define*

$$K = \left\{ v \in H_0^1(\Omega) : v \geq \psi \quad \text{a.e. in } \Omega \right\}.$$

a) *Check that K is convex and closed in $H_0^1(\Omega)$.*
b) *Show there is a unique function $u \in K$ that minimises the functional*

$$J(v) = \frac{1}{2} \int_\Omega |\nabla u|^2 \, d\mathbf{x}$$

on K, characterised by the following variational inequality:

$$\int_\Omega (\nabla v - \nabla u) \cdot \nabla u \, d\mathbf{x} \geq 0, \quad \text{for every } v \in K. \tag{6.55}$$

c) *Deduce that, if $u \in H^2(\Omega)$, then u solves the obstacle problem if and only if*

$$-\Delta u \geq 0, \quad u - \psi \geq 0 \quad \text{and} \quad \Delta u (u - \psi) = 0 \qquad \text{a.e. on } \Omega.$$

In particular, u is a harmonic function on every open set where $u > \psi$.

---

[14] A concave function is always Lipschitz.

6.3.15. Let $B_1 \subset \mathbb{R}^2$ be the unit disc centred at the origin, $Q_T = B_1 \times (0, T)$, Γ_D and Γ_N open subsets in $\partial\Omega$ with $\overline{\Gamma}_D \cup \overline{\Gamma}_N = \partial\Omega$. Consider the mixed problem:

$$\begin{cases} u_t - \operatorname{div}(xe^y \nabla u) + (\operatorname{sign}(xy)u)_y = f & \text{in } Q_T, \\ u(x, y, 0) = u_0(x, y) & \text{in } B_1, \\ u = g & \text{on } \Gamma_D, \\ \partial_\nu u + u = 0 & \text{on } \Gamma_N, \end{cases}$$

where $f = f(x, y, t)$, $u_0 = u_0(x, y)$ and $g = g(t)$ are given functions.

a) Using appropriate function spaces, write a weak formulation.
b) Study existence and uniqueness and give a stability estimate.

6.3.16. Let Ω be a bounded domain in $\mathbb{R}^n$ and $Q_T = \Omega \times (0, T)$. Let Γ_D and Γ_N be open in $\partial\Omega$ with $\overline{\Gamma}_D \cup \overline{\Gamma}_N = \partial\Omega$. Consider the mixed problem:

$$\begin{cases} u_t - \operatorname{div}(a(\mathbf{x})\nabla u - \mathbf{b}u) - c(\mathbf{x})u = 0 & \text{in } Q_T, \\ u(\mathbf{x}, 0) = g(\mathbf{x}) & \text{in } \Omega, \\ u = 0 & \text{on } \Gamma_D, \\ a(\mathbf{x})\partial_\nu u + hu = 0 & \text{on } \Gamma_N, \end{cases}$$

where $g \in L^2(\Omega)$ and $h \in \mathbb{R}$ is constant.

a) Using suitable function spaces, write the weak formulation.
b) Under which conditions on the coefficients a, $\mathbf{b}$, c and h is the problem well posed?

6.3.17. The potassium concentration $c(x, y, z, t)$ in a cell of spherical shape Ω and radius R, with boundary Γ, satisfies the evolution problem

$$\begin{cases} c_t - \operatorname{div}(\mu\nabla c) - \sigma c = 0 & \text{in } \Omega \times (0, T) \\ \mu\nabla c \cdot \nu + \kappa c = \kappa c_{\text{ext}} & \text{on } \Gamma \times (0, T) \\ c(x, y, z, 0) = c_0(x, y, z) & \text{on } \Omega \end{cases} \tag{6.56}$$

where c_{ext} is the given external concentration which is constant, σ and κ are positive numbers and μ is a strictly positive function. Write the weak formulation and analyse the well-posedness, providing suitable assumptions on the coefficients and on the data.

6.3.1 Solutions

Solution 6.3.1. **a)** The form is always continuous (the setting is finite-dimensional), and is coercive if and only if the matrix A is positive definite (with coercivity constant equal to the smallest eigenvalue of A). Let us remind that in finite dimensions weak coercivity coincides with coercivity, for there are no dense subspaces in $\mathbb{R}^n$ other than $\mathbb{R}^n$ itself.

b) The form is continuous ($L^p = L^p(0, 1)$):

$$|a(u, v)| \leq \|A\|_{L^\infty}\|u'\|_{L^2}\|v'\|_{L^2} + \|B\|_{L^\infty}\|u'\|_{L^2}\|v\|_{L^2} + \|C\|_{L^\infty}\|u\|_{L^2}\|v\|_{L^2} \leq$$
$$\leq (\|A\|_{L^\infty} + \|B\|_{L^\infty} + \|C\|_{L^\infty})\|u\|_{H^1}\|v\|_{H^1}.$$

To prove the coercivity, we have to estimate from below

$$a(u, u) = \int_0^1 A(x)(u')^2 dx + \int_0^1 B(x)u'u \, dx + \int_0^1 C(x)u^2 \, dx.$$

Here we can provide different conditions. A necessary one is $A \geq A_0 > 0$. In fact, if A is negative on a subinterval we can construct functions u with $\|u\|_{L^2}$ arbitrarily small and u' concentrated where A is negative, making $A(u, u)$ negative.

There remains to estimate the mixed term. Let $|B(x)| \leq B_0$, $B'(x) \leq B_1$; then by Problem 5.2.24, Chap. 5 (page 301):

$$\int_0^1 B(x)u'u \, dx \geq \left[\frac{B(x)}{2} u^2(x) \right]_0^1 - \int_0^1 B'(x)u^2 \, dx$$

$$\geq -B_0\|u\|_{L^\infty}^2 - B_1 \|u\|_{L^2} \geq -B_0\|u\|_{H^1}^2 - B_1 \|u\|_{L^2}.$$

If $C(x) \geq C_0$, we can write

$$a(u, u) \geq (A_0 - B_0) \int_0^1 (u')^2 dx + (C_0 - B_1) \int_0^1 u^2 \, dx.$$

Finally, if $A_0 - B_0 > 0$ and $C_0 - B_1 > 0$ the form is coercive. Assuming $A_0 - B_0 > 0$ we can only guarantee weak coercivity (for the Hilbert triple $\{H^1(0, 1), L^2(0, 1), (H^1(0, 1))'\}$).

c) The form is continuous, for

$$|a(u, v)| \leq \|\alpha\|_{L^\infty} \|\nabla u\|_{L^2} \|\nabla v\|_{L^2} \leq \|\alpha\|_{L^\infty} \|\nabla u\|_{H^1} \|\nabla v\|_{H^1}$$

by Hölder's inequality. The form is coercive if and only if $\alpha \geq \alpha_0 > 0$ a.e. on Ω.

Solution 6.3.2. **a)** The boundary conditions are inhomogeneous, so it is convenient to define $w(t) = u(t) - t - 1$. As $w'(t) = u'(t) - 1$, $w''(t) = u''(t)$, the function w solves

$$\begin{cases} -w'' - e^{-t}w' = 4 + e^{-t} & 0 < t < 1 \\ w(0) = w(1) = 0. \end{cases}$$

The weak formulation is:

$$\int_0^1 \left[w'v' - e^{-t}w'v \right] dt = \int_0^1 \left[4 + e^{-t} \right] v \, dt \quad \text{for every } v \in H_0^1(0, 1).$$

Arguing as in Problem 6.2.1 (page 336) we can apply Lax-Milgram, and obtain that the problem is well posed.

b) Choose $V = H^1(0, 1)$. For $v \in V$

$$\int_0^1 -(4 + t^2)u''v \, dt = -(4 + t^2)u'v|_0^1 + \int_0^1 \left[(4 + t^2)u'v' + 2tu'v \right] dt =$$

$$= -5u'(1)v(1) + \int_0^1 \left[(4 + t^2)u'v' + 2tu'v \right] dt =$$

$$= 5u(1)v(1) + 4v(0) + \int_0^1 \left[(4 + t^2)u'v' + 2tu'v \right] dt.$$

Hence the weak formulation

$$\int_0^1 \left[(4+t^2)u'v' + 2tu'v + 3uv\right] dt + 5u(1)v(1) = \int_0^1 \sin t\, v\, dt - 4v(0), \quad \forall v \in H^1(0,1).$$

Arguing as in Problems 5.2.25 (page 302) and 5.2.26 (page 303), one can prove that the right-hand side is the sum of two continuous linear functionals on $H^1(0,1)$. Combining the arguments of Exercise 6.3.1.b) with Problem 5.2.24 (page 301), the bilinear form $B(u,v)$ is continuous. Finally

$$B(u,u) = \int_0^1 \left[(4+t^2)(u')^2 + 2tu'u + 3u^2\right] dt + 5u^2(1)$$

$$= \int_0^1 \left[(4+t^2)(u')^2 + 2u^2\right] dt + 5u^2(1) \geq \int_0^1 \left[4(u')^2 + 2u^2\right] dt,$$

hence B is coercive and the Lax-Milgram theorem can be applied.

c) V will be a suitable subspace of $H^1(0,1)$, adapted to the boundary (periodicity) conditions. To find V let us multiply the equation by $v \in H^1(0,1)$ and integrate by parts. Formally,

$$\int_0^1 -u''v\, dt = -u'(1)v(1) + u'(0)v(0) + \int_0^1 (u')^2\, dt.$$

Now if u is in $H^1(0,1)$, we cannot, in general, make pointwise sense of u'. Hence a correct choice for test functions must annihilate the first two terms on the right. Since we want $u'(0) = u'(1)$, it suffices to take $v(0) = v(1)$. So set

$$V = H^1_{per} \equiv \{v \in H^1(0,1): v(0) = v(1)\}.$$

Then one can see[15] that V is closed in $H^1(0,1)$, hence a Hilbert space. The weak formulation is

$$B(u,v) \equiv \int_0^1 \left[u'v' + u'v + uv\right] dt = 0, \qquad \text{for every } v \in V.$$

Supposing u regular and a weak solution, it is enough to integrate by parts the other way aroundăto deduce that u is a strong solution, too. The bilinear form $B(u,v)$ is continuous on V, and as

$$\int_0^1 uu' = \frac{1}{2}\left[u^2(1) - u^2(0)\right] = 0,$$

it is also coercive. Hence the problem has only the trivial solution.

Solution 6.3.3. a) The problem is well posed thanks to the Lax-Milgram theorem (proceed as in Exercise 6.3.2.b). Let us find the classical formulation, supposing $u \in C^2(0,1) \cap C^1([0,1])$. Integrating by parts,

$$(x+1)u'v|_0^1 - \int_0^1 \left[(x+1)u''v + u'v\right] dx + u(0)v(0) = v(1), \quad \forall v \in H^1(0,1),$$

---

[15] To prove it use the argument of Problem 5.2.28 on page 306.

i.e.

$$\int_0^1 \left[(x+1)u'' + u'\right] v \, dx = (2u'(1) - 1)v(1) + (-u'(0) + u(0))v(0), \qquad \forall v \in H^1(0, 1).$$

In particular, the previous identity must hold for every $v \in H_0^1(0, 1)$, which forces

$$(x+1)u'' + u' = 0, \quad \text{in } (0, 1).$$

Then

$$(2u'(1) - 1)v(1) + (-u'(0) + u(0))v(0) = 0 \qquad \text{for every } v(0), v(1) \in \mathbb{R}.$$

By choosing v zero at one endpoint and non-zero at the other, and then swapping, we see that u solves the mixed (Robin-Neumann) problem

$$\begin{cases} (x+1)u'' + u' = 0 & \text{on } (0, 1) \\ u'(0) = u(0) \\ u'(1) = 1/2. \end{cases}$$

We can explicitly solve the problem: integrating once we get

$$(x+1)u' = C_1$$

and the Neumann condition gives $C_1 = 1$; integrating again, we find

$$u = \log(x+1) + C_2$$

and now the Robin condition gives $C_2 = 1$. Overall,

$$u(x) = \log(x+1) + 1.$$

b) On V we use the standard norm of $V = H^1(0, 2)$. It is not hard to see (see Exercise 6.3.1.b)) that B is continuous. Also F is continuous as soon as (for instance) $f \in L^2(0, 2)$. For the coercivity, we preliminarily observe that, as $u(2) = 0$

$$\int_0^2 x^2 u' u \, dx = \frac{1}{2} \int_0^2 x^2 (u^2)' \, dx = \left[\frac{1}{2} x^2 u^2\right]_0^2 - \int_0^2 x u^2 \, dx = -\int_0^2 x u^2 \, dx.$$

Then

$$B(u, u) = \int_0^2 [(u')^2 - 2xu^2] \, dx.$$

It is easy to see B is not coercive: for example, by choosing $u = 2 - x$ (which belongs to V), we get $B(u, u) = -2/3$. At the same time

$$B(u, u) + 5\|u\|_{L^2}^2 \geq \|u\|_V^2,$$

so the form is weakly coercive. By Fredholm's alternative, whether the problem has solutions depends on the solutions of the homogeneous problem:

$$B(v, w) = \int_0^2 \left[w'v' - 2x^2 v'w - 4xwv\right] dx = 0, \qquad \text{for every } v \in V.$$

To solve this let us write the classical formulation. Assuming w regular,

$$\int_0^2 \left[w' - 2x^2 w\right] v' \, dx = \left[w'v - 2x^2 wv\right]_0^2 - \int_0^2 \left[w'' - 2x^2 w' - 4xw\right] v \, dx$$

so the adjoint problem (formally) reads

$$w'(0)v(0) + \int_0^2 \left[w'' - 2x^2 w'\right] dx = 0 \qquad \text{for any } v \in V.$$

As is customary, choosing $v \in H_0^1(0, 2)$ first, and then $v \in V$, we obtain

$$w'' - 2x^2 w' = 0 \quad \text{in } (0, 2)$$

and the condition $w'(0) = 0$. Altogether the classical form is

$$\begin{cases} w'' - 2x^2 w' = 0 & \text{in } (0, 2), \\ w'(0) = 0, \\ w(2) = 0. \end{cases}$$

Integrating once gives $e^{-x^2} w' = C_1$; the Neumann boundary condition implies $C_1 = 0$, whence $w = C_2$ and then $w = 0$. As the adjoint problem has only the zero solution, by Fredholm's alternative the original problem is well posed for every f.

There remains to compute the classical form. Proceeding along the previous lines, we find

$$\begin{cases} u'' + 2x^2 u' + 4xu = -f & \text{in } (0, 2) \\ u'(0) = 0 \\ u(2) = 0. \end{cases}$$

Solution 6.3.4. Note first that the differential equation can be written as

$$-\left(\cos x \, u'\right)' + xu = -f.$$

To write a variational formulation, we follow Problem 6.2.3 on page 340, which actually corresponds to the choice $L = \pi/6$, $f \equiv 1$. More precisely, we choose the Hilbert space

$$V := \left\{ v \in H^1(0, L) : v(L) = 0 \right\}$$

with norm $\|v\|_V = \|v'\|_{L^2(0,1)}$ (recall that in V a Poincaré inequality holds), and proceeding as usual we obtain the following variational formulation: *to find $u \in V$ such that*

$$a(u, v) = \int_0^L [\cos x \, u'v' + xuv] dx - u(0)v(0) = -\int_0^L fv \, dx \qquad \forall v \in V.$$

To study well-posedness, we first check the hypotheses of the Lax-Milgram theorem. Observe that, since $u(0) = -\int_0^L u' dx$, we can write, using Schwarz's inequality,

$$|u(0)| \leq \int_0^L |u'| \, dx \leq \sqrt{L} \, \|u\|_V \tag{6.57}$$

so that

$$|a\,(u,v)| \leq (1+L)\,\|u\|_V\,\|v\|_V + L\,\|u\|_{L^2(0,L)}\,\|v\|_{L^2(0,L)} \leq (1+L+C_P^2 L)\,\|u\|_V\,\|v\|_V.$$

Therefore a is continuous. Now, using (6.57), we have

$$a\,(u,u) = \int_0^L [\cos x\,(u')^2 + xu^2]dx - u^2\,(0) \geq (\cos L - L)\,\|u\|_V^2$$

and therefore a is coercive if $L < L^*$, where $L^* \simeq 0.739\ldots$ is the solution of the equation

$$\alpha \equiv \cos L - L = 0.$$

Case $L < L^*$. Since clearly $f \in V^*$, the Lax-Milgram theorem applies and we get existence, uniqueness and the stability estimate

$$\|u\|_V \leq \frac{1}{\alpha}\,\|f\|_{L^2(0,L)} \tag{6.58}$$

(in particular, for $L = \frac{\pi}{6}$ and $f \equiv 1$, we recover Problem 6.2.3).

Case $L \geq L^*$. In this case a is only weakly coercive; to see it, we estimate $u\,(0)$ in a different way, writing

$$u^2\,(0) = -\int_0^L \frac{d}{dx}u^2 dx = -2\int_0^L uu'dx$$

and using Young's inequality to get, for any $\varepsilon > 0$:

$$u^2\,(0) \leq \varepsilon\,\|u\|_V^2 + \frac{1}{\varepsilon}\,\|u\|_{L^2(0,L)}^2.$$

Then, choosing (for instance) $\varepsilon = \frac{1}{2}\cos L$ we can write

$$a\,(u,u) \geq \frac{1}{2}\cos L\,\|u\|_V^2 - 2\,\|u\|_{L^2(0,L)}^2.$$

Therefore

$$\tilde{a}\,(u,v) \equiv a\,(u,v) + 2\,(u,v)_{L^2(0,L)}$$

is coercive and a is weakly coercive.

We can apply the Fredholm alternative. We have to examine the homogeneous adjoint problem, which is, since a is symmetric:

$$\begin{cases} \mathcal{L}w = \cos x\,w'' - \sin x\,w' - xw = 0 \quad 0 < x < L \\ w'\,(0) + w\,(0) = 0,\ w\,(L) = 0. \end{cases} \tag{6.59}$$

We want to show that there exists only one value $L = L_0$ such that problem (6.59) has a nontrivial solution and this vanishes at $x = L_0$. It is convenient to consider the following related Cauchy problem:

$$\begin{cases} \mathcal{L}y = \cos x\,y'' - \sin x\,y' - xy = 0 \quad -\frac{\pi}{2} < x < \frac{\pi}{2} \\ y\,(0) = r \\ y'\,(0) = -r. \end{cases} \tag{6.60}$$

From general ODE theory we know that, for any $r \in \mathbb{R}$, there exists a unique solution y_r of problem (6.60), analytic in $\left(-\frac{\pi}{2}, \frac{\pi}{2}\right)$. Let us examine the properties of y_r.

1. *We have*

$$y_r = ry_1.$$

In fact both y_r and ry_1 satisfy (6.60). By uniqueness they coincide. Therefore it is enough to consider y_1.

2. *Let* $z(x) = 1 - x - y_1$. *Then:*

$$z(x) \geq 0 \quad \text{for every } x \in \left(0, \frac{\pi}{2}\right).$$

To prove it, note first that, since $\sin x - x + x^2 \geq 0$ for every $x \in \mathbb{R}$,

$$\mathcal{L}z = \sin x - x + x^2 - \mathcal{L}y_1 \geq 0 \quad \text{in } \left(0, \frac{\pi}{2}\right). \tag{6.61}$$

As a consequence, z *cannot have a positive maximum at a point in* $(0, \frac{\pi}{2})$. Indeed, if $x_0 \in \left(0, \frac{\pi}{2}\right)$ and $z(x_0) = \max_{[0,1]} z$, then $z(x_0) > 0$, $z'(x_0) = 0$, $z''(x_0) \leq 0$. Thus,

$$\mathcal{L}z(x_0) = \cos x_0\, z''(x_0) - x_0 z(x_0) < 0$$

in contradiction with (6.61).

Moreover, $z(0) = z'(0) = 0$ and $z > 0$ near $x = 0$. Indeed, differentiating twice the differential equation for y_1 and using $y_1(0) = 1$, $y_1'(0) = -1$, we have

$$y_1(x) = 1 - x - \frac{1}{4!}x^4 + \frac{1}{5!}x^5 + \dots .$$

so that

$$z(x) = \frac{1}{4!}x^4 - \frac{1}{5!}x^5 + \dots > 0$$

in a positive neighbourhood of $x = 0$.

As a consequence, z cannot become negative in $\left(0, \frac{\pi}{2}\right)$, otherwise there would be a point in $\left(0, \frac{\pi}{2}\right)$ at which z would attain a positive maximum.

3. From 2. we deduce that $y_1 < 1 - x$ in $\left(0, \frac{\pi}{2}\right)$. Therefore y_1 *must have a zero at some point* $L_0 \in (0, 1]$.

Moreover y_1 is *strictly decreasing in* $\left(0, \frac{\pi}{2}\right)$ since it is easy to check that y_1 cannot have a maximum or a minimum at a point inside $\left(0, \frac{\pi}{2}\right)$. Therefore $y_1 > 0$ in $[0, L_0)$ and $y_1 < 0$ in $\left(L_0, \frac{\pi}{2}\right)$. Numerical calculations give $L_0 = 0.915\dots$.

4. From what we have proved so far, we deduce that there exists only one value $L = L_0 \in \left(0, \frac{\pi}{2}\right)$ such that problem (6.59) has a non-trivial solution. Actually, in this case, *the family of solutions to (6.59) is a 1–dimensional vector space generated by* y_1.

Going back to the original problem, we can draw the following conclusions:

Case $0 < L < L^*$. For any $f \in L^2(0, 1)$, there exists a unique solution $u \in V$ and

$$\|u\|_V \leq \frac{1}{\alpha} \|f\|_{L^2(0,L)} \qquad (\alpha = \cos L - L). \tag{6.62}$$

Case $L^* \leq L < \frac{\pi}{2}, L \neq L_0$. For any $f \in L^2(0,1)$, there exist a unique solution $u \in V$ and a constant C independent of u and f (but depending on L) such that

$$\|u\|_V \leq C \|f\|_{L^2(0,L)} . \tag{6.63}$$

Case $L = L_0$. The problem has a solution only if $(f, y_1)_{L^2(0,L)} = 0$. In this case the family of solutions is given by

$$u = \bar{u} + c y_1 \quad c \in \mathbb{R}$$

where $\bar{u}$ is a particular solution of (6.3.4).

Note that if f is constant, $(f, y_1)_0 \neq 0$, since $y_1 > 0$ in $(0, L_0)$, so that the problem has no solution.

Solution 6.3.5. a) We can mimic Problem 6.2.5 (page 343), with minor changes.

b) Set

$$\|v\| = \left(\int_{\mathbb{R}} v^2 e^{-x^2} dx \right)^{1/2}$$

and denote by $\|v\|_V$ the norm induced by (6.51). As

$$|Fv| = \left| \int_{\mathbb{R}} f v \, e^{-x^2} dx \right| \leq \|f\| \|v\| \leq \|f\| \|v\|_V ,$$

the functional F defines an element of V^*. Riesz's theorem implies there is a unique solution u to (6.52) and

$$\|u\|_V \leq \|f\| .$$

For the classical interpretation let us integrate formally, by parts, the second term of the inner product; then (6.52) reads

$$[e^{-x^2} u' v]_{-\infty}^{+\infty} + \int_{\mathbb{R}} [u e^{-x^2} - (e^{-x^2} u')'] v \, dx = \int_{\mathbb{R}} f v \, e^{-x^2} dx \qquad \text{for every } v \in V$$

and as v was arbitrary, if we choose v with compact support we find

$$-(e^{-x^2} u')' + u e^{-x^2} = f e^{-x^2} \quad \text{on } \mathbb{R},$$

equivalent to (*Hermite's equation*)

$$u'' - 2x u' - u = f \quad \text{on } \mathbb{R},$$

and then

$$\lim_{x \to \pm\infty} e^{-x^2} u'(x) = 0$$

by picking first v equal to zero at $\pm\infty$, and then equal to 1 at $\mp\infty$. The boundary conditions are thus homogeneous Neumann conditions, weighted by the Gaussian function.

Solution 6.3.6. a) Assuming u and v regular we can write

$$\int_{\Omega} \nabla u \cdot \nabla v \, d\mathbf{x} = \int_{\partial\Omega} \partial u_{\nu} \, v \, d\sigma - \int_{\Omega} \Delta u \, v \, d\mathbf{x}.$$

In particular, since $H_0^1(\Omega) \subset H^1(\Omega)$, the weak formulation implies

$$\int_\Omega [-\Delta u + u - f] v \, d\mathbf{x} = \int_{\partial\Omega} [g - \partial_\nu u] v \, d\mathbf{x} = 0, \quad \text{for every } v \in H_0^1(\Omega),$$

from which we deduce $-\Delta u + u - f = 0$ a.e. on Ω. Substituting back in the variational formulation, we find

$$\int_{\partial\Omega} [g - \partial_\nu u] v \, d\mathbf{x} = 0, \quad \text{for every } v \in H^1(\Omega),$$

from which $\partial_\nu u = g$ a.e. on $\partial\Omega$. In conclusion (6.53) is the weak formulation of the problem

$$\begin{cases} -\Delta u + u = f & \text{in } \Omega \\ \partial_\nu u = g & \text{on } \partial\Omega. \end{cases}$$

b) Observe that the left-hand side in the weak formulation is the standard scalar product in $H^1(\Omega)$. At the same time if $u \in H^1(\Omega)$, by the trace inequality

$$\|u\|_{L^2(\partial\Omega)} \leq C\|u\|_{H^1(\Omega)}.$$

Using Schwarz's inequality we deduce that the linear functional on the right of (6.53) is continuous on $H^1(\Omega)$, so Riesz's theorem implies the problem is well posed.

Solution 6.3.7. It is immediate to see that the problem is of the type

$$B(u, v) = Fv,$$

where B is the scalar product in V and F (for $f \in L^2(B_1)$) is continuous on $H^1(B_1)$, so *a fortiori* on V. In this case we can apply the Lax-Milgram theorem or, directly, Riesz's theorem.

If f is regular, i.e. $f \in C^\infty(\overline{B}_1)$, we know[16] that $u \in C^\infty(\overline{B}_1)$, so we may integrate by parts,

$$\int_{\partial B_1} \partial_\nu u \, v \, d\sigma + \int_{B_1} [-\Delta u + u - f] v \, d\mathbf{x} = 0 \quad \text{for any } v \in V.$$

Since $H_0^1(B_1) \subset V$, we deduce, in particular,

$$\int_{B_1} [-\Delta u + u - f] v \, d\mathbf{x} = 0 \quad \text{for every } v \in H_0^1(B_1).$$

Consequently $-\Delta u + u - f = 0$. Substituting in the weak formulation, we get

$$\int_{\partial B_1} \partial_\nu u \, v \, d\sigma = 0 \quad \text{for any } v \in V. \tag{6.64}$$

Let us show that this relation holds if and only if $\partial_\nu u$ is constant on ∂B_1. On one hand, if $\partial_\nu u$ is constant, the integral of the product with a function of zero average is clearly null. On the other hand, suppose $g = \partial_\nu u$ is not constant. Then there exist points σ_1 and σ_2 on ∂B_1 where $g(\sigma_1) = a$ and $g(\sigma_2) = b$ with (say) $a < b$. By continuity we can find arcs Γ_1, Γ_2 in $\partial\Omega$ so that

$$g|_{\Gamma_1} \leq a, \quad g|_{\Gamma_2} \geq b.$$

---

[16] [18, Chap. 8, Sect. 6].

Let w be a $C^1(\overline{B}_1)$ function, positive on Γ_1, negative on Γ_2, null on $\partial\Omega \setminus (\Gamma_1 \cup \Gamma_2)$ and such that

$$\int_{\Gamma_1} w \, d\sigma = -\int_{\Gamma_2} w \, d\sigma.$$

Then w belongs to V and can be used as test function in (6.64):

$$\int_{\partial\Omega} g \, w \, d\sigma = \int_{\Gamma_1} g \, w \, d\sigma + \int_{\Gamma_2} g \, w \, d\sigma \leq a \int_{\Gamma_1} w \, d\sigma + b \int_{\Gamma_2} w \, d\sigma$$

$$= (a - b) \int_{\Gamma_1} w \, d\sigma < 0.$$

This contradicts (6.64). In conclusion, the problem is

$$\begin{cases} -\Delta u + u = f & \text{in } B \\ \partial_{\nu} u = \text{constant} & \text{on } B_1 \\ \int_{\partial B_1} u = 0. \end{cases}$$

Solution 6.3.8. Multiplying the first equation by $v_1 \in H_0^1(\Omega)$ and integrating by parts, we obtain

$$\int_{\Omega} [\nabla u_1 \cdot \nabla v_1 + u_1 v_1 - u_2 v_1] \, d\mathbf{x} = \int_{\Omega} f_1 v_1 \, d\mathbf{x}, \quad \text{for any } v_1 \in H_0^1(\Omega).$$

Similarly,

$$\int_{\Omega} [\nabla u_2 \cdot \nabla v_2 + u_1 v_2 + u_2 v_2] \, d\mathbf{x} = \int_{\Omega} f_2 v_2 \, d\mathbf{x}, \quad \text{for any } v_2 \in H_0^1(\Omega).$$

The previous two equations can be combined into the following one

$$\int_{\Omega} [\nabla u_1 \cdot \nabla v_1 + \nabla u_2 \cdot \nabla v_2 + u_1 v_1 - u_2 v_1 + u_1 v_2 + u_2 v_2] \, d\mathbf{x} =$$

$$= \int_{\Omega} [f_1 v_1 + f_2 v_2] \, d\mathbf{x}, \quad \text{for every } (v_1, v_2) \in H_0^1(\Omega) \times H_0^1(\Omega)$$

(just choose $v_i = 0$ to get either). Define the Hilbert space

$$V = H_0^1(\Omega) \times H_0^1(\Omega),$$

with product

$$(\mathbf{u}, \mathbf{v})_V = (\nabla u_1, \nabla v_1)_{L^2(\Omega)} + (\nabla u_1, \nabla v_1)_{L^2(\Omega)}$$

(where $\mathbf{u} = (u_1, u_2)$, $\mathbf{v} = (v_1, v_2)$). Set $\mathbf{f} = (f_1, f_2)$,

$$B(\mathbf{u}, \mathbf{v}) = \int_{\Omega} [\nabla u_1 \cdot \nabla v_1 + \nabla u_2 \cdot \nabla v_2 + u_1 v_1 - u_2 v_1 + u_1 v_2 + u_2 v_2] \, d\mathbf{x}$$

and

$$F\mathbf{v} = \int_{\Omega} \mathbf{f} \cdot \mathbf{v} \, d\mathbf{x}.$$

This gives the variational formulation

$$B(\mathbf{u}, \mathbf{v}) = F\mathbf{v} \quad \text{for every } \mathbf{v} \in V.$$

Using Schwarz's and Poincaré's inequalities we can write

$$|F\mathbf{v}| \leq \|f_1\|_{L^2}\|v_1\|_{L^2} + \|f_2\|_{L^2}\|v_2\|_{L^2} \leq C_P^2\|\mathbf{u}\|_V\|\mathbf{v}\|_V$$

and

$$\begin{aligned}
|B(u,v)| &\leq \|\nabla u_1\|_{L^2}\|\nabla v_1\|_{L^2} + \|\nabla u_2\|_{L^2}\|\nabla v_2\|_{L^2} \\
&\quad + (\|u_1\|_{L^2} + \|u_2\|_{L^2})(\|v_1\|_{L^2} + \|v_2\|_{L^2}) \\
&\leq (1 + C_P^2)\|\mathbf{u}\|_V\|\mathbf{v}\|_V.
\end{aligned}$$

Moreover,

$$B(\mathbf{u}, \mathbf{u}) = \int_\Omega \left[|\nabla u_1|^2 + |\nabla u_2|^2 + u_1^2 + u_2^2\right] d\mathbf{x} \geq \|\mathbf{u}\|_{L^2}^2,$$

and therefore B is coercive. It is then possible to use Lax-Milgram and infer existence, uniqueness and a stability estimate for the solution.

Solution 6.3.9. Let us first check that the operator is uniformly elliptic, i.e. that the coefficients are bounded (true) and there exists $a_0 > 0$ such that

$$\sum_{j,k=1}^{2} a_{jk}(x,y) z_j z_k \geq a_0 \left(z_1^2 + z_2^2\right), \qquad \text{for every } (z_1, z_2) \in \mathbb{R}^2. \tag{6.65}$$

In fact,

$$\sum_{j,k=1}^{2} a_{jk}(x,y) z_j z_k = (2 + \sin(xy)) z_1^2 + 3z_1 z_2 - 3z_1 z_2 + (4 - \sin(xy)) z_2^2$$

$$\geq z_1^2 + 3z_2^2 \geq z_1^2 + z_2^2,$$

thus (6.65) holds with $a_0 = 1$. Let $V = H_{0,\Gamma_D}^1(\Omega)$ be the space of functions in $H^1(\Omega)$ vanishing on Γ_D. Poincaré's inequality holds, so we choose $\|\nabla u\|_{L^2(\Omega)}$ as norm on V. For the variational formulation, multiply the equation by $v \in V$ and integrate by parts:

$$\int_\Omega [A(x,y)\nabla u \cdot \nabla v + u\mathbf{b} \cdot \nabla v]\, dx dy - \int_{\Gamma_N} [A(x,y)\nabla u \cdot \mathbf{v} + u\mathbf{b} \cdot \mathbf{v}]v\, d\sigma = \int_\Omega fv\, dxdy. \tag{6.66}$$

On Γ_N we have $\mathbf{v} = (0, -1)$ and

$$A(0,y)\nabla u \cdot \mathbf{v} = 3u_x(0,y) - 4u_y(0,y) = -hu(0,y)$$
$$u\mathbf{b} \cdot \mathbf{v} = -yu(0,y).$$

Substituting in (6.66) produces

$$\int_\Omega [A(x,y)\nabla u \cdot \nabla v + u\mathbf{b} \cdot \nabla v]dxdy + \int_0^1 [h+y]u(0,y)v(0,y)\, dy = \int_\Omega fv\, dxdy.$$

If we denote by $B(u, v)$ the bilinear form on the left, the variational formulation is: determine a function $u \in V$ such that

$$B(u, v) = (f, v), \qquad \text{for every } v \in V.$$

As $f \in L^2(\Omega)$, the functional on the left is in V^*. Let us verify the continuity of B. By the trace and Poincaré inequalities:

$$|B(u, v)| \leq 4 C_*^2 (h + 2) \|\nabla u\|_{L^2} \|\nabla v\|_{L^2}$$

so B is continuous. As for coercivity, recalling that(6.65) holds with $a_0 = 1$,

$$\int_\Omega A(x, y) \nabla u \cdot \nabla u \, dx dy \geq \|\nabla u\|_{L^2}^2$$

while, as $\operatorname{div} \mathbf{b} = -3$,

$$\int_\Omega u \mathbf{b} \cdot \nabla u = \frac{1}{2} \int_\Omega \mathbf{b} \cdot \nabla(u^2) = -\frac{1}{2} \int_0^1 y u^2 (0, y) \, dy + \frac{3}{2} \int_\Omega u^2$$

hence $(h > 0)$

$$B(u, u) \geq \|\nabla u\|_{L^2}^2 + \frac{3}{2} \int_\Omega u^2 + \int_0^1 [h + \frac{y}{2}] u^2 (0, y) \, dy \geq \|\nabla u\|_{L^2}^2$$

and coercivity of B follows, with constant 1. The assumptions of the Lax-Milgram theorem hold. Therefore the given problem has a unique solution; what is more,

$$\|\nabla u\|_{L^2(\Omega)} \leq C_P \|f\|_{L^2(\Omega)},$$

where C_P is a Poincaré constant.

Solution 6.3.10.   a) Recall that if φ is a scalar field and $\mathbf{F}$ a vector field, both regular, we have

$$\operatorname{div}(\varphi \mathbf{F}) = \varphi \operatorname{div} \mathbf{F} + \nabla \varphi \cdot \mathbf{F}.$$

Since

$$\nabla e^{-\mathbf{b} \cdot \mathbf{x}} = -e^{-\mathbf{b} \cdot \mathbf{x}} \mathbf{b}$$

we obtain

$$-e^{-\mathbf{b} \cdot \mathbf{x}} \Delta u + e^{-\mathbf{b} \cdot \mathbf{x}} \mathbf{b} \cdot \nabla u = -\operatorname{div}\left(e^{-\mathbf{b} \cdot \mathbf{x}} \nabla u\right).$$

Hence u solves the divergence problem

$$\begin{cases} -\operatorname{div}\left(e^{-\mathbf{b} \cdot \mathbf{x}} \nabla u\right) = e^{-\mathbf{b} \cdot \mathbf{x}} f & \mathbf{x} \in \Omega \\ \partial_\nu u = 0 & \mathbf{x} \in \partial\Omega. \end{cases}$$

b) There is a standard way for proving (e.g. following Problem 6.2.12 on page 355) that the bilinear form appearing in the weak formulation is (beside continuous) weakly coercive, so Fredholm's theory applies. The operator is self-adjoint, and the adjoint homogeneous problem is

$$\begin{cases} -\operatorname{div}\left(e^{-\mathbf{b}\cdot\mathbf{x}}\nabla u\right) = 0 & \mathbf{x} \in \Omega \\ \partial_{\nu} u = 0 & \mathbf{x} \in \partial\Omega. \end{cases}$$

It is easy to show[17] that the solutions are only constants. The starting problem has solutions if and only if

$$\int_{\Omega} e^{-\mathbf{b}\cdot\mathbf{x}} f(\mathbf{x})\, d\mathbf{x} = 0$$

in which case the solution is defined up to an additive constant.

Solution 6.3.11.   The weak formulation is

$$B(u, v) = \int_{Q} [\nabla u \cdot \nabla v - 2uv] = \int_{Q} \mathbf{f}\cdot\nabla v = Fv, \qquad \text{for every } v \in H_0^1(Q).$$

The functional F belongs to $H^{-1}(Q)$ if and only if $\mathbf{f} \in L^2(Q;\mathbb{R}^2)$. We have that $\log t$ is in $L^2(0,1)$. Now as

$$\sin x \sim x \text{ as } x \to 0, \qquad \sin x \sim \pi - x \text{ for } x \to \pi,$$

we obtain the required integrability for $\mathbf{f}$.

As seen in Problem 6.2.13 (page 356), B is continuous and weakly coercive in $H_0^1(Q)$ and the adjoint homogeneous problem has only the solution $\bar{u}(x, y) = c \sin x \sin y$, with $c \in \mathbb{R}$. Fredholm's theory implies that the problem can be solved if and only if

$$F\bar{u} = 0.$$

We have $\nabla\bar{u} = (c \cos x \sin y, c \sin x \cos y)$ and

$$F\bar{u} = \left(\int_0^\pi c \sin y\, dy\right)\left(\int_0^\pi \log(\sin x)\cos x\, dx\right) = 0.$$

The problem has infinitely many solutions $u(x, y) + c \sin x \sin y$.

Solution 6.3.12.   Define $V = H^1(\Omega)$ and

$$V_1 = \left\{u \in H^1(\Omega_1), u = 0 \text{ in } \Omega \setminus \Omega_1\right\} \text{ and } V_2 = \left\{u \in H^1(\Omega_2), u = 0 \text{ in } \Omega \setminus \Omega_2\right\}.$$

Given $u_0 \in H^1(\Omega)$, let u_{2n+1} $(n \geq 0)$ and u_{2n} $(n \geq 1)$ be the solutions of the mixed problems

$$-\Delta u_{2n+1} + u = f \quad \text{in } \Omega_1, \qquad u_{2n+1} = u_{2n} \text{ on } \partial\Omega_1 \cap \Omega, \ \partial_\nu u = 0 \text{ on } \partial\Omega_1 \cap \partial\Omega$$

and

$$-\Delta u_{2n} + u = f \quad \text{in } \Omega_2, \qquad u_{2n} = u_{2n-1} \text{ on } \partial\Omega_2 \cap \Omega, \ \partial_\nu u = 0 \text{ on } \partial\Omega_2 \cap \partial\Omega$$

respectively. Let us follow the solution of Problem 6.2.15 on page 361 to obtain

$$u - u_{2n+1} = P_{V_1^{\perp}}(u - u_{2n}) \quad \text{and} \quad u - u_{2n} = P_{V_1^{\perp}}(u - u_{2n-1}),$$

---

[17] For example as in Problem 6.2.10 (page 351).

where the projections are intended with respect to the inner product $(u, v) = \int_\Omega [uv + \nabla u \cdot \nabla v]$ on V. The function u_n converges in $H^1(\Omega)$ to the solution $u \in H^1(\Omega)$ of problem (6.54) if (see Problem 5.2.9, Chap. 5, page 288) V_1 and V_2 are closed in V and

$$V_1^\perp \cap V_2^\perp = \{0\}$$

i.e. $V_1 + V_2$ is dense in V. Let $v \in V$. To verify the closure of V_1 and V_2, we may resort to the technique of Exercise 5.3.26 on page 313, Chap. 5. Details are left to the reader. To check the density of $V_1 + V_2$ in V, let w_1 and w_2 be regular with $w_1 = 0$ on $\Omega \setminus \Omega_1$ and $\Omega \setminus \Omega_2$, positive elsewhere. Setting

$$v_1 = \frac{w_1}{w_1 + w_2} v \quad \text{and} \quad v_1 = \frac{w_2}{w_1 + w_2} v,$$

we have $v_1 \in V_1, v \in V_2$ and $v_1 + v_2 = v$. Then $V_1 + V_1$ is not only dense, it coincides with V.

Solution 6.3.13.   a) First, on ∂B_1 the outward unit normal is $\nu = x\mathbf{i} + y\mathbf{j}$, while $\tau = y\mathbf{i} - x\mathbf{j}$ is tangential. The unit vector

$$\sigma = \frac{(x+y)\mathbf{i} + (y-x)\mathbf{j}}{\sqrt{2}} = \frac{\nu + \tau}{\sqrt{2}}$$

is *never* tangent, for $\sigma \cdot \nu = 1/\sqrt{2}$. Hence we are dealing with an oblique-derivative problem. In relationship to Problem 6.2.16, with $b = 1$,

$$A(x, y) = \begin{pmatrix} 1 & 1 \\ -1 & 1 \end{pmatrix} \quad \text{and} \quad \mathbf{b}(x, y) = (by, -bx) = (0, 0).$$

The variational formulation reads

$$a(u, v) = \int_{B_1} A(x, y) \nabla u \cdot \nabla v \, dxdy = \int_{B_1} fv \, dxdy + \int_{\partial B_1} gv \, ds.$$

Explicitly, we have

$$\int_{B_1} (u_x v_x + u_y v_x - u_x v_y + u_y v_y) \, dxdy = \int_{\partial B_1} \left(1 + \alpha x^2\right) v \, ds, \quad \text{for every } v \in H^1(B_1)$$

which can be solved if and only if

$$\int_{\partial B_1} \left(1 + \alpha x^2\right) ds = 2\pi + \alpha\pi = 0$$

i.e. for $\alpha = -2$.
   b) We have

$$\int_{B_1} (u_x v_x + u_y v_x - u_x v_y + u_y v_y) \, dxdy = \int_{B_1} \left(x^2 - y^2\right)^n v \, dxdy, \quad \text{for every } v \in H^1(B_1)$$

which can be solved if and only if

$$\int_{B_1} \left(x^2 - y^2\right)^n dxdy = 0$$

i.e. if and only if n is *odd*.

Solution 6.3.14. **a)** If u and $v \in K$ and $s \in [0, 1]$, then

$$(1 - s)u + sv \in K,$$

so K is convex. To check that K is closed in $H_0^1(\Omega)$ consider a sequence $\{v_n\} \subset K$ converging in $H_0^1(\Omega)$ to v. In particular, there exists a subsequence $\{v_{n_k}\}$ convergent to v a.e. on Ω. Therefore $v_{n_k} \geq \psi$ a.e. on Ω, which implies $v \geq \psi$ a.e. on Ω and hence K is closed in $H_0^1(\Omega)$.

Let us view this in terms of projections: $J(v)$ represents, up to the factor $1/2$, the square of the distance in $H_0^1(\Omega)$ between v and $w = 0$, and to minimise J is the same as minimising the distance of K from the origin. The existence of a minimiser for J amounts thus to the existence of the *projection of the origin on the closed, convex set K*. So we can use Problem 5.2.10 on page 289 (Chap. 5), and conclude that there is a unique $u \in K$ such that

$$J(u) = \min_{v \in K} J(v).$$

This element is characterised by

$$(u, u - v)_{H_0^1(\Omega)} \geq 0, \qquad \text{for every } v \in K$$

corresponding precisely to (6.55).

c) Set $u - v = w$ in (6.55). Then $w \geq 0$ on Ω. If $u \in H^2(\Omega)$ we can integrate by parts so to have the derivatives on u:

$$\int_\Omega w \Delta u \leq 0, \qquad \text{for every } w \in H_0^1(\Omega), w \geq 0 \text{ a.e. on } \Omega,$$

which forces (prove it) $\Delta u \leq 0$ a.e. on Ω. Additionally, if $D \subset \Omega$ is an open set over which $u > \psi$, take $\eta \in C_0^1(D)$. If $h \in \mathbb{R}$, positive or negative, is small enough,

$$v = u + hv > \psi$$

on D and $v \geq \psi$ on Ω. So we can take it as test function in (6.55) and deduce

$$h \int_D \nabla u \cdot \nabla \eta = 0 \qquad \text{for every } \eta \in C_0^1(D).$$

But this means $\Delta u = 0$ on D.

Remark. When $n = 2$ we can view the graph of u as an elastic membrane fixed to the boundary of Ω. $J(u)$ is proportional to the potential energy of the membrane's deformation. The problem asks to find the configuration of least energy (equilibrium) under the constraint that the membrane cannot move *below* ψ, which acts as an *obstacle*.

Solution 6.3.15. **a)** Set

$$w(x, y, t) = u(x, y, t) - g(t)$$

so to have zero Dirichlet datum on Γ_D. The problem for w is

$$\begin{cases} w_t - \operatorname{div}(xe^y \nabla w) + |x|(\operatorname{sign}(y)w)_y = F(x, y, t) & \text{in } Q_T, \\ w(x, y, 0) = u_0(x, y) - g(0) & \text{in } B_1, \\ w = 0 & \text{on } \Gamma_D, \\ \partial_\nu w + w + g = 0 & \text{on } \Gamma_N, \end{cases} \qquad (6.67)$$

where
$$F(x, y, t) = f(x, y, t) - \dot{g}(t), U_0(x, y) = u_0(x, y) - g(0).$$

For the weak formulation we use $V = H^1_{0,\Gamma_D}(B_1)$, the space of functions in $H^1(B_1)$ with zero trace on Γ_D. The Poincaré inequality holds

$$\|v\|_{L^2(B_1)} \leq C_P \|\nabla v\|_{L^2(B_1)}$$

so we choose

$$\|v\|_V = \|\nabla v\|_{L^2(B_1)}.$$

Multiply the equation by $v \in V$, integrate on B_1 and use Gauss's formula, keeping in mind the mixed boundary conditions; as $\boldsymbol{\nu} = (x, y)$ is the unit normal to ∂B_1,

$$\int_{B_1} [w_t v + xe^y \nabla w \cdot \nabla v - |x| \operatorname{sign}(y) w v_y] \, dx dy + \int_{\Gamma_N} |xy| \, wv \, d\sigma = \int_{B_1} Fv \, dx dy.$$

Set

$$B(w, v) = \int_{B_1} [xe^y \nabla w \cdot \nabla v - |x| \operatorname{sign}(y) w v_y] \, dx dy + \int_{\Gamma_N} |xy| \, wv \, d\sigma$$

and let us pass from the notation $w(x, y, t)$ to

$$\mathbf{w}(t) : t \to w(\cdot, t).$$

As usual $(\cdot, \cdot)$ is the inner product on $L^2(B_1)$. The weak formulation of (6.67) is: determine $\mathbf{w} \in L^2(0, T; V) \cap C([0, T]; L^2(B_1))$ such that $\dot{\mathbf{w}} \in L^2(0, T; V^*)$ and:

i)   For every $v \in V$,
$$\frac{d}{dt}(\mathbf{w}(t), v) + B(\mathbf{w}(t), v) = (F, v)$$

in $\mathcal{D}'(B_1)$ for almost every $t \in (0, T)$.

ii) $\mathbf{w}(t) \to U_0$ in $L^2(B_1)$ as $t \to 0^+$.

To see if the problem is well posed we check the *continuity* and *weak coercivity* of B on V. On $\overline{B}_1$
$$e^{-1} \leq xe^y \leq e, \quad |xy| \leq 1.$$

So for every $v, z \in V$:

$$|B(v, z)| \leq e\|v\|_V \|z\|_V + \|v\|_{L^2(B_1)} \|z_y\|_{L^2(B_1)} + \|v\|_{L^2(\Gamma_N)} \|z\|_{L^2(\Gamma_N)}.$$

By the Poincaré and trace inequalities[18]

$$|B(v, z)| \leq (e + C_P + C_*^2) \|v\|_V \|z\|_V,$$

and B is continuous on V. Moreover,

$$B(v, v) = \int_{B_1} [xe^y |\nabla v|^2 - |x| \operatorname{sign}(y) v v_y] \, dx dy + \int_{\Gamma_N} |xy| \, v^2 \, d\sigma$$
$$\geq e^{-1} \|v\|_V^2 - \|v_y\|_{L^2(B_1)} \|v\|_{L^2(B_1)}.$$

---

[18] $\|v\|_{L^2(\Gamma_N)} \leq C_* \|v\|_V$, see [18, Chap. 7].

By the Poincaré and trace inequalities, plus the elementary fact

$$ab \leq \frac{1}{2e}a^2 + \frac{e}{2}b^2$$

we can write

$$B(v,v) \geq \frac{1}{e}\|v\|_V^2 - \frac{1}{2e}\|v\|_V^2 - \frac{e}{2}\|v\|_{L^2(B_1)}$$

$$= \frac{1}{2e}\|v\|_V^2 - \frac{e}{2}\|v\|_{L^2(B_1)}.$$

Hence the bilinear form

$$\widetilde{B}(v,z) = B(v,z) + \frac{e}{2}\|v\|_{L^2(B_1)}^2$$

is coercive with constant $1/2e$, so B is weakly coercive.

By assuming

$$f(x,y,t) \in L^2(Q_T), g \in H^1(t) \text{ and } U_0(x,y) = u_0(x,y),$$

the problem has a unique weak solution, and

$$\max_{[0,T]} \|u(\cdot,t)\|_H^2 + \int_0^T \|\nabla u(\cdot,t)\|_H^2 \, dt$$

$$\leq ce^T \left\{ \|u_0 - g(0)\|_{H_0^1(B_1)}^2 + \int_0^T \left\{ \|f(\cdot,\cdot,s)\|_{L^2(B_1)}^2 + (\dot{g}(s))^2 \right\} ds \right\}.$$

Solution 6.3.16. **a)** Set

$$B(u,v) = \int_\Omega [a(\mathbf{x})\nabla u \cdot \nabla v - u\mathbf{b}\cdot\nabla v + c(\mathbf{x})uv]\,dxdy + \int_{\Gamma_N}[h+\mathbf{b}\cdot\mathbf{v}]uv\,d\sigma$$

and let us pass from using $u(\mathbf{x})$ to

$$\mathbf{u}(t): t \to u(\cdot,t).$$

The symbol $(\cdot,\cdot)$ is the inner product on $L^2(\Omega)$. The variational formulation goes as usual. Take $V = H_{0,\Gamma_D}^1(\Omega)$, space of $H^1(\Omega)$ functions with zero trace on Γ_D, normed by $\|v\|_V = \|\nabla v\|_{L^2(B_1)}$. We want to find $\mathbf{u} \in L^2(0,T;V) \cap C([0,T];L^2(\Omega))$ such that $\dot{\mathbf{u}} \in L^2(0,T;V^*)$ and:

i)   For every $v \in V$,

$$\frac{d}{dt}(\mathbf{u}(t),v) + B(\mathbf{u}(t),v) = 0$$

in $\mathcal{D}'(B_1)$ and for almost every $t \in (0,T)$.

ii) $\mathbf{u}(t) \to U_0$ in $L^2(\Omega)$ as $t \to 0^+$.

b) Let us assume:

$$|a(\mathbf{x})|, |\mathbf{b}(\mathbf{x})|, |c(\mathbf{x})| \leq M, \quad a(\mathbf{x}) \geq a_0 > 0, \quad |\text{div }\mathbf{b}(\mathbf{x})| \leq M', \quad \text{a.e. on } \Omega,$$

$$h + \mathbf{b}\cdot\mathbf{v} \geq 0 \text{ a.e. on } \Gamma_N \quad (\text{if } h = 0, \text{ the flow of } \mathbf{b} \text{ through } \Gamma_N \text{ is } \textit{outgoing}).$$

By the Poincaré and trace inequalities the bilinear form B is continuous (the reader should fill in the details). As for weak coercivity, using repeatedly the assumptions on the coefficients:

$$B(u, u) \geq \int_\Omega [a(\mathbf{x}) |\nabla u|^2 - u\mathbf{b} \cdot \nabla u + c(\mathbf{x}) u^2] + \int_{\Gamma_N} [h + \mathbf{b} \cdot \mathbf{v}] u^2 \, d\sigma$$

$$\geq a_0 \int_\Omega [|\nabla u|^2 - \frac{1}{2} \mathbf{b} \cdot \nabla(u^2) + c(\mathbf{x}) u^2] + \int_{\Gamma_N} [h + \mathbf{b} \cdot \mathbf{v}] u^2 \, d\sigma$$

$$= a_0 \int_\Omega |\nabla u|^2 + \int_\Omega [\frac{1}{2} \operatorname{div} \mathbf{b} + c(\mathbf{x})] u^2 + \int_{\Gamma_N} [h + \frac{1}{2} \mathbf{b} \cdot \mathbf{v}] u^2 \, d\sigma$$

$$\geq a_0 \|\nabla u\|_V^2 - \left(\frac{1}{2} M' + M\right) \|u\|_{L^2}^2 \equiv a_0 \|\nabla u\|_V^2 - \lambda_0 \|u\|_{L^2}^2.$$

The form $\widetilde{B}(u, v) = B(u, v) + \lambda_0(u, v)$ is thus coercive on V, which means B weakly coercive on V. Under the given hypotheses, the general theory applies and the problem has one weak solution.

Solution 6.3.17.   A Robin problem is assigned. Since μ is a positive function, we write

$$\mu(x) \geq \mu_0 > 0, \quad \text{a.e. } (\mathbf{x}, t) \text{ in } Q_T = \Omega \times [0, T]$$

and this inequality ensures the uniform ellipticity of the equation.

We consider $V = H^1(\Omega)$ and $v \in V$, with the usual norm. We recall that in $H^1(\Omega)$ the following trace inequality holds

$$\|v\|_{L^2(\Gamma)} \leq C_T \|v\|_V.$$

Writing $\mathbf{c}(t) = c(\cdot, t)$ and $\dot{\mathbf{c}}$ for c_t, let us proceed formally and multiply the first equation of system (6.56) by v; integrating on Ω, we find

$$\langle \dot{\mathbf{c}}, v \rangle_* + \int_\Omega \mu \nabla \mathbf{c} \cdot \nabla v \, d\mathbf{x} - \int_\Gamma \mu \nabla \mathbf{c} \cdot \mathbf{v} \, v \, d\sigma - \int_\Omega \sigma c v \, d\mathbf{x} = 0$$

namely

$$\langle \dot{\mathbf{c}}, v \rangle_* + \int_\Omega \mu \nabla \mathbf{c} \cdot \nabla v \, d\mathbf{x} + \int_\Gamma \kappa c v \, d\sigma - \int_\Omega \sigma c v \, d\mathbf{x} = \int_\Gamma \kappa c_{ext} v \, d\sigma.$$

Let us introduce the bilinear form and the functional

$$\overline{B}(w, v; t) = \int_\Omega \mu \nabla w \cdot \nabla v \, d\mathbf{x} + \int_\Gamma \kappa w v \, d\sigma - \int_\Omega \sigma w v \, d\mathbf{x}, \qquad F v = \int_\Gamma \kappa c_{ext} v \, d\sigma.$$

The problem can be written in weak form as follows. *Find $\mathbf{c} \in L^2(0, T; V)$ such that $\mathbf{c}' \in L^2(0, T; V')$ and:*

1. *For every $v \in V$ and for a.e. $t \in [0, T]$*

$$\langle \dot{\mathbf{c}}(t), v \rangle_* + \overline{B}(\mathbf{c}(t), v) = F v.$$

2. $\mathbf{c}(0) = c_0$ *in Ω.*

In order to use the theory for the well-posedness of the problem we have to show that $\overline{B}$ is continuous and weakly coercive and that F is continuous. Using the trace and the Schwarz inequalities we find that the functional F is continuous:

$$|F v| \leq \kappa c_{ext} |\Gamma|^{1/2} \|v\|_{L^2(\Gamma)} \leq 2R \sqrt{\pi} \kappa c_{ext} C_T \|v\|_V.$$

On the other hand, since $\mu \in L^\infty(\Omega)$, the bilinear form $\overline{B}$ is V-continuous, in fact ($L^p = L^p(\Omega)$)

$$|\overline{B}(c,v)| \leq \|\mu\|_{L^\infty}\|\nabla c\|_{L^2}\|\nabla v\|_{L^2} + \sigma\|c\|_{L^2}\|v\|_{L^2} + \kappa C_T^2\|c\|_V\|v\|_V$$

$$\leq \left(\|\mu\|_{L^\infty} + \sigma + \kappa C_T^2\right)\|c\|_V\|v\|_V.$$

Moreover

$$\overline{B}(v,v) + \lambda\|v\|_{L^2} = \int_\Omega \mu|\nabla c|^2 d\mathbf{x} + \int_\Gamma v^2\, d\sigma + \int_\Omega (\lambda - \sigma)v^2 d\mathbf{x} \geq \min(\mu_0, \lambda - \sigma)\|v\|_V^2,$$

hence $\overline{B}$ is weakly coercive if we choose, say, $\lambda = \sigma + \mu_0/2$. The constant of weak coercivity is then $\mu_0/2$.

Appendix A

Sturm-Liouville, Legendre and Bessel Equations

A.1 Sturm-Liouville Equations

A.1.1 Regular equations

The eigenfunctions of a large class of ODEs form complete orthonormal systems in suitable Hilbert spaces. To this class belong equations of the form

$$- \left(p \left(x \right) u' \right)' + q \left(x \right) u = \lambda w \left(x \right) u \left(x \right), \qquad a < x < b \tag{A.1}$$

under the assumption that p, p', q, and w are *continuous and positive on* $[a,b]$. In such a case (A.1) is called a **regular Sturm-Liouville equation**. We associate to (A.1) the boundary conditions:

$$\begin{aligned} \alpha u \left(a \right) - \beta p \left(a \right) u' \left(a \right) = 0 \\ \gamma u \left(b \right) - \delta p \left(b \right) u' \left(b \right) = 0, \end{aligned} \tag{A.2}$$

where the coefficients $\alpha, \beta, \gamma, \delta$ are real numbers. To avoid trivial situations we shall assume the following normalization condition:

$$\alpha^2 + \beta^2 = \gamma^2 + \delta^2 = 1.$$

In general, problem (A.1), (A.2) has nontrivial solutions only for special values of the parameter λ, called **eigenvalues**. The corresponding solutions are called **eigenfunctions**, and they form the **eigenspace** associated to λ. Let us introduce the (Hilbert) space $L_w^2 \left(a, b \right)$ of weighted square-integrable functions u on (a, b) with respect to the *weight* function w:

$$L_w^2 \left(a, b \right) = \left\{ u : \int_a^b u^2 \left(x \right) w \left(x \right) dx < \infty \right\}.$$

Then the following theorem holds:

© Springer International Publishing Switzerland 2015
S. Salsa, G. Verzini, *Partial Differential Equations in Action. Complements and Exercises*,
UNITEXT – La Matematica per il 3+2 87, DOI 10.1007/978-3-319-15416-9_A

Theorem A.1. *There exists an increasing sequence of positive numbers* $\{\lambda_j\}_{j\geq 1}$ *such that* $\lambda_j \to +\infty$ *and:*

a) *Problem (A.1), (A.2) admits a non-trivial solution if and only if* λ *equals one of the* λ_j.

b) *For every* j, λ_j *is simple, that is the associated eigenspace has dimension 1.*

c) *The eigenfunction system* $\{\varphi_j\}_{j\geq 1}$ *(suitably normalised) is an orthonormal basis in* $L_w^2\,(a,b)$.

A.1.2 Legendre's equation

When the coefficient p is zero, e.g. at a or b, the equation is **irregular** and the study becomes more complicated. A classical case is that of Legendre's equation

$$\left[\left(1-x^2\right)u'\right]' + \lambda u = 0 \qquad -1 < x < 1. \tag{A.3}$$

In the applications this is coupled with boundary conditions of the type

$$u\,(-1)\ \text{finite},\quad u\,(1)\ \text{finite}. \tag{A.4}$$

Particular solutions to (A.3), (A.4) are **Legendre's polynomials**, defined by the *Rodrigues* formula:

$$L_n\,(x) = \frac{1}{2^n n!}\frac{d^n}{dx^n}\left(x^2-1\right)^n \qquad (n \geq 0).$$

Each polynomial L_n corresponds to the eigenvalue $\lambda_n = n\,(n+1)$. The first four Legendre polynomials are

$$L_0\,(x) = 1,\, L_1\,(x) = x,\, L_2\,(x) = \frac{1}{2}\left(3x^2-1\right),\, L_4\,(x) = \frac{1}{2}\left(5x^3-3x\right).$$

The following theorem holds:

Theorem A.2. *In relationship to problem (A.3), (A.4):*

a) *There exists a non-zero solution if and only if*

$$\lambda = \lambda_n = n\,(n+1)\,, \qquad n = 0,1,2,\dots\,.$$

b) *For every* $n \geq 0$ *the solution corrisponding to* λ_n *is unique up to a constant factor, and coincides with Legendre's polynomial* L_n.

c) *The normalised polynomials*

$$\left\{\sqrt{\frac{2n+1}{2}}L_n\right\}_{n\geq 0}$$

form an orthonormal system in $L^2\,(-1,1)$.

Theorem A.2 allows to expand any $f \in L^2(-1, 1)$ in **Fourier-Legendre series**:

$$f(x) = \sum_{n=0}^{\infty} f_n L_n(x), \qquad \text{where } f_n = \frac{2n+1}{2} \int_{-1}^{1} f(x) L_n(x), \, dx$$

with $L^2(-1, 1)$-convergence. We also have a result about pointwise convergence, in perfect analogy with Fourier series.

Theorem A.3. *If f and f' have at most a finite number of jump points in $(-1, 1)$, then*

$$\sum_{n=0}^{\infty} f_n L_n(x) = \frac{f(x+) + f(x-)}{2}$$

for every $x \in (-1, 1)$.

A.2 Bessel's Equation and Functions

A.2.1 Bessel functions

Here is a short summary of the main properties of *Bessel functions*. First, though, we need to introduce a function interpolating the values of the factorial $n!$. The *gamma function* $\Gamma = \Gamma(z)$ is

$$\Gamma(z) = \int_0^{\infty} e^{-t} t^{z-1} dt \qquad (A.5)$$

for z complex with $\mathrm{Re}\, z > 0$. The function Γ is analytic for $\mathrm{Re}\, z > 0$ and satisfies the following relationships:

$$\Gamma(z+1) = z\Gamma(z)$$
$$\Gamma(z)\Gamma(1-z) = \frac{\pi}{\sin \pi z} \qquad (z \neq 0, 1, 2, \ldots).$$

In particular,

$$\Gamma(n+1) = n! \qquad (n = 0, 1, 2, \ldots)$$

and

$$\Gamma\left(n + \frac{1}{2}\right) = \frac{1 \cdot 3 \cdot 5 \cdots 2n - 1}{2^n} \sqrt{\pi} \qquad (n = 1, 2, \ldots).$$

One can define $\Gamma(z)$ for z **real, negative** and **not integer,** using

$$\Gamma(z) = \frac{\Gamma(z+1)}{z}.$$

In fact, we know how to compute Γ on $(0, 1)$, and the formula allows to find Γ on $(-1, 0)$. In general, once we know Γ on $(-n, -n+1)$, we can compute it on $(-n-1, -n)$. Finally,

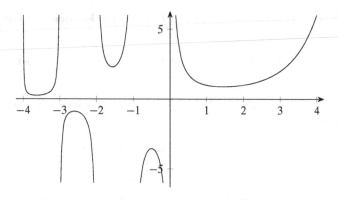

Fig. A.1 Graph of the *gamma* function on the real axis

coherently with (A.5) we define

$$\Gamma\left(-2n\right) = -\infty \quad \text{and} \quad \Gamma\left(-2n - 1\right) = +\infty.$$

In this way Γ is defined on the entire real axis (Fig. A.1).

Bessel's function of the first kind and order p, p real, is

$$J_p\left(z\right) = \sum_{k=0}^{\infty} \frac{(-1)^k}{\Gamma\left(k + 1\right)\Gamma\left(k + p + 1\right)} \left(\frac{z}{2}\right)^{p+2k}.$$

In particular, if $p = n \geq 0$ is an integer (Fig. A.2):

$$J_n\left(z\right) = \sum_{k=0}^{\infty} \frac{(-1)^k}{k!\left(k + n\right)!} \left(\frac{z}{2}\right)^{n+2k}.$$

When $p = -n$ is a *negative* integer, the first n terms of the series vanish and

$$J_{-n}\left(z\right) = (-1)^n J_n\left(z\right).$$

Hence $J_n\left(z\right)$ and $J_{-n}\left(z\right)$ are *linearly dependent*.

If p **is not integer**, for $z \to 0$ we have asymptotic behaviours:

$$J_p\left(z\right) = \frac{1}{\Gamma\left(1 + p\right)} \left(\frac{z}{2}\right)^p + O\left(z^{p+2}\right), \qquad J_{-p}\left(z\right) = \frac{1}{\Gamma\left(1 - p\right)} \left(\frac{z}{2}\right)^{-p} + O\left(z^{-p+2}\right)$$

so $J_p\left(z\right)$ and $J_{-p}\left(z\right)$ are *linearly independent*.

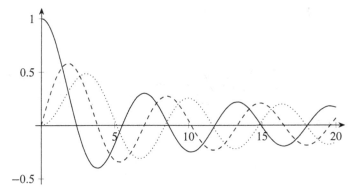

Fig. A.2 Graphs of J_0 (*solid*), J_1 (*dashed*) and J_2 (*dotted*)

Functions of the first kind satisfy a number of identities:

$$\frac{d}{dz}\left[z^p J_p(z)\right] = z^p J_{p-1}(z), \qquad \frac{d}{dz}\left[z^{-p} J_p(z)\right] = -z^{-p} J_{p+1}(z). \tag{A.6}$$

In particular

$$J_0'(z) = -J_1(z).$$

From these we also infer that for $p = n + \frac{1}{2}$ (and only in that case), the corresponding Bessel functions are elementary. For instance,

$$J_{\frac{1}{2}}(z) = \sqrt{\frac{2}{\pi z}}\sin z, \quad J_{-\frac{1}{2}}(z) = \sqrt{\frac{2}{\pi z}}\cos z.$$

Particularly important are the **zeroes** of J_p. For any p, there is an infinite increasing sequence $\{\alpha_{pj}\}_{j\geq 1}$ of positive numbers such that

$$J_p(\alpha_{pj}) = 0 \quad (j = 1, 2, \ldots).$$

When p is **not** an integer, *every linear combination*

$$c_2 J_p(z) + c_2 J_{-p}(z)$$

is a Bessel function of the second kind. **The (standard) function of the second kind** is

$$Y_p(z) = \frac{\cos p\pi\, J_p(z) - J_{-p}(z)}{\sin p\pi}.$$

When $p = n$ is integer, one defines[1] (Fig. A.3)

$$Y_n(z) := \lim_{p\to n} Y_p(z)$$

Note that $Y_p(z) \to -\infty$ when $z \to 0^+$.

---

[1] One can prove that the limit exists.

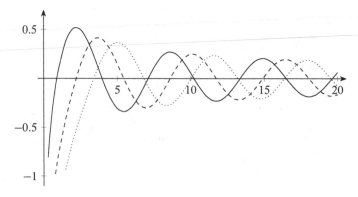

Fig. A.3 Graphs of Y_0 (*solid*), Y_1 (*dashed*) and Y_2 (*dotted*)

A.2.2 Bessel's equation

The Bessel functions J_p, Y_p are solutions of the so-called **Bessel equation of order** $p \geq 0$:

$$z^2 y'' + z y' + \left(z^2 - p^2\right) y = 0.$$

The general integral is given, for any $p \geq 0$, by

$$y(z) = c_1 J_p(z) + c_2 Y_p(z).$$

In the most important applications, one is typically led to solve the (**parametric**) equation (**with parameter** λ)

$$z^2 y'' + z y' + \left(\lambda^2 z^2 - p^2\right) y = 0 \tag{A.7}$$

on a bounded interval $(0, a)$, with boundary conditions of the sort

$$y(0) \text{ finite}, \quad y(a) = 0. \tag{A.8}$$

For these, the following theorem holds.

Theorem A.4. *Problem (A.7), (A.8) has nontrivial solutions if and only if*

$$\lambda = \lambda_{pj} = \left(\frac{\alpha_{pj}}{a}\right)^2.$$

In that case the solutions are

$$y_{pj}(z) = J_p\left(\frac{\alpha_{pj}}{a} z\right)$$

up to multiplicative constants. Moreover, the normalised functions

$$\frac{\sqrt{2}}{a J_{p+1}\left(\alpha_{pj}\right)} y_{pj}$$

form an orthonormal basis in $(w(z) = z)$

$$L_w^2 (0, a) = \left\{ u : \|u\|_{2,w}^2 = \int_0^a u^2 (z) \, z \, dz < \infty \right\}$$

by virtue of the orthogonality relations:

$$\frac{2}{a^2 J_{p+1}^2 (\alpha_{pj})} \int_0^a z J_p (\lambda_{pj} z) J_p (\lambda_{pk} z) \, dz = \begin{cases} 0 & j \neq k \\ 1 & j = k. \end{cases}$$

With Theorem A.4 we can expand any $f \in L_w^2 (0, a)$ in **Fourier-Bessel series**:

$$f (z) = \sum_{j=1}^\infty f_j J_p (\lambda_{pj} z), \qquad \text{where } f_j = \frac{2}{a^2 J_{p+1}^2 (\alpha_{pj})} \int_0^a z f (z) J_p (\lambda_{pj} z) \, dz,$$

with $L_w^2 (0, a)$-convergence.

Let us compute, for example, the expansion of $f (x) = 1$ on the interval $(0, 1)$, with $p = 0$:

$$f_j = \frac{2}{J_1^2 (\alpha_{pj})} \int_0^1 z J_0 (\alpha_{0j} z) \, dz.$$

Using (A.6),

$$\frac{d}{dz} [z J_1 (z)] = z J_0 (z)$$

so we may write

$$\int_0^1 z J_0 (\alpha_{0j} z) \, dz = \left[\frac{1}{\lambda_{0j}} z J_1 (\alpha_{0j} z) \right]_0^1 = \frac{J_1 (\alpha_{0j})}{\lambda_{0j}}.$$

Finally

$$1 = \sum_{j=1}^\infty \frac{2}{\lambda_{0j} J_1 (\alpha_{0j})} J_0 (\alpha_{0j} z)$$

with convergence in norm $L_w^2 (0, 1)$.

Also in this case one can insure pointwise convergence.

Theorem A.5. *If f and f' have at most finitely many jump discontinuities on $(0, a)$, then*

$$\sum_{j=1}^\infty f_j J_p (\lambda_{pj} z) = \frac{f (z+) + f (z-)}{2}$$

at every point $z \in (0, a)$.

Appendix B

Identities

Here is a compilation of significant formulas and identities of common use.

B.1 Gradient, Divergence, Curl, Laplacian

Let $\mathbf{F}, \mathbf{u}, \mathbf{v}$ be vector fields and f, φ scalar fields, all assumed regular on $\mathbb{R}^3$.

Orthogonal Cartesian coordinates

1. *gradient*:

$$\nabla f = \frac{\partial f}{\partial x}\mathbf{i} + \frac{\partial f}{\partial y}\mathbf{j} + \frac{\partial f}{\partial z}\mathbf{k}$$

2. *divergence*:

$$\operatorname{div} \mathbf{F} = \frac{\partial}{\partial x}F_x + \frac{\partial}{\partial y}F_y + \frac{\partial}{\partial z}F_z$$

3. *Laplacian*:

$$\Delta f = \frac{\partial^2 f}{\partial x^2} + \frac{\partial^2 f}{\partial y^2} + \frac{\partial^2 f}{\partial z^2}$$

4. *curl*:

$$\operatorname{curl} \mathbf{F} = \begin{vmatrix} \mathbf{i} & \mathbf{j} & \mathbf{k} \\ \partial_x & \partial_y & \partial_z \\ F_x & F_y & F_z \end{vmatrix}$$

Cylindrical coordinates

$$x = r\cos\theta, \ y = r\sin\theta, \ z = z \qquad (r > 0, 0 \le \theta \le 2\pi)$$

$$\mathbf{e}_r = \cos\theta\mathbf{i} + \sin\theta\mathbf{j}, \ \mathbf{e}_\theta = -\sin\theta\mathbf{i} + \cos\theta\mathbf{j}, \ \mathbf{e}_z = \mathbf{k}.$$

© Springer International Publishing Switzerland 2015
S. Salsa, G. Verzini, *Partial Differential Equations in Action. Complements and Exercises*,
UNITEXT – La Matematica per il 3+2 87, DOI 10.1007/978-3-319-15416-9_B

1. *gradient*:

$$\nabla f = \frac{\partial f}{\partial r}\mathbf{e}_r + \frac{1}{r}\frac{\partial f}{\partial \theta}\mathbf{e}_\theta + \frac{\partial f}{\partial z}\mathbf{e}_z$$

2. *divergence*:

$$\operatorname{div}\mathbf{F} = \frac{1}{r}\frac{\partial}{\partial r}(rF_r) + \frac{1}{r}\frac{\partial}{\partial \theta}F_\theta + \frac{\partial}{\partial z}F_z$$

3. *Laplacian*:

$$\Delta f = \frac{\partial^2 f}{\partial r^2} + \frac{1}{r}\frac{\partial f}{\partial r} + \frac{1}{r^2}\frac{\partial^2 f}{\partial \theta^2} + \frac{\partial^2 f}{\partial z^2} = \frac{1}{r}\frac{\partial}{\partial r}\left(r\frac{\partial f}{\partial r}\right) + \frac{1}{r^2}\frac{\partial^2 f}{\partial \theta^2} + \frac{\partial^2 f}{\partial z^2}$$

4. *curl*:

$$\operatorname{curl}\mathbf{F} = \frac{1}{r}\begin{vmatrix} \mathbf{e}_r & r\mathbf{e}_\theta & \mathbf{e}_z \\ \partial_r & \partial_\theta & \partial_z \\ F_r & rF_\theta & F_z \end{vmatrix}$$

Spherical coordinates

$$x = r\cos\theta\sin\psi,\; y = r\sin\theta\sin\psi,\; z = r\cos\psi \qquad (r > 0, 0 \le \theta \le 2\pi, 0 \le \psi \le \pi)$$

$$\mathbf{e}_r = \cos\theta\sin\psi\,\mathbf{i} + \sin\theta\sin\psi\,\mathbf{j} + \cos\psi\,\mathbf{k}$$
$$\mathbf{e}_\theta = -\sin\theta\,\mathbf{i} + \cos\theta\,\mathbf{j}$$
$$\mathbf{e}_z = \cos\theta\cos\psi\,\mathbf{i} + \sin\theta\cos\psi\,\mathbf{j} - \sin\psi\,\mathbf{k}.$$

1. *gradient*:

$$\nabla f = \frac{\partial f}{\partial r}\mathbf{e}_r + \frac{1}{r\sin\psi}\frac{\partial f}{\partial \theta}\mathbf{e}_\theta + \frac{1}{r}\frac{\partial f}{\partial \psi}\mathbf{e}_\psi$$

2. *divergence*:

$$\operatorname{div}\mathbf{F} = \underbrace{\frac{\partial}{\partial r}F_r + \frac{2}{r}F_r}_{\text{radial part}} + \underbrace{\frac{1}{r}\left[\frac{1}{\sin\psi}\frac{\partial}{\partial \theta}F_\theta + \frac{\partial}{\partial \psi}F_\psi + \cot\psi\,F_\psi\right]}_{\text{spherical part}}$$

3. *Laplacian*:

$$\Delta f = \underbrace{\frac{\partial^2 f}{\partial r^2} + \frac{2}{r}\frac{\partial f}{\partial r}}_{\text{radial part}} + \frac{1}{r^2}\underbrace{\left\{\frac{1}{(\sin\psi)^2}\frac{\partial^2 f}{\partial \theta^2} + \frac{\partial^2 f}{\partial \psi^2} + \cot\psi\frac{\partial f}{\partial \psi}\right\}}_{\text{spherical part (Laplace-Beltrami operator)}}$$

4. *curl*:

$$\operatorname{curl}\mathbf{F} = \frac{1}{r^2\sin\psi}\begin{vmatrix} \mathbf{e}_r & r\mathbf{e}_\psi & r\sin\psi\,\mathbf{e}_\theta \\ \partial_r & \partial_\psi & \partial_\theta \\ F_r & rF_\psi & r\sin\psi\,F_z \end{vmatrix}.$$

B.2 Formulas

Gauss's formulas

The following formulas hold on $\mathbb{R}^n$, $n \geq 2$, and we denote by:

- Ω a bounded domain with regular boundary $\partial\Omega$ and outward normal $\mathbf{v}$.
- $\mathbf{u}, \mathbf{v}$ vector fields that are regular[1] up to the boundary of Ω.
- φ, ψ regular scalar fields up to the boundary of Ω.
- $d\sigma$ the infinitesimal surface element of $\partial\Omega$.

We have the following formulas:

1. $\int_\Omega \operatorname{div} \mathbf{u}\, d\mathbf{x} = \int_{\partial\Omega} \mathbf{u} \cdot \mathbf{v}\, d\sigma$ (divergence formula)

2. $\int_\Omega \nabla\varphi\, d\mathbf{x} = \int_{\partial\Omega} \varphi\mathbf{v}\, d\sigma$

3. $\int_\Omega \Delta\varphi\, d\mathbf{x} = \int_{\partial\Omega} \nabla\varphi \cdot \mathbf{v}\, d\sigma = \int_{\partial\Omega} \partial_\mathbf{v}\varphi\, d\sigma$

4. $\int_\Omega \psi \operatorname{div} \mathbf{F}\, d\mathbf{x} = \int_{\partial\Omega} \psi\mathbf{F} \cdot \mathbf{v}\, d\sigma - \int_\Omega \nabla\psi \cdot \mathbf{F}\, d\mathbf{x}$

5. $\int_\Omega \psi\Delta\varphi\, d\mathbf{x} = \int_{\partial\Omega} \psi\partial_\mathbf{v}\varphi\, d\sigma - \int_\Omega \nabla\varphi \cdot \nabla\psi\, d\mathbf{x}$ (integration by parts)

6. $\int_\Omega (\psi\Delta\varphi - \varphi\Delta\psi)\, d\mathbf{x} = \int_{\partial\Omega} (\psi\partial_\mathbf{v}\varphi - \varphi\partial_\mathbf{v}\psi)\, d\sigma$

7. $\int_\Omega \operatorname{curl} \mathbf{u}\, d\mathbf{x} = -\int_{\partial\Omega} \mathbf{u} \times \mathbf{v}\, d\sigma$

8. $\int_\Omega \mathbf{u} \cdot \operatorname{curl} \mathbf{v}\, d\mathbf{x} = \int_\Omega \mathbf{v} \cdot \operatorname{curl} \mathbf{u}\, d\mathbf{x} + \int_{\partial\Omega} (\mathbf{v} \times \mathbf{u}) \cdot \mathbf{v}\, d\sigma$

Identities

$$\operatorname{div} \operatorname{curl} \mathbf{u} = 0$$
$$\operatorname{curl} \operatorname{grad} \varphi = \mathbf{0}$$
$$\operatorname{div} (\varphi\mathbf{u}) = \varphi \operatorname{div} \mathbf{u} + \nabla\varphi \cdot \mathbf{u}$$
$$\operatorname{curl} (\varphi\mathbf{u}) = \varphi \operatorname{curl} \mathbf{u} + \nabla\varphi \times \mathbf{u}$$
$$\operatorname{curl}(\mathbf{u} \times \mathbf{v}) = (\mathbf{v} \cdot \nabla)\mathbf{u} - (\mathbf{u} \cdot \nabla)\mathbf{v} + (\operatorname{div} \mathbf{v})\mathbf{u} - (\operatorname{div} \mathbf{u})\mathbf{v}$$
$$\operatorname{div}(\mathbf{u} \times \mathbf{v}) = \operatorname{curl} \mathbf{u} \cdot \mathbf{v} - \operatorname{curl} \mathbf{v} \cdot \mathbf{u}$$
$$\nabla (\mathbf{u} \cdot \mathbf{v}) = \mathbf{u} \times \operatorname{curl} \mathbf{v} + \mathbf{v} \times \operatorname{curl} \mathbf{u} + (\mathbf{u} \cdot \nabla)\mathbf{v} + (\mathbf{v} \cdot \nabla)\mathbf{u}$$
$$(\mathbf{u} \cdot \nabla)\mathbf{u} = \operatorname{curl} \mathbf{u} \times \mathbf{u} + \tfrac{1}{2}\nabla |\mathbf{u}|^2$$
$$\operatorname{curl} \operatorname{curl} \mathbf{u} = \nabla(\operatorname{div} \mathbf{u}) - \Delta\mathbf{u} \qquad (\operatorname{curl} \operatorname{curl} = \operatorname{grad} \operatorname{div} - \text{Laplacian}).$$

---

[1] $C^1(\overline{\Omega})$ is enough.

B.3 Fourier Transforms

$$\widehat{u}(\xi) = \int_{\mathbb{R}} u(x)\, e^{-i\xi x}\, dx$$

General formulas

u	$\widehat{u}$
$u(x-a)$	$e^{-ia\xi}\widehat{u}(\xi)$
$e^{iax}u(x)$	$\widehat{u}(\xi-a)$
$u(ax),\, a>0$	$\dfrac{1}{a}\widehat{u}\left(\dfrac{\xi}{a}\right)$
$u'(x)$	$i\xi\widehat{u}(\xi)$
$xu(x)$	$i\widehat{u}'(\xi)$
$(u*v)(x)$	$\widehat{u}(\xi)\,\widehat{v}(\xi)$
$u(x)v(x)$	$(\widehat{u}*\widehat{v})(\xi)$

Special transforms

u	$\widehat{u}$
$e^{-a|x|},\, a>0$	$\dfrac{2a}{a^2+\xi^2}$
$\dfrac{1}{a^2+x^2}$	$\dfrac{\pi}{a}e^{-a|\xi|}$
$e^{-ax^2},\, a>0$	$\sqrt{\dfrac{\pi}{a}}\,e^{-\frac{\xi^2}{4a}}$
$\dfrac{\sin x}{x}e^{-|x|}$	$\arctan\dfrac{2}{\xi^2}$
$\chi_{[-a,a]}(x)$	$2\dfrac{\sin a\xi}{\xi}$
$\delta(x)$	1
1	$2\pi\delta(\xi)$

B.4 Laplace Transforms

$$\widetilde{u}\,(s) = \int_0^{+\infty} u\,(t)\,e^{-st}\,dt$$

General formulas ($u(t) = 0$ for $t < 0$)

u	$\widetilde{u}$
$u\,(t-a),\, a > 0$	$e^{-as}\widetilde{u}\,(s)$
$e^{at}u\,(t),\, a \in \mathbb{C}$	$\widetilde{u}\,(s-a)$
$u\,(at),\, a > 0$	$\dfrac{1}{a}\widetilde{u}\left(\dfrac{s}{a}\right)$
$u'\,(t)$	$s\widetilde{u}\,(s) - u(0^+)$
$u''\,(t)$	$s^2\widetilde{u}\,(s) - u'(0^+) - su(0^+)$
$tu\,(t)$	$-\widetilde{u}'\,(s)$
$\dfrac{u\,(t)}{t}$	$\int_s^{+\infty}\widetilde{u}\,(\tau)\,d\tau$
$\int_0^t u\,(\tau)\,d\tau$	$\dfrac{\widetilde{u}\,(s)}{s}$
$(u * v)\,(t)$	$\widetilde{u}\,(s)\,\widetilde{v}\,(s)$

Special transforms

u	$\widetilde{u}$
$\mathcal{H}(t)e^{at},\, a \in \mathbb{C}$	$\dfrac{1}{s-a}$
$\mathcal{H}(t)\sin at,\, a \in \mathbb{R}$	$\dfrac{a}{s^2+a^2}$
$\mathcal{H}(t)\cos at,\, a \in \mathbb{R}$	$\dfrac{s}{s^2+a^2}$
$\mathcal{H}(t)\sinh at,\, a \in \mathbb{R}$	$\dfrac{a}{s^2-a^2}$
$\mathcal{H}(t)\cosh at,\, a \in \mathbb{R}$	$\dfrac{s}{s^2-a^2}$
$\mathcal{H}(t)t^n,\, n \in \mathbb{N}$	$\dfrac{n!}{s^{n+1}}$
$\mathcal{H}(t)t^{\alpha},\, \operatorname{Re}\alpha > -1$	$\dfrac{\Gamma(\alpha+1)}{s^{\alpha+1}}$
$\mathcal{H}(t)e^{-t^2}$	$e^{s^2/4}\int_{s/2}^{+\infty} e^{-\tau^2}\,d\tau$

References

Partial differential equations

[1] E. DiBenedetto, *Partial differential equations*, Cornerstones, Birkhäuser Boston, Inc., Boston, MA, 2nd ed., 2010.

[2] L. C. Evans, *Partial differential equations*, vol. 19 of Graduate Studies in Mathematics, American Mathematical Society, Providence, RI, 2nd ed., 2010.

[3] A. Friedman, *Partial differential equations of parabolic type*, Prentice-Hall, Inc., Englewood Cliffs, N.J., 1964.

[4] G. P. Galdi, *An introduction to the mathematical theory of the Navier-Stokes equations*, Springer Monographs in Mathematics, Springer, New York, 2nd ed., 2011. Steady-state problems.

[5] D. Gilbarg and N. S. Trudinger, *Elliptic partial differential equations of second order*, Classics in Mathematics, Springer-Verlag, Berlin, 2001. Reprint of the 1998 edition.

[6] P. Grisvard, *Elliptic problems in nonsmooth domains*, vol. 24 of Monographs and Studies in Mathematics, Pitman (Advanced Publishing Program), Boston, MA, 1985.

[7] R. B. Guenther and J. W. Lee, *Partial differential equations of mathematical physics and integral equations*, Dover Publications, Inc., Mineola, NY, 1996. Corrected reprint of the 1988 original.

[8] L. L. Helms, *Introduction to potential theory*, Robert E. Krieger Publishing Co., Huntington, N.Y., 1975. Reprint of the 1969 edition, Pure and Applied Mathematics, Vol. XXII.

[9] F. John, *Partial differential equations*, vol. 1 of Applied Mathematical Sciences, Springer-Verlag, New York, 4th ed., 1991.

[10] O. D. Kellogg, *Foundations of potential theory*, Reprint from the first edition of 1929. Die Grundlehren der Mathematischen Wissenschaften, Band 31, Springer-Verlag, Berlin New York, 1967.

© Springer International Publishing Switzerland 2015
S. Salsa, G. Verzini, *Partial Differential Equations in Action. Complements and Exercises*,
UNITEXT – La Matematica per il 3+2 87, DOI 10.1007/978-3-319-15416-9

[11] G. M. LIEBERMAN, *Second order parabolic differential equations*, World Scientific Publishing Co., Inc., River Edge, NJ, 1996.

[12] J.-L. LIONS AND E. MAGENES, *Non-homogeneous boundary value problems and applications. Vol. 1–2*, Springer-Verlag, New York-Heidelberg, 1972. Translated from the French by P. Kenneth, Die Grundlehren der mathematischen Wissenschaften, Band 181.

[13] R. MCOWEN, *Partial Differential Equations: Methods and Applications*, Prentice-Hall, New Jersey, 1964.

[14] P. J. OLVER, *Introduction to partial differential equations*, Undergraduate Texts in Mathematics, Springer, Cham, 2014.

[15] M. H. PROTTER AND H. F. WEINBERGER, *Maximum principles in differential equations*, Springer-Verlag, New York, 1984. Corrected reprint of the 1967 original.

[16] J. RAUCH, *Partial differential equations*, vol. 128 of Graduate Texts in Mathematics, Springer-Verlag, New York, 1991.

[17] M. RENARDY AND R. C. ROGERS, *An introduction to partial differential equations*, vol. 13 of Texts in Applied Mathematics, Springer-Verlag, New York, 2nd ed., 2004.

[18] S. SALSA, *Partial Differential Equations in Action. From Modelling to Theory*, vol. 86, UNITEXT – La Matematica per il 3+2, Springer International Publishing, Cham, 2nd ed., 2015.

[19] J. SMOLLER, *Shock waves and reaction-diffusion equations*, vol. 258 of Grundlehren der Mathematischen Wissenschaften [Fundamental Principles of Mathematical Sciences], Springer-Verlag, New York, 2nd ed., 1994.

[20] W. A. STRAUSS, *Partial differential equations*, John Wiley & Sons, Ltd., Chichester, 2nd ed., 2008. An introduction.

[21] D. V. WIDDER, *The heat equation*, Academic Press [Harcourt Brace Jovanovich, Publishers], New York-London, 1975. Pure and Applied Mathematics, Vol. 67.

Mathematical modelling

[22] D. J. ACHESON, *Elementary fluid dynamics*, Oxford Applied Mathematics and Computing Science Series, The Clarendon Press, Oxford University Press, New York, 1990.

[23] J. BILLINGHAM AND A. C. KING, *Wave motion*, Cambridge Texts in Applied Mathematics, Cambridge University Press, Cambridge, 2000.

[24] R. COURANT AND D. HILBERT, *Methods of mathematical physics. Vol. 1–2*, Interscience Publishers, Inc., New York, N.Y., 1953.

[25] R. DAUTRAY AND J.-L. LIONS, *Mathematical analysis and numerical methods for science and technology. Vol. 1–5*, Springer-Verlag, Berlin, 1985.

[26] C. C. LIN AND L. A. SEGEL, *Mathematics applied to deterministic problems in the natural sciences*, vol. 1 of Classics in Applied Mathematics, Society for Industrial and Applied Mathematics (SIAM), Philadelphia, PA, 2nd ed., 1988.

[27] J. D. MURRAY, *Mathematical biology. I–II*, vol. 17–18 of Interdisciplinary Applied Mathematics, Springer-Verlag, New York, 3rd ed., 2002–03.

[28] H.-K. RHEE, R. ARIS, AND N. R. AMUNDSON, *First-order partial differential equations. Vol. 1–2*, Dover Publications, Inc., Mineola, NY, 2001.

[29] O. SCHERZER, M. GRASMAIR, H. GROSSAUER, M. HALTMEIER, AND F. LENZEN, *Variational methods in imaging*, vol. 167 of Applied Mathematical Sciences, Springer, New York, 2009.

[30] L. A. SEGEL, *Mathematics applied to continuum mechanics*, vol. 52 of Classics in Applied Mathematics, Society for Industrial and Applied Mathematics (SIAM), Philadelphia, PA, 2007. Reprint of the 1977 original.

[31] G. B. WHITHAM, *Linear and nonlinear waves*, Pure and Applied Mathematics (New York), John Wiley & Sons, Inc., New York, 1999. Reprint of the 1974 original.

ODEs, analysis and functional analysis

[32] R. A. ADAMS AND J. J. F. FOURNIER, *Sobolev spaces*, vol. 140 of Pure and Applied Mathematics (Amsterdam), Elsevier/Academic Press, Amsterdam, 2nd ed., 2003.

[33] H. BREZIS, *Functional analysis, Sobolev spaces and partial differential equations*, Universitext, Springer, New York, 2011.

[34] E. A. CODDINGTON AND N. LEVINSON, *Theory of ordinary differential equations*, McGraw-Hill Book Company, Inc., New York-Toronto-London, 1955.

[35] I. M. GEL' FAND AND G. E. SHILOV, *Generalized functions. Vol. I: Properties and operations*, Translated by Eugene Saletan, Academic Press, New York-London, 1964.

[36] V. G. MAZ' JA, *Sobolev spaces*, Springer Series in Soviet Mathematics, Springer-Verlag, Berlin, 1985.

[37] W. RUDIN, *Principles of mathematical analysis*, McGraw-Hill Book Co., New York-Auckland-Düsseldorf, 3rd ed., 1976. International Series in Pure and Applied Mathematics.

[38] W. RUDIN, *Real and complex analysis*, McGraw-Hill Book Co., New York, 3rd ed., 1987.

[39] L. SCHWARTZ, *Théorie des distributions*, Hermann, Paris, 1966.

[40] A. E. TAYLOR AND D. C. LAY, *Introduction to functional analysis*, John Wiley & Sons, New York-Chichester-Brisbane, 2nd ed., 1980.

[41] A. E. TAYLOR AND D. C. LAY, *Introduction to functional analysis*, Robert E. Krieger Publishing Co., Inc., Melbourne, FL, 2nd ed., 1986.

[42] K. Yosida, *Functional analysis*, Classics in Mathematics, Springer-Verlag, Berlin, 1995. Reprint of the sixth (1980) edition.

[43] W. P. Ziemer, *Weakly differentiable functions*, vol. 120 of Graduate Texts in Mathematics, Springer-Verlag, New York, 1989. Sobolev spaces and functions of bounded variation.

Numerical analysis

[44] R. Dautray and J.-L. Lions, *Mathematical analysis and numerical methods for science and technology. Vol. 4, 6*, Springer-Verlag, Berlin, 1985.

[45] E. Godlewski and P.-A. Raviart, *Numerical approximation of hyperbolic systems of conservation laws*, vol. 118 of Applied Mathematical Sciences, Springer-Verlag, New York, 1996.

[46] A. Quarteroni and A. Valli, *Numerical approximation of partial differential equations*, vol. 23 of Springer Series in Computational Mathematics, Springer-Verlag, Berlin, 1994.

Stochastic processes and finance

[47] M. Baxter and A. Rennie, *Financial Calculus An introduction to derivative pricing*, Cambridge University Press, 1996.

[48] B. Øksendal, *Stochastic differential equations*, Universitext, Springer-Verlag, Berlin, 6th ed., 2003. An introduction with applications.

[49] P. Wilmott, S. Howison, and J. Dewynne, *The mathematics of financial derivatives*, Cambridge University Press, Cambridge, 1995. A student introduction.

Collana Unitext – La Matematica per il 3+2

Series Editors:
A. Quarteroni (Editor-in-Chief)
L. Ambrosio
P. Biscari
C. Ciliberto
M. Ledoux
W.J. Runggaldier

Editor at Springer:
F. Bonadei
francesca.bonadei@springer.com

As of 2004, the books published in the series have been given a volume number.
Titles in grey indicate editions out of print.
As of 2011, the series also publishes books in English.

A. Bernasconi, B. Codenotti
Introduzione alla complessità computazionale
1998, X+260 pp, ISBN 88-470-0020-3

A. Bernasconi, B. Codenotti, G. Resta
Metodi matematici in complessità computazionale
1999, X+364 pp, ISBN 88-470-0060-2

E. Salinelli, F. Tomarelli
Modelli dinamici discreti
2002, XII+354 pp, ISBN 88-470-0187-0

S. Bosch
Algebra
2003, VIII+380 pp, ISBN 88-470-0221-4

S. Graffi, M. Degli Esposti
Fisica matematica discreta
2003, X+248 pp, ISBN 88-470-0212-5

S. Margarita, E. Salinelli
MultiMath – Matematica Multimediale per l'Università
2004, XX+270 pp, ISBN 88-470-0228-1

A. Quarteroni, R. Sacco, F.Saleri
Matematica numerica (2a Ed.)
2000, XIV+448 pp, ISBN 88-470-0077-7
2002, 2004 ristampa riveduta e corretta
(1a edizione 1998, ISBN 88-470-0010-6)

13. A. Quarteroni, F. Saleri
    Introduzione al Calcolo Scientifico (2a Ed.)
    2004, X+262 pp, ISBN 88-470-0256-7
    (1a edizione 2002, ISBN 88-470-0149-8)

14. S. Salsa
    Equazioni a derivate parziali - Metodi, modelli e applicazioni
    2004, XII+426 pp, ISBN 88-470-0259-1

15. G. Riccardi
    Calcolo differenziale ed integrale
    2004, XII+314 pp, ISBN 88-470-0285-0

16. M. Impedovo
    Matematica generale con il calcolatore
    2005, X+526 pp, ISBN 88-470-0258-3

17. L. Formaggia, F. Saleri, A. Veneziani
    Applicazioni ed esercizi di modellistica numerica
    per problemi differenziali
    2005, VIII+396 pp, ISBN 88-470-0257-5

18. S. Salsa, G. Verzini
    Equazioni a derivate parziali – Complementi ed esercizi
    2005, VIII+406 pp, ISBN 88-470-0260-5
    2007, ristampa con modifiche

19. C. Canuto, A. Tabacco
    Analisi Matematica I (2a Ed.)
    2005, XII+448 pp, ISBN 88-470-0337-7
    (1a edizione, 2003, XII+376 pp, ISBN 88-470-0220-6)

20. F. Biagini, M. Campanino
    Elementi di Probabilità e Statistica
    2006, XII+236 pp, ISBN 88-470-0330-X

21. S. Leonesi, C. Toffalori
Numeri e Crittografia
2006, VIII+178 pp, ISBN 88-470-0331-8

22. A. Quarteroni, F. Saleri
Introduzione al Calcolo Scientifico (3a Ed.)
2006, X+306 pp, ISBN 88-470-0480-2

23. S. Leonesi, C. Toffalori
Un invito all'Algebra
2006, XVII+432 pp, ISBN 88-470-0313-X

24. W.M. Baldoni, C. Ciliberto, G.M. Piacentini Cattaneo
Aritmetica, Crittografia e Codici
2006, XVI+518 pp, ISBN 88-470-0455-1

25. A. Quarteroni
Modellistica numerica per problemi differenziali (3a Ed.)
2006, XIV+452 pp, ISBN 88-470-0493-4
(1a edizione 2000, ISBN 88-470-0108-0)
(2a edizione 2003, ISBN 88-470-0203-6)

26. M. Abate, F. Tovena
Curve e superfici
2006, XIV+394 pp, ISBN 88-470-0535-3

27. L. Giuzzi
Codici correttori
2006, XVI+402 pp, ISBN 88-470-0539-6

28. L. Robbiano
Algebra lineare
2007, XVI+210 pp, ISBN 88-470-0446-2

29. E. Rosazza Gianin, C. Sgarra
Esercizi di finanza matematica
2007, X+184 pp, ISBN 978-88-470-0610-2

30. A. Machì
Gruppi – Una introduzione a idee e metodi della Teoria dei Gruppi
2007, XII+350 pp, ISBN 978-88-470-0622-5
2010, ristampa con modifiche

31 Y. Biollay, A. Chaabouni, J. Stubbe
    Matematica si parte!
    A cura di A. Quarteroni
    2007, XII+196 pp, ISBN 978-88-470-0675-1

32. M. Manetti
    Topologia
    2008, XII+298 pp, ISBN 978-88-470-0756-7

33. A. Pascucci
    Calcolo stocastico per la finanza
    2008, XVI+518 pp, ISBN 978-88-470-0600-3

34. A. Quarteroni, R. Sacco, F. Saleri
    Matematica numerica (3a Ed.)
    2008, XVI+510 pp, ISBN 978-88-470-0782-6

35. P. Cannarsa, T. D'Aprile
    Introduzione alla teoria della misura e all'analisi funzionale
    2008, XII+268 pp, ISBN 978-88-470-0701-7

36. A. Quarteroni, F. Saleri
    Calcolo scientifico (4a Ed.)
    2008, XIV+358 pp, ISBN 978-88-470-0837-3

37. C. Canuto, A. Tabacco
    Analisi Matematica I (3a Ed.)
    2008, XIV+452 pp, ISBN 978-88-470-0871-3

38. S. Gabelli
    Teoria delle Equazioni e Teoria di Galois
    2008, XVI+410 pp, ISBN 978-88-470-0618-8

39. A. Quarteroni
    Modellistica numerica per problemi differenziali (4a Ed.)
    2008, XVI+560 pp, ISBN 978-88-470-0841-0

40. C. Canuto, A. Tabacco
    Analisi Matematica II
    2008, XVI+536 pp, ISBN 978-88-470-0873-1
    2010, ristampa con modifiche

41. E. Salinelli, F. Tomarelli
    Modelli Dinamici Discreti (2a Ed.)
    2009, XIV+382 pp, ISBN 978-88-470-1075-8

42. S. Salsa, F.M.G. Vegni, A. Zaretti, P. Zunino
Invito alle equazioni a derivate parziali
2009, XIV+440 pp, ISBN 978-88-470-1179-3

43. S. Dulli, S. Furini, E. Peron
Data mining
2009, XIV+178 pp, ISBN 978-88-470-1162-5

44. A. Pascucci, W.J. Runggaldier
Finanza Matematica
2009, X+264 pp, ISBN 978-88-470-1441-1

45. S. Salsa
Equazioni a derivate parziali – Metodi, modelli e applicazioni (2a Ed.)
2010, XVI+614 pp, ISBN 978-88-470-1645-3

46. C. D'Angelo, A. Quarteroni
Matematica Numerica – Esercizi, Laboratori e Progetti
2010, VIII+374 pp, ISBN 978-88-470-1639-2

47. V. Moretti
Teoria Spettrale e Meccanica Quantistica – Operatori in spazi di Hilbert
2010, XVI+704 pp, ISBN 978-88-470-1610-1

48. C. Parenti, A. Parmeggiani
Algebra lineare ed equazioni differenziali ordinarie
2010, VIII+208 pp, ISBN 978-88-470-1787-0

49. B. Korte, J. Vygen
Ottimizzazione Combinatoria. Teoria e Algoritmi
2010, XVI+662 pp, ISBN 978-88-470-1522-7

50. D. Mundici
Logica: Metodo Breve
2011, XII+126 pp, ISBN 978-88-470-1883-9

51. E. Fortuna, R. Frigerio, R. Pardini
Geometria proiettiva. Problemi risolti e richiami di teoria
2011, VIII+274 pp, ISBN 978-88-470-1746-7

52. C. Presilla
Elementi di Analisi Complessa. Funzioni di una variabile
2011, XII+324 pp, ISBN 978-88-470-1829-7

53. L. Grippo, M. Sciandrone
Metodi di ottimizzazione non vincolata
2011, XIV+614 pp, ISBN 978-88-470-1793-1

54. M. Abate, F. Tovena
Geometria Differenziale
2011, XIV+466 pp, ISBN 978-88-470-1919-5

55. M. Abate, F. Tovena
Curves and Surfaces
2011, XIV+390 pp, ISBN 978-88-470-1940-9

56. A. Ambrosetti
Appunti sulle equazioni differenziali ordinarie
2011, X+114 pp, ISBN 978-88-470-2393-2

57. L. Formaggia, F. Saleri, A. Veneziani
Solving Numerical PDEs: Problems, Applications, Exercises
2011, X+434 pp, ISBN 978-88-470-2411-3

58. A. Machì
Groups. An Introduction to Ideas and Methods of the Theory of Groups
2011, XIV+372 pp, ISBN 978-88-470-2420-5

59. A. Pascucci, W.J. Runggaldier
Financial Mathematics. Theory and Problems for Multi-period Models
2011, X+288 pp, ISBN 978-88-470-2537-0

60. D. Mundici
Logic: a Brief Course
2012, XII+124 pp, ISBN 978-88-470-2360-4

61. A. Machì
Algebra for Symbolic Computation
2012, VIII+174 pp, ISBN 978-88-470-2396-3

62. A. Quarteroni, F. Saleri, P. Gervasio
Calcolo Scientifico (5a ed.)
2012, XVIII+450 pp, ISBN 978-88-470-2744-2

63. A. Quarteroni
Modellistica Numerica per Problemi Differenziali (5a ed.)
2012, XVIII+628 pp, ISBN 978-88-470-2747-3

74. A. Bermúdez, D. Gómez, P. Salgado
    Mathematical Models and Numerical Simulation in Electromagnetism
    2014, XVIII+430pp, ISBN 978-3-319-02948-1

75. A. Quarteroni
    Matematica Numerica. Esercizi, Laboratori e Progetti (2a Ed.)
    2013, XVIII+406pp, ISBN 978-88-470-5540-7

76. E. Salinelli, F. Tomarelli
    Discrete Dynamical Models
    2014, XVI+386pp, ISBN 978-3-319-02290-1

77. A. Quarteroni, R. Sacco, F. Saleri, P. Gervasio
    Matematica Numerica (4a Ed.)
    2014, XVIII+532pp, ISBN 978-88-470-5643-5

78. M. Manetti
    Topologia (2a Ed.)
    2014, XII+334pp, ISBN 978-88-470-5661-9

79. M. Iannelli, A. Pugliese
    An Introduction to Mathematical Population Dynamics.
    Along the trail of Volterra and Lotka
    2014, XIV+338pp, ISBN 978-3-319-03025-8

80. V. M. Abrusci, L. Tortora de Falco
    Logica. Volume 1
    2014, X+180pp, ISBN 978-88-470-5537-7

81. P. Biscari, T. Ruggeri, G. Saccomandi, M. Vianello
    Meccanica Razionale (2a Ed.)
    2014, XII+390pp, ISBN 978-88-470-5725-8

82. C. Canuto, A. Tabacco
    Analisi Matematica I (4a Ed.)
    2014, XIV+508pp, ISBN 978-88-470-5722-7

83. C. Canuto, A. Tabacco
    Analisi Matematica II (2a Ed.)
    2014, XII+576pp, ISBN 978-88-470-5728-9

84. C. Canuto, A. Tabacco
    Mathematical Analysis I (2nd Ed.)
    2015, XIV+484pp, ISBN 978-3-319-12771-2

85. C. Canuto, A. Tabacco
   Mathematical Analysis II (2nd Ed.)
   2015, XII+550pp, ISBN 978-3-319-12756-9

86. S. Salsa
   Partial Differential Equations in Action. From Modelling to Theory (2nd Ed.)
   2015, XVIII+688, ISBN 978-3-319-15092-5

87. S. Salsa, G. Verzini
   Partial Differential Equations in Action. Complements and Exercises
   2015, VIII+422, ISBN 978-3-319-15415-2

The online version of the books published in this series is available at
SpringerLink.
For further information, please visit the following link:
http://www.springer.com/series/5418

Printed in the United States
By Bookmasters